The Law of the Seabed

Publications on Ocean Development

A SERIES OF STUDIES ON THE INTERNATIONAL, LEGAL, INSTITUTIONAL AND POLICY ASPECTS OF OCEAN DEVELOPMENT

General Editors

Robin Churchill
Alex Oude Elferink

VOLUME 90

The titles published in this series are listed at *brill.com/pood*

The Law of the Seabed

Access, Uses, and Protection of Seabed Resources

Edited by

Catherine Banet

BRILL
NIJHOFF

LEIDEN | BOSTON

Library of Congress Cataloging-in-Publication Data

Names: Banet, Catherine, editor.
Title: The law of the seabed : access, uses, and protection of seabed resources / edited by Catherine Banet.
Description: Leiden ; Boston : Brill Nijhoff, 2020. | Series: Publications on ocean development, 0924–1922 ; volume 90 | Includes bibliographical references.
Identifiers: LCCN 2019048170 (print) | LCCN 2019048171 (ebook) | ISBN 9789004391550 (hardback) | ISBN 9789004391567 (ebook)
Subjects: LCSH: Ocean bottom—Law and legislation. | Ocean mining—Law and legislation. | Marine resources conservation—Law and legislation.
Classification: LCC KZA1390 .L395 2020 (print) | LCC KZA1390 (ebook) | DDC 341.4/55—dc23
LC record available at https://lccn.loc.gov/2019048170
LC ebook record available at https://lccn.loc.gov/2019048171

Typeface for the Latin, Greek, and Cyrillic scripts: “Brill”. See and download: brill.com/brill-typeface.

ISSN 0924-1922
ISBN 978-90-04-39155-0 (hardback)
ISBN 978-90-04-39156-7 (e-book)

This book is printed on acid-free paper and produced in a sustainable manner.

Contents

PART 3
Exploitation of Non-Living Resources from the Seabed beyond National Jurisdiction

PART 4
Exploitation of Marine Biodiversity and Living Resources on the Seabed beyond National Jurisdiction

PART 5
Principles Applicable to Sovereign States When Exploiting Seabed Resources within National Jurisdiction

PART 7
Conflicting Uses or Coexistence, Resolving Mechanisms and Protection Regimes: Towards a More Integrated Approach

Preface

A significant part of global trade, value creation, energy security and international peace and security depends on the seas and oceans, which cover around 71% of the planet's surface.[1] The same applies to the bulk of internet communications, which rely heavily on submarine cables, and, no less so, to most of the world's unknown biological resources and biodiversity. A number of challenges and possibilities related to the oceans are highlighted in governmental strategies, including notably the Norwegian Government's White Paper submitted to the Norwegian parliament in 2017 and the Government's Updated Ocean Strategy published in 2019.[2]

There is, in fact, good news as regards the toolbox provided by international law for policy-makers. A comprehensive, unified and universal legal framework exists for ocean and maritime spaces. This framework establishes a system of competencies and possibilities of proactive coordination, further regulation where necessary, and dispute settlement, together with a number of key principles and rules. It successfully promotes and integrates a variety of important goals, which previously could seem irreconcilable. Moreover, any further regulations must be fully consistent with this legal framework and not undermine the existing architecture.[3] Instead of lawless black holes, issues of lack of compliance or implementation are unfortunately frequent.

This framework did not appear overnight. It builds on a monumental series of building-blocks, imagine layer upon layer of legal mortar, stemming mainly from State practice and legislation, treaty-making and case-law. The resulting masonry is solid. Its most sophisticated expression is found in the United

---

1 The author is solely responsible for this contribution, which was written before the author assumed his current position.

2 'The place of the oceans in Norway's foreign and development policy' – Meld. St. 22 (2016–2017) Report to the Storting, <https://www.regjeringen.no/contentassets/1b21c0734b5042e489c24234e9927b73/en-gb/sved/stm201620170022000engpdfs.pdf>; and 'Blue opportunities – The Norwegian Government's Updated Ocean Strategy', 3 June 2019, <https://www.regjeringen.no/globalassets/departementene/nfd/dokumenter/strategier/nfd_havstrategi_2019_engelsk.pdf>.

3 See article 311 (2) and (3) of the *United Nations Convention on the Law of the Sea*, concluded 10 December 1982, entered into force 16 November 1994, 1833 UNTS 396 (UNCLOS). See also, for example, paras. 6–7 of resolution 72/249 adopted by the United Nations General Assembly on 24 December 2017 on an 'International legally binding instrument under the United Nations Convention on the Law of the Sea on the conservation and sustainable use of marine biological diversity of areas beyond national jurisdiction', UN Doc. A/RES/72/249.

Nations Convention on the Law of the Sea (UNCLOS).[4] Among its stated aims is to make "an important contribution to the maintenance of peace, justice and progress for all peoples of the world," The number of parties bound by this convention has reached 168 at the time of writing.

Even the relatively few States that have not ratified or acceded to the convention recognize that a number of its provisions reflect international customary law, and thus universally binding norms. This significantly includes the United States of America, which is the State that set in motion modern law-making relating to the continental shelf. This happened on 28 September 1945 with the seminal adoption of what has later consistently been referred to as the Truman Proclamation.[5] Subsequently, the United States continued to be at the forefront in spearheading more predictable rules in this domain, in line with momentous technological and economic developments. Key steps included the adoption in 1958 of the Geneva Convention on the continental shelf and, thereafter, the negotiations on the definition and the extent of the continental shelf in UNCLOS article 76 and Annex II. The latter consecrates a method originally suggested by the American geologist Hollis Dow Hedberg, with other additions. Today, the Hedberg formula is most frequently used for defining the outer limits of the continental shelf. The existence of an extensive legal framework applicable to all Arctic waters was confirmed in the Ilulissat Declaration on 28 May 2008 by the coastal States surrounding the central part of the Arctic Ocean, who confirmed their commitment to this extensive legal framework and to the orderly settlement of any possible overlapping claims in the Arctic.

Indeed, there is good news. The broad normative unity and coherence of the international law of the sea is striking. In many areas and on subjects which could be described as potential hotspots of contention or colliding interests, the universal and unified nature of this body of law is regularly demonstrated. This fact is actually often overlooked or, wrongly, just taken for granted. The yearly resolutions of the United Nations General Assembly on the topic of "Oceans and the Law of the Sea" are incidentally the longest of all resolutions adopted by the plenary of the General Assembly. They arguably contain some of the most concrete operative paragraphs for timely adoption of national

4 *Ibid.*

5 Proclamation No. 2667, 'Policy of the United States with Respect to the Natural Resources of the Subsoil and Sea-Bed of the Continental Shelf', 28 September 1945, 10 Fed. Reg. 12303; XIII Bulletin, Dept. of State, No. 327, 30 September 1945, p. 485.

measures and concerted international action. These resolutions merit careful study and a broader readership.[6]

As already indicated, some words of caution are nevertheless due. The existence of a considerable number of applicable principles and rules, either as treaty obligations or as customary international law, does not automatically entail that all necessary or adequate regulations and arrangements have been adopted or implemented. Nor does it signify that legal rules are being complied with thanks to appropriate controls and enforcement.

The first word of caution has to do with the fact that many rules of the international law of the sea confer powers – and may be distinguished from rules imposing obligations or duties. In many cases they could also be seen as a road map or a work plan that, in turn, requires adequate action to be taken by the competent national authorities or international organizations, notably in the form of appropriate regulations.

Such rules might to some extent be compared to the operating system of a computer. The latter manages hardware and software resources, but requires additional software or applications to actually enable operations essential for the user. The user would usually not have to bother too much about the operating system, but would instead focus on this "secondary" software. Let us here use a simple yet classic illustration. The law of the sea empowers States to establish exclusive economic zones of 200 nautical miles. Whether to establish such a zone is a sovereign prerogative of the coastal State to decide. The coastal State may choose to establish or keep less comprehensive fisheries or ecological zones, as long as the limits and constraints established in UNCLOS article 311 (2) and (3) are respected, as these require consistency with the key system of the convention.

UNCLOS is often referred to as the international constitution of the oceans. A constitution will still require adoption of legislation and systems of compliance, enforcement and adjudication. Within this framework, there is a considerable potential for creativity, innovation and adaptation to evolving conditions as regards policy formulation. This is also the case for the law of the sea. Moreover, cross-fertilization, helped by comparative studies or various other phenomena of reception of norms, may be particularly useful for States

6 Notably two resolutions adopted in December 2018, respectively 'Oceans and the Law of the Sea' (UN Doc. A/RES/73/124) and 'Sustainable fisheries, including through the 1995 Agreement for the Implementation of the Provisions of the United Nations Convention on the Law of the Sea of 10 December 1982 relating to the Conservation and Management of Straddling Fish Stocks and Highly Migratory Fish Stocks, and related instruments' (UN Doc. A/RES/73/125).

and the competent international organizations considering the adoption of relevant measures.

More good news is that control and compliance may be aided by new advances in technology, science and various forms of know how building on best practices. Satellite tracking and digital advances related to big data are merely catchwords to illustrate the potential for quantum leaps in efficiency, which also have to be accompanied by thoughtful regulators. Yet an example of innovation is the approach taken by Norway in promoting development assistance in the context of definition of the outer limits of the continental shelf.[7] Mr. Harald Brekke, member of the Commission of the Limits of the Continental Shelf from its inception in 1997 until 2012, played a key role also in this regard. He is among the contributors to this book on the law of the sea-bed.

This book addresses a need. It concerns a vast number of rapidly developing new challenges and possibilities. It contributes cross-disciplinary perspectives and various academic "angles of attack" on issues that merit further analysis, cross-fertilization and critical inquiry. A long chain of individuals, including practitioners and academics, e.g. geologists, policy-makers, negotiators and arbitrators, has contributed to the constant weaving of a less open textured law of the sea – yet striving at the same time for preserving due flexibility in light of differing geographical and local contexts and conditions. The contributors to this book are participants in this international endeavour. We are therefore sincerely grateful to Catherine Banet for this valuable initiative.

Rolf Einar Fife

7 R.E. Fife, 'A Perspective on Development and the Law of the Sea: How to Provide Support for the Establishment of the Outer Limits of the Continental Shelf, in Sainz-Borgo et al (eds.), *Liber Amicorum in Honour of Gudmundur Eiriksson*, University for Peace, San José, Costa Rica & O.P. Jindal Global University, New Dehli, 2017, 51–67.

Illustrations

Abbreviations

ABNJ	Area Beyond National Jurisdiction
ABS	Access and Benefit Sharing
APEIS	Area of Particular Environmental Interest
ATS	Antarctic Treaty System
AUV	Autonomous Underwater Vehicle
BAT	Best Available Techniques
BCC	Benguela Current Convention
BIT	Bilateral Investment Treaty
BREF	BAT Reference Document
CBD	Convention of Biological Diversity
CCAMLR	Commission for the Conservation of Atlantic Marine Living Resources
CCS	carbon capture and storage
CO_2	carbon dioxide
CDM	Clean Development Mechanism
CJEU	Court of Justice of the European Union
CLCS	Commission on the Limits of the Continental Shelf
COMRA	China Ocean Mineral Resource R&D Association
CTD	Conductivity-Temperature-Depth
DOSI	Deep-Ocean Stewardship Initiative
DSM	deep seabed mining
DSTD	deep-sea tailing disposal
ECT	Energy Charter Treaty
EEA	European Economic Area
EHR	Enhanced Hydrocarbon Recovery
EOR	Enhanced oil recovery
EIA	Environmental Impact Assessment
EU	European Union
ESA	EFTA Surveillance Authority
EEZ	exclusive economic zone
FAO	United Nations Food and Agriculture Organization
FSA	United Nations Fish Stocks Agreement
GAIRS	generally accepted international rules and standards
GFCM	General Fisheries Commission for the Mediterranean
GHG	greenhouse gases
ICJ	International Court of Justice

ICPC	International Cable Protection Committee
ICZM	integrated coastal zone management
IEA	International Energy Agency
INDEEP	International Network for Scientific Investigations of the Deep Sea
IUU fishing	Illegal, Unreported and Unregulated fishing
ILC	International Law Commission
IMO	International Maritime Organization
IPCC	Intergovernmental Panel on Climate Change
IPRS	Intellectual Property Rights
ISA or ISBA	International Seabed Authority
ITLOS	International Tribunal for the Law of the Sea
IUCN	International Union for Conservation of Nature
JDA	Joint Development Agreement
JDZ	joint development zone
km	kilometres
LDC Protocol	Protocol to the London Dumping Convention
LOSC	Law of the Sea Convention
LTC	Legal and Technical Commission (of the International Seabed Authority)
MPA	Marine Protected Area
MGR	Marine genetic resource
MOU	Memorandum of Understanding
MSP	Marine Spatial Planning
NAFTA	North American Free Trade Agreement
NAFO	Northwest Atlantic Fisheries Organisation
NATO	North Atlantic Treaty Organization
NEAFC	North-East Atlantic Fisheries Commission
NGO	Non-Governmental Organisation
nm	nautical miles
NPD	Norwegian Petroleum Directorate
OAU	Organisation for African Unity
OSD	Offshore Safety Directive (EU)
OSPAR	Convention for the Protection of the Marine Environment of the North-East Atlantic
PLDC	Protocol to the London Dumping Convention
PSSA	Particularly Sensitive Sea Area
PSV	Production Support Vessel
REES	Rare Earth Elements
RFMO	regional fisheries management organization

RSO	Regional Seas Organisation
ROV	Remote Operated Vehicle
SEAFO	Southeast Atlantic Fisheries Organisation
SIODFA	Southern Indian Ocean Deepsea Fishers Association
SIOFA	Southern Indian Ocean Fisheries Agreement
SMS	Seafloor massive sulphides
SPRFMO	South Pacific Regional Fisheries Management Organisation
STD	submarine tailing disposal
TPP	Trans-Pacific Partnership
TSO	Transmission System Operator
UK	United Kingdom
UN Charter	Charter of the United Nations
UNCLOS	United Nations Convention on the Law of the Sea
UNEP	United Nations Environmental Programme
UNFCCC	United Framework Convention on Climate Change
UNGA	United Nations General Assembly
USA	United States of America
USGS	U.S. Geological Survey
VME	Vulnerable marine ecosystem
WMD	Weapons of Mass Destruction

Notes on Contributors

Håkon With Andersen

Professor of modern history at the Department of Historical Studies at the Norwegian University of Science and Technology (NTNU) since 1992. History of science and technology as specialty. Leader of the research project *3ROceans: The high seas and deep ocean: Representation, Resources and regulatory governance* (https://www.ntnu.edu/3roceans).

Finn Arnesen

has been Professor of Law at the University of Oslo, Norway since 1998.

Lars Olav Askheim

General Counsel at Statnett SF, the Norwegian transmission system operator for electricity.

Catherine Banet

(Ph.D.) is Associate Professor at the University of Oslo, Faculty of Law, Scandinavian Institute of Maritime Law, Department of Petroleum and Energy Law, Norway. Her core fields of legal expertise include energy law, environmental law, competition law (state aids), EU/EEA law and law of the sea. She is member of the Advisory Academic Group to the International Bar Association (IBA), Section for Energy, Environment and Natural Resources and Infrastructure Law (SEERIL). Banet is one of the four academic directors of the professional LL.M programme North Sea Energy Law Partnership (NSELP) and co-editor of the annual European Energy Law Reports (EELR). She is Academic Fellow at the Center on Regulation in Europe (CERRE).

Nigel Bankes

is a professor law at the University of Calgary where he holds the chair in natural resources law. He is also an adjunct professor with the KG Jebsen Centre for the Law of the Sea at UiT, the Arctic University of Norway.

Alvar Braathen

is Professor in geology at the Department of Geosciences of the University of Oslo, Norway.

Yana Evgenyevna Brazovskaya
Associate Professor of the International and Maritime Law Department, Admiral Makarov State University of Maritime and Inland Shipping, Saint-Petersburg, Russia.

Harald Brekke
is Project Coordinator and Senior Geologist in the Norwegian Petroleum Directorate (NPD), in Stavanger, Norway. In 1996, after more than 10 years as an exploration geologist, he became the project coordinator for the technical part of establishing the outer limits of the continental shelf of Norway in accordance with the UN Convention of the Law of the Sea. He was member of the Commission on the Limits of the Continental Shelf (CLCS) from 1997 to 2012. He has given advice to many coastal states in the preparation of their submissions to the CLCS. He has been member of the Legal and Technical Commission (LTC) of the International Seabed Authority since 2013; currently holding the position as Vice-chair of the Commission. In parallel with his work in the CLCS and the LTC, he has been the NPD coordinator of international Arctic research and mapping projects in the Barents Sea and Arctic Ocean.

Richard Caddell
Lecturer in Law, Cardiff University, UK.

Chelsea Zhaoxi Chen
is a PhD candidate from China Tsinghua University School of Law researching on energy regulatory law. She obtained her undergraduate law degree from the London School of Economics in the UK and her Master's degree (JSM) from Stanford Law School in the USA. Before doing her PhD, she worked in London as an English Solicitor specializing in cross-border energy project finance.

Aldo Chircop
is Professor of Law and Canada Research Chair in Maritime Law and Policy. He is based at the Marine and Environmental Law Institute, Schulich School of Law, Dalhousie University, Canada. His current principal fields of research are the international law of the sea and international maritime law. Professor Chircop was consultant to the International Seabed Authority and rapporteur at the 2008 Chatham House and 2012 Beijing meeting of experts on Article 82. He has published extensively and is co-editor of the *Ocean Yearbook* (Brill).

Joanna Dingwall
PhD, University of Glasgow; LL.M. (International Legal Studies), New York University; DLP, Glasgow Graduate School of Law; LL.B. (First Class Honours),

University of Glasgow, is admitted to practice law in Scotland, New York, England and Wales. The author is a public international lawyer currently lecturing international law at the University of Glasgow, School of Law. She is also a lawyer to the Scottish Government concerning legal aspects of Scotland's offshore renewable energy industry and the marine environment. Previously, the author worked as a London-based public international law and international dispute resolution practitioner for several years.

Edwin Egede
is a Reader in International Law & International Relations at the Cardiff University School of Law & Politics and obtained his PhD in International Law of the Sea from the Cardiff University Law School. Recently, he has been appointed as Expert Consultant on the African Blue Economy by various International Organisations, including the United Nations and the African Union. In addition, he is a Barrister and Solicitor of the Supreme Court of Nigeria with over 25 years practice experience. Currently, he is a member of the executive of the International Bar Association (IBA) Public Law Committee.

Rolf Einar Fife
Ambassador of Norway to the European Union.

Rosa Greaves
is a Professor II at the University of Oslo, Norway. She is Professor Emeritus (formerly Professor of European Commercial Law) at the University of Glasgow, UK and a Visiting Professor at the Catholic University of Lisbon, Portugal. She has been a Visiting Professor at Universities in Australia, UK, Europe and the UK. She is a Barrister and member of Inner Temple (London). In 2015 she was elected to membership of the Academia Europaea.

James Harrison
is Senior Lecturer in International Law, University of Edinburgh School of Law, UK. He has been a member of staff at Edinburgh Law School since 2007, where he teaches and researches on a wide range of international law topics, including international law of the sea, international environmental law, and international dispute settlement. He has written broadly on these subjects, including two monographs: *Making the Law of the Sea: A Study in the Development of International Law* (Cambridge University Press, 2011) and *Saving the Oceans through Law: The International Legal Framework for the Protection of the Marine Environment* (Oxford University Press, 2017). Alongside his academic work, he has acted as a legal adviser or consultant for a number of governments, intergovernmental organizations and non-governmental organizations.

Catherine Horsfield
Attorney and Programme Head: Mining, Centre for Environmental Rights, Cape Town, South Africa.

David Langlet
is Professor of Ocean Governance Law, University of Gothenburg, Sweden. Langlet is the first incumbent of the Chair in Ocean Governance Law at University of Gothenburg. He has previously held positions as e.g. senior lecturer at Stockholm University and research fellow at the Faculty of Law at Oxford University. Langlet's research has engaged with a range of topics in the fields of environmental law, law of the sea, energy law, and international economic law. Recent books include Langlet D. and Rayfuse R. (eds), *The Ecosystem Approach in Ocean Planning and Governance – Perspectives from Europe and Beyond* (Brill Nijhoff 2018) and Langlet D. and Mahmoudi S. *EU Environmental Law and Policy* (OUP 2016).

Alla Pozdnakova
Professor of Law and Chair of the Research Group on International Law and Governance at the University of Oslo, Norway.

Eva Ramirez-Llodra
is a deep-sea ecologist with a PhD from the University of Southampton (UK, 2001). From 2002 to 2010, she was the project manager of ChEss-Census of Marine Life (NOC, UK) and a postdoctoral researcher at ICM-CSIC (Spain, 2005–2013). Since January 2014, she is a Research Scientist at NIVA (Norway) and from June 2019, she is the Science Coordinator at REV Ocean. Her main expertise is in marine biodiversity and early life histories of deep-sea fauna in relation to anthropogenic stressors. She has co-edited 3 special volumes, published over 50 scientific papers and book chapters. She is co-Principal Investigator of the International Network for Scientific Investigations of Deep--Sea Ecosystems (INDEEP) and member of the Advisory Board for the Deep Ocean Stewardship Initiative (DOSI).

Rosemary Rayfuse
is Scientia Professor in International Law at UNSW Sydney, Australia and a Fellow of the Academy of the Social Sciences in Australia (FASSA). Since 2011 she has been a Conjoint Professor in the Faculty of Law at Lund University, Sweden. From 2014–2017 she was Visiting Professor at the University of Gothenburg. Her research focuses on issues of oceans governance, including

polar oceans governance, high seas fisheries, protection of the marine environment in areas beyond national jurisdiction, and the normative effects of climate change on the law of the sea and international law in general.

Martha M. Roggenkamp
is professor of Energy at the University of Groningen and Director of the Groningen Centre of Energy Law and Sustainability, The Netherlands.

Saul Roux
Science and Policy Specialist, Centre for Environmental Rights, Cape Town, South Africa.

Gulnara Flurovna Ruchkina
Dr., is professor of the International and Maritime Law Department, Admiral Makarov State University of Maritime and Inland Shipping, Saint-Petersburg, Russia. He is also Head of the Business and Corporate Law Department, Financial University of the RF Government, Moscow.

Erik Røsæg
is formerly the Director of the Scandinavian Institute of Maritime Law, University of Oslo, Norway. He is now at UiO PluriCourts – Centre for the Study of the Legitimate Roles of the Judiciary in the Global Order. He teaches and writes in the fields of law of the sea, maritime law, and third party interests in commercial law. He has been much involved in the negotiations of liability Conventions in the IMO, chairs the Norwegian Maritime Law Commission, and has chaired other committees preparing draft legislation. He has published extensively in national and international journals and books, and he has been a speaker and consultant in a number of jurisdictions.

Tullio Scovazzi
is Professor of International Law, University of Milano-Bicocca, Milan, Italy. He occasionally participates, as legal expert, in negotiations and meetings relating to the sea or cultural properties.

Alexander S. Skaridov
CAPT RN (ret), Professor, Head of the International and Maritime law department of the Russian Admiral Makarov State University of Maritime and Inland Shipping, St. Petersburg, Russia.

Kristoffer Svendsen
LL.B. (Bond), LL.M. (Bond), LL.M. (MGIMO-University of MFA), and Ph.D. (UiT – the Arctic University of Norway) is a post doctoral fellow at the K.G. Jebsen Centre for the Law of the Sea, Faculty of Law, UiT – the Arctic University of Norway.

Seline Trevisanut
Professor of International Law and Sustainability, Utrecht Law School, The Netherlands.

Morten Walløe Tvedt
is Senior Research Fellow at the Fridtjof Nansen Institute and Associate Professor in law at Molde University College. He has published extensively in the area of law and biotechnology. He has a particular research interest in how to translate international law into legally binding obligations on private actors, like multinational companies and university users as intermediaries for product development.

Jaap J.A. Waverijn
is a lawyer at Stibbe and a PhD researcher at the Groningen Centre of Energy Law of the University of Groningen, The Netherlands. His PhD research focuses on legal questions relating to the ownership of and establishment of security rights concerning offshore energy installations, licenses and subsidies. The aim is to increase legal certainty and thus decrease the cost of capital and attract additional investments. Jaap regularly publishes on competition law and has consulted the Fijian Consumer and Competition Commission. He is a fellow of the Platform for International Energy Governance and associated member of the Transatlantic Maritime Emissions Research Network TRAMEREN. He is a founder of Energy Law Research Community ENERLEX. Jaap holds a Bachelor of Laws in Dutch law from the University of Groningen and an LLM in international and European law from the University of Cambridge.

INTRODUCTION

The Law of the Seabed

Catherine Banet

1 Focus of the Book

Humankind has always been fascinated by the seabeds for their mysteries, their perils and their riches, along with the beauty of their biological diversity. Moreover, the ocean floor comprises a constellation of components that are, at the same time, fragile and vital to sustainability and the balance of the rest of the world's interdependent eco-systems.

For an area which is barely accessible to most human beings, the ocean floor plays a major role in the Earth ecological balance. Oceans cover 70 percent of the Earth surface and constitute more than 95 percent of the planet living space. The deep seabed beyond national jurisdiction covers half of the Earth's surface, and 50 per cent of is below 3,000 meter depth. The seabeds sustain important functions and ecosystem services, including nutrient regeneration, carbon sequestration, biological and mineral resources.

Both the exploitation and the conservation of seabeds pose fundamental questions for consideration by jurists, policy makers and negotiators, issues associated with the man-made rule of law but with consequences surpassing purely legal considerations and that require great qualities of judgement and care. Short-term commercial aspirations compete with national strategic and technological goals as well as global ecological concerns. There is, as often in the case, a balance to strike between enabling seabed resources use and preserving fragile environments. The question for lawyers is how to best ensure that known and potential impacts are taken into account when taking decisions as to the use of the seabed and its resources, and how to ensure that access to seabed resources benefit all stakeholders in an equitable and sustainable manner. Legal principles such as the precautionary principle or common heritage of mankind must serve as guidance. Procedural tools, such as environmental impact assessments (EIAs), aim to make sure that consequences are assessed and interested parties heard. Marine management tools aim to ensure long-term coherence between activities and interests. This book aims to offer a new perspective on the juridical aspects raised by the use and protection of natural resources on and underneath the world's seabeds. Several chapters

DOI:10.1163/9789004391567_002

also carry a strong call for science-based decisions and the development of sound ecosystem-based management rules.

There have been numerous works on the law of sea with relevance for seabed uses and protection, in particular in relation to the deep seabed.[1] However, there are to our knowledge few contributions which look at the seabed in such a comprehensive and transversal manner as the present book, covering several sectors, actors and jurisdiction areas.

This book aims to make several contributions to the literature. First, a common scientific knowledge basis is necessary to correctly design legal rules, appraise their consequences – known and potential –, and reach evidence-based decisions.[2] Therefore, the book starts with a multidisciplinary definition of the seabed that is rarely provided. Second, the book presents an up-to-date analysis of the most pressing and fundamental legal questions related to the use and protection of the seabed. It does this by juxtaposing sectoral regimes and comparing the regulatory approach in areas within and beyond the national jurisdiction. In doing so, the book argues for a more consistent and cross-sectoral approach, identifying some common management principles and tools. Third, many chapters offer a critical analysis and make suggestions for improvement of the applicable legal regime, the manner to fill legal gaps or to advance current treaty negotiations. In that sense, the book

1 See notably: K. Zou (ed.), *Global Commons and the Law of the Sea*, Maritime Cooperation in East Asia, Volume: 5 (Brill, 2018); A.G. Elferink and E.J. Molenaar (eds.), *The International Legal Regime of Areas beyond National Jurisdiction: Current and Future Developments* (Brill, 2010); J.M. Van Dyke, D. Zaelke and G. Hewison (eds.), *Freedom for the Seas in the 21st Century – Ocean Governance and Environmental Harmony* (Island Press, 1993); Rosemary Rayfuse, *Research Handbook on International Marine Environmental Law* (Edward Elgar Publishing Ltd, 2015); D. Vidas (ed.), *Law, Technology and Science for Oceans in Globalisation* (Brill, 2010); N. Bankes and S. Trevisanut, *Energy from the Sea: An International Law Perspective on Ocean Energy* (Brill, 2015); E. Egede, *Africa and the Deep Seabed Regime: Politics and International Law of the Common Heritage of Mankind* (Springer, 2011).

Some earlier publications are also of notable value, such as: *Our Seabed Frontier: Challenges and Choices*, National Research Council (The National Academies Press, Washington, 1989); *The Ocean: Our Future*, Independent World Commission on the Oceans (Cambridge University Press, 1998); M.G. Schmidt, *Common Heritage or Common Burden?: The United States Position on the Development of a Regime for Deep Sea-Bed Mining in the Law of the Sea Convention* (Oxford University Press, 1990).

2 As a matter of example, see the Treaty obligation for the European Union to take account of available scientific and technical data when preparing its policy on the environment (Art. 191.3, Treaty on the Functioning of the European Union). See, more generally, on evidence-based law making, but still in the case of EU law: Š. Majcen, 'Evidence based policy making in the European Union: the role of the scientific community' (2017) 24 *Environmental Science and Pollution Research* 7869. The European Commission refers itself to evidence-base decision making as part of its 'Better Regulation Agenda'.

aims to contribute actively to several of the current processes at international, European or national levels.

Providing a new comprehensive and critical perspective on seabed regulation is also timely. The issues and concerns inherent to the seabed have been known for a long time, but new uses, new technologies, new knowledge about seabed ecosystems, higher tensions and potential disputes due to competing uses and interests, urge to reflect on which regulatory approaches to pursue.

In 2019, the United Nations (UN) celebrated the 25th anniversary of the entry into force of the 1982 United Nations Convention on the Law of the Sea (UNCLOS)[3] and the establishment of the International Seabed Authority (ISA).[4] The regime for deep seabed mining beyond national jurisdiction – in the Area –, which is under the competence of ISA, has developed towards a 'Mining Code', with the prospect of commercial activities getting increasingly realistic. So far, ISA has entered into 29 15-year contracts with contractors for the exploration of deep sea minerals, and the draft regulations on exploitation should be finalised in 2020.

In parallel, UN negotiations have started on an international legally-binding instrument under UNCLOS on the conservation and sustainable use of marine biological diversity of areas beyond national jurisdiction.[5] Marine genetic resources, including questions on the sharing of benefits, measures such as area-based management tools and marine protected areas, EIAS, and technology transfer are among the topics discussed.

Within national jurisdiction, states have been adopting legislation to implement UNCLOS and fulfil their obligations thereunder, but have also indicated looking at the exploitation of new minerals – beyond oil and gas – on their continental shelf, and have enacted new legislation in that sense. For example, in March 2019, Norway enacted a new Seabed Minerals Act, opening for exploration for and extraction of seabed minerals on its continental shelf.[6] This new

3 United Nations Convention on the Law of the Sea (UNCLOS) adopted and opened for signature 10 December 1982, entered into force 16 November 1994, 1833 UNTS 3.

4 UNCLOS, Art. 156.

5 A Preparatory Committee was established by the United Nations General Assembly (UNGA) Resolution 69/292 of 19 June 2015 with the view of developing an international legally binding instrument under UNCLOS on the conservation and sustainable use of marine biological diversity of areas beyond national jurisdiction. In its resolution 72/249 of 24 December 2017, the UNGA decided to convene an Intergovernmental Conference to consider the recommendations of the Preparatory Committee. The first session of the Intergovernmental Conference took place on 4–17 September 2018 and the second session on 25 March–5 April 2019. The third session was convened from 19 to 30 August 2019. The fourth session is scheduled for the first half of 2020.

6 *Lov om mineralvirksomhet på kontinentalsokkelen* (*havbunnsmineralloven*) of 22.03.2019, entered into force on 1 July 2019.

quest for minerals is fuelled by the needs of the digitalisation of the economy and the energy transition. It requires secure, stable and sufficient supply of metals and rare earth elements (REEs) for electronic devices like smart-phones, or energy technologies like solar panels, wind turbines and electric vehicles.[7]

Since the laying of the first telegraph cable across the seafloor of the Atlantic between Britain and the United States in 1866,[8] continents have been increasingly interconnected. Interconnectivity is another feature of today's world which sets its mark on the seabed, whether it is for telecommunication purposes (cables for transfer of digital data including telephone and the Internet) or energy transport purposes between coastal states or for bringing energy to shore (oil and gas pipelines from production field, power cables from offshore wind farms, interconnectors).[9] Those cables have developed to form a global network of undersea infrastructures which, in closed-basin seas such as the Mediterranean Sea, the Baltic Sea and the North Sea, results in a high number of crossings. Since each crossing is governed by a separate 'crossing agreement' entered into by the owners, the laying of a new cable or pipeline in the above mentioned sea areas usually involve the conclusion of a high number of those agreements. The laying of new undersea cables and pipelines may also involve crossing the continental shelf of other coastal states, a situation which may trigger the conclusion of a specific agreement between the states concerned, and/or the application of relevant provisions from UNCLOS[10] and regional agreements. The two Nord Stream gas pipelines from Russia to Germany via the Baltic sea are good examples of such crossing situation, where delineation of the course for the laying, permitting procedures, due consideration to marine protected areas and national security interests have been among the legal issues at stake.

Those are only few examples of the new context for seabed resources management.

---

7 On the role of minerals and REEs in the digital economy and the energy transition, see for example: G. Pitron, *La guerre des métaux rares : La face cachée de la transition énergétique et numérique* (Broché, 2018); 'Substitution of critical raw materials in low-carbon technologies: lighting, wind turbines and electric vehicles', European Commission, Joint Research Centre, Öko-Institut e.V., 2016.

8 Other submarine telegraph cables had been laid down previously over more limited distances, like across the English Channel between France and Great Britain in 1852. Source: R. Salvador, G. Fouchard, Y. Rolland et A.P. Leclerc, *Du morse à l'Internet, 150 ans de télécommunications par câbles sous-marins* (AAcsM, 2006).

9 For a world map of submarine cables see for example <https://www.submarinecablemap.com/>.

10 UNCLOS, Art. 79.

2 Access, Uses and Protection of Seabed Resources as Cross-Cutting Themes

Three cross-cutting themes structure the book: access, uses and protection of seabed resources. A short introduction to those three themes follows below.

2.1 *The Legal Status of the Seabed and Access to Its Resources*

Once jurisdiction over seabed areas has been established through a process of delineation, coastal states can exercise their sovereign rights within their national jurisdiction. This results in a definition by the negative of the Area, which is 'the seabed and ocean floor and subsoil thereof, beyond the limits of national jurisdiction'.[11] It is notable that the question of delimitation of jurisdiction and of whom should have access to seabed resources, is relatively recent, and had mostly been a matter for private companies until the mid-20th century.[12] Technological innovation has increased interest in the use of the seabed for purposes of control over resources and maritime spaces, and claiming jurisdiction has become the primary formal step for coastal states at the international level when the presence of seabed resources is proven.

Defining rules for delimitating the different maritime spaces according to a zonal approach, and defining the rights attached to those different zones, have been the primary focus of negotiators at the three UN Conferences on the law of the sea. Still today, defining the limits of national jurisdiction can be a contentious matter for coastal states aiming at harnessing seabed resources.[13]

Besides the tensions between coastal state's sovereign rights and maritime nations' interest in ensuring free access to oceans and straits, the voice of Arvid Pardo, the then Maltese ambassador to the United Nations, raised in 1967, calling for the adoption of a common status and regime for 'the seabed and the subsoil thereof' beyond national jurisdiction, which will secure 'the use of their resources in the interests of mankind'.[14] The final text of UNCLOS defines the Area itself and its mineral resources as 'common heritage of mankind'. The status of common heritage of mankind involves the non-appropriation of the seabed by states or private entities, the sharing of benefits for the common

11 UNCLOS, Art. 1(1)(1).

12 See Chapter 3 of this book, H.W. Andersen, 'A short human history of the ocean floor'.

13 See notably the ITLOS Special Chamber, Judgment in Dispute Concerning Delimitation of the Maritime Boundary Between Ghana and Côte d'Ivoire in the Atlantic Ocean, 23 September 2017.

14 Arvid Pardo's speech, UNGA 22nd session, 1 November 1967, Agenda Item 92, full text available at <http://www.un.org/depts/los/convention_agreements/texts/pardo_ga1967.pdf>.

good of humankind, and the use of the seabed for peaceful purposes only.[15] The deep seabed and its resources is to be administered for the benefit of mankind as a whole, by ISA as an autonomous international organisation, on behalf of the international community.

This book discusses both the specific case of the Area and its mineral resources as well as spaces and resources falling outside that definition. For example, which legal status should be given to newly discovered resources such as marine genetic resources (MGR)? Do they pertain to global commons when located beyond national jurisdiction? Could the concept of common heritage of mankind apply to them? One recurrent theme in several chapters involves a discussion of the consequences of the legal status of the seabed on the types of exploitation and protection regimes we choose for seabed resources and ecosystems.

This can translate into a series of obligations for state and non-state parties. States in particular are bound by general principles defined in both treaties and international customary law, such as the precautionary principle or the prevention principle. More specific obligations can derive from the legal status of the seabed, such as in Article 194(2) UNCLOS which provides for due diligence in the respect of marine environment protection.[16] This comes in addition to Article 192 UNCLOS which imposes on all States a general obligation to protect and preserve the marine environment. It can also translate into a more detailed legal regime for access and benefit sharing.

2.2 *Multiplication of Seabed Uses*

The oceans and the deep sea bed are increasingly coveted and are becoming ever more crowded. Innovations in marine technologies open for even more uses of the seabed than anticipated. There has been a multiplication and diversification of human uses of the seabed and its resources, and perspectives have differed between seeing the seabed as a *medium* (notably for the laying of infrastructures) and seeing it as a *resource* in and of itself (where seabed resources are both geological and biological). The spectrum of actual and potential uses has been constantly expanding and includes, notably, the laying of undersea cables and pipelines, marine research, extraction of petroleum and other mineral resources, exploitation of the renewable energy resources of

15 UNCLOS, Art. 137, 138 and 140(1).

16 See Advisory Opinion, *Responsibilities and obligations of States with respect to activities in the Area,* Seabed Disputes Chamber of the International Tribunal for the Law of the Sea, 1 February 2011, ITLOS Reports 2011, p. 10 (Case No. 17); UNCLOS Articles 208(3) and 209(2) 01.

the sea (wind, wave, tidal, storage), ancillary construction of islands and platforms (still mostly anchored), bottom fisheries, exploitation of marine genetic resources, military uses and even carbon dioxide storage beneath the seabed.

Tensions and legal issues are emerging out of the conflict of perspectives between those who define the seabed and its resources as the common heritage of mankind and those who seek to convert the ocean floor into a 'seabed factory'.[17] The discussion of the conflicting interests between exploitation and conservation of the seabed and its resources is another leitmotiv in this book.

2.3 *Conflicting Uses, Coexistence, Resolving Mechanisms and Protection Regimes*

The multiplication of seabed uses leads to a discussion on their coexistence. The predominance of a sectoral approach is highlighted, and different solving mechanisms discussed.

Indeed, despite the overarching regime defined in UNCLOS, the regulation of ocean activities is essentially a sectoral matter, both within and beyond national jurisdiction. The same applies to the work of the different competent organisations, which often lacks close coordination and prevents further consistency between the regimes.[18] Private parties are often relying on bilateral agreements to solve possible conflicts of uses of seabed space and resources. Finally, following a sectoral approach also means that gaps exist between the different regimes, bringing legal uncertainty and increasing the risk of fragmentation. The issue is known, but, as activities develop, there is a clear need for closer coordination at the overarching level, in the form of common principles, common tools and standardised practices.

While uses of the seabed are diversifying and the prospects of a 'blue economy' are expanding,[19] sustainable use and protection of the marine environment should act as an overarching goal, as reflected in UN Sustainable Development Goal 14 (SDG 14). SDG 14 aims to the conservation and

---

17 On the concept of 'seabed factory', see 'Getting to the bottom of it', Norwegian Continental Shelf 2–2016, pp. 14–17, available at <https://www.npd.no/globalassets/1-npd/publikasjoner/norsk-sokkel-en/arcive/ncs-2-2016.pdf>.

18 See, as concerns ISA's works, Section XXII – 'Relationship with the United Nations and other relevant international organizations and bodies' in *Report of the Secretary-General of the International Seabed Authority under article 166, paragraph 4, of the United Nations Convention on the Law of the Sea*, ISBA/25/A/2, 3 May 2019, available at <https://www.isa.org.jm/document/isba25a2>.

19 The sustainable exploitation of seabed resources is full part of the visions developed for the 'blue economy'. As a matter of example, see *The EU Blue Economy Report 2019*, European Commission, available at <http://periscope-network.eu/analyst/annual-report-eu-blue-economy>.

sustainable use of the oceans, seas and marine resources for sustainable development.[20] This necessary common approach is justified by the essential place the seabed has in what can be called 'the oceans commons'.

3 Content of the Book

The book is divided into seven parts, following a systematic approach to the different issues of access, uses and protection of seabed resources, building on a distinction between the regimes applicable to areas beyond national jurisdiction and within national jurisdiction.

The book also balances theoretical chapters with case studies. Norway and the North Sea are at several occasions taken as examples due to the extent of seabed issues currently taking place in that region, but several chapters also present studies of the legislation in Australia, Canada, China, Namibia, New Zealand, Mexico, Papua New Guinea, Russia, South Africa and the United States of America (USA) in addition to the European Union and specific EU countries.

3.1 *Part I: Defining the Seabed*

Part I of the book offers a multidisciplinary definition of the seabed and its ecosystems. Indeed, it is necessary to combine knowledge of geoscience and marine biology to capture the richness and intrinsic characteristics of the seabed and its biodiversity. This science-based understanding of how the ocean floors have been formed, and what they contain and represent as ecosystems for species is put in perspective with a legal history analysis of its regulation over time.

Braathen and Brekke (Chapter 1), both geologists, provide readers with a fundamental understanding of the geoscience of the seabed. They describe the natural processes impacting the structure of the seabed, i.e. both its topography and subsurface geology. A detailed understanding of the seabed geology is a pre-condition to any exploitation decision and aims to ensure the good management of the resources, or, on the contrary, to preserve the seabed from any activity. The authors also address the manner how human activity impacts the physical seabed and could change it over time, affecting its physical consistency. They carry out an important message towards decision makers and the

20 For an update on the progress of SDG 14, see <https://sustainabledevelopment.un.org/sdg14>.

law community in general, making some recommendations as to the regulation of activities on the seabed.

Ramirez-Llodra (Chapter 2) describes the deep-sea ecosystems from a marine biology perspective. The chapter gives a unique picture of the high biodiversity and variety of faunal communities and their habitat which remain hardly accessible to most human beings. The deep seafloor has shown to support one of the highest biodiversities in the planet in a wide array of interconnected habitats. The chapter starts by presenting the general biological features and key ecosystem functions of the main deep-sea ecosystems. The chapter continues by describing the main human activities that affect, directly or indirectly, deep-sea habitats and their ecosystems. These anthropogenic impacts are grouped into 2 main categories: waste dumping and resource exploitation. As Braathen and Brekke did, this chapter asks for science-based decisions, with a call for a better understanding of deep-sea ecosystem composition, diversity and functioning, and the manner they response to stressors when considering resource utilisation regimes.

In his chapter (Chapter 3), **Andersen** gives us a short human history of the ocean floor. He reviews the long history of how western societies have perceived and to some extent experienced the deep ocean floor from the perils of the seafarers in the sixteenth century to UNCLOS III. The historical approach reveals how much influence the representation of the seafloor and its resources has had on their regulation. Importantly, Andersen makes clear the shift operated from conceiving the seafloor as a medium to looking at it as a resource. This shift in perception and interest also explains the shift in interest representation between private companies and national governments, between flag states and coastal states. The chapter ends by raising the question of the suitability of UNCLOS to answer the new challenges.

3.2 *Part II: Delimitation of Jurisdiction over the Seabed and Right to Its Resources*

Part II of the book is dedicated to the principles applying to the delimitation of jurisdiction over the seabed as a necessary prerequisite to access to its resources.

Brekke (Chapter 4) reviews the main delineation principles currently applied in order to determine maritime jurisdiction of the coastal state over the seabed, as defined in Article 76 of the 1982 UNCLOS, and as interpreted by the courts. The chapter continues by reviewing the Norwegian experiences with setting the country's maritime boundaries, as a concrete application of the principles identified. The elements related to the organisation of the work for the purpose of submissions to the Commission on the Limits of the Continental Shelf (CLCS), the mapping of the outer limits and to the cooperation with neighbouring states are successively analysed. The chapter ends with an overview of the currently identified resources on the Norwegian continental shelf.

Skaridov (Chapter 5) explores the question of delimitation of jurisdiction in the Arctic as a fundamental starting point for discussing the regime of exploration for and exploitation of seabed resources in an area of the world which is subject to competing state and commercial interests, and so a source of potential conflicts. The author starts by pointing out the difficulty of establishing a precise scientific knowledge basis as to the volumes of resources contained in the Arctic seabed. Most part of the chapter is then dedicated to the question of competing claims and the analysis of Arctic states agreements and disputes regarding the establishment of baselines, the delimitation of the state territories and the applications in the view of establishing the outer limits of continental shelf submitted to the CLCS. The chapter also reviews the national Arctic strategy of the five Arctic coastal states. The author advances several proposals on possible ways of resolving disputes.

Brazovskaya and Ruchkina (Chapter 6) offer a study of the international regulation of the seabed in the Antarctic, and address the two topics of delineation of maritime spaces and use of seabed mineral resources. In the absence of a conventional delimitation of Antarctic spaces, the authors start by addressing the challenging task of establishing the boundaries in the polar regions, which are constantly covered with ice and, accordingly, the difficulty of measuring the width of the territorial sea and the EEZ. Second, the authors note that despite the fact that this continent was discovered already in 1819, its geographical and climatic features do not allow its full use. Meanwhile, a series of legal principles defined in the Antarctic Treaty System put limits on the use of the territory, limiting it to peaceful purposes and scientific research. In that context, the authors discuss the international legal regime which would apply to the exploration for and exploitation of Antarctic mineral resources, since, as they state, sooner or later, the question of industrial extraction of Antarctic natural resources will be raised. Given the 'frozen' territorial claims of countries, the authors argue that this will require the adoption of a special legal regime for the continental Antarctic shelf as an alternative to the continental shelf regime in the sense of the 1982 UNCLOS.

3.3 *Part III and IV: Exploitation of Non-Living, Living and Marine Biodiversity Resources in Areas beyond National Jurisdiction*

The next two parts of the book look at the areas beyond national jurisdiction (ABNJ), first for the exploitation of non-living resources (Part III) and then for the living and marine biodiversity resources (Part IV). Putting the regime for living and non-living resources in parallel enables to draw conclusions in terms of suitability of a common approach, or, on the contrary, the need to design different models, notably for the purpose of benefit sharing.

Part III of the book dedicated to non-living resources starts with **Dingwall**'s chapter (Chapter 7), which provides a comprehensive analysis of the international legal framework applicable to deep seabed mining beyond national jurisdiction, i.e. in the Area. This framework was elaborated under UNCLOS, leading to the establishment of ISA. ISA governs deep seabed mining activities on behalf of humanity, as a reflection of the qualification of the deep seabed as 'common heritage of mankind'. The Authority is responsible for regulating the Area and granting mining contracts to allow States and other entities to explore for and exploit deep seabed minerals. The Chapter reviews the key elements of this regime, the current level of activity and the role of the different actors – ISA, contractors, state sponsors, the Enterprise –, and discusses as well the case of actors still outside the UNCLOS deep seabed mining regime (non-States Parties to UNCLOS, such as the USA).

Røsæg (Chapter 8) continues the discussion on deep seabed mining by identifying gaps in the currently applicable regime that prevent moving forward towards commercial activities in the Area. The unclear applicable regime for mortgaging the equipment involved in mining operations is seen as a problem. Other examples involve patent infringements, labour protection, and extra contractual liabilities. Balancing the pros and cons of acting outside or within the scope of the Authority, the chapter discusses different possible alternatives, and which role sponsoring states and flag states could play for filling the legislative gaps and offering a more sound framework legislation for commercial activities in the Area than just adding conditions to licenses.

Activities in the deep seabed area beyond national jurisdiction raise important maritime security issues such as the emplacement of weapons of mass destruction, prospects of piracy and terrorism against ships engaged in deep seabed mining activities, as well as the challenge that deep seabed mining could actually raise as regards environmental security. Yet, relatively little has been written on maritime security and the Area. **Egede** makes therefore an important contribution (Chapter 9) by investigating possible maritime security issues that could arise in the Area. In doing so, the chapter takes both a rather traditional state-centric maritime security approach and a more non-State centric viewpoint. When discussing the notion of maritime security in the law of the sea, Egede points out the increasing shift from a sole focus on States actors to an increasing engagement with diverse non-State actors. Non-state actors which include international organizations, private maritime security companies, pirates and armed robbers at sea, private fishing trawlers engaged in Illegal, Unreported and Unregulated (IUU) fishing, terrorists groups and victims of maritime crimes. This shift is accompanied by new interactions between State actors and non-State actors involved in peaceful and less-peaceful activities. This triggers a discussion of the applicable international legal framework, including the one under the responsible of ISA.

Part IV of the book is dedicated to the exploitation of genetic resources and living resources, a topic which has concentrated much attention lately, with the ongoing negotiations on the use of marine biological diversity in areas beyond national jurisdiction (BBNJ).[21] All three authors come with critical analysis of the current regime, the existing gaps and make suggestions for improvement.

Scovazzi (Chapter 10) discusses rights to genetic resources in ABNJ, which is one of the central issues of the ongoing negotiations at the United Nations. Among the still pending questions are whether the new regime will be based on freedom of the sea or common heritage of mankind or a third *sui generis* approach. Other notable questions relate to access to the resources, benefit-sharing (both non-monetary and monetary benefits), inclusion of genetic resources of the water column, intellectual property rights, role of traditional knowledge, and 'straddling' genetic resources.

Tvedt (Chapter 11) continues the discussion started by Scovazzi, looking at the potential elements in regulating rights and use of genetic resources in ABNJ, as it appears that the existing Access and Benefit Sharing (ABS) regimes cannot just be copy-pasted. The author follows a practical approach and explores different models for how to make an ABS-system for ABNJ legally binding on users. This means that all suggestions are tested against whether it would be possible to oblige private parties to a system of both access and benefit sharing. For example, the chapter provides a clear view on how the proposed regulatory model can be aligned with the exclusive rights awarded by patents when the activity results in an invention. The chapter discusses open and semi-open source options for using repositories as a practical means of securing access to the resources for the many. It highlights the potential role of private contracts for such a global regime to become functional.

Caddell (Chapter 12) looks at the regime for exploitation of living resources directly on the seabed in the ABNJ. While limited regulation existed in relation to bottom fisheries two decades ago, the UN General Assembly has played an essential role, calling for the protection of vulnerable marine ecosystems (VMEs) on the seabed in a series of highly influential Resolutions. This political impetus has prompted the adoption of complementary conservation and management measures by regional fisheries management organisations

---

21 See above (n5).

(RFMOs), flag states and other actors. The chapter reviews the progress made in the development of standards for deep-sea bottom fishing. It also identifies a series of shortcomings in the implementation of those standards, and advances a series of recommendations as to the future of the regulatory regime.

3.4 *Part V: Principles Applicable to Sovereign States When Exploiting Seabed Resources within National Jurisdiction*

Part V of the book looks at the manner sovereign states regulate or intend to regulate the exploration for and exploitation of seabed resources within their national jurisdiction, and which constraints international and European law put on the development of national legislation.

Roux and Horsfield (Chapter 13) offer a series of case studies of national legislations applicable to seabed mining within the national jurisdiction of coastal states. Interest in exploration and exploitation activities has been increasing over the past decade and has largely been concentrated in the EEZ of five states namely: New Zealand, Australia, Namibia, Mexico and Papua New Guinea. The approach, policy positions or decisions adopted by these countries in relation to seabed mining within their jurisdictions has generally been cautious, with due regard to the precautionary principle. These have ranged from permanent bans, moratoria, strategic environmental assessments and reviews by environmental agencies. Through a comparative analysis of these cases, the chapter identifies a structure for decision-making on seabed mining that can be used by coastal States.

Arnesen, Greaves and Pozdnakova (Chapter 14) look at the example of the European Union, and how the latter intends to regulate Member States' activities on the seabed. The authors consider the question of EU competences with respect to the seabed, namely whether the EU has competence; what kind of competence it has (shared or exclusive), and the territorial limits of this competence (functional approach). Then, the authors examine how the EU deals with two specific sectors in light of new challenges, namely, seabed mining and environmental protection and liability.

Chen (Chapter 15) offers another case study, reviewing the 2016 Law of the People's Republic of China on Exploration for and Exploitation of Resources in the Deep Seabed Area (the Deep Seabed Law). The enactment of the law and its implementing regulations represents China's national implementation of UNCLOS. The chapter raises two central questions: why does China need this law and why 2016 represented a positive context for its adoption. The author undergoes a systematic review of the provisions of the Deep Seabed Law, commenting specifically the three core intents, namely controlling, securing and preparing. The author also compares the Law with the equivalent legislation of other countries and concludes with an assessment of the limitations and impact on China's other areas of law. This chapter provides readers with a better understanding of China's strategy of building a 'deep sea maritime power' by 2020. By pointing out the legal innovation of the Deep Seabed Law, the author argues that it could serve as a reference for other countries that intend to incorporate UNCLOS obligations on deep seabed mining into their domestic legal system.

Chircop (Chapter 16) examines Article 82 of UNCLOS, which is a novel provision introducing the first-ever international royalty on production from non-living resources within national jurisdiction, specifically from the extended continental shelf as defined in Article 76. The author points out that Article 82 has several textual ambiguities that could pose a challenge for its interpretation and implementation by both affected coastal States and ISA. The Authority plays a particular role in the implementation of Article 82, since it is responsible for receiving payments or contributions in kind and for effecting their distribution to States Parties to the Convention, especially developing countries, in accordance with equitable criteria. Article 82 is expected to be first activated on Canada's extended continental shelf off Newfoundland in the Northwest Atlantic. In implementing Article 82 Canada faces domestic political, economic and legal challenges, in addition to the textual ambiguities. This chapter discusses the major issues and underscores the important leadership and precedential role played by Canada in domesticating Article 82 and developing a relationship with the Authority on this matter.

3.5 *Part VI: Building, Operating and Removing Installations on and beneath the Seabed*

Part VI of the book looks at the applicable regimes for the building, operating, re-moving and even re-use of installations on and beneath the seabed. The use and re-use of installations in connection to storage of carbon dioxide (CO_2) under the seabed is also part of the upcoming topics which need further legal scrutiny.

Bankes (Chapter 17) examines the legal issues associated with the use of sub-seabed transboundary geological structures, including saline aquifers, for storage or disposal purposes focusing on the geological sequestration of CO_2. Underground CO_2 storage represents the final stage of the carbon capture and storage (CCS) chain. The chapter reviews the existing law requirements and guidelines that frame the operations of injection of CO_2 as part of enhanced oil recovery (EOR) as well as CO_2 disposal or storage operations (CO_2/EOR), such as the Protocol to the London Dumping Convention. It also discusses the implications of Articles 74(3) and 83(3) of UNCLOS for the situations of transboundary geological storage sites, including in areas of overlapping maritime entitlements. The author examines how these issues have been or could be dealt with in delimitation agreements, framework agreements and joint development agreements. As the CO_2 storage projects worldwide get closer to operating phase, the legal issues raised by Bankes become increasingly imperative to consider.

Trevisanut (Chapter 18) addresses the issue of decommissioning of installations placed on the seabed. The number of offshore installations, reaching the end of their life-cycle, are increasing and many of them can represent an environmental hazard or area threat for other legitimate uses of the sea and the seabed. This time constraint put on decommissioning decisions calls for an analysis of the applicable international legal framework, as undergone by the author who also highlights some of its ambiguities. The chapter offers a systematic review of the legal regime incorporated into UNCLOS, and continues by focusing on generally accepted international rules and standards (GAIRS) at both a global and regional level. The author makes important conclusions as to the manner the lack of a binding instrument at the universal level, and the geographical fragmentation of the existing legal regime act as an obstacle to the development of clear global GAIRS, and thus to a more coherent and effective legal framework for the decommissioning of offshore installations.

Roggenkamp's chapter (Chapter 19) makes the link between the two topics of decommissioning of offshore installations and CO_2 storage and disposal. The author examines the possibility of using depleted oil and gas fields in the North Sea area for CO_2 storage. An important issue in this regard is the relationship with the removal obligations that exist offshore. Instead of removing the installations after the oil/gas production has ceased, the possibility to re-use the depleted reservoir for CO_2 storage could entail that the offshore installations and pipelines have to be kept in place. The chapter looks into this possibility from the perspective of both international law and the national law of three North Sea coastal States: the Netherlands, United Kingdom and Norway. This review reveals legal uncertainties for the transitional period between production ceases and a CO_2 storage permit is awarded. As the exploitation of oil and gas fields in the North Sea area is ageing and maturing, and although first steps have been made to facilitate re-use of installations, the author calls for legislators and policy makers to act and seize this window of opportunity.

In his chapter (Chapter 20), **Harrison** explores the extent to which the investment treaty framework can be applied to seabed investments and, if so, the scope of protection that may be expected by investors. It addresses the geographical scope of international investment treaties, and considers the substantive protections that are available if a seabed investment is found to fall within the scope of those treaties. It also questions whether the treaties can be applied to seabed investments made within the jurisdiction of coastal states. The analysis demands an understanding of how international investment law interacts with relevant rules in the international law of the sea and international environmental law. Given the intensification in seabed activity around the world, there is a need to clarify key issues that arise in the interpretation and application of investment standards in this context.

Waverijn (Chapter 21) questions to which extent seabed installations can be mortgaged as a way of reducing financing risks. To conduct his analysis, he takes energy installations in the North Sea as study case. Indeed, offshore oil and gas, offshore wind and ocean energy projects are capital intensive with billions of euros required for their construction. It is common business practice to raise debt to finance them, which is more expensive when risks are greater. Security rights such as mortgage and pledge reduce the risks of lenders, however restrictions exist under national law. In his chapter, the author analyses the solutions provided under Dutch property law, as it is currently impossible to mortgage installations situated on the seabed further than 22.2 kilometres off the Dutch coast. In his view, allowing for such mortgages could benefit the development of the offshore energy sector as this would reduce risks.

3.6 *Part VII: Conflicting Uses or Coexistence, Resolving Mechanisms and Protection Regimes: Towards a More Integrated Approach*

The final part of the book, **Part VII**, is dedicated to the question of coexistence between activities competing for the same seafloor area and which regime exists for solving conflicts in uses, such as resolving mechanisms. It contains a call for more consistency in the approach of coexistence of activities, but also in the manner divergent interests, including environmental protection and sustainable use of the oceans, can be conciliated through management tools.

Rayfuse (Chapter 22) examines the general principles which lay the basis of normative frameworks and management approaches in the efforts to promote cross-sectoral cooperation and coordination, and avoid inter-sectoral conflicts and negative decisions for the conservation and protection of the marine environment. Her chapter reviews the new cross-sectoral management approaches and tools that are emerging. It then turns to a discussion of the key tools that have been developed to assist environmental management of marine and seabed uses and an examination of emerging frameworks for cross-sectoral management aimed specifically at avoiding, minimising or resolving conflicting uses on the seabed, particularly in ABNJ.

Askheim's chapter (Chapter 23) deals with agreements between owners/operators of crossing pipelines, power cables and telecom cables (all termed connectors) on the seabed. It reviews the legal basis for such agreements as well as their main provisions. The manner in which liability and indemnity clauses are designed is given attention throughout the entire implementation of the project (pre-completion, construction and post-completion). The chapter makes clear the need to complete the provisions of UNCLOS in specific crossing agreements as a way by which the owners of subsea transportation assets can organise themselves in the most balanced and predictable manner.

Using the Nord Stream gas pipelines in the Baltic Sea as a case study, **Langlet** (Chapter 24) investigates how competing interests relating to submarine pipelines are handled by the law of the sea. Particular attention is given to the role of transit states – i.e. States over whose continental shelf a pipeline is laid without it entering their territorial waters – and the limits to their right to regulate the laying of pipelines and thus act as arbiters of potentially competing interests such as the right of transport, the effective protection of the marine environment, and national security considerations. Only some few interests are recognized by UNCLOS as legitimate bases for a coastal State to adopt measures, but it could be tempting to use such grounds to pursue other objectives. It is concluded that although the pertinent rules in UNCLOS are complex and partly vague, the States concerned have in most cases diligently avoided pushing the limits of coastal State jurisdiction as set out in the Convention.

Svendsen (Chapter 25) looks at the regime for liability and compensation, and discusses which manner can best compensate damage caused by deep seabed mining activities in the Area. The author takes the view that the draft regulations for exploitation, in their present form, do not construct an adequate model for liability and compensation for damage as a result of these activities. The chapter sketches the current parallel system of the sponsoring State's responsibility for damage caused in the Area and the sponsored contractor's liability for damage caused in the Area. The chapter attempts to draw a model for an improved liability and compensation system for damage caused by deep-sea mining in ABNJ.

4 Acknowledgments

This book is the result of international collaboration between researchers gathered under the Oslo Law of the Sea Forum, 'OSLOS', a research network of the Scandinavian Institute of Maritime Law at the University of Oslo.[22] The success of the project is a tribute to all the authors, each of whom has contributed greatly in delivering state-of-the-art research in his or her specialized area of research.

The gathering of authors at the preparatory meeting in Oslo was made possible by funding received from the MARINFORSK programme of the Research Council of Norway (project number 255761).

The present editor would also like to thank Prof. Erik Røsæg and Prof. Henrik Ringbom for the opportunity to lead this project.

Finally, this book would not have been realized without the continual support of the publisher. Special thanks go to BRILL and their team for steadfast confidence in the project and vital support throughout the entire process.

---

22 Website of the Oslo Law of the Sea Forum (OSLOS): <https://www.jus.uio.no/nifs/english/research/projects/oslos/index.html>.

PART 1

Defining the Seabed

CHAPTER 1

Characterizing the Seabed: a Geoscience Perspective

Alvar Braathen and Harald Brekke

1 Introduction

The sea with its seabed has assets that have become extensively exploited as a source for food, energy and transport. Traditionally, the advancements of mankind on land have been more successful than in the oceans, mostly due to a physical setting that challenges easy-access and low-cost operations. However, this is gradually changing under the combined factors of the depletion of the required new resources as assets on land, and an increasing demand, resulting in an increased attention for the oceans. Meanwhile, the oceans remain mostly unknown, could be hostile and are for sure fragile.

In this contribution, our aim is not to explore the possibilities laying in the oceans and its seabed as other chapters in this book extensively cover them. Our objective is to inform about the seabed itself and define it, by asking some key questions that geoscientists could answer: What is the seabed made up of? Is the seabed a static entity or will the bed change over time? How could human activity influence the physical consistency of the seabed? These questions are basic but compulsory if one wish to implement a regulatory regime to seabed activities. It may be obvious, but decision-makers have to know what they regulate.

2 What Characterizes the Seabed?

The term *seabed* refers to the top-surface of earth in seas and oceans, also known as the seafloor or ocean floor. This surface has a *topography*, which is directly related to the nature of its *subsurface* geology, in places modified by ocean currents and sedimentary processes. Both the topography and the subsurface are important factors in the use of the seabed by humankind. From a resource perspective, humankind determines how resources from the seabed may be exploited and its resources distributed or protected by preventing exploitation. In all circumstances, the nature and accessibility of the seabed will be a crucial preliminary consideration. To understand the characteristics of the topography and subsurface of the seabed, we will have to look at the *formation* of the seabed, including the outer parts of the Earth's crust below oceanic water (2.1), and then the nature and processes of the interface between solid or semi-solid earth materials and water column in the sea (2.2). These are subjects covered extensively in many textbooks.[1] Our approach is therefore that of a summary.

2.1 *The Earth's Crust*

The main geological difference between the continents and the deep oceans is the nature of the Earth's crust. In geology, one speaks of *continental crust* as opposed to *oceanic crust*. The continental crust is very thick (20–40 km) and relatively light, while oceanic crust is thin (5–15 km) and considerably denser, as shown in figures 1 and 2. This is because continental crust consists of light minerals rich in silica and aluminium, while oceanic crust consists of heavy, dense minerals rich in iron and magnesium. The effect of this is that the continents float high on the earth's mantle, like a cork on water, while the deep ocean floor is barely afloat, like a piece of heavy wood. This marked difference in buoyancy, thickness and relative elevation, is actually the underlying reason for the general global distribution of land and water. The thick buoyant continents support the vast areas of emergent land, while the heavy and low-lying oceanic crust forms the floor of the deep parts of the huge ocean basins that accommodate the world's seawater.

The differences between the continental and oceanic crust relates to the processes by which they are formed. The outer part of the Earth, including the crust, is divided into several large segments called lithospheric plates, which fit together like pieces in a jig saw puzzle. These plates 'float' and drift slowly in different directions upon the deeper mantle of the Earth. This implies that along some boundaries the plates separate and move away from each other (divergent boundaries), while along other boundaries they collide (convergent boundaries). In geology, this process is called *plate tectonics*.

Along the boundaries where plates separate, space is continuously created as the plates move away from each other at the speed of 1–10 cm/yr. This space is immediately filled by molten rock from the mantle in the form of intrusions and lavas, which solidify and form new crust. This is the way the dense and heavy oceanic crust is formed. The process is called *ocean spreading* since it

1 See notably: J.P. Kennett, *Marine Geology* (Prentice Hall, 1982); E. Seibold and W. Berger, *The sea floor: an introduction to marine geology* (Springer, 2017).

implies that the ocean crust on both sides of the plate boundary is growing, causing the ocean to widen. Such widening oceans have all originated in break-up zones along which old continents were rifted apart by forces in Earth's mantle. These break-up zones then become new divergent plate boundaries between which the new ocean will form by ocean spreading. The Atlantic Ocean of today is such a spreading ocean, which accommodates increasing separation between the American continent on one side and Europe and Africa on the other.

Ocean spreading along some plate boundaries implies that other boundaries experience plate collision, making up convergent plate boundaries. At such boundaries, the crust of one plate will be forced under the other plate, with the contact termed a subduction zone. The heavy oceanic crust tends to be forced down. As the down-pressed plate moves down into the mantle in the subduction zone, the oceanic crust will melt again, and the molten rock will rise to the surface where it forms chains of volcanoes on the overriding plate. These volcanoes in many places form chains of islands, so called *island arcs*. Where the overriding plate is that of a continent the volcanoes becomes part of the margin of that continent. Water and sediments will be involved in the melting process within the subduction zone giving rocks enriched in light minerals akin to continental crust. As millions of years pass, such islands and/or continental margins at convergent boundaries of plates will collide with islands or continents on other plates. Since they are formed of light rocks they will not be subducted into the mantle again, but tend to become accreted to the colliding islands or continent forming mountain chains. That is the way continents grow. For instance, the Pacific Ocean of today is surrounded by subduction zones where its oceanic crust disappears beneath the American and Asian continents. Millions of years into the future, the Pacific Ocean will close and the two continents collide.

Thus, the plate tectonic process results in a cycle where oceans come and go while continents collide and grow (Fig. 1.1 in Annex). The plate tectonic process started more than 4 billion years ago when the Earth had cooled sufficiently for molten rock to solidify and begin forming crust on the planet's surface.

2.2 *Seabed Topography*

The general topography of the seabed of the Earth may be described in terms of the *continental shelf*, the *continental slope*, the *continental rise*, the *continental margin*, and the *deep ocean floor* (Fig. 1.1). These terms are used partly in a legal sense by Article 76 of the United Nations Convention of the Law of the Sea (UNCLOS) to define the outer limits of coastal state's jurisdiction over the continental shelf.

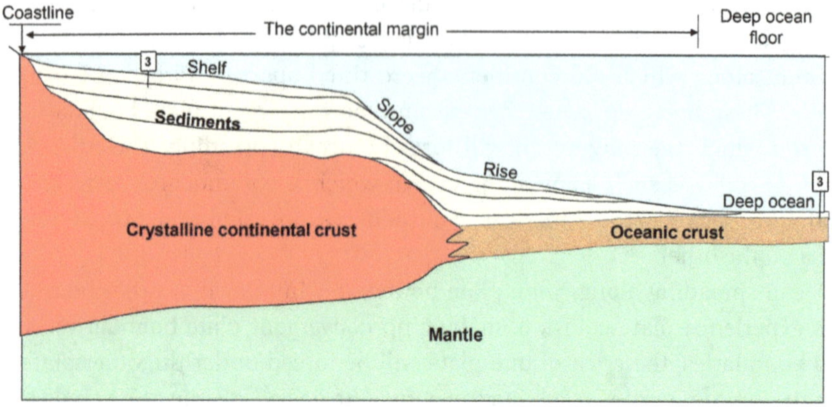

FIGURE 1.1 Profile showing a cross section of the elements of the general topography of the seabed extending from the coastline to the deep ocean floor

The topography of the seabed is related to the lateral distribution of the two types of crust, as shown in Figure 1.2. All the continents, which make up nearly all of the earth's land areas, have a continental shelf underlain by continental crust dominated by light components (minerals such as quartz and feldspar). This shelf is the part of the continent that is submerged below the sea, and make up vast areas of sea floor at a water depth starting at the beach and descending to some hundred meters. The shelf is bound seaward by the continental slope of similar light composition, dipping gently towards the deep ocean. In general, the continental slope is formed near the edge of the continental mass where the continental crust thins considerably and merges with the oceanic crust. At the base of the slope, at water depths below 3,000–4,000 meters, we reach the realm of the deep ocean floor that is underlain by oceanic crust made up of dark, heavy components. In many parts of the world one also finds a continental rise, which is an area of very gentle dip between the base of the slope and the deep ocean floor. The classical continental rise is a wedge shaped pile of sediments derived from the shelf areas and accumulated next to the base of the slope – in many places the sedimentary wedge partly overly oceanic crust.

In geology, the underwater areas of the planet are subdivided into two parts: the continental margins and the deep ocean floor. The continental margin extends from the shoreline to the end of the continental rise or to the base of the continental slope where no rise exists. The area seaward of the continental margin is the deep ocean floor.

The continental shelf is relatively flat and shallow. The widest continental shelves are found in the continental margins that were formed at divergent plate margins; also termed passive margins (see Fig. 1.1). Passive margins form

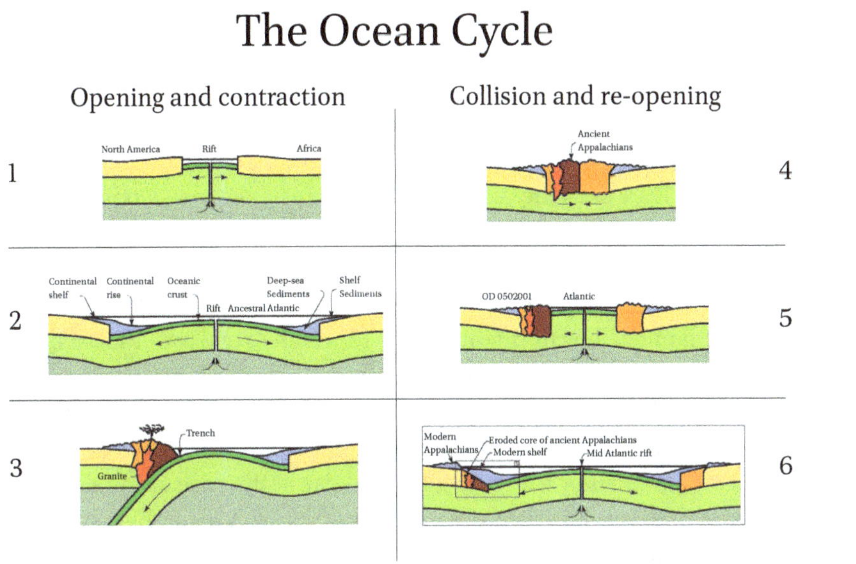

FIGURE 1.2 The Ocean Cycle. Schematic illustration of six progressive steps in the ocean cycle exemplified by the geological evolution of the Atlantic margins.

BASED ON PRESS AND SIEVER, 1974

1) The evolution starts with break-up, splitting North America from Africa, leading to formation of passive continental margins separated by oceanic crust gradually formed by intrusion of molten rock and extrusion of lavas at the mid-oceanic ridge (steps 1–2). This rifting and onset of ocean floor spreading of the Proto-Atlantic Ocean (the Iapetus Ocean) happened ca. 700 million years ago.
2) The mature stage of the ocean floor spreading in the Iapetus Ocean in Cambrian times (ca. 500 million years ago). In stages 1) and 2) the margins of the continents are tectonically and magmatically relatively passive while they are being progressively separated by the ocean floor spreading process. Continental margins in this geological setting are therefore termed 'passive margins'.
3) The contraction stage of the Iapetus Ocean in Ordovician times by the establishment of subduction zones along the continental margins. Such continental margins are tectonically and magmatically very active, and are therefore termed 'active margins'.
4) The collision between the old continents resulting in the final closure and destruction of the Iapetus Ocean and the consequent formation of the Appalachian/Caledonian Orogen (mountain chain) about 420 million years ago.
5) Renewed rifting along the axis of the orogen and re-establishment of seafloor spreading in the Atlantic Ocean through Jurassic and Cretaceous times. Note that continents at this stage have grown through the crustal material accreted at the continental margins during the continental collision and the previous ocean contractional stage.
6) Present spreading stage of the Atlantic Ocean where the passive continental margins still grow by the deposition of continentally derived sediments in the continental slopes along the outer parts of the margins.

Note: F. Press and R. Siever. *Earth* (W.H. Freeman and Company, San Francisco, 1974).

by the extension and rifting of the continental crust prior to the break-up of the continent, ultimately leading to ocean floor and spreading. Thus, the widest continental shelves in the world are developed in the continental margins of the Atlantic Ocean. Narrow continental shelves are typical in continental margins associated with subduction zones at convergent plate boundaries: the active margins. They are generally narrow, as for instance seen along the western margin of the American continent.

All continental shelves end at the continental slope. The shape and nature of the continental slope vary considerably around the world. Along most passive continental margins, the continental slope forms an overall even surface with low gradients. Several of these slopes are underlain by large accumulations of sediment derived from rivers or glaciers that has draped and smoothed earlier topography. Such slopes are dominated by sedimentary processes and are generally accompanied by a large continental rise at the base of the slope. For other passive margins, however, the continental slope is a complex system of plateaus, ridges and steep escarpments formed by fault movements and from volcanic activity. For the active margins, the continental slope of the overriding plate typically constitutes the steep, landward slope towards the subduction trench (Fig. 1.2). Such trenches are the deepest places on Earth.

The topography of the deep ocean floor seaward of the continental slope is formed by the oceanic crust and later modified by sediment deposition. The oceanic crust is produced by volcanism (sea-floor spreading) along the diverging plate boundary. Hot rock is less dense than cold rock. Therefore, the young crust along the ocean spreading boundaries is relatively elevated to about 2,500 meters depth because the rocks are heated by the volcanic processes. Away from the spreading axis, the crust becomes denser as it cools with time and gradually subsides to large depths of about 4,000–6,000 meters. Thus, the spreading boundaries stand up as a global system of broad mountain chains (volcanoes) of the deep oceans. These mountain chains are called *mid-ocean ridges*. The floor of these mid-ocean ridges is very irregular reflecting the original volcanic terrain of hills and mountains. Away from the elevated parts of the mid-ocean ridges, the crust becomes covered by sediment through time. In the deepest, older parts of the oceans the rugged terrain of the crust is totally buried by sediments, which forms the vast, flat abyssal plains of the world's oceans.

The oceans are locally punctuated by a different type of volcanic edifices, seen as seamounts that rise several thousand meters from the deep ocean floor. These volcanoes are geologically different from those of the ocean spreading process as they form above so-called hot-spots in the deep mantle. Many of these seamounts are surmounted by islands like those of Hawaii and Polynesia.

3 Seabed Composition and Its Mapping

3.1 *Seabed Composition*

Earth materials in the surface seabed of the continental shelf is made up of a veneer of material falling out of suspension, such as dust from continents, biological material, sometimes ash and, more recently, human waste. Seafloor biological activity creates debris, for instance from reefs. Sand and mud is mostly supplied by rivers entering the ocean at given points. From their entry point, biological debris, sand and mud are washed around by waves and currents in the sea before settling in positions where wave and current action have limited impact. Typically, this process of deposition will create distinct layering, or stacking of beds, of which the younging-up order, or sequence, of distinct beds make up a lithostratigraphy (Fig. 1.3).

With increased burial by progressive burial, or chemical reactions between grains in sediments and fluids, initially unconsolidated biological debris, sand, mud and clay will start to advance mechanical strength by creating bonds in a process termed diagenesis. This is the path towards lithification that ultimately leads to rocks, such as limestone, sandstone, mudstone and shale underlying the unconsolidated to poorly consolidated upper beds underneath the seafloor.

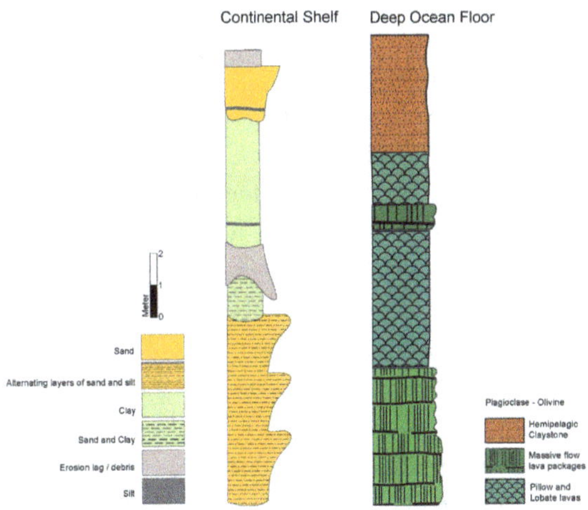

FIGURE 1.3 Typical lithostratigraphy of the continental shelf compared to similar stratigraphy of the deep ocean floor, shown in stratigraphic columns in which the total thickness would be between 10 and 100 meters in vertical section down from the sea-floor (on the top). The distribution of sediments such as sand and clay is directly linked to the source for various sediments and subsequently the physical conditions (waves, currents, composition) of the water column and its impact on the seafloor.

The seabed of the deep ocean floor shows similar traits to beds of the continental shelf. However, with increasing distance to land (source of sand), limited seafloor biological activity, and depths excluding wave impact, a significant part of the seabed material has been falling out of suspension (ash, windblown dust, biological material). Local patches of sand and mud relates to infrequent currents down the continental slope. With the volcanic activity at the mid oceanic ridges, hard rocks from volcanic flows of the oceanic crust make up the units below the topside veneer of poorly consolidated sediments. Furthermore, cooling of volcanic material triggers hydrothermal activity, with seafloor groundwater venting revealed as black smokers.

3.2 *Mapping the Seabed and Its Composition*

In many parts of the world, the seafloor has been subject to mapping. For large areas, fairly detailed seafloor morphology maps are available, based in campaigns by national authorities, military activity, and compilation of datasets from mostly commercial ships equipped with advanced echo-sounders. The motivation for this work has been multifold, spanning form protection of reefs to placement of infrastructure. In Figure 1.4 we present one example, based on the Norwegian Mapping authority's (*Kartverket*[2]) efforts to manage the seafloor of the Barents shelf and nearby fjords of North Norway.[3]

In the case of the Barents Shelf seafloor, the mapping campaigns show that it is composed of sand, mud and shale, with some areas of concentrated biological activity displaying deep-water patch reefs. This distribution of materials on the seafloor is controlled by former or contemporaneous wave and current energy at the base of the marine water column. The basic physics behind moving sand or clay grains dictate that sand reflect most energy and shale lower energy. Noticeable, the energy level in areas with sand could be sufficient to create net seafloor erosion, hence causing gradual removal of the topmost part of the seabed.

In fjord areas like in Norway, the sediment distribution is closer linked to land, as rivers convey the bulk of the sediments to the fjord basin. Accordingly, sand is found at the mouth of rivers and streams, whereas mud and clay settle from suspension further out in the fjord. As discharge by river can be very high and even episodic during floods, seabed thickness varies greatly from thick deltas to thin deep fjord deposits.

---

2 Norwegian Mapping Authority (*Kartverket*): <https://www.kartverket.no/>.

3 For further insight, information is given on the Geological Survey of Norway's website at <www.ngu.no>.

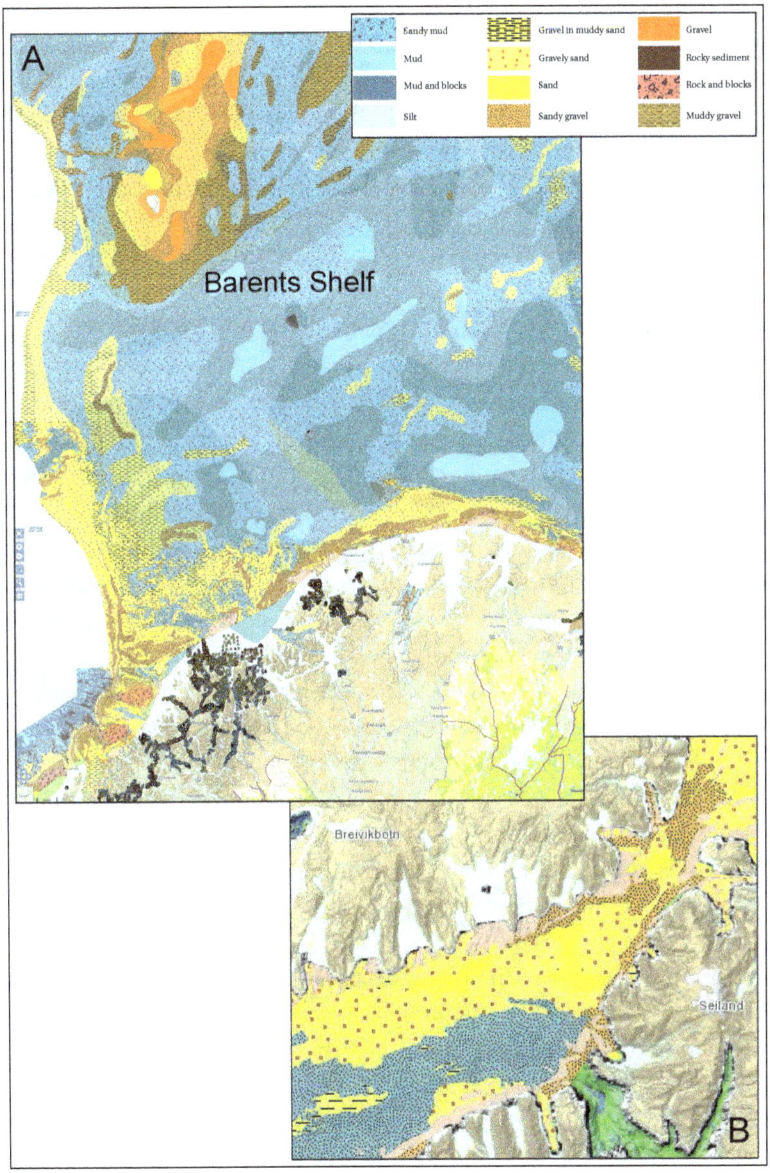

FIGURE 1.4 (A) Seafloor composition of the Barents Sea continental shelf, and (B) a similar map showing the seafloor composition in a fjord of North Norway.

EXAMPLES FROM THE DATABASE OF THE GEOLOGICAL SURVEY OF NORWAY (NGU) (WW.NGU.NO). WITH THE CONSENT OF NGU.

Maps like those presented in Figure 1.4, directly describing seafloor composition, allow insight to temporal stability. The derived information is crucial for placement of infrastructure, as further outlined below. Such detailed data are nonetheless not common on a global scale. They mostly exist in ocean basins where there are significant national economic or other interests at stake. Hence, some areas of the world are already well mapped, while others are far from being mapped.

4 Natural Processes in and on the Seabed

As mentioned, in most places the seafloor experiences net deposition or erosion of sediments, although major or catastrophic events (e.g., storms, tsunamis, landslides) can temporarily change this status. From the moment a bed is building up, biological activity will modify its consistence. Most common is bioturbation by sediment eating organisms, consuming whatever organic material that has been captured in the sediment. These forms of life thrive in given positions, with deeper areas below the photogenic zone less attractive. Further, in shallow waters with extensive wave action, or in locations exposed to fresh water from streams, living conditions are uninviting. In areas where organic material is not fully consumed, there will be bacterial degradation creating gas (hydrocarbons). This gas is biogenic, and should not be mistaken as thermogenic gas. The latter relates to chemical reactions in organic shales (and sometimes limestones) that survived the biological activity and experienced deeper burial. As described above, deeper burial instigate the transition to rock that, combined with general heating, mobilize hydrocarbons in organic beds. Byproducts are gas and oil.

The transition from loose sediment to rock linked to compaction from burial and chemical reactions reduces the pore space available for fluids. Fluids will typically be captured sea water or groundwater that migrate in from other beds. General compaction reduces available space for fluids, forcing them to migrate. As basically all fluids are lighter than rock and earth materials, this migration follows paths that eventually will bring fluids to the surface, unless trapped. The consequence is that large parts of the seafloor experience expulsion of fluids such as fresh or salt (brine) ground water, biogenic and thermogenic gas and even oil. Much of this relaxed flow is bypassing the seabed during migration from deeper geological units to the surface.

There are numerous examples of the fluid expulsion at the Earth's surface, with two illustrations presented in Figures 1.5 and 1.6. This phenomenon is not unique to the oceans, and excellent natural examples of it can be found

FIGURE 1.5
A CO_2 leak in Utah, USA
PHOTOGRAPH BY ALVAR BRAATHEN

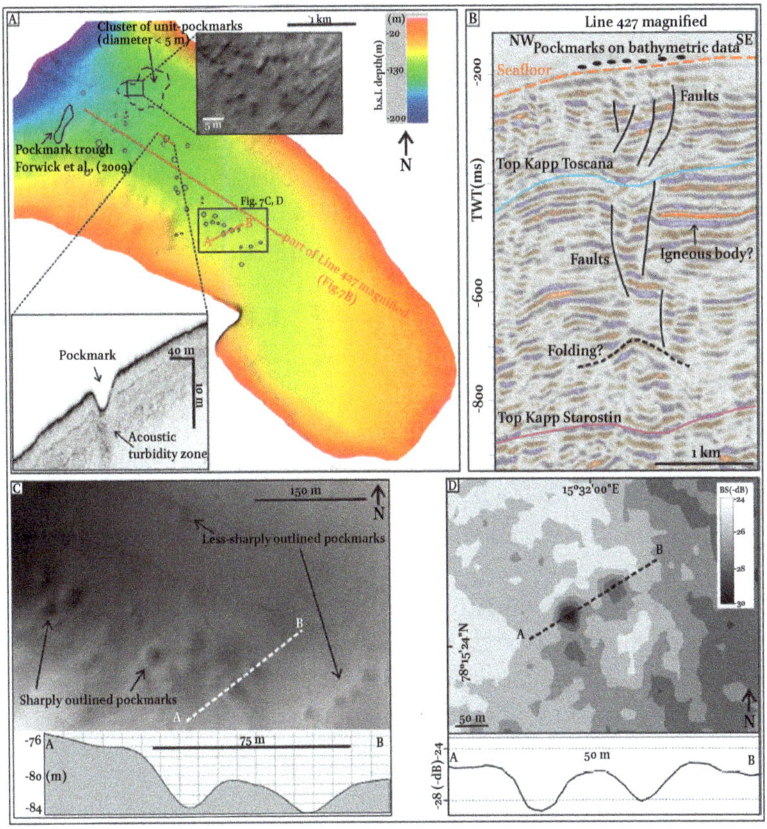

FIGURE 1.6 Pockmarks of Isfjorden, Svalbard, Norway. High-resolution multibeam bathymetric data from SA1 (Adventfjorden) showing the distribution of pockmarks. (A) The dashed-line area has c. 25–30 unit-pockmarks (top inset: magnified unit-pockmarks on bathymetric data; bottom inset: acoustic turbidity zone beneath a pockmark on the sub-bottom profile. (B) Structural interpretation of 2D seismic data (Line 427) in deeper successions beneath the pockmarks. (C) Sharply and less-sharply outlined circular pockmarks. Profile A-B across the pockmarks is illustrated beneath. (D) Multibeam backscatter data and corresponding A-B profile across the pockmarks in Fig. C.
MODIFIED FROM ROY ET AL. 2016

onshore, such as in Utah (Southwest USA). This region display numerous sites that leakage natural CO_2 to the Earth's surface from deeper reservoirs/aquifers. Extensive geological investigations suggest fluids migrate up conduits along narrow zones in the crust that have experienced earthquakes and which are called faults. This migration has lasted for at least 400,000 years.[4] Not surprisingly, drilling in 1937 next to a CO_2 charged spring punctured one fluid migration route, triggering significant CO_2 outbursts.

Another example of fluid expulsion from the seafloor can be found in the pristine High Arctic of Norway (see Figure 1.6). In Svalbard, the Longyearbyen CO_2 Lab project undertook mapping of the fjord seafloor as a baseline study. The concept was to map pre-existing leaks before starting injection campaigns into rocks nearby. Investigations found hundreds of small depressions with circular rims on the seafloor, so-called pockmarks.[5] They are formed by fluids or gas rising out of the seabed, lifting away sediments and gradually forming a 1–10 m deep crater. Mobilized sediments settle along the rim of the crater. As the wilderness of Svalbard has experienced minimal impact by human activity, this illustrates how fluid migration to the seafloor is a natural process, with pockmarks as the physical evidence. For the CO_2 Lab, key questions were attached to timing of fluid migration and flow rate. A central question was notably to know whether these pockmarks were the result of ongoing flow. Most pockmarks turned out to be dormant, reflecting former fluid expulsion episodes. However, a few pockmarks in the fjord are active, showing mainly methane gas mixed with ground, where thermogenic gas is sourced from the deep subsurface. The organic black shales of the Svalbard bedrock emit natural gas.

For the broader audience, the two examples of Utah and Svalbard discussed above offer a higher learning value, which is that, basically, all geological materials will leak, given sufficient time. However, these leaks are overall slow and will seldom be catastrophic (in geology terms this means shorter than thousand years).

4 B. Dockrill, and Z. Shipton, 'Structural controls on leakage from natural CO_2-geologic storage site: Central Utah', U.S.A., *Journal of Structural Geology*, v. 32, no. 11 (2010) 1768–1782.

5 S. Roy, M. Hovland, & A. Braathen, 'Evidence of fluid seepage in Grønfjorden, Spitsbergen: Implications from an integrated acoustic study of seafloor morphology, marine sediments and tectonics,' *Marine Geology*, no. 380 (2016), 67–78.

5 Seabed Resources and Human Activity

The resources of the seabed are both geological and biological. Geological resources are found on the surface – mainly as sand and gravel, and deep sea minerals –, as well as in the subsurface of the seabed – mainly as oil and gas. Biological resources in the form of sedentary organisms live on or just below the surface as part of the biodiversity, with species specialized according to available resources, depending on the composition of the surface and subsurface of the seabed, and the associated geological processes.

The floor of the continental shelf is usually underlain by thick sequences of sediments that may host hydrocarbon resources. This is especially the case for continental shelves of passive continental margins where the sediments may date back to long before the break-up of the continent. Therefore, these margins host most of the offshore oil and gas industry of the world.

Continental shelfs worldwide is the site for anchored infrastructures, for instance drill holes that bypass the seabed on their way to the deeper subsurface and which are placed there to extract hydrocarbons. Other drill holes are used to inject (waste-)water, gas or, in a few places, even anthropogenic CO_2. Decades with learning around these operations make them familiar to the general public, although technical challenges around infrastructure placement such as wells are resolved by experts. Similar knowledge exists around harvesting of biota from the seafloor (algae, shell fish, etc.) and, to a lesser extent, mining biota in the uppermost seabed for instance by seafloor trawling. These undertakings are well regulated in most waters, with national and bilateral agreements regulating undertakings.[6]

The deep ocean floor at 1000's of meters depth has been regarded economically less interesting as it is nearly barren of life that can be harvested. However, a growing global population facing the transition to a society increasingly dependent on digital processes and renewable energy has given rise to a general agreement that there will be an increasing demand for metals in the future, both in volume and diversity. This notion is reflected in the current significantly growing interest in deep sea mining. Exploration for deep sea polymetallic minerals is taking place both in the continental shelf areas of coastal states and in the international seabed (the Area). At present, there are 29 exploration contracts in the Area, and the activity is expected to move

6 See R. Caddell, Chapter 12 of this book, 'Deep-Sea Bottom Fisheries and the Protection of Seabed Ecosystems: Problems, Progress and Prospects.'

to exploitation in a near or not too far future.[7] There are two main types of polymetallic deposits on the seafloor of the deep oceans: ferro-manganese minerals and seafloor massive sulphides.[8] The ferro-manganese mineral deposits are formed by oxyhydroxide minerals of iron and manganese, which also carry minor, but economically interesting quantities of nickel, copper, cobolt, rare earth elements (REE) and other metals. They form two types of deposits: nodules and crusts. The individual nodules are formed as concentric layers of minerals that are precipitated from the seawater above the seafloor or from the pore water of the sediments just below the seafloor. Nodules grow in areas of very sparse sedimentation and form large nodule fields in the abyssal plains that lie far from the sediment sources of the continents. The largest nodules fields are found in the Pacific Ocean. The ferro-manganese crusts grow as mineral lamina precipitated onto surfaces of bare rock on the seafloor (e.g. escarpments, ridges and seamounts). Such crusts are found in all oceans.

The seafloor massive sulphides are formed by hydrothermal processes associated with volcanic activity, mainly along the mid-ocean ridges of the oceans (see above Section 2.2). The heat of the volcanic activity sets up a circulation system of hot water within the ocean crust, leaching metallic elements from the rock and carrying them up to the seafloor in hydrothermal vents. In contact with the cold seawater, the metals will precipitate as a black cloud of sulphides; hence the name 'black smokers'. Such deposits are known from all oceans, and some of them are already the object of industry activity. The Canadian company Nautilus is currently developing a sulphide deposit called 'Solwara' within the Exclusive Economic Zone (EEZ) of Papua New Guinea.[9] The Japanese agency JOGMEC in August 2017 carried out a successful test of mining equipment on a sulphide deposit in the Okinawa Trough.[10] It seems that deep sea mining will take place as affordable technology opens new avenues.

How will human activity on the seabed impact the local environment? Infrastructure on the shelf such as oil platforms are anchored by traditional

---

7 See website of the International Seabed Authority at <https://www.isa.org.jm/>. For an analysis of the applicable legal regime, see J. Dingwall, Chapter 7 of this book, 'Commercial Mining Activities in the Deep Seabed Beyond National Jurisdiction: The International Legal Framework.'

8 See R. Sharma (ed.), *Deep-Sea Mining. Resources Potential, Technical and Environmental Considerations*, (Springer International Publishing, 2017).

9 See further information on the website of the company Nautilus Minerals Inc at <http://www.nautilusminerals.com/IRM/content/default.aspx>.

10 'Japan successfully undertakes large-scale deep-sea mineral extraction,' Japan Times, 26 September 2017 <https://www.japantimes.co.jp/news/2017/09/26/national/japan-successfully-undertakes-large-scale-deep-sea-mineral-extraction/#.XA19JGyWyhc>.

methods (chain and anchor) or, for some installations, stands on the seafloor by their own weight. Others stand on pillars knocked into the seabed. Future production facilities will however to a larger degree lay on the sea floor or be dug into the seabed. Common for all these semi-permanent facilities is that they cause changes in the physical conditions for the seafloor, thereby impacting the seabed with time. Waves amplify when hitting obstacles such as pillars, drill hole casings, storage tanks or tubes/cables, increasing the likelihood for erosion of the seafloor. On the contrary, infrastructure dampening the sea could cause deposition. Infrastructure also has a load that could cause subsidence. Further, infrastructure could generate heat and/or light impacting the biological environment. All these effects are in most cases technically manageable if considered when it comes to design. Similarly, environmental impacts caused by local sea floor modifications, or even spills, have been extensively analyzed, as has the ownership and obligations to acreage. The environmental impacts from local seafloor modifications are also regarded a major issue to be considered with regard to the future deep sea mining for minerals; or similar mining for biota in the seabed of the shelfs.

6 Geoscience Posts to the Regulatory Community

Our considerations around the seabed, based on our background as geologists, bring forward many aspects that are well known, and well regulated, but we feel that a reminder is in any case valuable. In order to reach science-based decisions reflecting a precautionary approach, lawmakers must get access to sufficient insight to understand the implications and impacts that the envisaged new laws may cause.

CHAPTER 2

Deep-Sea Ecosystems: Biodiversity and Anthropogenic Impacts

Eva Ramirez-Llodra

1 Introduction

1.1 *Planet Ocean*

We call our planet 'Earth', but 70 per cent of the Earth is covered by oceans, with oceans constituting more than 95 per cent of Earth's living space. Fifty per cent of these oceans are below 3000 meter (m) depth and the average ocean depth is 3800 m. The largest biome on Earth is, thus, composed by deep marine ecosystems of about 1 billion km^3 of deep water and 326 million km^2 of deep seafloor. The deep sea is considered to start at 200 m depth, where solar energy cannot support primary productivity through photosynthesis. This depth changes regionally depending on water turbidity, but it often coincides with the shelf break where the seafloor transitions to the continental slope and is marked by a significant increase of the slope angle.[1] Although large-scale bathymetry (depth topographic maps) exists for the whole ocean floor, deep-sea ecosystems are still of the least explored on Earth, with less than 0,0001% physically sampled or visually observed.[2] In the last 170 years, twenty two new deep-sea habitats and associated fauna have been discovered. The deep seafloor, long believed to be a featureless and stable environment, has been shown to support one of the highest biodiversities in the planet in a wide array of interconnected habitats.[3] These ecosystems sustain important functions and derived ecosystem services, spanning from nutrient regeneration and carbon

---

1 Tyler PA, Baker MC, Ramirez-Llodra E 'Deep-Sea Benthic Habitats', in Clark MR, Consalvey M, Rowden AA (eds) *Biological sampling in the deep sea* (John Wiley & Sons Ltd., West Sussez, 2016), pp. 1–15.

2 Ramirez Llodra E, Brandt A, Danovaro R, De Mol B, Escobar E, German CR, Levin LA, Martinez-Arbizu P, Menot L, Buhl-Mortensen P, Narayanaswamy BE, Smith CR, Tittensor DP, Tyler PA, Vanreusel A, Vecchione M (2010) 'Deep, Diverse and Definitely Different: Unique Attributes of the World's Largest Ecosystem', *Biogeosciences* 7: 2851–2899 doi doi:10.5194/bgd-7-2361-2010.

3 Tyler PA, *Ecosystems of the Deep Oceans Ecosystems of the World* (Elsevier, Amsterdam, 2003) p. 569.

 | DOI:10.1163/9789004391567_004

sequestration to biological and mineral resources, not to mention cultural and educational services,[4] many of which are key to the health of the planet. Deep-sea research is rapidly progressing in parallel to technological development, in parallel to an increase in the exploration for and exploitation of deep-sea resources.[5] However, the limited understanding of the composition, diversity and functioning of many deep-sea ecosystems restricts our capacity to develop robust ecosystem-based management measures that are necessary if we are to balance resource use and ecosystem conservation.[6]

1.2 *Aim and Structure*

The aim of this chapter is to provide a general overview of deep-sea ecosystems, their faunal composition and the key functions and services that they provide. This will allow for a better understanding of the current and potential impacts derived from an increasing number of human activities and indirect stressors.

The first part of the chapter (Section 2) briefly describes the habitat, general biological features and key ecosystem functions of the main deep-sea ecosystems (Figure 2.1), from the shelf break to the deepest trenches. Starting from the shelf break (ca. 250 m deep), where the deep sea starts, the key ecological aspects of the different habitats found on continental margins are discussed, highlighting the heterogeneity of a system that was before supposed to be featureless and poor in life. The chapter then describes the vast abyssal plains that support a very high biodiversity of small fauna and the rich underwater mountains, or seamounts, where filter feeders such as corals and sponges thrive. The composition and functioning of hydrothermally active and inactive habitats of the mid-ocean ridges, back-arc basins and some active seamounts are explained, finishing with a short introduction to the deepest habitats on Earth, the hadal trenches.

The chapter then briefly introduces the main human activities that affect, directly or indirectly, deep-sea habitats and their ecosystems (Section 3). These anthropogenic impacts are grouped into two main categories: waste dumping and resource exploitation. Climate change, although a critical issue globally

4 Thurber AR, Sweetman AK, Narayanaswamy BE, Jones DOB, Ingels J, Hansman RL 'Ecosystem function and services provided by the deep sea', *Biogeosciences* (2014) 11: 3941–3963.

5 Ramirez-Llodra E, Tyler PA, Baker MC, Bergstad OA, Clark M, Escobar E, Levin LA, Menot L, Rowden AA, Smith CR, Van Dover CL 'Man and the last great wilderness: human impact on the deep sea' *PLoS ONE* (2011) 6(8) e22588 doi doi:10.1371/journal.pone.0022588.

6 Mengerink KJ, Van Dover CL, Ardron J, Baker MC, Escobar-Briones E, Gjerde K, Koslow A, Ramirez-Llodra E, Lara-Lopez A, Squires D, Sutton T, Sweetman AK, Levin LA, 'A call for deep-ocean stewardship', *Science* (2014) 344: 696–698.

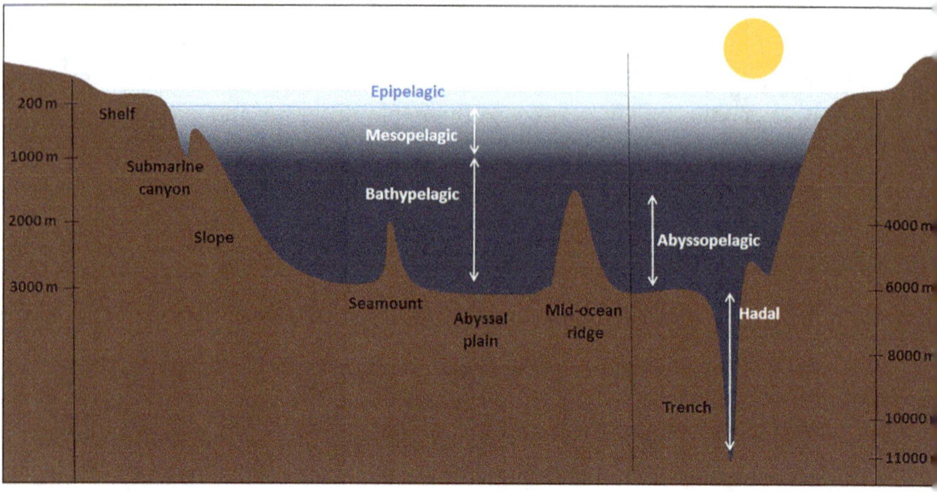

FIGURE 2.1 Diagram showing seafloor habitats and the water column biome, from the coast to the deepest trenches. Note different depth scales at each side of the dotted line.

that affects the oceans as a whole, including deep-sea ecosystems,[7] has not been included in these discussions. However, the importance of cumulative impacts on deep-sea ecosystems and the role played by climate change is discussed in the last paragraph.

2 Deep-Sea Ecosystems

2.1 *A History of Exploration*

The development of deep-sea research as a science is associated with the development of new techniques of navigation, sampling and measuring, and follows the path of great oceanic expeditions. It was during the last two centuries that these expeditions obtained the first scientific results, which would fascinate and inspire a whole new branch of oceanographers. The cruise of H.M.S *Beacon* to the Aegean (1841–1842) could be considered one of the first biological deep-sea cruises. Prof. Edward Forbes, from Edinburgh University, joined the ship as a naturalist and made around 100 dredge hauls down to a depth of 420 m. As the *Beacon* crew dredged deeper, fewer species were found,

7 Levin LA, Le Bris N, 'The deep ocean under climate change', *Science* (2015) 350: 766–768.

leading to Forbes' 'Azoic Theory'[8] where he proposed that no life existed at great depths. However, the British admiral Sir John Ross had already collected a deep-sea echinoderm while dredging at 1460 m during his exploration for the Northwest Passage in 1818.[9] Later, the Norwegian Michael Sars (1850) published a list with 19 species from waters deeper than 550 m, and his son, George Ossian Sars extended the list to 92 species. With evidence accumulating of a diverse deep-water fauna, C.W. Thomson and W.B. Carpenter encouraged the Royal Society and the Admiralty to organise a deep-sea expedition, resulting in the H.M.S. *Lightning* cruise in 1868 to the NE Atlantic, and the H.M.S. *Porcupine* cruise (1869) to NE Atlantic and Mediterranean Sea (Rice 1986). With the important discoveries of the *Lightning* and *Porcupine*, W.B. Carpenter's application for a scientific circumnavigation expedition was accepted in April 1872. H.M.S. *Challenger* set sail from Sheerness on December 7th 1872 for her three and a half years cruise with C.W. Thomson as chief scientist. The *Challenger* expedition was set up to study the physical, chemical and biological processes in the deep ocean. This circumglobal oceanographic voyage has been considered by many to be the true birth of modern oceanography.[10]

The *Challenger* expedition was followed by an era of pioneering deep-sea research, involving numerous ships from several countries. But it was in the 1960s and 1970s, that an important change in the approach of deep-sea biological research took place. Descriptive biology was complemented with a more ecological, evolutionary and experimental approach, led by North American researchers such as Profs. Hessler, Sanders and Grassle.[11] However, the conquest of the oceans would not have been complete if humans had not developed the ways of entering the deep-sea environment, to observe, explore and experiment *in situ*. Therefore, parallel to the remarkable developments in navigation and oceanographic technologies, there is the history of diving, deeper and longer. Beebe's Bathysphere in 1930 was the first deep-water vehicle for observation of the seabed. From there, in little more than 50 years, the advances in deep-sea technology have led to a variety of novel instruments. Some of these include maned submersibles, Remote Operated

---

8 Forbes E, 'Report on the Mollusca and Radiata of the Aegean Sea, and on their distribution, considered as bearing on geology', Report of the British Association for the Advancement of Science for 1843, British Association for the Advancement of Science (1844) pp. 129–193.

9 Menzies RJ, George RY, Rowe GT, *Abyssal Environment and Ecology of the World Oceans* (Wiley-Interscience, New York, 1973).

10 See Tyler (n3).

11 See Hessler RR, Sanders HL, 'Faunal diversity in the deep-sea', *Deep-Sea Research* (1967) 14: 65–78. See as well Grassle FJ, Sanders HL, 'Life histories and the role of disturbance', *Deep-Sea Research* (1973) 20: 643–659.

FIGURE 2.2 State-of-the art equipment for deep-sea research. A: Autonomous Underwater Vehicle Hugin from Kongsberg Maritime (Norway). B: human occupied vehicle Alvin from WHOI (USA). C: Remote Operated Vehicle Triton XLR (Norway). D: Remoted Operated Vehicle Isis from the UK.
A: COPYRIGHT, E. RAMIREZ-LLODRA/MARMINE. B: COPYRIGHT, C. GERMAN, WHOI. C: COPYRIGHT, E. RAMIREZ-LLODRA/ MARMINE. D: COPYRIGHT P. TYLER, UNI. SOUTHAMPTON (UK).

Vehicles (ROV), Autonomous Underwater Vehicles (AUVs), new chemical and physical sensors and cabled observatories (Figure 2.2). These instruments are used in combination with other sampling gear, such as multicorers, boxcorers, trawls and sledges to sample benthic fauna; plankton nets with several opening and closing mechanisms; Conductivity-Temperature-Depth devices (CTDs) that measure conductivity, temperature and depth; multibeam echosounders and sidescan sonars to map the seafloor, etc. The use and continuous development of these technologies provides a wealth of novel information on the composition, structure and functioning of deep-sea ecosystems.[12] This comprehensive knowledge is essential for the development of robust management

12 Clark MR, Consalvey M, Rowden AA, *Biological sampling in the deep sea* (John Wiley & Sons Ltd., West Susex, 2016), p. 451.

and conservation measures to be applied to deep-sea ecosystems. Below, the main characteristics of the major deep-sea habitats and their communities are briefly considered, to set the scene for the discussion on anthropogenic impacts upon these ecosystems.

2.2 *Continental Margins*

The continental shelf expands from the coastline to the shelf break and is an area of relatively shallow water, mostly less than 250 m, with the exception of some large shelves like the Norwegian shelf, that has depth down to 500 m. The start of the deep sea is often considered to coincide with the shelf break, from which the continental margin descends along a slope from about 250 m to 3000 m depth (Figure 2.1). The continental margins cover about 11% of the ocean floor (ca. 40 million km^2) and can be passive or active. Passive margins are found where an ocean rift has split a continent in two, generating an ocean basin in between, while active margins are found where the ocean floor is so dense that it sinks back into the Earth forming trenches along subduction zones.[13] Continental margins are characterised by high habitat heterogeneity, including sedimentary slopes, submarine canyons, cold-water corals, cold seeps, mud volcanoes, pockmarks and oxygen minimum zones.[14] These habitats support a variety of faunal communities that support a wide array of functions.

2.2.1 Sedimentary Slopes

Sedimentary slopes are often characterised by high biodiversity of small meiofauna (organisms retained on a 32 micron sieve, such as nematodes) and macrofauna (organisms retained on a 0.3 to 0.5 millimetre sieve, mostly small crustaceans and polychaete worms). This infauna (organisms that live in the surface layers of the sediment), together with the microorganisms in the seafloor, play a key role in the biological pump, where carbon fixed by shallow-water organisms through photosynthesis and subsequently falling to the seafloor is remineralised and carbon and nutrients that are upwelled fuel again primary productivity in the surface layers. The margin megafauna (animals identifiable from seafloor videos and photos) are often dominated

13 Menot L, Sibuet M, Carney RS, Levin LA, Rowe GT, Billett DSM, Poore G, Kitazato H, Vanreusel A, Galeron J, Lavrado HP, Sellanes J, Ingole B, Krylova E, 'New perceptions of continental margin biodiversity', in McIntyre AD (ed) *Life in the World's Oceans: Diversity, Distribution and Abundance* (Blackwell Publishing Ltd, Oxford, 2010), pp. 79–102.

14 See Levin LA, Sibuet M, Gooday AJ, Smith CR, Vanreusel A, 'The roles of habitat heterogeneity in generating and maintaining biodiversity on continental margins: an introduction', *Marine Ecology* (2010) 31: 1–5. See also Menot et al. (n13).

FIGURE 2.3 Examples of faunal communities from continental margins. A: herd of the echinoid *Linopheuses*. B: stalked crinoids on a rocky submarine canyon wall. C: Cold water corals from the Gulf of Mexico. D: Community of *Escarpia laminate* from the Gulf of Mexico cold seeps.
PHOTOS A & B: COPYRIGHT, P. TYLER, UNI. SOUTHAMPTON (UK)
PHOTOS C & D: COPYRIGHT, C. FISHER, PSU (USA).

by echinoderms and crustaceans as well as fish, depending on the region (Figure 2.3A). Some of these groups include valuable commercial species and, thus, sedimentary slopes are subjected to increasingly intense fisheries in certain regions of the world.[15]

2.2.2 Submarine Canyons

Submarine canyons are large geomorphological features covering 11.2% of continental margins globally.[16] The topography of canyons intercepts regional hydrographic patterns resulting in modified local currents that trap particles.

15 Koslow J.A., Boehlert G.W., Gordon J.D.M., Haedrich R.L., Lorance P., Parin N., 'Continental slope and deep-sea fisheries: implications for a fragile ecosystem', *ICES J. Mar. Sci.* (2000) 57:548–557.

16 See Harris P, Macmillan-Lawler M, Rupp J, Baker E, 'Geomorphology of the oceans', *Marine Geology* (2014) 352: 4–24. See also Fernandez-Arcaya U, Ramirez Llodra E, Allcock AL, Davies JS, Dissanayake A, Harris P, Howell K, Huvenne VA, Macmillan-Lawler M, Martin

Canyons thus act as conduits for particles from the fertile coast and shelves to the deep basins, fueling the deep faunal communities.[17] Canyons provide also a variety of habitats that support diverse faunal types. The canyon head and walls are characterised by rocky outcrops that provide substratum for filter feeders such as crinoids, gorgonians or corals that use the currents in the canyon to filter seawater and capture food (Figure 2.3B). The axis of the canyon is filled with fine sediment that support rich benthic communities like the ones found on the sedimentary slopes. Canyons have been described as 'essential habitats'[18] because they can provide refuge and habitat for spawning species and juveniles, as well as feeding grounds for certain species.[19] The rough topography of canyons has limited fisheries, but technological developments are opening new fishing grounds in areas that were before difficult to access. Additionally, the modified currents in canyons enhance the transportation of chemical pollutants and litter that can accumulate at the base of the canyon.

2.2.3 Cold Water Corals

Cold-water corals are found at temperatures ranging from 4 to 13 °C and depths between 50 and 6000 m depths.[20] Most of the reef-forming cold-water corals, such as *the Lophelia pertusa* reefs in the NE Atlantic, are found on the upper part of the continental slope and on seamounts. Reef-forming corals are estimated to cover an area of ca. 280 000 km^2 worldwide. The 3-dimensional structure of cold-water corals can form long-lived reefs or gardens, providing habitat and refuge to a large variety of organisms, both in the adult and juvenile stages. These ecosystems support a high biodiversity and high biomass along continental margins (Figure 2.3C). Extensive damage on cold-water corals from trawling has occurred, resulting in highly productive systems being transformed into coral rubble. The recovery of damaged cold-water corals is

---

J, Menot L, Nizinski M, Puig P, Rowden AA, Sanchez F, Van den Beld IM, 'Ecological role of submarine canyons and need for canyon conservation: a review', *Frontiers in Marine Science* (2017) 4: DOI=10.3389/fmars.2017.00005.

17 Masson DG, Huvenne VAI, de Stigter HC, Wolff GA, Kiriakoulakis K, Arzola RG, al., 'Efficient burial of carbon in a submarine canyon', *Geology* (2010) 38: 831–834.

18 Company JB, Ramirez-Llodra E, Sardà F, Aguzzi J, Puig P, Canals M, Calafat A, Palanques A, Solé M, Sànchez-Vidal A, Martín J, Lastras G, Tecchio S, Koenig S, Fernandez-Arcaya U, Mechó A, Fernández P, 'Submarine canyons in the Catalan Sea (NW Mediterranean): megafaunal biodiversity patterns and anthropogenic threats', in Würts (ed) *Mediterranean submarine canyons: ecology and governance* (IUCN, Málaga, 2012), pp. 133–145.

19 See Fernandez-Arcaya et al (n16).

20 Roberts JM, Wheeler AJ, Freiwald A, 'Reefs of the deep: the biology and geology of cold-water coral ecosystems', *Science* (2006) 312: 543–547.

likely to be slow (decades to centuries) and when the habitat has been altered and the corals eliminated, recovery is unlikely.[21]

2.2.4 Cold Seeps

Cold seeps are found both at active and inactive margins and the estimated global area is 10 000 km^2. These habitats are characterised by the cold seepage of fluid with high concentrations of methane and hydrogen sulphide. These reduced chemicals are used by microorganisms as source of energy to produce organic matter, in a process called chemosynthesis. Chemosynthctic-based ecosystems, such as cold-seeps or hydrothermal vents, are the only communities in the deep-ocean where the faunal communities are supported by in situ primary productivity. But here, this productivity is based on chemical energy instead of solar energy used by plants in the sunlit zone. These chemoautotrophic microorganims in cold seeps are found both free living and in symbiosis with benthic fauna.[22] The primary productivity at cold seeps supports communities of relatively low biodiversity but high biomass of highly specialised fauna. Some of the key organisms often found at cold seeps include bivalves, gastropods, siboglinid tubeworms, decapod crustaceans and cladorhizid sponges[23] (Figure 2.3D).

2.3 *Abyssal Plains*

Abyssal plains are vast regions of relatively flat seafloor extending from 3000 to 6000 m depth (Figure 2.1), covered by a layer of fine sediment that can reach thousands of meters in thickness. The abyssal plains cover a total area of 245 million km^2, about 75% of the deep seafloor, representing one of the largest ecosystems on Earth. Their vastness and remoteness makes abyssal plains one of the least explored regions of the oceans.[24] As for the rest of the deep-sea fauna, excluding chemosynthetically-based ecosystems, the lack of light to fuel

21 Rogers A, 'The Biology, Ecology and Vulnerability of Deep-Water Coral Reefs', IUCN-Rep-2004-002 (IUCN, 2004).

22 See Tunnicliffe V, Juniper KS, Sibuet M, 'Reducing environments of the deep-sea floor', in Tyler PA (ed) *Ecosystems of the World*, Vol 28 Ecosystems of the deep oceans (Elsevier, London, 2003) pp. 81–110. See also Baker MC, Ramirez-Llodra E, Tyler PA, German CR, Boetius A, Cordes E, Dubilier N, Fisher C, Levin LA, Metaxas A, Rowden A, Santos RS, Shank TM, Van Dover CL, Young CM, Waren A, 'Biogeography, Ecology and Vulnerability of Chemosynthetic Ecosystems in the Deep Sea', Chapter 9 in McIntyre AD (ed) *Life in the World's Oceans: Diversity, Distribution, and Abundance* (Wiley Blackwell Oxford, 2010) pp. 161–183.

23 Levin LA, 'Ecology of cold seep sediments: interactions of fauna with flow, chemistry and microbes', *Oceanography and Marine Biology Annual Review* (2005) 43: 1–46.

24 See Ramirez-Llodra et al. (n2).

photosynthesis results in the abyssal fauna being heterotrophic. This means that the organisms rely fully on the arrival of organic matter from the surface layers, falling as 'marine snow' through the water column or advected along the margin. Abyssal plains are thus often food limited,[25] but these habitats support one of the highest biodiversities on Earth. This high biodiversity is mostly composed of small organisms, from microbes to meiofauna and macrofauna.[26] Abyssal plains are subjected to relative extreme ecosystem parameters, including very high pressures (1 atmosphere for each 10 m depth), low temperatures (about 2 °C), usually very slow bottom currents and usually very low annual organic matter input.[27] The quantity and quality of this flux of organic matter varies seasonally depending on the geographic region and the productivity of the surface oceanic layers. Thus, ecosystem composition, structure and function vary regionally at abyssal plains. A major characteristic of abyssal fauna is that rare is common. This means that most organisms collected from abyssal depths have been recorded as a few individuals (typically less than 5) from one or two sampling sites.[28] Technological development has greatly increased our sampling activity, providing a wealth of samples with a high number of species new to science, most of them represented by small, single individuals. The rate at which potentially new species are being collected together with the decrease in expert taxonomists (specialists in species identification and naming) have led to what has been termed 'taxonomic impediment'.[29] This results in a significant delay between the discovery of a new species (when it is collected and identified as new) and the scientific description of the species (when it is given a name and published, thus becoming available).[30] Addressing this issue is thus essential if we are to obtain a thorough understanding of abyssal community composition, structure and function.

Although remote, abyssal plains are subjected to different environmental stressors. In particular, some abyssal plains (e.g. Pacific Ocean) include

25 Smith C, De Leo FC, Bernardino AF, Sweetman AK, Martinez-Arbizu P, 'Abyssal food limitation, ecosystem structure and climate change', *Trends in Ecology and Evolution* (2008) 23: 518–528.

26 Ebbe B, Billett DSM, Brandt A, Ellingsen K, Glover A, Keller S, Malyutina M, Martinez Arbizu P, Molodtsowa T, Rex M, Smith C, Tselepides A, 'Diversity of Abyssal Marine Life', Chapter 8 in McIntyre AD (ed) *Life in the World's Oceans: Diversity, Distribution, and Abundance* (Wiley Blackwell, Oxford, 2010) pp. 139–160.

27 Smith CR, Demopoulos AWJ, 'Ecology of the deep Pacific Ocean floor', in Tyler PA (ed) *Ecosystems of the World*, Volume 28: Ecosystems of the Deep Ocean (Elsevier, Amsterdam, 2003) pp. 179–218.

28 See Ebbe et al. (n26).

29 Ibid.

30 See Ramirez-Llodra et al. (n2).

important mineral resources in the form of polymetallic manganese nodules (see below Section 3.5.1) which are currently under exploration licenses. Thus, improving scientific understanding of the structure and function of these ecosystems at the local and regional scales is essential prior to the signature of exploitation contracts. Climate change will also have an impact on abyssal faunal word wide, mainly related to changes in organic matter fluxes caused by changes in surface primary productivity, as well as potential water column stratification and changes in global circulation.[31]

2.4 *Seamounts*

Seamounts and knolls are underwater mountains rising from 100 to over 1000 m from the surrounding seafloor (Figure 2.1). The number of seamounts and knolls has been estimated to be ca. 100 000, covering an area of 8.5 million km^2, which represents 2.6% of the seafloor.[32] However, the biological communities of only less than 300 seamounts have been studied with enough detail to provide a thorough description of their composition, let alone functioning. The topography of seamounts modifies locally the prevailing currents and results in the retention of particles above the seamount, providing an enhanced food supply to the seamount fauna. The available rocky substratum, elevation from the seafloor and modified hydrography of seamounts support high abundances and biomass of often distinct faunal communities.[33] The dominant fauna includes sessile, filter-feeder organisms such as corals and sponges, which in turn provide habitat for a variety of other species, such as fish, echinoderms and crustaceans. Seamounts have often been described as isolated habitats supporting hot spots of species richness with high degrees of endemism. However, knowledge is still scarce and recent evidence does not support these widely accepted paradigms.[34] They are also proposed to serve as stepping stones for dispersal of species across the abyssal plains. The high abundance of commercially-valuable fishes that may aggregate over seamounts has attracted industrial interest to these distinctive topographic habitats, with, in some

31 See Levin and Le Bris (n7).

32 See Consalvey M, Clark MR, Rowden AA, Stocks KI, 'Life on Seamounts', Chapter 7, in McIntyre AD (ed) *Life in the World's Oceans: diversity, distribution and abundance* (John Wiley & Sons, West Sussez, 2010), pp. 123–138. See also: Ramirez Llodra et al. 2010 (n2) ; Yesson, C., Clark, M., Taylor, M.L. and Rogers, A., 'The global distribution of seamounts based on 30 arc seconds bathymetry data', *Deep Sea Research Part I* (2011) 58: 442–453.

33 Schlacher TA, Rowden AA, Dower JF, Consalvey M, 'Seamount science scales undersea mountains: new research and outlook', *Marine Ecology Progress* (2010) Series 31: 1–13.

34 Rowden AA, Dower JF, Schlacher TA, Consalvey M, Clark MR, 'Paradigms in seamount ecology: fact, fiction and future', *Marine Ecolgy* (2010) 31: 226–241.

cases, devastating impacts on the sessile fauna and the long-lived populations of target fish (see below Section 3.3).

2.5 *Mid-Ocean Ridges and Hydrothermal Vents*

2.5.1 Mid-Ocean Ridges

Mid-ocean ridges form a 65 000 km long, semi-continuous, linear range of volcanic mountains where new oceanic crust is being formed and hydrothermal vents are found (Figure 2.1). Mid-ocean ridges support a wealth of habitats, from rocky substratum that includes hills and seamounts to deep axial valleys that can reach 4000 m depth and are covered with fine sediment.[35] The rocky seafloor supports communities dominated by filter feeders such as crinoids, sponges, corals and gorgonians and attracts motile fauna such as fish, galatheid crustaceans and cephalopods. This fauna contrast with the sediment communities, which are like those found in abyssal plains.[36]

2.5.2 Hydrothermal Vents

Hydrothermal vents and their associated fauna, discovered in 1977 in the Galapagos Rift (Pacific Ocean), are one of the major discoveries of the last decades.[37] A total of ca. 2000 vents has been estimated to occur globally,[38] although recent models have suggested a number 3 to 6 times higher.[39] Vents are found on mid-ocean ridges and back-arc basins where cold oxygenated deep seawater penetrates through the cracks of the ocean crust and reacts with the hot rock close to the magma chamber underlying the ridge.

35 Bergstad OA, Falkenhaug T, Astthorsson O, Byrkjedal I, Gebruk AV, Piatkowski U, Priede IG, Santos RS, Vecchione M, Lorance P, Gordon JDM, 'Towards improved understanding of the diversity and abundance patterns of the mid-ocean ridge macro- and megafauna', *Deep-Sea Research* (2008) II 55: 1–5. See also Vecchione M, Bergstad OA, Byrkjedal I, Falkenhaug T, Gebruk AV, Godø OR, Gislason A, Heino M, Høines Å, Menezes GMM, Piatkowski U, Priede IG, Skov H, Søiland H, Sutton T, de Lange Wenneck T, 'Biodiversity Patterns and Processes on the Mid-Atlantic Ridge', Chapter 6 in McIntyre AD (ed) *Life in the World's Oceans: Diversity, Distribution and Abundance* (Blackwell Publishing Ltd, Oxford, 2010) pp. 103–121.

36 See Vecchione et al. (n35).

37 Corliss JB, Dymond J, Gordon LI, Edmond JM, von Herzen RP, Ballard RD, Green K, Williams D, Bainbridge A, Crane K, van Andel TH, 'Submarine thermal springs on the Galapagos Rift', *Science* (1979) 203: 1073–1083.

38 Baker ET, German CR, 'On the Global Distribution of Hydrothermal Vent Fields Mid-Ocean Ridges', *American Geophysical Union* (2013) pp. 245–266.

39 Baker ET, Resing JA, Haymon RM, Tunnicliffe V, Lavelle JW, Martinez F, Ferrini V, Walker SL, Nakamura K, 'How many vent fields? New estimates of vent field populations on ocean ridges from precise mapping of hydrothermal discharge locations', *Earth and Planetary Science Letters* (2016) 449: 186–196.

During this process, the fluids can exceed 350 °C, dissolving metals and sulphur from the rocks. The heated fluid rises back to the surface of the seafloor and, when it mixes with the cold oxygenated water, the dissolved metals and sulphides precipitate, appearing as black smokers. The deposition of these particles forms the vent chimneys and can accumulate as massive sulphide deposits. Hydrothermal vents support unique faunal communities based on chemosynthetic primary productivity. As in cold seeps (see Section 2.2.4), chemoautotrophic microbes use the reduced chemicals (e.g. hydrogen sulphide) from the vent fluid as source of energy to produce organic matter.[40] These microorganisms are found free living forming bacterial mats over the vent chimneys, but also in tight symbiosis with benthic fauna. The availability of primary productivity on the seafloor supports high abundance and biomass of highly specialised megafauna communities. At the same time, the extreme environmental conditions found at vents (high temperature gradients, high levels of toxic chemicals, dynamism of vents) result in a low biodiversity with a high proportion of endemic species[41] (see Figure 2.4). The deposition of metals from the vent fluids can result in large accumulations of commercially-interesting minerals, in what is known as seafloor massive sulphide deposits (see Section 3.5.3.).

2.6 *Trenches*

The trenches are the deepest areas of the seafloor, extending from 6000 m to 11 km, in what is known as the hadal zone (Figure 2.1). The deepest point on Earth is in the Marianas Trench, in the western Pacific, with a maximum-recorded depth of 11 033 m in the Challenger Deep. There are 33 trenches around the world, covering an area of 0.2% of the seafloor.[42] Trenches are covered with fine sediment and their main characteristic is the very high hydrostratic pressure (600 to 1100 atmospheres), while temperature and oxygen variables are similar to those found on abyssal plains. The trench macro- and megafauna communities are composed by diverse fauna with a high degree of endemism, including hadal fish, large amphipods, shrimp, polychaetes, bivalves and

40 Van Dover CL, 'Mining seafloor massive sulphides and biodiversity: what is at risk?, *ICES Journal of Marine Science* (2010) doi:10.1093/icesjms/fsq086. See also Baker et al. (n22).

41 Tunnicliffe et al. (n22).

42 Blankenship-Williams LE, Levin LA, 'Living Deep: a synopsis of hadal trench ecology', *Marine Technology Society Journal* (2009) 43: 137–143.

FIGURE 2.4 Examples of hydrothermal vent ecosystems. A: black smokers from the Mid-Atlantic Ridge; B: the vent shrimp *Rimicaris exoculata* from the Mid-Atlantic Ridge; C: *Bathymodiolus* mussel bed from the Mid-Atlantic Ridge; D: *Riftia pachyptila* from the East Pacific Rise.
PHOTOS A, B & C COPYRIGHT MISSAO SEHAMA, 2002 (FUNDED BY FCT, PDCTM 1999/MAR/15281)
PHOTO D COPYRIGHT C. VAN DOVER, DUKE UNI. (USA)

holothurians.[43] The smaller faunal fraction, the meiofauna (32–63 microns) is dominated by soft-bodied foraminifera.[44]

3 Anthropogenic Impacts to the Deep Seafloor

Technological development in the last half century has facilitated access to deep-sea ecosystems. This has provided evidence of a wealth of undiscovered

43 Jamieson A, *The hadal zone: life in the deepest oceans* (Cambridge University Press, Cambridge, 2015).

44 Todo Y, Kitazato H, Hashimoto J, Gooday AJ, 'Simple foraminifera flourish at the ocean's deepest point', *Science* (2005) 307: 689–689.

biodiversity and ecosystem functions as well as important resources, both mineral (hydrocarbons, minerals) and biological (fisheries, genetic resources). Interest in the exploration for and exploitation of these resources is rapidly increasing, paralleling the increasing demand for raw materials and the depletion of resources on land and in the coastal area.[45] Additionally, the remoteness of the deep seafloor has promoted for centuries the disposal of waste and, even under the current restrictive regulations on dumping waste in the seas and oceans, the issue of marine litter continues to increase. Below, we briefly describe the major activities that can have a significant impact on deep-sea ecosystems.

3.1 *Marine Litter*

Marine litter is defined by the United Nations Environmental Programme (UNEP) as 'any persistent, manufactured or processed solid material discarded, disposed of or abandoned in the marine and coastal environment'. Although dumping litter in the sea was banned by the Convention on the Prevention of Marine Pollution by Dumping of Wastes and Other Matter 1972 (known as the London Convention) and subsequent London Protocol (1996), litter continues to be a major threat to marine ecosystems. Major sources of marine litter are heavily populated coastal areas and rivers, as well as illegal dumping from boats.[46] About 6,4 Million tonnes of litter have been reported to enter the oceans each year.[47] Litter can float in the surface or water column, eventually sinking and accumulating on the seafloor. Plastics are the most abundant litter type observed on the deep seabed, followed by metal and glass[48] (Figure 2.5). The impacts of marine litter on the benthic fauna have not been studied in detail, but effects such as suffocation, entanglement, physical damage, ghost fishing of discarded/lost nets and chemical pollution from decomposing materials (e.g. plastics additives, microplastics, paints) have been suggested as major issues requiring further investigation.

---

45 Ramirez-Llodra et al. (n5).

46 Ibid.

47 UNEP, *Marine Litter: A Global Challenge* (UNEP, Nairobi, 2009).

48 Ramirez-Llodra E, De Mol B, Company JB, Coll M, Sardà F, 'Effects of natural and anthropogenic processes in the distribution of marine litter in the deep Mediterranean Sea', *Progress in Oceanography* (2013) 118: 273–287. See also Pham C, Ramirez-Llodra E, Alt C, Amaro T, Bergmann M, Canals M, Company JB, Davies J, Duinvevald G, Galgani F, Howell KL, A.I. HV, Jones DOB, Lastras G, Morato T, Gomes-Pereira JN, Purser A, Stewart H, Tojeira I, Tubau X, Van Rooij D, Tyler PA, 'Marine litter distribution and density in European Seas, from the shelves to deep basins', *PLoS ONE* (2014) 9(4): e95839.

FIGURE 2.5 Marine litter collected in the deep Mediterranean Sea with an otter trawl. A: plastic litter from 1200 m in the Central Mediterranean. B: oil drum collected at 2000 m on the Western Mediterranean. C: plastics collected from 3000 m in the Western Mediterranean. D: glass bottles collected at 1750 m from the Western Mediterranean.
PHOTOS A, B, C COPYRIGHT E. RAMIREZ-LLODRA/ICM-CSIC/BIOFUN. PHOTO D COPYRIGHT A. MECHO/ICM-CSIC/BIOFUN

3.2 *Submarine Tailing Disposal*

Tailings are the fine waste produced by mining activities after extraction of the target metals from the ore. Most industrial mines dispose the vast amounts of tailing waste in land-based dams. However, in countries were the topography or climate do not allow for safe management of dams (e.g. Norway, Indonesia, Papua New Guinea), the disposal of tailings in the sea is used as a suitable option. There are currently two main types of tailing disposal in the sea.[49] In submarine tailing disposal (STD), tailings are disposed through an underwater pipeline at relatively shallow depths (<100 m). Tailings create a gravity flow that deposits the waste on the seafloor. In deep-seat tailing disposal (DSTD), tailings are disposed via a submerged pipeline below the mixing zone (>100 m). The tailings create a gravity flow that deposits the waste on the deep seafloor below 1000 m depth.

The main impacts of STDs and DSTDs, reviewed in[50] include: 1) smothering of the benthic communities by hyper-sedimentation at the local scale; 2) potential toxic effects from heavy metals or added chemicals (flocculants, floatation); 3) impact of changes in grain, which can modify the organic content in the sediment, and grain structure, with some tailing particles having very sharp edges that can physically damage feeding structures or the settlement of larvae/juveniles; and 4) plume dispersal, upwelling and slope failure, which can re-distribute tailings far from the original settling area, thus affecting communities at the regional scale.

Acknowledging the urgent need for further research and robust management measures, the International Maritime Organisation (IMO), together with the Deep-Ocean Stewardship Initiative (DOSI), the International Network for Scientific Investigations of the Deep Sea (INDEEP) and the Norwegian Research Council (NRC) funded-project MITE-DEEP, co-organised a workshop to discuss current knowledge on DSTD processes and environmental impacts. The discussions and conclusions have been synthesised in a report to be discussed by the parties of the London Convention/London Protocol for future action.[51] In parallel, the European Commission is in the process of updating the first 'Reference Document on Best Available Techniques for

49 Shimmield TM, Black KD, Howe JA, Hughes DJ, Sherwin T, *Final report: Independent Evaluation of Deep-Sea Mine Tailings Placement (DSTP) in PNG*, SAMS, Oban, UK, 2010. See also: Hughes DJ, Shimmield TM, Black KD, Howe JA, 'Ecological impacts of large-scale disposal of mining waste in the deep sea', *Nature Scientific Reports* (2015) 5:09985 doi 10.1038/srep09985; and Ramirez-Llodra E, Trannum HC, Evenset A, Levin LA, Andersson M, Finne TE, Hilario A, Flem B, Christensen G, Schaanning M, Vanreusel A, 'Submarine and deep-sea mine tailing placements: a review of current practices, environmental issues, natural analogs and knowledge gaps in Norway and internationally', *Marine Pollution Bulletin* (2015) 97: 13–35.

50 Reichelt-Brushett A, 'Risk assessment and ecotoxicology. Limitations and recommendations for ocean disposal of mine waste in the Coral Triangle', *Oceanography* (2012) 25: 40–51. See also: Ramirez-Llodra et al. (n49); and Morello EB, Haywood MDE, Brewer DT, Atpe SC, Asmunda G, Kowng YTJ, Dennis D, 'The ecological impats of submarine tailings placement', *Oceanography and Marine Biology: An Annual Review* (2016) 54: 315–366.

51 Vogt C (in press) Proceedings of the GESAMP International Workshop on the Impacts of Mine Tailings in the Marine Environment, 10–12 June 2015, Lima, Peru.

Management of Tailings and Waste-Rock in Mining Activities'. The revised 'Best Available Techniques Reference Document for the Management of Waste from Extractive Industries' (MWEI BREF) is subject to the EU Directive on the management of waste from extractive industries (2006/21/EC) and has been published in December 2018.[52]

3.3 *Fishing*

Increased demand for marine biological resources and technological development have promoted the continuous increase of deep-sea fisheries exploitation,[53] with fishing grounds commonly found below 1500 m depth. Bottom trawling (i.e. the towing of a trawl net along the seafloor) has the highest impact, both on the fauna (target and non-target species) and the habitat. Deep-sea target species are often long-lived and have delayed maturity, so the exploitation of such communities, which depletes the population of reproductively-active adults, has rarely proven sustainable.[54] The trawling gear has also a major impact on the seafloor and the benthic communities. In sedimentary slopes, where most trawling takes place, recent studies in the Mediterranean have shown that the regular trawling of the seabed triggers sediment flows downslope, with unknown effects on the faunal communities.[55] Furthermore, long-term trawling activity in a region can modify the shape of the submarine landscape, reducing the original complexity of the seafloor in the same way that agriculture does on land.[56] Fishing over seamounts has resulted in significant impacts, again on the target species and the ecosystem.[57] Seamounts are characterised by rich communities of sessile fauna, including sponges and corals that provide habitat to other fauna (see Section 2.3). These communities are heavily impacted by bottom trawling and their recovery is in the order of decades or centuries (see Section 2.2.3).

Fishing regulations are implemented by coastal states and RFMOs (Regional Fisheries Management Organisations). These regulations may include quota managements, licensing systems and protection of specific habitats

52 Available at < https://susproc.jrc.ec.europa.eu/activities/MWEI/documents/jrc109657_mwei_bref_-_for_pubsy_online.pdf>.

53 Morato T, Watson R, Pitcher TJ, Pauly D, 'Fishing down the deep', *Fish and Fisheries* (2006) 7: 24–34.

54 See Clark et al. (n12).

55 Martín J, Puig P, Palanques A, Masqué P, García-Orellana J, 'Effect of commercial trawling on the deep sedimentation in a Mediterranean submarine canyon', *Marine Geology* (2008) 252: 150–155.

56 Puig P, Canals M, Company JB, Martin J, Amblas D, Lastras G, Palanques A, Calafat AM, 'Ploughing the deep sea floor', *Nature* (2012) 489: 286–290.

57 Pitcher TJ, Clark MR, Morato T, Watson R, 'Seamount Fisheries: Do They Have a Future?' *Oceanography* (2010) 23: 134–144.

(e.g. seamounts, ridges, cold-water corals, upper slope) by closing the areas to fishing activities.[58] In recent years, certain RFMOs have extended their regulations to the protection of vulnerable benthic marine ecosystems in international waters.[59]

3.4 *Oil and Gas Exploitation*

The decrease in land-based resources and developing technology has promoted the increase of oil and gas exploitations in deep waters, with routine drilling below 200 m depth in many regions. In well explored areas, such as the Gulf of Mexico, ultra-deep water drilling (>1000 m depth) activities, which reach 3000 m depth, are expanding.[60] Impact of oil and gas exploitation can come from various activities related to offshore oil and gas development. Some of the major direct impacts are relatively local, including the physical damage to the benthic habitat and community caused by the installation of the drilling infrastructure (ca. 100 m radius), and the discharge of drilling muds and produced water that can affect benthic communities at distances of about 300 m from the source.[61] Effects of drill muds on all size classes of the benthic community (meio-, macro- and megafauna) include changes in density, biomass and diversity, but little is known on the effects on the microbial community. These potential effects of oil and gas exploitation activities are particularly important in operations close to cold-water corals, where coral polyps mortality can increase by burial from thin layers (6.5 mm) of drill cuttings.[62] Additionally, large environmental impacts can occur during accidental oil spills, such as the Deepwater Horizon blowout accident of the Macondo well in the Gulf of Mexico in 2010.[63] Impacts to the deep benthic fauna have

58 Bensch A, Gianni M, Gréboval D, Sanders J, Hjort A, 'Worldwide review of bottom fisheries in the high seas', *FAO Technical Paper* (2008) 522: 1–145.

59 For a review of the applicable legal regime, see R. Caddell, 'Deep-Sea Bottom Fisheries and the Protection of Seabed Ecosystems: Problems, Progress and Prospects', Chapter 12 of this book.

60 Reviewed in Cordes EE, Jones DOB, Schlacher TA, Amon DJ, Bernardino AF, Brooke S, Carney R, DeLeo DM, Dunlop KM, Escobar-Briones EG, Gates AR, Génio L, Gobin J, Henry L-A, Herrera S, Hoyt S, Joye M, Kark S, Mestre NC, Metaxas A, Pfeifer S, Sink K, Sweetman AK, Witte U, 'Environmental Impacts of the Deep-Water Oil and Gas Industry: A Review to Guide Management Strategies', *Frontiers in Environmental Science* (2016) 4: 1–26 doi 10.3389/fenvs.2016.00058.

61 Ibid.

62 Larsson AI and Purser A, 'Sedimentation on the cold-water coral *Lophelia pertusa*: cleaning efficiency from natural sediments and drill cuttings', *Marine Pollution Bulletin* (2011) 62: 1159–1168.

63 Joye SB, Bracco A, Ozgokmen T, Chanton JP, Grosell M, MacDonald IR, Cordes EE, Montoya JP, Passow U, 'The Gulf of Mexico ecosystem, six years after the Macondo Oil Well Blowout', *Deep Sea Res.* (2016) II 129, 4–19.

been detected in an area of 300 km², with significant impacts to the cold-water coral communities 22 km away from the well and at depths of 1950 m.[64]

Although experiments of the toxic effects on deep-sea fauna of chemical dispersants used during oil spills are limited, there is evidence that dispersants can affect larval development, cause tissue degradation in invertebrates and damage cold-water corals. Management of oil and gas exploration and exploitation licenses falls under national regulations and should include activity management, where some processes or technologies are restricted, temporal management, where temporal variations in feeding, breeding or migration of key species is considered, and spatial management, where exploitation may be restricted in relation to the proximity of sensitive species or habitats.[65]

3.5 *Deep-Sea Mining*

The commercial exploitation of deep-sea mineral resources has not started yet. However, in the last 15 years, interest in exploration for and exploitation of these resources has greatly increased Currently, there are four major resource types that are being considered for commercial exploitation from habitats deeper than 200 m depth: manganese nodules, cobalt-rich crusts, seafloor massive sulphides and phosphorite nodules. Each of these resources is found in a specific habitat with particular geochemical and biological characteristics, which will define the significance of the mining impact and the ecosystem recovery potential.[66] Below, we briefly describe each of these mineral resources, their associated ecosystems and main expected impacts and recovery potential form mining activities.

3.5.1 Manganese Nodules

Manganese nodules are polymetallic concretions made of manganese and iron sulphides which form by precipitation from the ambient sea-water over millions of years.[67] Manganese nodules are rich in manganese, copper, cobalt and nickel and are found on abyssal plains, particularly in the Pacific Ocean. The sediments support rich communities of meio- and macrofauna, with

64 Fisher CR, Hsing P-Y, Kaiser CL, Yoerger DR, Roberts HH, Shedd WW, Cordes EE, Shank, Timothy M., Berlet SP, Saunders MG, Larcom EA, Brooks JM, 'Footprint of deepwater horizon blowout impact to deep-water coral communities', Proceedings of the Natural Academy of Scienes of the USA (2014) 111: 11744–11749.

65 Cordes et al. (n60).

66 Ramirez-Llodra et al. (n5).

67 SPC, 'Deep Sea Minerals: Manganese Nodules, a physical, biological, environmental, and technical review', Vol. 1B, in Baker E, and Beaudoin, Y. (ed), *Secretariat of the Pacific Community* (2013) p. 52.

larger animals such as holothurians, sea urchins, sea stars, polycahetes and octocorals also present, but in lower abundance. The nodules are colonised by large single-celled foraminifera.[68] The processes in these abyssal plains are very slow, with very slow sedimentation rates and very weak bottom currents. Additionally, nodules are formed at geological-time scales. Thus, the recovery and recolonization of these ecosystems will be extremely slow and not at the ecological time-scales that mining-licenses will operate, making robust spatial management plans more valuable than possible restoration measures.[69]

3.5.2 Cobalt-Rich Ferromanganese Crusts

Cobalt-rich ferromanganese crusts form by precipitation from the seawater over millions of years over all rocky surfaces free of sediment in the deep oceans. Potentially exploitable crusts are found on the flanks of seamounts, knolls and ridges at depths of 800–2500 m.[70] These crusts are rich in cobalt, nickel and platinum. Although little is known of the fauna specifically on cobalt-rich crusts (in comparison to that of seamounts), these geomorphological structures provide substrate for a variety of sessile filter feeders, such as corals and sponges, and other motile fauna including crustaceans and echinoderms.

3.5.3 Seafloor Massive Sulphides

Seafloor massive sulphides (SMS) form through the precipitation of metals from the fluids at hydrothermal vents, typically at depths between 1000 and 3000 m. SMS are sources of copper, gold, silver, zinc and lead.[71] Vent communities are characterised by very high abundances and biomass of highly adapted species, with a high degree of endemism, supported by microbial chemoautotrophy (see Section 2.4.2).[72] These systems are very dynamic and subjected to sporadic volcanic eruptions, particularly in fast-spreading ridges, as well as changes in the activity of individual chimneys and sources of diffuse flow. There are two

68 Smith and Demopoulos (n27).

69 SPC (n67).

70 SPC, 'Deep Sea Minerals: Cobalt-rich Ferromanganese Crusts, a physical, biological, environmental, and technical review', Vol. 1C, In Baker E, Beaudoin Y (eds). *Secretariat of the Pacific Community* (2013).

71 Van Dover, C.L., Arnaud-Haond S., Gianni, M., Helmreich, S., Huber, J.A., Jaeckel, A.L., Metaxas, A., Pendleton, L.H., Peterseni, S., Ramirez-Llodra, E., Steinberg, P.E., Tunnicliffe, V. & Yamamoto, H., 'Scientific rationale and international obligations for protection of active hydrothermal vent ecosystems from deep-sea mining', *Marine Policy* (2018) 90: 20–28. See also SPC, 'Deep Sea Minerals: Sea-Floor Massive Sulphides, a physical, biological, environmental, and technical review', Vol. 1A, in Baker E, Beaudoin Y (eds). *Secretariat of the Pacific Community* (2013).

72 Van Dover et al. (n71).

scientifically documented cases where naturally impacted vent communities from volcanic eruptions recovered one decade after the eruption.[73] However, these processes took place in fast-spreading ridges, while the major SMS identified to date are on slow-spreading ridges, which are much less dynamic systems. The recovery of such ecosystems from mining depends, thus, on the habitat itself, as well as on the availability of larvae, juveniles or mobile adults from intact populations that are able to disperse to and colonise the new vents systems post-mining.[74] However, mining will add on to the existing natural loss of critical habitat, and cumulative impacts may result in significant changes in the abundance and distribution of vent species.[75] Because of the rarity of active hydrothermal vent systems, their unique fauna and the challenges of identifying representative systems for area-based management, it has been proposed that active hydrothermal vents are protected legally from direct and indirect mining impacts.[76]

3.5.4 Phosphorite Nodules

Phosphorite nodules are formed from limestone deposits following chemical reactions in areas with upwelling and high surface productivity on upper continental slopes (200–400 m). Phosphorite nodules contain products used to make phosphate fertiliser and they have recently been explored off New Zealand and Namibia. In these regions, the dominant fauna includes echinoderms, galatheid crabs, sponges, corals and bryozoans, and abundant amphipods in the sediment. However, the impacts of potential mining of the mineral resources on the upper continental margin have been little investigated.

3.5.5 Impacts of Deep-Sea Mining

The main impacts of deep-sea mining on the seafloor include the depletion or physical damage to the habitat and fauna by the mining equipment, changes in seafloor topography and geochemical characteristics, creation of sediment plumes and potential toxicity from metal and/or process chemicals release (Figure 2.6). Additionally, light and noise may be an issue for deep-water fauna and sediment plumes may impact pelagic life, including larvae and juveniles. These processes will affect the composition, structure and functioning of the faunal communities in different ways depending on the ecosystem considered.

---

73 Tunnicliffe et al. (n22).

74 Boschen RE, Rowden AA, Clark MR, Gardner JPA, 'Mining of deep-sea seafloor massive sulfides: A review of the deposits, their benthic communities, impacts from mining, regulatory frameworks and management strategies', *Ocean & Coastal Management* (2013) 84: 54–67.

75 Van Dover et al. (n71) and Boschen et al. (n74).

76 Boschen et al. (n74).

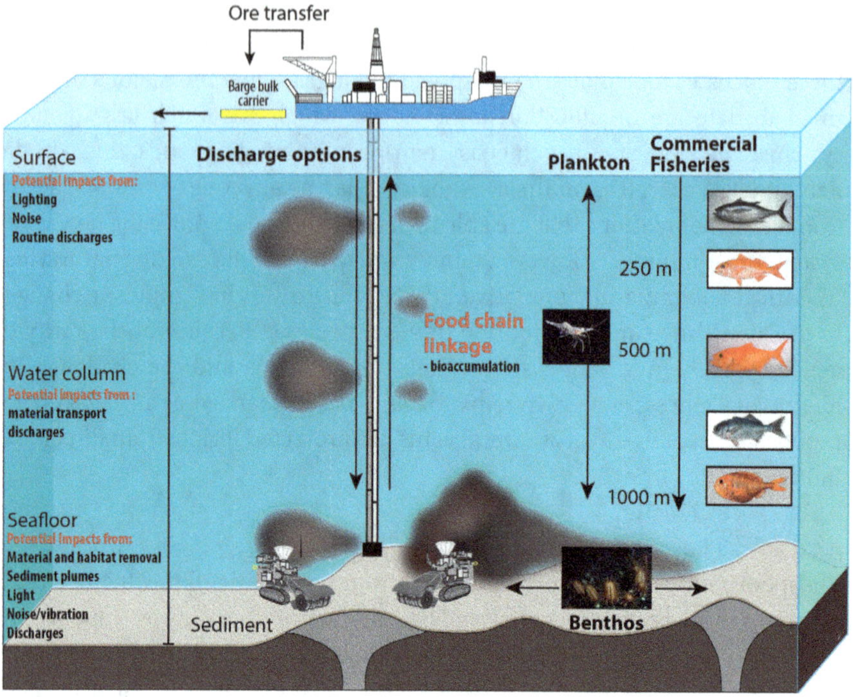

FIGURE 2.6 Deep-sea mining system and associated impacts on the pelagic and benthic ecosystems
IMAGE COURTESY OF DR MALCOLM CLARK, NIWA (NZ) AND IUCN

For example, mining manganese nodules at abyssal plains, where processes such as nodule formation and sedimentation are extremely slow (millennia), will have a very significant and long-lasting impact on the ecosystem.

With a new deep-sea mining industry emerging, regulatory bodies, both for areas within and beyond national jurisdiction, need to develop regulations and licenses where potential economic gains need to be balanced against impacts on the environment, other ocean users and civil society. The need to ensure the protection of the environment requires a robust scientific understanding of what can cause a significant adverse change to deep-sea biodiversity, ecosystem structure and function that will cause serious harm to the affected ecosystem.[77] The International Seabed Authority (ISA) is responsible for the

77 Levin LA, Mengerink K, Gjerde KM, Rowden AA, Van Dover CL, Clark MR, Ramirez-Llodra E, Currie B, Smith CR, Sato KN, Gallo N, Sweetman AK, Lily H, Armstrong CW, Brider J, 'Defining "serious harm" to the marine environment in the context of deep-seabed mining', *Marine Policy* (2016) 74: 245–259 doi http://dx.doi.org/10.1016/j.marpol.2016.09.032.

regulations and license contracts for exploration and exploitation of minerals on the seabed beyond national jurisdiction (The Area), under the principle that The Area and its mineral resources are 'common heritage of mankind'.[78] Within territorial waters, regulations are often lacking, but interested nations are currently developing such regulations, which, for parties of the UN Law of the Sea Convention, must be at least as restringing as the ISA regulations.

3.6 *Cumulative Impacts*

The deep ocean is experiencing increasing pressure from human activities targeting its resources or receiving and accumulating synthetic waste and chemical pollution. These different impacts may have synergies on single ecosystems if acting together, with a magnified effect on the structure and functioning of the faunal communities.[79] In particular, climate change-related stressors such as warming water masses, de-oxygenation, changes in primary productivity and ocean stratification, can affect the oceans globally.[80] These global climatic stressors will add to direct impacts from other human activities, such as fishing or mining (Figure 2.7), possibly reducing resilience and recovery potential of the affected ecosystems. Different extractive industries may also be in spatial conflict. For example, in New Zealand and Namibia, phosphorite nodule reserves on the upper continental margin coincide with existing fishing grounds.[81] Based on the still limited scientific understanding of the composition and functioning of many deep-sea ecosystems, several stakeholders recommend the development of precautionary and ecosystem-based management systems. These measures should balance the use of mineral and biological resources with the maintenance of healthy marine systems and the ecosystems services they provide.[82]

---

78 See: Jaeckel A, 'An environmental management strategy for the International Seabed Authority? The legal basis.', *The International Journal of Marine and Coastal Law* (2015) 30: 1–27; Jaeckel A, Ardron JA, Gjerde KM, 'Sharing benefits of the common heritage of mankind – Is the deep seabed mining regime ready?', *Marine Policy* (2016) 70: 198–204.

79 Ramirez-Llodra et al. (n5).

80 Levin and Le Bris (n7).

81 Environmental Protection Authority (EPA) – Te Mana Rauhi Taiao. Decision on marine consent application. Chatham Rock Phosphate Limited, To mine phosphorite nodules on the Chathman Rise (2015).

82 Mengerink et al. (n6).

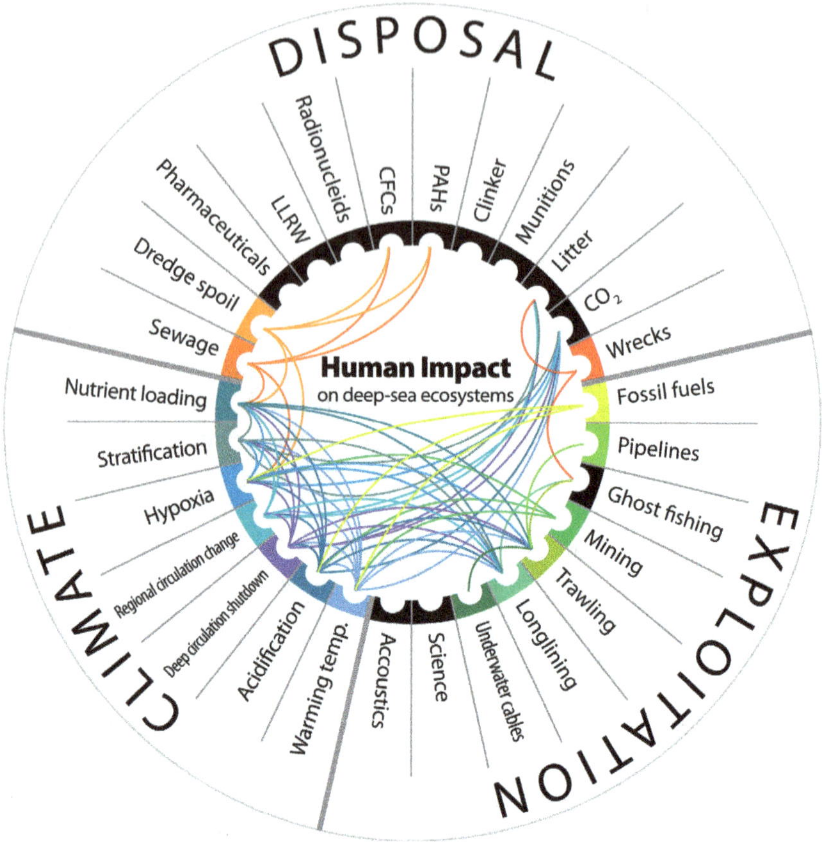

FIGURE 2.7 Interactions amongst waste disposal, exploitation of resources and climate change that may have synergistic effects in deep-sea ecosystems
FROM RAMIREZ-LLODRA ET AL., 2011. PLOS ONE: 6(8) E22588

Acknowledgements

ERLL was funded by the MarMine project (grant no. 247626/O30) and with support from the Norwegian Institute for Water Research, NIVA, Oslo.

CHAPTER 3

A Short Human History of the Ocean Floor

Håkon With Andersen

1 Introduction

It could be argued that representations are the key to understanding human actions. It is our inner picture of the sea floor that makes things happen – whatever picture that is. The more so since the ocean floor is not directly accessible to us in any way – we depend on representations. So let us start this human history of the ocean floor reminding ourselves that our object of study is not direct accessible and that our impressions of the sea floor is always mediated in one way or another: by different technologies, by science or by literature or cultural traditions.

Science came to play an important role in overtaking earlier guesswork and anecdotes about the sea floor. But even scientific views were changing. It suffices to remember the ridicule Alfred Wegener (1880–1930) was subjected to with his theory of continental drift from 1912. Not to mention the fascinating story of the cartographer Marie Tharp (1920–2003) and her detailed drawings of the ocean floor that finally contributed to the breakthrough of plate tectonics in the late 1960s and restored Wegener's ideas. The representations developed afterwards combined with all sort of technological devices have made the ocean floor a place for a great variety of claims and hunt for resources.

In this chapter, I will try to establish something that could be called a human history of the sea floor. A place so inaccessible requires other means and ways to figure out the relation between humans and the deep sea. It is important to acknowledge that the representation of the seafloor is the most important element in this history. Secondly that the resources and their regulations always have been based on these representation. As time flows these have shifted and varied. Science has come to play an important part as have real examination of the sea floor. This chapter is an overview, too short of details and modifications, but it might be an introduction to an area very few have seen, but still covers almost 5⁄7 of the Earth's surface.[1]

---

1 The literature in the field is particular centered on the history of oceanography. Some main references here are the classic study by Deacon, Margaret B. *Scientists and the sea 1650–1900: a study of marine science* (London: Academic Press, 1971). It was followed by a newer anthology

© HÅKON WITH ANDERSEN, 2020 | DOI:10.1163/9789004391567_005

The chapter consists of four major parts as a parallel to four major historical phases of the cultural appropriation of the sea floor. After an introduction (Section 1), it starts with the *Carta Marina* from 1539 describing the dangers of crossing the surface of the sea and the first attempts for soundings and measuring of tidal waters (Section 2). In the second major part (Section 3), the first attempts to measure the deep sea and to use the seafloor as a place for cables are discussed. The third period (Section 4) introduces the powerful alliance between science and navies leading up to important research project as 'The Challenger' and 'Meteor' really increasing the knowledge of the seafloor. The last major section (Section 5) is devoted to the regulation of the seafloor ending with UNCLOS III and the scientific development underpinning it.

2 The Dangerous Sea

The deep ocean has always been a mystery to mankind. The idea of something bottomless, a void, is frightening. Even more frightening was the idea of what this bottomless void could hold. Rumors and ideas were circulated and also collected by intellectuals in the renaissance and later.

2.1 *Representing the Unknown Deep Ocean Floor*

It seems appropriate that we start this decent to the deep ocean floor with a famous map, *Carta Marina,* from 1539 (Figure 3.1). The map was made by a clergy, Olaus Magnus (1490–1557).[2] It covered most of the northern part of coastal Europe and the North Sea and the Atlantic. Olaus Magnus used years to gather knowledge and experiences from merchants, sailors, fishermen and whalers to be able to draw the map. To us the interesting part is not only the land masses and coasts that are drawn but also what is to be found in the deep ocean and that come to the surface of the map. Monsters, large as mountains, lived in the unfathomable depths, threatening every seaman who dared to sail across open sea. Olaus Magnus had, of cause, not seen these monsters himself,

from 2001: Deacon, Margaret, Rice, Tony and Summerhayes, C.P. *Understanding the oceans: a century of ocean exploration* (London: UCL Press, 2001). For a newer general introduction, H.M. Rozwadowsky is highly recommendable: Rozwadowski, H.M. *Fathoming the Ocean* (Harvard University Press, 2009); Helen M. Rozwadowsky, 'Focus: knowing the oceans: a role for the history of science', *ISIS: Journal of the History of Science in Society* (2014) *105*(2), 335–337 (see as well her edited focus group of paper in *Isis* 2014). A good introduction to the historical development of plate tectonics can be found in Lawrence, David M., *Upheaval from the abyss: ocean floor mapping and the Earth science revolution.* (New Brunswick, N.J: Rutgers University Press, 2002).

2 Richter, Herman and Olaus, Magnus. *Olaus Magnus: Carta marina, 1539* (Vol. 11:2) (Lund: Lärdomshistoriska samfundet, 1967).

FIGURE 3.1 Carta Marina, by Olaus Magnus
HTTP://WWW.NPM.AC.UK/RSDAS/PROJECTS/CARTA_MARINA/ "CARTA MARINA SATELLITE IMAGES", PUBLIC DOMAIN, HTTPS://COMMONS.WIKIMEDIA.ORG/W/INDEX.PHP?CURID=558827

he had to rely on reports as he relied on reports of coasts and lands, weather and winds. Hence, what we see in the map are not fantasies, but first attempts to empirically say something about what the deep ocean in fact contained as it had been witnessed by sea folks. It was a representation of what the oceans concealed.

2.2 *Addressing the Perils of the Open Sea by Exploring the Seafloor*

The sea monsters and ocean storms were perils of the open sea. There were, however, other more concrete and always threatening ways that the sea and the sea bottom could be a menace for the seamen and the ship masters. One thing was storms and bad weather, even more dangerous was the treacherous sea bottom threatening to ground the ship and destroy it. As late as the end of 19th century the largest cause of averages and losses of ship was 'grounding' or 'stranded and abandoned'. As much as ⅔ to ¾ of all ship losses was caused by this.[3]

3 DnV accident statistics, from Annual report. See Paulsen, Gard, Andersen, Håkon With, Collett, John Peter and Stensrud, Iver Tangen *Building Trust. The history of DNV 1864–2014* (Dinamo Forlag, 2014).

The danger of grounding and the anxiety for being stuck on the bottom lead very early to measures to sound the depth of the sea under the ship. Sounding was also used to determine if land was not too far away. However, the main reason was to ensure safe travel. Still, accident happened. One of the more famous was for instance James Cook's (1728–1779) grounding at the Great Barrier reef in the summer of 1770 when exploring the east coast of Australia.[4]

Sounding also made another thing clear: the ocean was very, very deep. Usually sounding lines would only reach some hundreds fathom deep.[5] Below that, nobody knew and nobody cared too much. Interest turned to what was considered problems for shipping. Not only grounding was a threat, but in the same way knowledge of tides became important in the 18th century. Serious studies of tides and the behavioral of tides were undertaken, along with attempts to map coastlines with the level of tides.

Nevertheless, very few cared about the deep ocean except for superstition and rumors about what the deep ocean actually hid. It was still fathomless in the 18th century, even if Olaus Magnus′ creatures had disappeared and may be substituted by whales and other large sea animals with other stories connected to them and the few that had seen them.

A small note should be made about the real observers of the ocean: the growing whaling industry towards the end of the 17th century and through the 18th and early 19th century. Whalers were the only one that really crisscrossed the oceans on their restless hunt for the large animals. Hence, they also became the most important reporters of conditions at sea and the lives of 'monsters'.[6]

3 Mapping the Seafloor as a First Answer to Its Growing Strategic Importance

Towards the end of the 18th century and start of the 19th century, the ocean took on a more strategic importance to the larger naval countries. It became important to systematize the knowledge of wind, currents and sailing conditions around the globe. Knowledge about the sea floor, where it was a danger to ships and where it could contribute to different sailing conditions became important. The French navy had done this since the middle of the 18th century

4 M.B. De Deacon, Margaret B. *Scientists and the sea 1650–1900: a study of marine science* (London: Academic Press, 1971).

5 A fathom is 6 feet or 1,83m. Older measures varies from 1,5m to 1,8m.

6 Philbrick, Nathaniel, *In the heart of the sea: the epic true story that inspired Moby Dick* (London: HarperCollins, 2001); Tønnessen, Johan Nicolay and Johnsen, Arne Odd, *The history of modern whaling* (Univ of California Press, 1982).

while the British navy established the British Admirality's Hydrographic Office in 1795. The first hydrographer of the admirality was Alexander Dalrymple (1737–1808), a fellow of the Royal Society. We immediately note the marriage between 'science' and the navy, a new combination that should mark the exploration of the seafloor for many decades to come.[7]

The hydrographic office should first and foremost gather intelligence about the sea and the oceans and improve maps and navigational manuals for both military and civil service. Maps had been made for commercial reasons for centuries, but particularly the use of chronometers and lunatic tables for more exact longitudinal positions had improved the quality of maps quite dramatically. The Admiralty and its hydrographic office had pioneered these resources since the last decades of the 18th century. But the task was even broader: the hydrographer should be the foremost advisor to the Board of Admiralty on all sort of intelligence about the ocean.[8] It strengthened the link between the navy and 19th century scientific activity at the same time as it also contributed to commercial activities, first and foremost through better maps and sailing manuals. Dalrymple, for example, came to prioritize international scientific cooperation and exchange of data and maps. He had obvious a desire to put hydrography before limited military gains.[9]

Matthew Fontaine Maury (1806–1873) came to play a parallel role in the United States in the antebellum period. In 1842, he was appointed director of the Navy's Depot of Charts and Instruments and soon after the head of the Naval observatory.[10] In connection with our study of the ocean floor Maury plays a prominent part as he was the first to make a crude map of the Atlantic Ocean floor and as such motivated the first use of the deep sea floor. We will come to that, but first we have to understand what was at stake for all the great powers hydrographic activities in the first part of the nineteenth century. We will use Maury as an example of this, even if more or less the same sort of work were conducted in Britain, France, Spain, Denmark and other European coastal states.

---

7 Webb, Adrian. 'More than just charts: hydrographic expertise within the Admiralty, 1795–1829', *Journal for Maritime Research* (2014) 16(1), 43–54; Clissold, P. *Chartering the Seas. The Admiralty Hydrographic Service 1795–1919*. Vice-Admiral Sir Archibald Day, K.B.E., C.B., D.S.O. (1968), Her Majesty's Stationery Office, London, 1967, 105s. *Journal of Navigation, 21*(03), 371–373.

8 Webb, Adrian (n7) p 45.

9 Webb, Adrian, 'Foundations for « International cooperation in the field of hydrography »: some contributions by British admirality hydrographers, 1795–1855', *International Hydrographic review* (2010) (4) p 8.

10 Rozwadowski, H.M (2009) (n1) p 44.

3.1 *Matthew Maury's Legacy*

In 2016, a whole section of the International Journal of Maritime History was devoted to Matthew F. Maury and his rather mixed legacy.[11] Mixed because he left his job to fight for the confederacy in 1861 and because he was not a real scientist according to historians of science. In the section, Maury's contribution is discussed along three dimensions: Maury, the pathfinder, the scientist and the reformer. These three roles characterize Maury quite well, as well as they characterized contemporary hydrography in general. However, Maury did it more intensely than most others hydrographers at the time. As a pathfinder, he was studying ship logs and created catalogues and maps of wind, weather and water in such a way that he could recommend the fastest way to travel.[12] He created in a way pathways for sailing ships, not only naval but also for merchant ships and whalers. This empirical and systematic work over many years did away with much of the superstition and personal preferences of the captains. As the merchant ship masters saw the benefit of their observations, they were easily persuaded to collect more data on their travels and in this way strengthen Maury's scientific work.

Maury's interests were much wider than just finding the best traveling routes in the days of sail. It included all parts of the oceans: winds, salinity, currents, weather and depths.[13] More or less unintentionally, Maury's work came to be important for the first serious use of the ocean floor, as a bed for communication technology, the first transatlantic telegraph cable.

3.2 *The First Submarine Telegraph Cables: the Seafloor as a Medium*

In a report to the secretary of the U.S. Navy from 1853, Maury had described the ocean bed as 'a plateau, which seems to have been placed there especially for the purpose of holding the wires of the submarine telegraph, and keeping them out of harm's way'.[14]

It was the American entrepreneur Cyrus Field, encouraged by Maury's report that initiated the attempt to lay such a cable. Moving to Britain he succeeded in rising sufficient funding for a first attempt to lay a cable. The cost was

11 See the introductory article: Rozwadowski, Helen M., 'Introduction: Reconsidering Matthew Fontaine Maury', *International Journal of Maritime History*, (2016) 28(2), 388–393.

12 Smith, Jason W., 'Matthew Fontaine Maury: Pathfinder', *International Journal of Maritime History*, (2016) 28(2), 411–420.

13 Hardy, Penelope K. 'Matthew Fontaine Maury: Scientist', *International Journal of Maritime History* (2016) 28(2), 402–410.

14 Headrick, Daniel R., Griset, Pascal, 'Submarine Telegraph Cables: Business and Politics, 1838–1939', *The Business History Review* (2001) 3/75, 543–578.

FIGURE 3.2
Landing of the Atlantic Cable of 1866, Heart's Content, Newfoundland, by Robert Charles Dudley
HTTPS://WWW.FLICKR.COM/PHOTOS/LAC-BAC/31367042642/IN/ALBUM-72157677492475586/, PUBLIC DOMAIN, HTTPS://COMMONS.WIKIMEDIA.ORG/W/INDEX.PHP?CURID=55655760

enormous, and the risk was high and became obvious when the first cable was lost in 1857. A second attempt was made in 1858. It succeeded temporarily so the address of the American president would reach Queen Victoria. However, a couple of weeks later it failed, never to work again. The Civil War prevented new attempts before finally the cable was laid in 1865. It broke again, but this time it was possible to repair it. In 1866, the cable finally worked and worked well: it served newspapers, business, politicians and administrations, even if it was expensive in use (see Figure 3.2 – Landing of the Atlantic Cable of 1866, Heart's Content, Newfoundland, by Robert Charles Dudley).

In the years from the 1850s onwards submarine telegraph cables were laid under many seas and oceans, all of them needed sufficient knowledge about the ocean bottom. British firms totally dominated the business, both for cable laying and for submarine telegraph. In 1892, British firms controlled ⅔ of all submarine cables (more than 160 000 km) with the US as a good second with 15%.[15] For cable laying, knowledge of the ocean bed's topology and the bottoms quality was of outmost importance. It was not any longer science for science sake, it was technology, it was strategy but foremost it was business.

It is difficult for us today to recognize the importance and the publicity that the submarine cables created. As communication technology, it was the first time that information and communication was separated from physical movement of persons or things. It is difficult to imagine the change that the cables brought about. Instead of using weeks and months for news, business instructions and letters, they could now be communicated within hours and minutes. The price was high, of course, but suddenly the globe became much smaller. It can be argued that the transatlantic cable across the abyss of the ocean was the real start of the global information society.

15 Ibid, p 560.

The Atlantic cable was not the first submarine cable, but it was the first to connect North America with Europe. Earlier submarine cables from the early 1850s was quickly destroyed by fishermen before they were properly protected and the insulation was destroyed by seawater. The use of guta perca as insulation and iron reinforcements proved to be effective. Hence, the possibility for larger distances opened up.

With the idea of a submarine cable, the seafloor was immediately brought to attention. The seafloor would be the medium the cable had to use, even the very deep seafloor. In addition, this was not a territory under national control, on the contrary, this was a transnational medium paralleled only by the ocean surface as used by international shipping. For the first time, technology demanded knowledge of the deep seafloor outside the fishing grounds on the shelfs and was in itself an impulse for further exploration.

4 Exploring the Seafloor

4.1 *Natural Science*

In the middle of the 19th century the interest for the sea bottom took on another new perspective. This time the interest came more directly from science, from both amateurs and more professionally inclined scientists. Science as a gentleman's 'sport' was well established in Great Britain with Royal Society as a core elite institution. Around 1850 the combination of yachting and bottom scraping came in fashion by gentlemen scientists.[16] Bottom scraping brought up a very new flora and fauna which could be described and discussed. In the second half of the 19th century, bottom scraping also became increasingly popular among scientists in other countries. This resulted in an increased knowledge about the sea floor at not too great depths.[17] The importance of these early scrapings was to open up for more serious studies of the deep ocean, were yachts with lines and scrapes would not be sufficient. The second part of the nineteen century represented in many ways the high tide for museums of natural history as the most prominent representation of modern science. As such, bottom scraping contributed substantially to the collections and, thus, to the representation of science.

A popular idea among scientists was the idea that there could be no life below 300 fathoms because of the loss of sunlight. The theses were put forward

16 Rozwadowski, H.M. (2009) (n1).

17 Andersen, Håkon With et al, *Aemula lauri : the Royal Norwegian Society of Sciences and Letters, 1760–2010* (Sagamore Beach: Science History Publications, 2009).

by the British naturalist Edward Forbes (1815–1854) in the early 1840s and got its own name: the 'azoic theory', i.e. the lifeless zone. Forbes had been dragging in the Irish sea and most important the Aegean sea and found less and less diversity of living beings the further down he dragged. It encouraged others to do the same and in the 1860s the azoic hypothesis obviously had problems. All these draggings increased the knowledge of life at the seafloor, even if it was not possible to drag very deep.[18]

Already the laying of the transatlantic cable found evidence to contradict the azoic theory when primitive lifeforms were found on broken cables that were brought up for repair. However, it would live on for a decade before it was finally rejected through the *Challenger* expedition in the early 1870s.

The seafloor had found a use and hence become a medium that required knowledge and interest. It was no longer a void only to be discussed or to be researched. Towards the end of the 19th century, both the U.S. navy and the British navy's Hydrological office became more and more involved in what we might call early oceanographic scientific work in a more systematic way.

The cost of doing this kind of research in the 19th century was staggering and that was one reason that only the larger navies were the ones that could contribute with ships and seamen. Hence a pragmatic alliance between the navy and scientists was formed. The ocean became in a way a new frontier, to quote Helen Rozwadowski, as there was no more land to be found.[19] New species, new charts, new details of shores and peoples were to be 'discovered'. In this the navy could be seen to continue an already old tradition from exploration and occupation of territories, but this time under the sea. For the scientist in the Royal Society or at the universities, it was inconceivable to fund large scale studies of the deep ocean without the cooperation of the navy.

The understanding of the ocean floor in the first part of the 19th century was based on the idea of a very rugged 'landscape'. Without the forces that would slowly tear and wear on mountains on land one thought that the seascape was even more dramatic that mountains on land. This led to the belief in so called 'vigias' that simply was summits that raised so high in general deep seas that it was a threat for ships and lives of men. So one reason for measuring depths of even very deep oceans was that there might be some vigias. Maurey became instrumental in the early 1850s to change the view of the deep sea bottom to something much more like a plateau, perfectly shaped for a submarine cable.

18 Corfield, R., *The Silent Landscape: The Scientific Voyage of HMS Challenger* (National Academies Press, 2003).

19 Rozwadowski, H.M. (2009) (n1).

During the 1850s and 1860s several expeditions were made both with American and British navy ships sounding and measuring temperature, salinity and currents in the oceans. With the steadily expanding telegraph cables, this was also useful to commercial companies.

4.2 *The HMS Challenger Expedition (1872–1876)*

However, no initiative could compare with the three and a half yearlong scientific expedition of an old British navy vessel, the *HMS Challenger* (Figure 3.3 – H.M.S. Challenger 1874).[20] From 1872 until 1876 she circumvented the earth three times and crisscrossed all large oceans, sounding depths, taking samples from the bottom, measuring salinity and temperature. The navy and scientists combined resources and intellect in a very fruitful mix. *Challenger* made around 400 deep soundings in all the larger oceans in the span of the three and a half years, each sounding was difficult to execute and took a fairly long time. In addition, they took samples of the seafloor and studied both the flora and fauna of the ocean (Figure 3.4 – Examining the 'haul' on board the *Challenger'*, W.H. Overend). The reports from the expedition filled 50 volumes and was not completed before 1895. They came to be the basis for all further investigations and new results from the findings did not stop before well into the 20th century.[21] The *Challenger* was equipped with special equipment for sounding, including a separate steam engine to enable sounding and dragging at extreme depths. She was also equipped with laboratories and storeroom for scientific specimen brought up from the deep.

The ship itself was a 2300-ton screw corvette built in 1857. It was a full square rigger with an additional 1200hp steam engine. With most of its cannons removed and the six scientists onboard, it still had a crew of about 200 men. In the more than three years she crisscrossed the Atlantic, the Pacific, the Indian Ocean and the Antarctic she regularly stopped every 200 miles to do soundings, scrapings and all sort of measurements, particularly temperature at different depths, salinity and currents. It was the grandest and most costly scientific expedition ever to have been carried through.[22]

The expedition was initiated by science, through the initiative of the vice president of The Royal Society, W.B. Carpenter (1813–1885). Charles Wyville Thomson (1830–1882) was to become the scientific leader of the *Challenger* expedition. He had led a couple of shorter expeditions to the North Atlantic in

20 Brunton, E.V., *The Challenger Expedition, 1872–1876: a visual index* (Natural History Museum, 1994). Corfield, R., *The Silent Landscape: The Scientific Voyage of HMS Challenger* (National Academies Press, 2003).

21 M. Deacon et al., 2001 (n1) p 31.

22 Ibid and Lawrence, David M. (2002) (n1).

FIGURE 3.3
H.M.S. Challenger 1874.
FROM NOAA ARCHIVE, PUBLIC DOMAIN, HTTPS://COMMONS.WIKIMEDIA.ORG/W/INDEX.PHP?CURID=7028789 HTTPS://EN.WIKIPEDIA.ORG/WIKI/CHALLENGER_EXPEDITION#/MEDIA/FILE:CHALLENGER.JPG REPORT ON THE SCIENTIFIC RESULTS OF THE VOYAGE OF H.M.S. CHALLENGER DURING THE YEARS 1873–76 UNDER THE COMMAND OF CAPTAIN GEORGE S. NARES, R.N., F.R.S. AND THE LATE CAPTAIN FRANK TOURLE THOMSON, R.N./PREPARED UNDER THE SUPERINTENDENCE OF THE LATE SIR C. WYVILLE THOMPSON, AND NOW OF JOHN MURRAY; PUBLISHED BY ORDER OF HER MAJESTY'S GOVERNMENT LIBRARY CALL NUMBER Q115. C4 *NOAA SOURCE INFORMATION* HTTP://WWW.PHOTOLIB.NOAA.GOV/SHIPS/SHIP3137.HTM

FIGURE 3.4 Examining the 'haul' on board the 'Challenger'.
W.H. OVEREND [PUBLIC DOMAIN], VIA WIKIMEDIA COMMONS, HTTPS://COMMONS.WIKIMEDIA.ORG/WIKI/FILE:EXAMINING_A_HAUL_ON_BOARD_THE_CHALLENGER.JPG

the smaller naval vessels, *HMS Lightening* and *HMS Porcupine* in the late 1860s. Both the navy and the Hydrographer of the Admiralty was positive to the project, and so was the Treasury. There were several reasons for this most unusual strong support for a very expensive and long expedition. One of them was the newly advance of subsea telegraph cables. Another was the treat of American deep sea explorations, following up on Maury's earlier work. A third reason was the way the government's financial structure was newly restructured.[23] Finally, it should be noted that Britain, as the total dominating sea power of

23 M. Deacon et al., 2001 (n1) p 28.

the world at that time, had to take on the responsibility of figuring out more about the seafloor than the other and much smaller maritime nations.

The leader of the expedition, Charles Wyville Thomson, died in 1882 and had to leave the completion of the reports to his younger colleague who also had participated in the whole expedition, John Murray (1841–1914). It was mainly due to his work that the result of the *Challenger* expedition came to put its mark on the next half hundred years of research and exploration of the sea floor.

The discoveries of *Challenger* were many and had profound consequences. First of all, one got to know how deep the ocean actually was and how the topography of the bottom varied with seamounts, ridges and large flat territories and also with extreme depths in deep trenches. Secondly, one got an idea of what the ocean floor consisted of, with the light oozes picked up all round the globe. With 200 miles between each sounding, the maps were not very good, but it sufficed to get an idea of the bottom. The extremely detailed reports that came from John Murray's hand were to be used for many decades into the future. The *Challenger* really shifted, or maybe better, created our view of the sea floor that to some extent still is valid today even as impressionistic as it was.

However, the number of soundings were rather small, compared with the enormous area of the oceans. To make profiles of the sea bottom for the laying of telegraph cables the soundings were not only suspicious few, the problems of correct soundings were also great. Already Maury in 1858 had shown how different soundings of the same route gave quite different profiles of the seafloor, mainly because the number of soundings were so few. Scientifically one could live with this uncertainty, but for telegraph cables it was another question. Here, the ocean floor as a medium for the cable was very important.

4.3 *Follow-up Expeditions: Meteor*

The *Challenger* expedition was also the breakthrough for oceanographic research internationally. However, no one had the resources or the patience to repeat such grandiose expedition as that of *Challenger*. Around the turn of the century, much interest turned away from the ocean floor and instead concentrated on the resources at sea: fish and fishing. With telegraph cables on the ocean floor now as quite ordinary business – even if it was very expensive and needed bottom surveying and fathoming, migrant fish schools and fish reproduction in the ocean became the new focus.[24]

---

24 Hamblin, Jacob Darwin, 'Seeing the Oceans in the Shadow of Bergen Values', *History of Science Society, Inc* (2014) Vol. 105, pp. 352–363. Rozwadowski, Helen M., *The sea knows no boundaries : a century of marine science under ICES*, International Council for the Exploration of the Sea (Seattle, Wash: ICES in association with University of Washington Press, 2002). Schwach, Vera, *Havet, fisken og vitenskapen : fra fiskeriundersøkelser til havforskningsinstitutt 1860–2000* (Havforskningsinstituttet, Bergen, 2000).

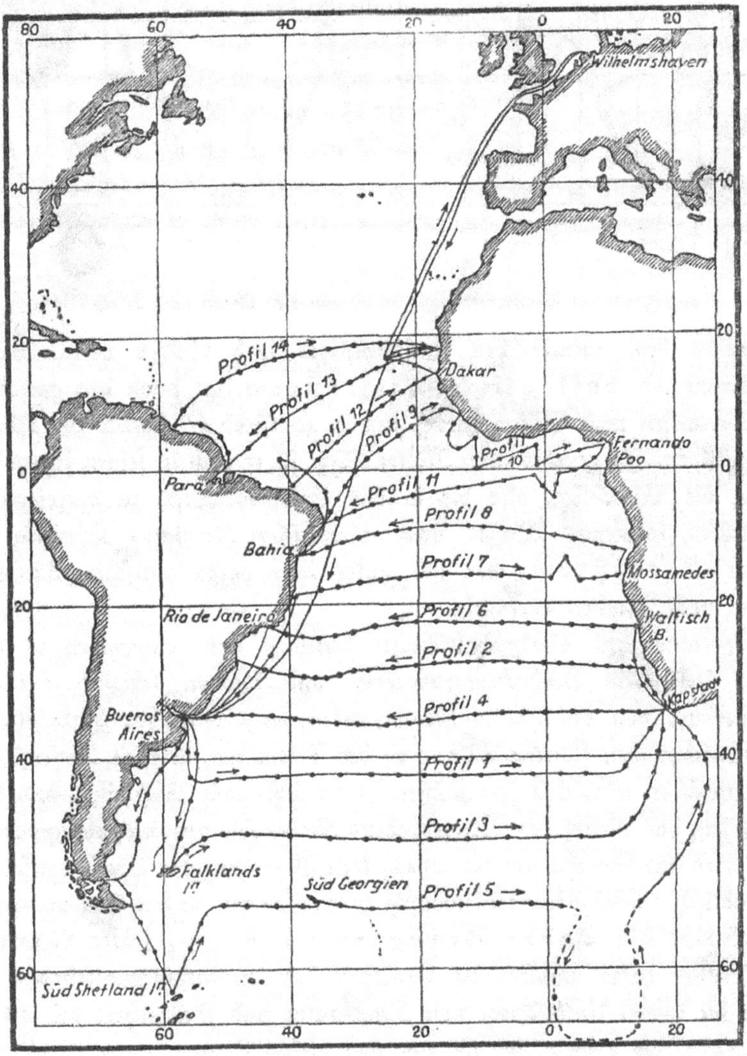

FIGURE 3.5 The Meteor expedition, original plan (1925–1927)
A. MERZ, HTTPS://COMMONS.WIKIMEDIA.ORG/WIKI/FILE:DAE_REISEPLAN_ORIG.JPG
PUBLIC DOMAIN VIA WIKIMEDIA COMMONS

Scientific expeditions did not stop even if interest turned to other problems. Of particular interest is the German *Meteor* expeditions in 1925–1927 (Figure 3.5 – The Meteor expedition, by A. Merz). For the first time sounding was carried out without line and heavy weights. The Meteor expedition pioneered acoustic waves as sounding method. Suddenly it was possible to do many more soundings and at the same time reduce the uncertainty connected to the real depth. While *Challenger* made 400 soundings, *Meteor* was able to do more than 6000 soundings in a much shorter time. From the 1920s we could say that it was possible to realistically profile the ocean floor.[25] *Meteor* was using an echo sounder based on a design by the Canadian inventor and professor Reginald A. Fessenden (1866–1932). Fessenden was originally known for his inventions in radio communication but had turned to echo sounding and 'iceberg warning' later in life. The *Meteor* expedition set the state for a much better knowledge of the subsea floor.[26]

5 Governing Deep Sea Resources

The 1930s saw in embryonic form another approach to the ocean and deep sea: the attempt to control resources in the sea much further away from land than until then had been customary law. Some South American nations were particularly eager to control what they considered their resources. Let us, however, take a small step back and consider the historical development of international law with respect to the ocean.

5.1 *The Shortcomings of the First International Law Response*

As many authors have pointed out, there was actually no law of the oceans aside for customary rules and, of course, private law regulating much of the shipping business. The reason was quite obvious, since no one needed such laws apart for the common conception of the free ocean and the limitation of the territorial sea to a cannon shot, usually interpreted as three nautical miles.

With expanding fisheries and the introduction of large steamer fishing vessels, conflicts escalated, particularly after the turn of the century. The contours of the frontlines that marked most of the 20th century became clearer: the large shipping nations (flag states) against the coastal states. The first ones fighting claims to increased national control over the coastal seas, the other ones wanted to protect what they considered their resources, also outside the rather narrow territorial waters.

Already in the early 1920s, the League of Nations started working finding out where there was a need for new international public law. In 1924, the Council of the League of Nations established a Committee of Experts for the Progressive Codification of International Law. The law of the sea was one topic considered

25 Höhler, Sabine, 'Depth records and ocean volumes: ocean profiling by sounding technology, 1850–1930', *History and technology* (2002) 18(2), 119–154.

26 Lawrence (2002) (n1) p 102–108. Höhler (2002) (n25).

for codification by the Committee of Experts. Through a series of debates, the Committee narrowed the questions concerning the seas down to the question of territorial waters.[27]

In 1930, the preparatory work of the Expert Committee was the basis for an international diplomatic conference: The Hague Codification Conference. It turned out that the second Committee that was charged with questions related to territorial waters was unable to conclude.[28] The conference was a failure and is often seen as irrelevant since it did not give any results on the territorial seas. Others might see it a bit differently. The conference was premature but the different view between the flag states and the coastal states became apparent and led several of the last to threaten to increase their own territorial waters unilaterally.[29]

To complicate the situation even further just after World War II, one of the most powerful shipping nations made two unilateral proclamations increasing both their fishing rights and, what interests us most here, their claim on the seafloor and the recourses in the subsoil of the continental shelf. This has later been known as the Truman declarations from September 28, 1945.[30]

The background was the search for oil that already before the war had lead oil companies into the Gulf of Mexico to search for oil in increasingly deeper water. Actually, this was not the first claim on the sea floor. In 1942, Britain signed an agreement with Venezuela on the Gulf of Paria (between Trinidad and Venezuela) on the rights to the seafloor and the water above.[31] In both cases, the agreement stated that the arrangement would not interfere with the right of shipping and would thus preserve the freedom of the seas, even if the resources in the water column and the continental shelf were claimed.

Until the mid-20th century, the ocean floor was interesting and experienced as a medium and as a scientific object of study. The question of resources or to whom it belonged was far away and outside anyone's actual claim. We should also note a particularity concerning the use of the ocean as a medium. It was largely dominated and regulated by private companies, not by states. We have already mentioned that shipping was mainly regulated by private laws and agreements, but so was also the telegraph cables. They were both laid,

---

27 Harrison, James, *Making the law of the sea: a study in the development of international law*, Cambridge studies in international and comparative law (New York: Cambridge University Press, 2011), p 29.

28 Ibid, p 30.

29 Gold, Edgar, *Maritime transport: the evolution of international marine policy and shipping law* (Lexington, Mass: Lexington Books, 1981).

30 Ibid, p 252.

31 Ibid, p 252.

controlled and operated by large, mainly private, companies, even if they of course were of strategic importance for states.

5.2 *New Attempts to Regulate the Seafloor through International Law in Changing Global Power Balance (1950s)*

In the 20th century, both fish and oil became important as it was possible to move further out from the shores. Hence, the quest for appropriation of both the water column and the ocean floor and its subsoil became an issue. Even if the League of Nation's conference failed in 1930, the United Nation tried again, both in 1958 and 1960 with the first and the second Law of the Sea Conferences, both a rather mixed success and partly failure.

For the seafloor, however, the conferences resulted in some constructive conclusions, even if they were rather temporary and incomplete. The 1958 conference was able to conclude with a regulation of the continental shelf. The shelf and its resources belonged to the coastal state until the depth of 200 meters or as deep as it was possible to extract resources. This was a rather unclear and technology dependent limit, but was probably the best that could be achieved at that time. As for the water column and its resources, none of the conferences were able to reach an agreement. Even if the ambition was reduced compared to the 1930 conference, the different views were still too antagonistic. The flag states opposed the claims of the coastal states. It should also be noted that with regard to the surface, the larger shipping nations blocked the establishment of a UN organization for regulating shipping businesses. The International Maritime Consultative Organization (IMCO, later IMO), which was established in 1949, did not get sufficient ratifications before 1959, mostly due to the opposition of the major flag states. In the light of this, one could say that the agreement on the continental shelf was an exception to be explained.[32]

Resources have to be controlled, one way or another. What we have seen through the first three quarters of the 20th century was a fight over the control of resources in the sea column, that is mainly fish. Large maritime nations wanted not only to have free access to oceans and straits, but also to be able to fish everywhere. Hence it was important to keep the limit of territorial waters to a minimum. Coastal states on the contrary wanted to lay their hands on larger areas of the sea that they argued was their rightful resources. As for the seafloor, it was not that much of a threat since, in most cases, it was not seen

32 Cafruny, Alan W. *Ruling the waves : the political economy of international shipping* (Vol. 17) (Berkeley: University of California Press, 1987).

as relevant for either fishing or shipping and only on very shallow waters was considered relevant for oil.

The cold war was another factor in this game. The US and Britons were opposed to larger territorial waters in light of the submarine warfare possibilities and threats even if this was not clearly stated. However, military and strategic concerns played a central role in the further development of the law of the sea, even if it was not clearly articulated.

It is important in this discussion to bear in mind a couple of important facts as for international law of the sea. The first fact of importance is to recognize Britain's immense dominating position as maritime power around 1900. Half of the world fleet was British. They also dominated both shipbuilding and subsea telegraph cable companies. Add to this the British empire and their domination is complete. As for shipping, another growing power became second only to the British: the US fleet or rather the fleet controlled by US owners. Rules set by the British Board of Trade was the closest one could come to international law of the sea.[33] Other large shipping nations, as the Scandinavians and Greece, followed closely the British policy and the rules of the Board of Trade, simply by necessity.

This situation could not last. The cod wars between Iceland and Britain towards the end of the 1950s was only a sign that times were changing. As the Empire shrank and countries got their independence one after the other, the demand of the coastal states grew stronger by the day. At the end of the 1960s, it was only a question of time before large exclusive economic zones (EEZ) would be a reality. At the same time the prospect of technological change had made the old limitation of the continental shelf irrelevant (200m).

5.3 *Changing Scientific and Political Parameters: the 1960s*

For the history of the seafloor, the 1960s was fascinating in two very different ways, scientific and political. It was a time of great upheaval in both dimensions. Let us start with the scientific one. Almost one hundred years after the *Challenger* expedition, a totally new perspective on the ocean floor would be found, a paradigmatic change almost without parallel. It started in a discredited place, with the geophysicist Alfred Wegener's (1880–1930) theses of continental drift from 1912/1915. Wegener argued that all the continents once had been one, the Pangea, and then drifted apart, colliding and turning.[34] One reason for the rejection was that Wegener was not able to give any mechanism

33 This might clearly be seen in the case of an international loadline for merchant ships, finally established in 1930.

34 Lawrence (2002) (n1), p 33 ff.

for the continental drift and his theory was regarded as rather odd. It was a common consensus that both the continents and the seafloor was of the same constitution and of the same age.

The 1960s marked the return of Wegener's ideas, but now in a totally different form, i.e. as a theory of plate tectonics and this time with explanations on mechanisms for plate drifts. The closer knowledge of oceanic ridges and the creation of new ocean floor at the same time as the oldest part moved slowly into the magma under the continents proved surprisingly to give a radical new understanding of the ocean floor's creation, young age and volcanic activity. The ocean floor was dramatically different from the continents: in age, structure, thickness and durability.

However, nobody had ever seen this part of the world, even if there were enormous amounts of soundings. An important contribution to the solution came to be a painstaking detailed drawing of the ocean floor that still today is quite amazing. Marie Tharp's (1920–2006) drawings are today considered to be major works in international cartography.[35] The seafloor's relatively young age, its thin structure and the volcanic activity gave rise to speculations of what was to be found on the seafloor. Already *Challenger* had noted the manganese nodules on the seafloor in the Pacific, but this new view from the late 1960s on the seafloor re-opened the question.

Hand in hand with the new scientific findings, a new political reality was also dawning. As the number of member states to the United Nations steadily increased during the de-colonialization process in the 1960s, the political balance and competition in the organization changed. The share of new states increased and the balance of power shifted from the old flag states to the new really international organization. The context at the end of the 1960s was quite different than before: a new scientific view on the ocean floor and a new political situation pushed for new ways of doing things.

If we add to this an increased technological and industrial pressure on the resources in the water column (fishing) and on the seabed (oil and all sorts of valuable minerals) there were every reason to reconsider the situation of the oceans. In particular, John Mero's book from 1965 on the *Mineral Resources of the Sea* came to play an important role for the understanding of the potential richness of the deep sea floor. The ocean as an open common needed some sort of regulation and control of access, at least to the resources that was limited.

35 Ibid. For a view of the refined map see 'Manuscript painting of Heezen-Tharp "World ocean floor" map by Berann (1977), available at <https://www.loc.gov/resource/g9096c.ct003148/>.

5.4 *Arvid Pardo's Legacy*

In 1967, the Maltese ambassador to the United Nation, Arvid Pardo (1914–1999), made a speech at the UN General Assembly that later has been hailed as exceptional and most important.[36] The title of the item submitted by the Maltese delegation and presented by Pardo was: 'Examination of the question of the reservation exclusively for peaceful purposes of the sea-bed and the ocean floor, and the subsoil thereof, underlying the high seas beyond the limits of present national jurisdiction, and the use of their resources in the interest of mankind'. Pardo's idea was basically that the ocean bottom and the subsoil with all its resources should belong to all mankind, regardless of their nation, land-locked or coastal, large or small. The seafloor should be declared the heritage of mankind and administrated by some new agency under the United Nations. It was a grand idea, since 70 percent of the Earth was ocean floor. Pardo's speech struck a note. The Maltese delegation left it to the Secretariat to further work on a resolution or to choose other means to forward the proposition.[37]

Pardo's argumentation and conclusion brought the full question of all the possible uses of the sea floor into the UN. In addition to the importance of the topic, the timing was perfect. The ocean and seabed was under attack in so many different ways: nuclear pollution, military appropriation and use of the deep seabed, the quest for the enormous amount of riches in the deep seafloor including mineral extraction, the unclear limit of the continental shelfs as for oil and gas extraction. Two international features made the period even more crucial: first, all the new developing states that from the end of the 1950s had been gaining power in UN; and, second, the contemporary technological optimism. It was an optimism not without reason. The space adventures had its high days, nuclear power promised low cost energy, nuclear bombs and missiles with several war heads, electronic computers, microelectronics, discoveries in biology (DNA) and medicine and so on, all made deep sea floor extraction seem realistic.

A policy to avoid a new run for the rich countries to carve up the seafloor like they did some decades earlier in Africa was welcomed in many quarters. Pardo was very explicit on the point that the developing countries should have a preference when the wealth of the seafloor was to be distributed.

Pardo's speech triggered a process that ended with calling a new conference on the law of the sea, the third in line. It might be debatable if there would

36 Arvid Pardo's speech, UNGA 22nd session, 1 November 1967, Agenda Item 92, full text available at <http://www.un.org/depts/los/convention_agreements/texts/pardo_ga1967.pdf>.

37 Ibid.

FIGURE 3.6 Arvid Pardo monument at the University of Malta
CONTINENTALEUROPE AT ENGLISH WIKIPEDIA, CC BY-SA 3.0, HTTPS://COMMONS.WIKIMEDIA.ORG/W/INDEX.PHP?CURID=47519850

have been a third conference anyway since more mundane problems like the extension of the continental shelfs and the still unsolved question of territorial sea and control over fish resources remained unsolved. However, no one can deny the visionary talk of Pardo as concerns the use of the deep sea floor, even if he was far ahead of his time and all too optimistic both on behalf of the volume of resources and of the appropriate technology.

Pardo's suggestion had a parallel that might have inspired the proposal. It was linked to the rather fresh agreement on de-militarization of outer space. An agreement had already been made and was open for signatures from January 1967. Here, the peaceful use of outer space to the benefit of all mankind was a central value.[38]

In 1973 started the United Nation Law of the Sea Conference number III, the UNCLOS III. It was the fourth attempt to make 'a constitution for the oceans', and this time, after several years of work, they succeeded. Meanwhile, it was not until 1994 that Pardo's international 'heritage of mankind' had been given an administration that would answer to its name. By then, it had shrunk substantially as the coastal states had expanded their legitimate continental shelfs well beyond anything that was thought as realistic in the 1950s. The old maritime powers had lost out on most issues, including on EEZs of 200 miles and territorial waters of 12 miles.

6 Conclusion

The old maritime powers, the old flag states, had the tendency to stretch the idea of control over the resources to control over the waters and the seafloor. Control over resources turns out to be similar to control over territory, as it might be argued it is at the seafloor.

Then, maybe Philip Steinberg is right: today we have very conflicting views of the ocean. As for the surface and for shipping, the ocean is a freeway, a medium owned by nobody, used by all. For the large continental shelves and the water column, the national ownership of the resources seems to drift in the direction of a nationally controlled territory. Lastly, the environmental aspects have become of central importance, with a stronger focus on the consequences of the common heritage regime, the responsibility for the ocean's ecology and on

38 Treaty on Principles Governing the Activities of States in the Exploration and Use of Outer Space, including the Moon and Other Celestial Bodies, opened for signature on 27 January 1967 and entered into force on 10 October 1967. Available at <http://disarmament.un.org/treaties/t/outer_space>.

the stewardship of the environment.[39] The environmental question might now have taken the place that a just distribution of wealth had in the 1970s. It raises the question of how well UNCLOS is designed to handle these kinds of questions or if amendments or new treaty approaches are needed.

39 Steinberg, Philip E., 'Of other seas: metaphors and materialities in maritime regions', *Atlantic Studies* (2013) *10*(2), 156–169.

PART 2

Delimitation of Jurisdiction over the Seabed and Right to Its Resources

CHAPTER 4

Setting Maritime Limits and Boundaries: Experiences from Norway

Harald Brekke

1 Introduction

Part VI of the 1982 UN Convention on the Law of the Sea (UNCLOS, the Convention) deals with the regime of the continental shelf of coastal states. Article 77 provides for the rights of the coastal state over the continental shelf, while Article 76 provides for the limits of that shelf. The coastal state has the obligation to delineate and establish those limits in accordance with Article 76 and Annex II of the Convention. Until now, 72 coastal states have made submissions in this respect. The background, principles and practical work involved in this procedure are described in this chapter.

After a review of the main delineation principles currently applied in order to determine maritime jurisdiction of the coastal state over the seabed, as defined in UNCLOS and interpreted by the courts (Section 2), the chapter continues by reviewing the Norwegian experiences with setting the country's maritime boundaries, as a concrete application of the principles identified. The elements related to the organisation of the work for the purpose of submissions to the Commission on the Limits of the Continental Shelf (CLCS) (Section 3), the mapping the outer limits (Section 4) and to the cooperation with neighbouring states (Section 5) are successively analysed. The chapter ends with an overview of the currently identified resources on the Norwegian continental shelf (Section 6).

2 Delineation Principles of Article 76 UNCLOS

In the 1958 Geneva Convention, the limit of the jurisdiction over the seabed, i.e. the continental shelf, is made dependent on exploitability so that it may extend 'to where the depth of the superjacent water admits of the exploitation

DOI:10.1163/9789004391567_006

of the natural resources.'[1] As technology developed in the 1960s, this criterion turned out not to imply any limit at all. Therefore, states agreed on provisions in the Convention for the precise definition and establishment of continental shelf limits as 'final and binding'.[2]

The agreement, however, is not about the specific limits for each state, but an agreement on the set of rules and provisions on how these limits are to be measured and generated. As regards the Territorial Sea, Contiguous Zone and the Exclusive Economic Zone (EEZ), their limits are generated directly from the coastal baselines of the coastal States.[3] The limits of the continental shelf beyond 200 nautical miles (nm) are different because they are not measured from the baselines.

The extent of the continental shelf of a coastal state is defined in Article 76 of the Convention. The principles of Article 76 are founded on the distinction between the terms *the continental shelf* and *the continental margin*. Both terms are defined in Article 76 for the purpose of the Convention by incorporating both legal and scientific aspects.

In geo-science, the continental margin is generally understood to be the submerged prolongation of the continent and is defined as the area of seafloor between the coast of a continent and the plains of the deep ocean floor at generally 4,000–6,000 meters (m) depth. Most definitions of the continental margin are based on the widely accepted morphological subdivision of its seabed into the continental shelf, the continental slope and the continental rise. In this context, the continental shelf is the relatively shallow seabed area (100–400 m depth) adjacent to the coast and landward of the continental slope.

Paragraph 1 of Article 76, however, defines *the continental shelf* of a coastal State as the seabed areas that extend beyond the territorial sea throughout *the natural prolongation* of the coastal State's land territory. Being *a natural prolongation of the territory* implies that, for the purposes of the Convention, the term *the continental shelf* is a juridical term meaning *the seabed areas over which the coastal State has jurisdiction*. Furthermore, Article 76 provides that this juridical continental shelf may extend up to the outer edge of the continental margin (or to 200 nm, whichever is the farther). That is, *the continental shelf* in its juridical sense may encompass the whole continental margin and not only its inner, shallow parts as in its scientific sense. In this way, the

1 United Nations Treaty Series No. 7302, vol. 499, pp. 312–321; Convention on the Continental Shelf. Done at Geneva, on 29 April 1958, Article 1.

2 UNCLOS, Article 76, paragraph 8.

3 'Norwegian Baselines, Maritime Boundaries and the UN Convention on the Law of the Sea', B.G. Harsson and G. Preiss, Arctic Review on Law and Politics, vol. 3, 1/2012 p. 108–129. ISSN 1891–6252.

Convention is using the concept of the continental margin as a means of measuring and delineating the limits to the prolongation of the continental shelf.

To this end, the Convention has introduced its own conceptual definition of the *continental margin.* Paragraph 3 of Article 76 states that:

> The continental margin comprises the submerged prolongation of the land mass of the coastal State, and consists of the sea-bed and subsoil of the shelf, the slope and the rise. It does not include the deep ocean floor with its oceanic ridges or the subsoil thereof.

In this definition, the starting point of the continental margin is not a continent as such, but the land mass of a coastal State, either it be hosted in a continent or in an oceanic island. This implies that the existence of a State is a prerequisite for the existence of a continental margin for the purposes of the Convention. Thus, paragraph 3 links the term the *continental margin* to the jurisdiction of a State but still incorporates elements that have kept their scientific, geological and morphological meaning, i.e. land mass, shelf, slope and rise.

Being the yardstick for the extent of the continental shelf, the submerged prolongation of the continental margin determines *the natural prolongation* of the land territory of a coastal State as referred to in paragraph 1 of Article 76. Paragraph 4 of Article 76, gives the operational definition for the outer edge of the continental margin for the purposes of the Convention. According to this paragraph, the outer edge of the continental margin is to be established at a certain distance from the foot of the continental slope by either of two methods: the Hedberg rule, or the Gardiner rule (Figure 4.1). This must imply that *the natural prolongation* referred to in paragraph 1 is not a separate property of the continental shelf itself but emanates from the application of paragraph 4.

This understanding is also supported by the International Tribunal of the Law of the Sea (ITLOS, the Tribunal) in the case between Bangladesh and Myanmar.[4] (That case also illustrated that the principles involved in the delineation of the outer limits of the continental shelf may be brought into a dispute on the delimitation between states.[5]) In its decision on the case, the Tribunal

---

4 International Tribunal for the Law of the Sea, 2012; Case No. 16, Dispute Concerning Delimitation of the Maritime Boundary between Bangladesh and Myanmar in the Bay of Bengal, adopted 14 March 2012.

5 C. Schofield, 'One step forwards, two steps back? Progress and challenges in the delimitation of maritime boundaries since the drafting of the United Nations Convention on the Law of the Sea' in G. Xue and A. White (ed), 30 Years of UNCLOS (1982–2012): Progress and Prospects (2013) 217–239.

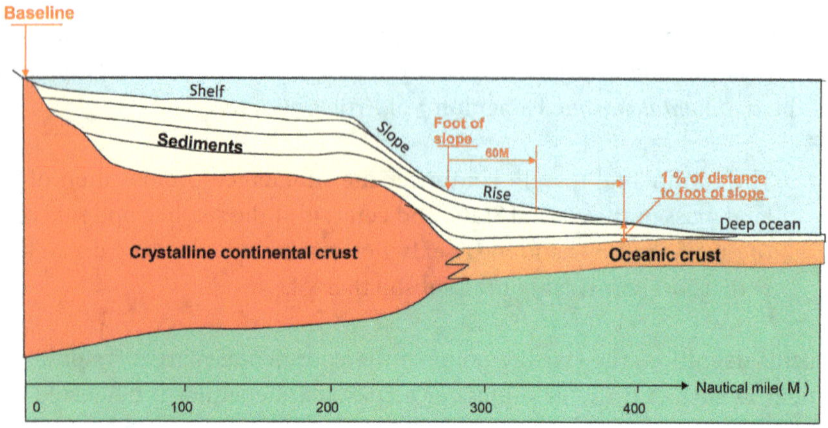

FIGURE 4.1 Schematic profile of the continental margin with geological terms in black and delineation terms of Article 76 in red. According to Article 76, the outer edge of the continental margin is to be set at a distance of either 60 nm from the foot of the slope (the Hedberg Rule) or at a location where the sediment thickness is not less than 1% of the distance to the floor of the slope (the Gardiner Rule)
THE NORWEGIAN PETROLEUM DIRECTORATE

spells out their view on the question of *the natural prolongation* in paragraphs 433 to 437 and concludes as follows:

> 437. For these reasons; the Tribunal is of the view that the reference to natural prolongation in article 76, paragraph 1, of the Convention, should be understood in light of the subsequent provisions of the article defining the continental shelf and the continental margin. Entitlement to a continental shelf beyond 200 nm should thus be determined by reference to the outer edge of the continental margin, to be ascertained in accordance with article 76, paragraph 4. To interpret otherwise is warranted neither by the text of article 76 nor by its object and purpose.

Thus, the application of Article 76, paragraph 4 is the basis for defining the outer edge of the continental margin and thereby identifying the entitlement to a continental shelf beyond 200 nm. Hence, the location of the foot of the continental slope is the critical factor, both for the entitlement to, and extent of, the continental shelf of a State.

Identifying the outer edge of the continental margin by such measurements deviate from the common geoscientific concept of the margin but has the advantage of defining a precise location of its outer edge. A coastal State may apply the two methods to suit its own interest, i.e. by applying only one, or alternating between the two, along its continental margin in order to establish the outer limits of its continental shelf beyond 200 nautical miles. A coastal State that intends to establish such limits, must submit the particulars of these limits to the Commission on the Limits of the Continental Shelf (CLCS, the Commission) along with supporting scientific and technical data.[6]

In order to apply the provisions and methods given in Article 76, the coastal State needs certain geophysical, geological and geodetic[7] data covering the critical parts of its continental margin. Bathymetric data is needed to map the topography of the seafloor and to identify the foot of the continental slope. Seismic data is needed to determine the thickness and, sometimes, the nature of sediments of the sub-surface. In many cases, geological data is required to characterize important seafloor features like submarine ridges and elevations. Geodetic methods and data are required to calculate the appropriate distances and construct the final limits. The acquisition and use of these data requires the expertise of geophysicists, geologists and hydrographers. This is also the reason why the Convention stipulates that the members of the Commission shall be experts in one of these fields.

The Commission gives recommendations to each coastal state separately on the location of outer limits of the continental shelf as it may be established according to Article 76, without regard to the outer limits of the continental shelf of neighbouring states. In several cases, this leads to areas of overlapping continental shelf of states with adjacent or opposite coasts. The settling of bilateral boundaries in such cases is up to the states involved only.[8] The final and binding limits of the continental shelf enters into force when their particulars are deposited with the Secretary-General of the UN and/or the Secretary General of the International Seabed Authority.[9]

3 Norway's Work with the Submissions to the CLCS

3.1 *Norwegian Organization of Work*

Norway signed the Convention on the day it was opened for signature and ratified it on 24 June 1996. It entered into force in Norway on 24 July 1996. The same year, the Norwegian Foreign Ministry initiated the work related to

---

6 UNCLOS, Article 76, paragraph 8, and Annex II, article 4.

7 Geodetic data means the formally correct data and information that are needed to produce a map, including datum, projection, coordinates and distance calculations.

8 UNCLOS, Article 83. See on that point M.D. Evans, 'Maritime Boundary Delimitation' in D. Rothwell, A. Oude Elferink, K. Scott and T. Stephens, 'The Oxford Handbook of The Law of the Sea' (Oxford, 2015), chapter 12, pp. 254–279.

9 UNCLOS, Article 76, paragraph 9, and Article 84.

the establishment the outer limits of the continental shelf of Norway beyond 200 nm in accordance with Article 76 and Annex II of the Convention.

The particulars of these limits along with supporting scientific and technical data were organized in two partial submissions to the Commission. The first submission was made on 27 November 2006 in respect of areas in the Arctic Ocean, the Barents Sea and the Norwegian Sea. The second submission was made on 4 May 2009 in respect of areas in the South Atlantic Ocean adjacent to the Bouvet Island, and the Southern Ocean adjacent to Dronning Maud Land in Antarctica.

Each submissions consisted of three parts as prescribed in the Scientific and Technical Guidelines of the Commission, i.e. an Executive Summary, a Main Body, and a third part with all the supporting scientific and technical data.[10] The Executive Summary was posted on the CLCS website[11] and contained a short description of the different parts of the continental shelf beyond 200 nm, charts and list of coordinates defining their outer limits, references to the provisions of Article 76 invoked, and a status of any delimitations matters with neighbouring states. The Main Body contained a detailed description of the hydrography and geology of the relevant continental margins, and the technical procedures and scientific methodologies applied in the implementation of Article 76 to delineate the continental margin. The results were presented by maps and coordinates and documented by all relevant data. The third part contained a copy of all data and computer-based analyses (e.g. GIS projects) referred to in the main body.

The submissions were prepared under the direction of the Norwegian Ministry of Foreign Affairs. The Norwegian Petroleum Directorate (NPD, the Directorate), an independent agency under the Ministry of Petroleum and Energy (MPE), was the technical coordinator and carried out the technical preparation of the submissions. The Directorate is Norway's expert body for offshore geology and geophysics. For the specific needs of the project, the NPD carried out an extensive marine programme for acquiring geophysical and geological data. These data were in general interpreted and analysed by the NPD. In addition, the data were made available to various other national agencies and institutions, including the Norwegian Mapping Agency, the Universities of Bergen, Oslo and Tromsø, and the Norwegian Polar Institute, which also made scientific and other contributions to the preparations of the submissions.

10 CLCS/11 The Scientific and Technical Guidelines of the Commission on the Limits of the Continental Shelf, paragraphs 9.1.4, 9.1.5 and 9.1.6.

11 Document available at: <http://www.un.org/Depts/los/clcs_new/clcs_home.htm>.

The NPD also established technical cooperation programmes with their peer institutions of the neighbouring coastal states with which Norway would have overlapping continental shelf areas beyond 200 nm, i.e. Russia, Iceland and Denmark with Greenland and the Faroe Islands. The cooperation included data exchange and joint mapping programmes. All of these activities were funded by the Government through the NPD.

3.2 *Commission Procedures and Recommendations regarding the Submission of 27 November 2006*

On 2 April 2007, the Commission established a sub-commission of seven of its members to examine and consider the details of the submission made by Norway. During the next two-year period, Norway provided clarifications and further data and information on request by the Sub-commission, made written responses to its preliminary considerations, and met regularly twice a year with the Sub-commission in New York. The Norwegian delegation was headed by the Ministry of Foreign Affairs with the NPD as technical assistant on a permanent basis. Experts from universities and the National Mapping Agency were included as needed. The Department of Geosciences at the University of Oslo was represented on an almost permanent basis because of their involvement in the previous joint Norwegian-Russian cooperation projects. During the two years of interaction with the Sub-Commission, Norway provided 34 documents with annexes, 25 PowerPoint presentations, and 31 CD/DVDs with supplementary information and data. On 13 March 2009, the sub-commission submitted its recommendations to the plenary of the Commission.

On 27 March 2009, the Commission adopted its recommendations concerning the Norwegian submission. The Commission agreed with the outer limits as submitted by Norway, except for recommending a minor reduction of area in the Norwegian Sea and a slight increase in the Nansen Basin in the Arctic Ocean (see Figure 4.2).

3.3 *Commission Procedures and Recommendations regarding the Submission of 4 May 2009*

In accordance with the request of Norway,[12] with reference to the Antarctic Treaty System as explained below, the Commission decided not to take any

12 Document available at: <https://www.un.org/Depts/los/clcs_new/submissions_files/nor30_09/nor2009_executivesummary.pdf>.

action on the part of the submission that related to the continental shelf appurtenant to Antarctica, i.e. the areas appurtenant to Dronning Maud Land.[13]

For the parts of the submission concerning the continental shelf appurtenant to the Bouvet Island in the South Atlantic Ocean, the Commission established a sub-commission of seven members at its 32nd session in August 2013. During the next three-year period, Norway provided further data and information on request by the Sub-commission, made written responses to its preliminary considerations, and met regularly thrice a year with the Sub-commission in New York. Again, the Norwegian delegation was headed by the Ministry of Foreign Affairs with the NPD as technical assistant on a permanent basis. Experts of the Department of Geosciences at the University of Oslo were represented on an almost permanent basis because of their expertise in geophysical modelling.

Norway was of the opinion that since the time the submission was made in 2009, the practice of the Commission had evolved in a way that made it relevant for Norway to revise the outer limits of the continental shelf in the area appurtenant to the Bouvet Island.

Norway agreed with the Sub-commission that the amendment of the outer limits proposed by Norway required the submission of a revised Executive Summary to be posted on the website of the Commission. Accordingly, Norway submitted a revised Executive Summary on 21 May 2015 that was posted on the website the same day. On the request of the Sub-commission, Norway collected and made available supplementary data to support the revised limits.

Based on all the data thus made available, the Sub-commission agreed with the revised outer limits as submitted by Norway and adopted its recommendations to this effect on 11 August 2016. The Sub-commission subsequently submitted its recommendations for the Commission's 41st plenary session during which Norway delivered its final presentation to the Commission on 16 August 2016. However, the term of the Commission members ended in June 2017 and the Meeting of States Parties elected members for the next five-year term on 14 June 2017. The consideration of the recommendations of the Sub-commission was then passed on to this new set of members. On request, Norway was given the opportunity to repeat its final presentation to the new Commission members on 7 August 2018.

On 8 February 2019, the Commission adopted its recommendations to Norway concerning the outer limits of the continental shelf in the area

---

13 Brekke, H. 2014. *Defining and recognizing the outer limits of the continental shelf in the polar regions*. In: R. Powell and K. Dodd (Editors), Polar Geopolitics? Knowledges, Resources and Legal Regimes. Edward Elgar Publishing Ltd., Celtenham, UK; Northampton, MA, USA.: 38–54.

appurtenant to the Bouvet Island. The Commission agreed with the outer limits as submitted by Norway, with minor changes.

The sub-commissions for the two submissions carried out very thorough examinations of the substantial amount of data and information submitted. The role of the Norwegian delegation was essentially to give assistance to the sub-commissions in their examinations by providing clarifications and supporting data. The sub-commissions put much of their time and work into satisfying themselves that the foot of the slope points were correctly defined and adequately documented. In this respect, the recommendations issued for the areas of the Arctic Ocean, the Barents Sea and the Norwegian Sea established a precedence for how to define the location of the foot of very low gradient continental slopes of sedimentary origin. The Sub-commission for the areas of the South Atlantic Ocean, put much emphasis also on satisfying itself that it could support the principle applied for the maximum constraint of the continental self. The recommendation on the area in the South Atlantic Ocean may become an important reference regarding the submerged prolongation of the landmass of oceanic islands and their maximum outer limits. The Norwegian delegation was dedicated to meeting the concerns of the sub-commissions with responses that would instil confidence in the basis for the recommendations on the final outer limits.

4 Mapping of the Outer Limits and Cooperation with Neighbour States

The technical work of the NPD consisted of four parts:
1. NPDs data acquisition program;
2. Data acquisition in cooperation with academic and peer government institutions;
3. NPD scientific cooperation with academic institutions and neighbour state's peer government institutions;
4. NPD analyses, compilation of all data and information, and preparation of final submission documents and supplementary material.

The technical work covered areas of continental shelf beyond 200 nm in five different oceans and seas (See Figure 4.2):
- The Barents Sea, in the area enclosed by the 200 nautical mile limits of Norway and Russia, also known as the 'Loop Hole';
- The Arctic Ocean, in an area north of Svalbard known as the 'Nansen Basin';
- The Norwegian Sea, in the area enclosed by the 200 nm limits of Norway, Iceland and Denmark with the Faroe Islands and Greenland, also referred to as the 'Banana Hole';

- The South Atlantic Ocean, in the area around the Bouvet Island in the South Atlantic;
- The Southern Ocean, in the area north of Dronning Maud Land in Antarctica.

The work progress and the conclusions reached for each for those five areas are reviewed successively in the sections below.

4.1 *The Barents Sea*

The Barents Sea is a vast shallow water shelf area situated north of mainland Norway and Russia, bounded in the east by Novaya Zemlya and the Kara Sea, and in the north and west by the archipelagos of Franz Josef Land and Svalbard and the deep waters of the Norwegian and Greenland Seas. The 'Loop Hole' is an area beyond, and entirely enclosed by, the 200 nm limits of Norway and Russia. This area forms part of the shallow waters of the Barents Sea, and all of it is situated landward of the foot of the continental slope and the 2,500 m isobath. With reference to the applicable rules in Article 76, paragraphs 3–5, of the Convention, Norway submitted that the area may be considered to be part of the continental shelf and that there was no need for further scientific or technical documentation.

4.2 *The Arctic Ocean*

The Arctic Ocean is enclosed by the shallow shelf areas of the five Arctic states: Canada, United States of America (USA), Russia, Norway, and Denmark with Greenland. A central topographic element of the Arctic Ocean is the submerged Lomonosov Ridge that extends right across the ocean between the Greenland and Siberian shelves. This ridge divides the ocean into two parts, the complex Amerasian Basin on the American side and the oceanic Eurasian Basin on the other. The Eurasian Basin in turn, is subdivided into the Amundsen Basin to the north and the Nansen Basin to the south.

The northward extension of the shallow shelf areas of the Barents and Kara Seas ends at the continental slope along the Nansen Basin. In the continental margin in the western part of this basin, north of the archipelagos of Svalbard and Franz Josef Land, it was clear that Norway and Russia would have overlapping areas of continental shelf. Thus, the two states agreed that it was in the interest of both to cooperate in the mapping and delineation of the outer limit of the continental shelf in this region. Accordingly, the NPD and its peer Russian institute, the All-Russia Research Institute of Geology and Mineral Resources of the World Ocean (VNIIOkeangeologia) in St. Petersburg entered into a joint research project in this regard. Besides the scientific results, the main purpose of the project was to establish a common view of the location of the outer limits of the continental shelf in the region and to provide the required supporting data and information needed by the two states to prepare their submissions to the Commission. For the purposes of the submissions, the project started in 1996 and ended in 2006. For scientific purposes, however, this cooperation between Norway and Russia has continued since.

The project included data exchange and the establishment of a common database of existing data. To cover gaps in the database, Norway in 2001 and 2005 acquired an additional 1,000 km of seismic and 2500 km of bathymetric data with the Swedish icebreaker 'Oden' and with the assistance of experts from the University of Bergen. All the data finally available, the analyses thereof, and results of the joint Norwegian/Russian research, formed the basis for the Norwegian submission in this region.

4.3 *The Norwegian Sea*

The Norwegian Sea is the part of the North Atlantic Ocean west of the coast of Norway between Stad (62° N) and Senja (70° N). To the west, it is bordered by the Greenland Sea along a submarine ridge system running from Iceland, via Jan Mayen to Bear Island. To the south, it is bordered by the North Sea and a submarine ridge connecting the Faroe Islands and Iceland. To the north, it is bordered by the shallow shelf areas of the Barents Sea.

Norway's continental margins in this region comprise two parts. One part is constituted by the continental margin of mainland Norway and the Svalbard archipelago. It extends from the North Sea in the south, through the Norwegian and Greenland Seas, and into the Arctic Ocean. The other part is the continental margin surrounding the island of Jan Mayen. The two parts face each other with the area beyond 200 nm, the 'Banana Hole', located in the middle.

This area comprised the potentially largest area of continental shelf beyond 200 nm of Norway in the northern hemisphere. The shallow shelf areas within 200 nm in this region was geologically well known because it had been subject to extensive exploration for, and exploitation of, oil and gas since the late 1970s. The deep ocean within the 'Banana Hole', however, needed a closer study in order to establish the outer limits of the continental shelf. Therefore, the NPD carried out three major marine campaigns on tender to acquire sufficient data for the purpose of the submission. This amounted to 10,500 km of seismic data and 270,000 km^2 of multi-beam bathymetric data. It was clear that in the southern part of the Banana Hole there would be a potential overlap with the continental shelf of Iceland and Denmark with the Faroe Islands. From the start, there was communication between the three states on both diplomatic and technical level. At the technical level, the NPD engaged in a data exchange programme and professional discussions with its peer institutions, the Faroese Earth and Energy Directorate (Jarðfeingi) of Denmark, and Iceland GeoSurvey (ISOR) of Iceland.

4.4 *The South Atlantic Ocean*

The Bouvet Island is located around 54° S, 4° E. The submerged prolongation of the landmass of the island consists of a wide submarine elevation that narrows north-eastwards into a long spur. The outer limits of the continental shelf beyond 200 nautical miles from the baseline, which are based on this submarine elevation extends north-eastwards to about 50° N, 15° E. The baseline of the Bouvet Island was established by Royal Decree of 25 February 2005, and deposited with the Secretary-General on 31 March the same year. The original submission in respect of the Bouvet Island was based mainly on publicly available data derived from previous Norwegian and international scientific expeditions.

However, in regard of the revision of the submission described in section 3.3, Norway carried out a new marine acquisition survey to provide further bathymetric data in critical parts of the area. This survey was carried out by the Norwegian Polar Institute in January 2014. In order to acquire further supplementary data as requested by the Sub-commission, Norway engaged the Russian State Research Navigation-Hydrographic Institute (GNINGI) to carry out an expedition to gather multi-beam bathymetric data and seabed rock samples for geochemical studies. For this purpose, GNINGI made available the research vessel R/V Akademik Fedorov during its annual visit to Antarctic waters in March 2016, and with VNIIOkeangeologia as operator for the rock sampling. The expedition resulted in the acquisition of 3,200 line km of MBES data, including an area of 8350 km^2 full coverage, and 88 rock samples by dredging.

4.5 *The Southern Ocean*

The Southern Ocean is the area of seas south of 60° S surrounding the Antarctic Continent. Within this ocean, the continental margin appurtenant to Dronning Maud Land extends beyond 200 nautical miles as measured from relevant basepoints established by Norway along the coastline of the land area. The seafloor of the continental margin in this area includes several large elevations and ridges. In order to establish a reliable outline of the outer edge of this continental margin in accordance with the provisions of article 76, Norway had to acquire more data. The whole area of the Southern Ocean is subject to the Antarctic Treaty System, and Norway recognized that any data acquisition should serve scientific purposes. In order to achieve this, Norway financed a three-year joint scientific program between the University of Bergen and the two research institutes VNIIOkeangeologia and PMGRE in St. Petersburg. The latter made available the research vessel R/V Akademik Karpinsky, which on an annual basis visit Antarctic waters, for acquiring bathymetric and seismic data. In addition, Norway also funded a data acquisition cruise with the Norwegian research vessel R/V G.O. Sars in the area. It all added up to 8000 km of seismic data and 9000 line km of multi-beam bathymetry data. The data acquired became the basis for several Master and PhD degrees and scientific publications. All data from the Southern Ocean have been submitted to the database of the Scientific Committee on Antarctic Research (SCAR).

5 Relationship with Neighbouring States

Upon receiving the submission from a coastal state, the Commission publishes the executive summary of the submission on its website.[14] A period of at least three months is then allowed for comments by other states before the Commission puts that submission on its agenda. This website has become a forum in which states make known their statements and views on the submissions of other states and any other related issues, in the form of *notes verbale*.

The submission that Norway made in respect of the Arctic Ocean, the Barents Sea and the Norwegian Seas implied actual and potential overlaps with the continental shelf limits of the neighbour states, Russia, Iceland and Denmark with Greenland and the Faroe Islands.

5.1 *Relationship with Russia*

Russia had made its first submission in 2001, in which it submitted a sector line as the western limit of the continental shelf in the Barents Sea and western Nansen Basin. This included almost the entire Loop Hole. Norway submitted the median line with Russia as its eastern limit of the continental shelf. Thus, there was an area of overlapping continental shelf area throughout the Barents Sea and into the Nansen Basin in the north. This overlapping area was not new to the two states; Norway and Russia had been negotiating their bilateral, marine boundary in this area since 1971. By mutual notes verbale, the two states made it clear that their continental shelf submissions did not prejudice anything in these negotiations.

14 Website of the CLCS: <www.un.org/Depts/los/clcs_new/clcs_home.htm>.

Following the submission by Russia in 2001, Norway submitted a note verbale to be posted on the website of the Commission. By this, Norway referred to the unresolved delimitation issue with Russia and specified the area under negotiations by giving the coordinates of each of the two state's positions. Norway, with reference to paragraph 5 of Annex I of the Rules of the Procedure of the Commission, declared that it had no objections to the Commission considering and making recommendations with regard to the Russian submission as these recommendations would not prejudice the future delimitation between the two states (now customary known as a note verbale of non-objection).[15]

Following the submission by Norway in 2006, Russia submitted a similar note verbale of non-objection with regard to Norway. The wording of the two notes verbale are identical, except for comments on details regarding coordinates, confirming prior, close consultations regarding these notes. In its note, Russia also made the reservations that the note itself should not prejudice the position of Russia 'towards the Spitsbergen archipelago and its continental shelf', and that the recommendations of the Commission should be without prejudice to 'the provisions of the Treaty concerning Spitsbergen of 1920 and, accordingly, to the regime of maritime areas adjacent to Spitsbergen'.[16]

On 27 June 2002, the Commission adopted its recommendations concerning the submission made by Russia in 2001. The Commission agreed to and recommended the outer limits in the Barents Seas as submitted by Russia. As regards the Central Arctic Ocean, the Commission recommended that Russia make a revised submission.[17] This implied that at the time Norway made its submission in 2006, the Loop Hole in the Barents Sea was already recommended as continental shelf area, while the area in the Nansen Basin was pending the outcome of the submission by Norway and the revised submission of Russia (See Figure 4.2).

After the Commission adopted its recommendations concerning the Norwegian submission in 2009, Norway and Russia also moved towards a final agreement in their delimitation negotiations. After 40 years of negotiations, the two states reached a maritime delimitation agreement that was signed in Murmansk on 15 September 2010 and entered into force on 7 July 2011. The agreed delimitation line extends to a point north of the outer limit as recommended by the Commission in order to accommodate a possible intersection with the future outer limit of the continental shelf of Russia.

15 Document available at: <http://www.un.org/Depts/los/clcs_new/submissions_files/rus01/CLCS_01_2001_LOS__NORtext.pdf>.

16 Document available at: http://www.un.org/Depts/los/clcs_new/submissions_files/nor06/rus_07_00325.pdf>.

17 Reports of the Secretary-General: A/57/57/Add. 1, paras 38–41, https://documents-dds-ny.un.org/doc/UNDOC/GEN/N02/629/28/PDF/N0262928.pdf?OpenElement.

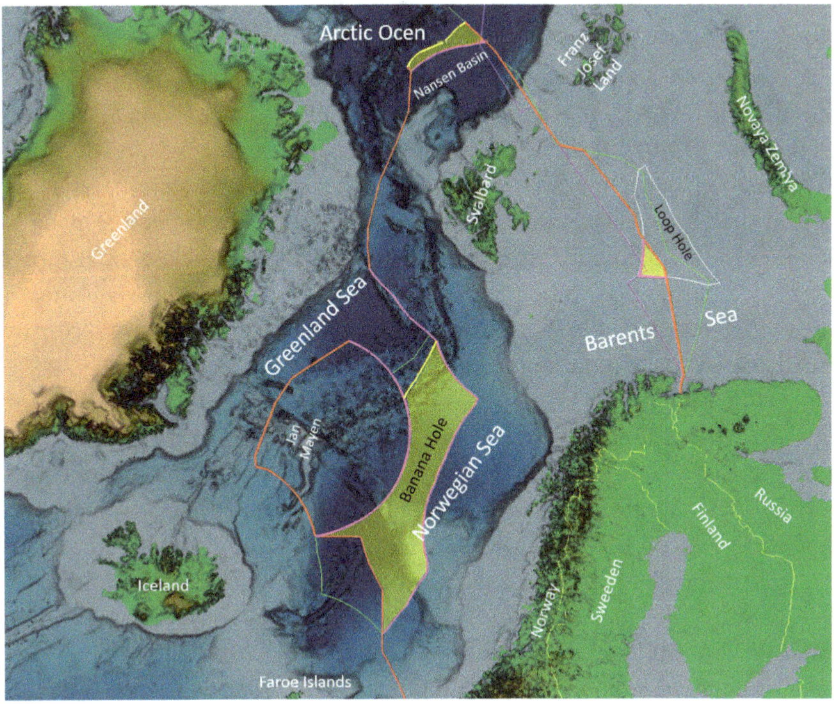

FIGURE 4.2 The continental shelf of Norway in the Arctic Ocean, Barents Sea and the Norwegian Sea. Red lines – agreed maritime delimitation with neighbour states, including the Agreed Minutes with Iceland and Denmark with Faroe Islands. Pink lines – 200 nautical mile lines of Norway. Yellow lines – outer limits of the continental shelf beyond 200 nautical miles as recommended for Norway by the Commission, but yet to be concluded by bilateral delimitation pending the future recommendations for Russia in the Arctic Ocean and Denmark with Greenland in the Norwegian Sea and the Arctic Ocean. The areas of continental shelf beyond 200 nautical miles are highlighted as yellow polygons. Thin green lines – outer limits of the continental shelf of Norway as originally submitted to the Commission by Norway in 2006. Thin purple line – western limit of the continental shelf of Russia in the Barents as submitted by Russia in 2001. Thin white line – delineation of the Russian part of the enclave of area beyond 200 nautical miles in the Barents Sea (the 'Loop Hole').
THE NORWEGIAN PETROLEUM DIRECTORATE

5.2 *Relationship with Denmark with Greenland*

On 20 February 2006, Norway and Denmark with Greenland signed an agreement concerning the delimitation of the continental shelf in the area between Greenland and Svalbard.[18] Upon the submission made by Norway to the

18 Document available at: <https://treaties.un.org/Pages/showDetails.aspx?objid=0800000280064a71>.

Commission later the same year, Denmark with Greenland submitted a note verbale of non-objection.[19]

At that time, the states anticipated a slight potential overlap of continental shelf areas in the northern border zone of the Banana Hole. Such overlap was confirmed in the submission made by Denmark with Greenland 26 November 2013. In 2006, no overlap of the continental shelf areas was foreseen in the Nansen Basin. However, at the time of the submission made by Denmark with Greenland in the Arctic Ocean on 15 December 2014, it turned out to be a substantial potential overlap with the continental shelf area of Norway already recommended by the Commission in this region. Norway issued notes verbale of non-objection on both of these occasions.[20]

5.3 *Relationship with Iceland and Denmark with the Faroe Islands*

During the preparation of their submissions in respect of the Banana Hole, Norway, Iceland and Denmark with the Faroe Islands developed the shared view that the whole area concerned consists of continuous continental shelf. It also became clear that there would be potential overlaps of continental shelf areas of the three states in this region, and that Norway would be the first state in position to make its submission to the Commission, while the two others would submit some years later. On this basis, Iceland and Denmark with the Faroe Islands approached Norway and proposed to discuss a possible future delimitation in the area before Norway lodged its submission. The Agreed Minutes that came out of these discussions were published on 21 September 2006.[21] These minutes include a Model Agreement and the coordinates for the points on which the future delimitations would be based. The final delimitation would be concluded at the time when all three states had received their recommendations from the Commission.

Iceland and Denmark with the Faroe Islands both made their submissions with respect to this area on 29 April 2009. In accordance with the Agreed Minutes all three states issued notes of non-objection with regard to

19 Document available at: <http://www.un.org/Depts/los/clcs_new/submissions_files/noro6/dnko7_00218.pdf>.

20 Documents available at: <http://www.un.org/Depts/los/clcs_new/submissions_files/dnk68_13/2014_01_21_NOR_NV_UN_001_14-00060.pdf>, and <http://www.un.org/Depts/los/clcs_new/submissions_files/dnk76_14/2014_12_17_nor_nv_dnk4_001.pdf>.

21 Document available at: <https://www.regjeringen.no/en/dokumenter/Agreed-Minutes/id446839/?q=Agreed minutes>.

each other's submissions.[22] The Commission adopted its recommendations for Denmark with the Faroe Islands, and Iceland in respect of this area on 11 March 2014 and 10 March 2016, respectively. Accordingly, the three states have now moved on to conclude their bilateral delimitation as agreed.

5.4 *The South Atlantic Ocean*

The outer limits of the continental shelf appurtenant to the Bouvet Island borders in its entirety with the international seabed, i.e. the Area. The submission made by Norway to the Commission in respect of the Bouvet Island of 4 May 2009, including its revision of 21 May 2015, did not attract any notes verbale from other states.

5.5 *The Southern Ocean*

In its submission of 4 May 2009, Norway noted that appurtenant to Antarctica there exist areas of continental shelf that has yet to be defined. In this context, Norway referred to two ways in which a submission in respect of such areas of continental shelf may be made:

- States concerned may submit information to the Commission, which would not be examined for the time being, or
- make a partial submission not including such areas of continental shelf, for which a submission may be made later, notwithstanding the provisions regarding the ten-year period established by article 4 of Annex II to the Convention.[23]

The outer limits of the continental shelf appurtenant to Dronning Maud Land on the Antarctic Continent as submitted by Norway are located entirely within the area of the Antarctic Treaty System. Consistent with the first option, Norway requested the Commission not to take any action for the time being with regard to this area.[24] Among the other claimant states in Antarctica, Australia and Argentina also referred to this option,[25] while United Kingdom,

22 Document available at: <http://www.un.org/Depts/los/clcs_new/commission_submissions.htm>.

23 http://www.un.org/Depts/los/clcs_new/submissions_files/nor30_09/nor2009_executivesummary.pdf.

24 http://www.un.org/Depts/los/clcs_new/submissions_files/nor30_09/nor2009_executivesummary.pdf.

25 http://www.un.org/Depts/los/clcs_new/submissions_files/aus04/Documents/aus_doc_es_attachment.pdf, http://www.un.org/Depts/los/clcs_new/submissions_files/arg25_09/arg_note_2009eng.pdf.

New Zealand and France referred to the second option.[26] Chile has announced that it will inform the Commission on this matter in due course.[27][28]

The submission by Norway attracted notes verbale from five states: USA, Russia, India, Netherlands and Japan, all of which confirmed that they do not recognize any State's claim to territory in Antarctica and that they acknowledged Norway's request to the Commission not to take any action on its submission concerning this area.[29]

6 Resources from the Seabed of the Continental Shelf

According to Article 77 of the Convention, the coastal state has the right to exploit the resources of the continental shelf, specified as the resources of the seabed and the subsoil thereof. These include the conventional geological resources as oil, gas and metallic seabed minerals, as well as biological resources in the form of sedentary species living on or within the seabed. As regards the Norwegian continental shelf, its oil and gas resources are located almost entirely within the 200 nautical miles limits. Establishing the outer limits in accordance the recommendations of the Commission, implies the inclusion of possible oil and gas resources only in the area of the Loop Hole in the Barents Sea and in some minor areas along the 200 nautical miles limits of the Banana Hole in the Norwegian Sea (Figure 4.3). The latter areas are in relatively deep waters and in part strongly invaded by subsurface magmatic rocks of low resource potential.

The areas of continental shelf beyond 200 nm in the Nansen Basin and in the Banana Hole consist mainly of deep ocean floor with no potential for oil and gas. These areas, however, have a good potential for metallic seabed minerals, namely polymetallic sulphides and polymetallic manganese crusts. Such

26 http://www.un.org/Depts/los/clcs_new/submissions_files/gbr08/gbr_nv_9may2008.pdf, http://www.un.org/Depts/los/clcs_new/submissions_files/nzl06/nzl_doc_es_attachment.pdf, http://www.un.org/Depts/los/clcs_new/submissions_files/fra09/fra_note_feb2009e.pdf.

27 http://www.un.org/Depts/los/clcs_new/submissions_files/preliminary/chl2009note_e.pdf.

28 H. Brekke, 'Defining and recognizing the outer limits of the continental shelf in the polar regions', in R.C. Powell and K. Dodds (eds) *Polar Geopolitics? Knowledges, Resources and Legal Regimes* (Edward Elgar Publishing Ltd., Cheltenham, UK, 2014), pp. 38–54.

29 http://www.un.org/Depts/los/clcs_new/submissions_files/submission_nor_30_2009.htm, https://documents-dds-ny.un.org/doc/UNDOC/GEN/N10/337/97/PDF/N1033797.pdf?OpenElement.

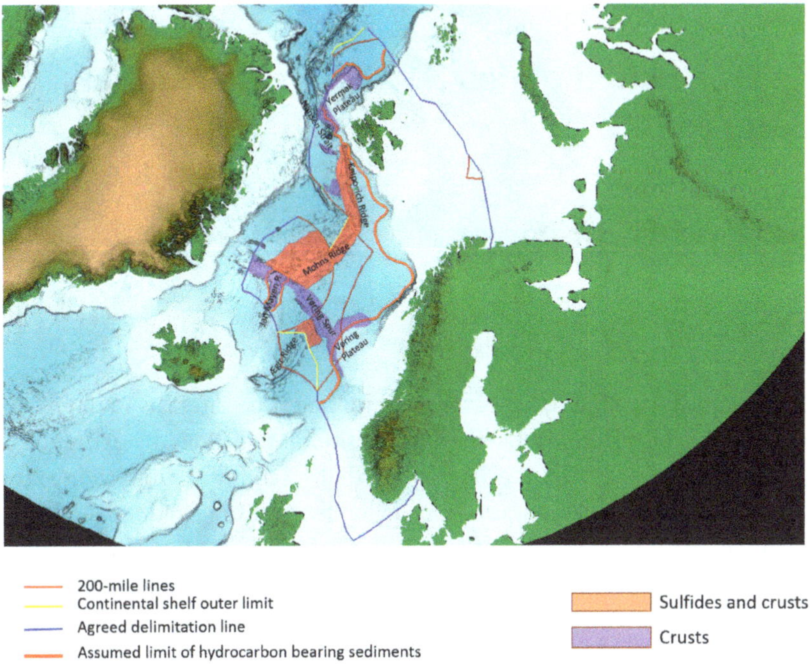

FIGURE 4.3 Map showing the possible distribution of petroleum and mineral resources of the Norwegian continental shelf
THE NORWEGIAN PETROLEUM DIRECTORATE

minerals are already discovered in the deep ocean areas of the Norwegian and Greenland Seas, both within and beyond 200 nautical miles (Figure 4.3). More knowledge about their distribution, volume or exploitability is needed for management purposes and estimation of their economic value.

The known biological resources of the continental shelf, like crustaceans and shellfish, are all associated with the shallow shelf areas, at depths down to 100–400 meters. All of these areas, except for the Loop Hole, are within the 200 nautical mile limits. The biology of the seabed of the deep ocean areas of the continental shelf is much less known. Current scientific research on the deep ocean seabed in the Norwegian Sea, however, has revealed the existence of a benthic fauna that may become an economic genetic resource in the future.

The seabed of the continental shelf around the Bouvet Island consists of magmatic rocks and thin sedimentary cover that cannot host oil or gas. There is a potential for mineral resources in this area, while the potential for biological resources is unknown.

CHAPTER 5

The Seabed in the High North – How to Address Conflicts?

Alexander S. Skaridov

1 Introduction

A general question underlying current competing state interests in the Arctic is whether claims to the Arctic continental shelf are justified by existence of potential resources. Indeed, the assumption that a significant proportion of the world's undiscovered oil and gas deposits lie beneath the Arctic seabed have turned the Arctic into a region of considerable geopolitical interest. However, geological investigations cover just small Arctic seabed areas; all other estimations are based on mathematic and methods of probabilistic modelling.

For the purpose of this chapter, we present estimations made by U.S. Geological Survey (USGS) which were based on a geological probabilistic methodology. The USGS estimated the deposits of undiscovered oil and gas in 33 geologic provinces that thought to be prospective for petroleum. The sum of the estimates for each province indicates that 90 billion barrels of oil, 1,669 trillion cubic feet of natural gas, and 44 billion barrels of natural gas liquids may remain to be found in the Arctic, of which approximately 84 % is expected to be found in offshore areas.[1] Another popular speculation regarding the Arctic is that, because of the changing climate of the Arctic, ice cover will reduce in the near future and most spaces of the Higher North will be available for normal navigation, which will dramatically influence the global shipping.[2]

Meanwhile, the USGS study, like other ones, is limited in its knowledge basis and method, also because of the geographical limitation of the seabed areas which were studied and their depth.[3] That is why, in our view, present

1 U.S. Department of the Interior, U.S. Geological Survey, 'Circum-Arctic Resource Appraisal: Estimates of Undiscovered Oil and Gas North of the Arctic Circle,' USGS Fact Sheet 2008–3049. Available at: <http://pubs.usgs.gov/fs/2008/3049/fs2008-3049.pdf>.

2 The third of the Arctic Ocean is covered with old ice, the surface of newly-formed ice is changing from year to year, and from measurement season to season; furthermore, there is a lot of data that the surface of newly-formed ice can enlarge.

3 In the USGS study, were included only those stocks that are considered to be recoverable using existing technology, even in conditions of permanent ice cover and depth of about 500 m.

aspirations of coastal countries to divide the Arctic spaces are not very much related to current scientific knowledge about seabed resources. All our further conclusions in this chapter will be made on the grounds that the above-mentioned statements do not have sufficient scientific proof and cannot be taken as a major challenge to the issue of the delimitation of the Arctic seabed. What is more important are the facts that only few nations[4] own the Arctic coast, and, because of this, those states have a much higher responsibility in the protection of Polar spaces than other states.

Another obvious consideration is that coastal economy attracts more and more resources, and its value is significantly increasing, which at the national level requires different conditions for investment and due consideration to the people living in the Arctic regions and who consider themselves indigenous.[5]

Certainly, the assumption that Arctic resources potential is exaggerated is not a strong enough argument to stop the process of establishing the boundaries and delimitation of the seabed. However, the desire to establish the boundaries far beyond the real possibilities of operation is in stark contrast with the simplicity and clarity of existing international legal norms regulating the delimitation and the legal regime of Arctic spaces.

This chapter explores the question of delimitation of jurisdiction in the Arctic as a fundamental starting point for discussing the regime for exploration for and exploitation of seabed resources in an area of the world which is subject to competing state and commercial interests, and so to potential conflicts. It starts by a brief historical sketch of countries claims to the Arctic seabed (Section 2), before considering the respective ambitions and so conflicting interests of the Arctic states (Section 3). The chapter ends with a reflection on possible ways of resolving disputes (Section 4).

2 Delimitation and Competing Claims over the Arctic Seabed

2.1 *Application of UNCLOS Provisions*

At least two key issues should be raised here: the applicability of the provisions of the 1982 UN Law of the Sea Convention (UNCLOS), and the legitimate access to Arctic resources by non-Arctic states.[6]

4 Russian Federation and Canada own 80% of the Arctic coastal area, Norway and Denmark about 16%, the USA 4%.

5 For example, for Russia comprises nearly 20,000 km of state border, about 11% of national income. Considerable amount of hydrocarbon and other minerals are already extracted there. It is the center of shipbuilding with ports and port stations.

6 In this chapter, we are not going to cover the last point since it needs a separate study.

From our point of view, the attempts of many scholars and official commentators to build up various concepts of extension of special rights (within national jurisdiction) on regulation the exploitation of the resources of the Arctic shelf and transport of goods based on the provisions of Article 234 UNCLOS (Ice covered areas) cannot be considered reasonable. This Article in UNCLOS does not deal with exclusive 'Arctic reference' (we can find 'ice covered areas' in other parts of the World ocean), and contains the terms which are not determined by the Convention, but which are key for the application of the UNCLOS. For instance, it is not obvious what to perceive by 'severe climate conditions', 'serious damage to ecological balance'.

It is the view of the author that the application of the UNCLOS provisions to the Polar regions without taking into consideration their peculiarities, will contradict the natural conditions of the environment. As a matter of fact, we cannot consider the icefield as an area of normal navigation, as well as define baselines to determine the outer limits of the territorial sea or other spaces if an island or a part of land which are surrounded by permanently frozen areas, such as icefront, grounded ice or grounded hummock. In other words, equating the Arctic Ocean without regard to the physical condition of the environment to any other area of the ocean does not add the natural grounds for the resolution of present and future delimitation. The preparatory documents analysis of UNCLOS gives evidence that the negotiators did not want to make all process of negotiations still more complicated and discuss such a difficult matter from the point of view of law.

2.2 *Position of the Five Arctic Coastal States (Ilulissat Declaration) and Continental Shelf Delineation*

On May 28, 2008 in Ilulissat in Greenland, the five Arctic coastal states (A-5)[7] adopted a declaration where they recall that an extensive international legal framework applies the law of the sea to the Arctic ocean which provides for important rights and obligations concerning the delineation of the outer limits of the continental shelf, the protection of the marine environment, including ice-covered areas, freedom of navigation, marine scientific research, and other uses of the sea. The countries stated that they remained committed to this legal framework and to the orderly settlement of any possible overlapping claims. The A-5 see no need to develop a new comprehensive international legal regime to govern the Arctic Ocean.[8]

7 A-5 is the short name for the five Arctic coastal states: Denmark, Canada, Norway, Russia and USA.

8 The Ilulissat Declaration, Arctic Ocean Conference, Ilulissat, Greenland, 27–29 May 2008. Available at: <http://www.arcticgovernance.org/the-ilulissat-declaration.4872424.html>.

Canada, Denmark,[9] Norway and Russia ratified the UNCLOS,[10] while the United States of America is still working on it. Each A-5 country claimed exclusive economic zones (EEZ) of 200 nautical miles (nm), and continental shelf areas adjacent to their coasts.

The definition of the continental shelf and the criteria by which a coastal State may establish the outer limits of its continental shelf are set out in UNCLOS, Article 76. In addition, the Third United Nations Conference on the Law of the Sea adopted on 29 August 1980 a Statement of Understanding which is contained in Annex II to the Final Act of the Conference.[11]

Pursuant to Article 76(7) UNCLOS, information on the limits of the continental shelf (CS) beyond 200 nm from the baseline from which the breadth of the territorial sea is measured shall be submitted by the coastal State to the Commission on the Limits of the Continental Shelf (CLCS) set up on the basis of equitable geographical representation. The Commission shall make recommendations to coastal States on matters related to the establishment of the outer limits of their continental shelf. It is essential to stress that the limits of the shelf established by a coastal State on the basis of CLCS recommendation will be final and binding only after adoption by the coastal states.

Article 76 uses two different terms in the meaning of the description of the boundaries: 'delineation' and 'delimitation' The first one refers to the line which should be determined by the coastal state as the outermost points of the continental margin to a distance drawn according to UNCLOS.[12]

The UNCLOS requirements concerning deposit of, and publication of the charts or lists of geographical coordinates showing the outer limits of the CS practically remained unfulfilled and experts, for visualization of the Arctic

9 Via Greenland. Denmark's sovereignty over all of Greenland was recognized by the United States in 1916 and by an international court in 1933. Denmark could also conceivably claim an Arctic sector (60°W to 10°W).

10 Norway ratifies the UNCLOS on June 24, 1996; Russian Federation ratifies the UNCLOS on March 12, 1997; Canada ratifies the UNCLOS on November 7, 2003 ; Denmark ratifies the UNCLOS November 16, 2004.

11 Where a State establishes the outer edge of its continental margin by applying the method set forth in the preceding paragraph of this statement, this method may also be utilized by a neighboring State for delineating the outer edge of its continental margin on a common geological feature, where its outer edge would lie on such feature on a line established at the maximum distance permissible in accordance with article 76, paragraph 4(a)(i) and (ii), along which the mathematical average of the thickness of sedimentary rock is not less than 3.5 kilometers.

12 In accordance article 76 incl. paragraph 4 (a)(i) and (ii).... either shall not exceed 200 nautical miles where the outer edge of the continental margin does not extend up to that distance ... or 350 nautical miles from the baselines from which the breadth of the territorial sea is measured or shall not exceed 100 nautical miles from the 2,500 meter isobaths, which is a line connecting the depth of 2,500 meters.

boundaries used drawn hypothetical boundaries which could be found in different publications and on the web. We also suggest an approximate perception of the Arctic borders position based on the maximum furthest outer boundaries from baselines, current agreements and claims, and on the principle of equidistance, accuracy of which are defined by the map scale. It is obvious that if we used maximum criteria of the width of CS delineation, it would leave the distinct polar enclave drawing on the map. However, coastal countries did not begin to define coordinates of the shelf outer boundary within 4–5 articles 76, but claimed submarine ridges that by UNCLOS are 'natural components of the continental margin' in the spaces of the seabed where they meet the criteria applied to submarine elevations through the second sentence of Article 76(6), in as much as the morphological ridgelike features are included in the definition of submarine elevations according to common and accepted formal definitions of submarine seafloor.

2.3 *Historic Claims over the Arctic and State Practice on Delimitation of the Continental Shelf*

2.3.1 Review of Historic Claims over the Arctic

On December 20, 2001 Russia delivered submission to the CLCS claiming that the Lomonosov Ridge was an extension of its continental shelf. Russia stood to potential acquisition of nearly one-half of the Arctic Ocean, including the North Pole. In 2002, the UN Commission neither rejected nor accepted the proposal, recommending an additional research was necessary. On August 4, 2015, Russia resubmitted its bid, containing new arguments based on the new scientific data collected in years of Arctic research.[13]

On November 27, 2006, Norway made an official submission into the CLCS and provided arguments to extend the Norwegian seabed claim beyond the 200 nm in three areas of the northeastern Atlantic and the Arctic: the 'Loop Hole' in the Barents Sea,[14] the Western Nansen Basin in the Arctic Ocean, and the 'Banana Hole' in the Norwegian Sea.[15] On March 27, 2009, Norway received recommendations from the CLCS with the advice to clarify territorial disputes in the Barents Sea, Arctic Ocean, Norwegian and Greenland Seas. The CLCS

13 Through this bid, Russia is claiming 1.2 million square kilometers (over 463,000 square miles) of Arctic sea shelf extending more than 350 nautical miles (about 650 kilometers) from the shore.

14 The delimitation of the continental shelf and the 200-mile zones in the Barents Sea is an essential basis for the exploration and exploitation of petroleum deposits in the area of overlapping claims, which covers an area of 175 000 square kilometers.

15 The submission also states that an additional submission for continental shelf limits in other areas may be posted later.

also acknowledged that the information for the Loop Hole contained in the Submission of Norway of 27 November 2006 fully satisfies the requirements of a submission for continental shelf beyond 200 M from the territorial sea baselines of Norway in accordance with article 76, paragraph 8, and article 4 of Annex II to the Convention. Only a bilateral delimitation between Norway and the Russian Federation remained to be carried out to delineate the extent of each coastal State's continental shelf in the Loop Hole.[16] The Commission recommended that Norway proceed with the delimitation of the continental shelf beyond 200 M in the Loop Hole by agreement with the Russian Federation with the assurance that both coastal States share entitlement to the seabed and subsoil located beyond 200 M in this part of the Barents Sea as the natural prolongations of their land territories.

The Commission recommended to Norway that, in accordance with article 84 of the Convention, upon entry into force of a maritime boundary delimitation agreement with the Russian Federation in the central Barents Sea, it deposit with the Secretary-General of the United Nations charts, or a list of geographical coordinates of points, showing the line of delimitation of the continental shelf beyond 200 nautical miles.

On December 15, 2014, the Kingdom of Denmark submitted to the CLCS in respect of the Northern Continental Shelf of Greenland, to attempt to prove that the Lomonosov Ridge is an extension of Greenland's land mass.

On 6 December 2013, Canada made its own submitted to the CLCS.

In sum, currently, Canada, the Kingdom of Denmark and the Russian Federation assert that the Lomonosov Ridge is a continuation of their own continental shelf, while the US claims it to be an oceanic ridge and thus not an extension of any State's continental shelf.

The provisions of Article 76 do not prejudice the question of delimitation of the continental shelf between States with opposite or adjacent coasts. In Arctic, continental shelves of coastal countries are adjacent and, according to Article 83 UNCLOS, delimitation should be carried out on the grounds of agreement based on international law in order to reach a fair decision. However, this formula has not always been working, and each country has its own history of polar claims and methods of delimitation.

16 The same approach has been taken to the delimitation of the continental shelf and the fisheries zones in the area between Greenland and Svalbard, which has now been settled, and to the delimitation of the continental shelf in the Norwegian Sea. It is also being used in dealing with the unresolved questions related to the delimitation line between Norway and Russia in the Barents Sea.

In February 1907, Canadian Senator Pascal Poirier proposed a resolution before the Canadian Senate to declare possession of the lands and islands between Canada and the North Pole. 18 years later, in June 1925, Canada became the first State to claim that its boundaries extended into to the Arctic and up to the North Pole by adoption Amendment to the Northwest Territories Act. Less than a year after President Truman established Proclamation 2667,[17] Canadian Ambassador to the US Lester Pearson, attempted to claim not only the islands, but the frozen sea north of the mainland between the meridians of Canada's east and west boundaries, extending up to the North Pole under Canadian sovereignty. Canada became the first country to extend its maritime boundaries based on the 'Sector principle' (between 60°W and 141°W).

In 1910–1915, Russia claimed territorial sea to 12 nm and organized Arctic Ocean Hydrographic Expedition, the first ever traverse of the Northern Sea Route. It represented a great effort by the Imperial Russian Navy to explore, survey, and chart the Northern Sea Route for commercial purposes. On 15 April 1926, Union of Soviet Socialist Republics (USSR) issued its 'Arctic Decree' where declared sovereignty over all lands and islands (including those that may be discovered in the future) between the USSR and the North Pole. The Decree was seen as a response to previous Canadian claim to the territory between its mainland and the North Pole the previous year. In 1926, Russia fixed its claim in Soviet law (between 32°04'35"E to 168°49'30"W).[18]

By a 2004 Norwegian legislation, the breadth of the territorial sea has been increased from 4 to 12 nm for mainland Norway, Spitsbergen and Jan Mayen. In 2005, this was also implemented for Bouvet Island. In connection with the extension of the territorial boundary in 2004, the EEZ extended to 200 nm beyond the baseline. Where there is less than 400 nm to another state's baseline, the zone is delimited by the agreed demarcation lines between the states.[19] In a legal sense, the Norwegian continental shelf is the seabed from the territorial boundary at 12 nm from mainland Norway, Svalbard and Jan Mayen

17 In his proclamation (July 1946), President Truman sought to preserve and utilize the natural resources of the subsoil and sea bed of the continental shelf (beyond the 3-mile limit) beneath the high sea that extended from the United States' shores, affirmed that the United States had exclusive jurisdiction over its continental shelf resources.

18 Norway (5°E to 35°E) made similar sector claims – as did the United States (170°W to 141°W), but that sector contained only a few islands so the claim was not pressed.

19 At Svalbard a fish protection zone has been established. This is calculated in the same manner from the baselines of the Svalbard archipelago. The zone is limited by the outer boundary of Norway's economic zone and agreed demarcation lines towards Greenland and Russia.

and out to the outer limits of the continental shelf or agreed demarcation line against another state.

2.3.2 The 2010 Delimitation Treaty between Norway and Russia

Norway and Russia have a longstanding delimitation history.[20] In the course of the fifteen to eighteenth centuries coastal border spaces between Norway and Russia did not have official delimitation and got their names 'common areas' (*fellesdistrikter*).[21] These areas remained the areas of common use for more than five centuries.[22]

In 1825, demarcation of the border was carried out in 'common area' which was 4,4 thousand of square km, according to Norwegian maps, and which was later owned by the Swedish-Norwegian Union for various reasons. Demarcation was acknowledged by the Convention 'Borders between Russia and Norway in the area of Lapland pogost', signed in 1826.

In 1871–1872, the united government of Sweden and Norway exchanged memoranda with the Russian government, where the equality of all states in the use of Spitsbergen natural resources was acknowledged. Reasonability of research work implementation in archipelago was acknowledged, the idea of exclusive ownership of the archipelago by one of the states was denied. International legal status of Spitsbergen as a *terranullius* was defined by the Convention. Actually, up to the adoption of the Convention about Spitsbergen in 1920 the stated territory was in common use by the states. Historically it resulted in the international practice of peaceful economic and scientific use of Spitsbergen transformed into custom and legally adopted in 1871–1872. This treaty was called 'Spitsbergen treaty of 1872'.[23]

In spite of the ratified agreement between the USSR and Norway about the marine borderlines between the two countries signed on February 15 1957, the negotiations about economic zones and continental shelf boundaries in the Barents Sea did not stop. From the very beginning of the official negotiations,[24]

20 The first reference of the contractual design of bilateral relations belong to the XIII century. In 1251, the first agreement on the settlement of relations in border areas between Russia (Novgorod) and Norway was concluded. In 1326, the agreement 'On Eternal Peace' was concluded between the Novgorod Republic and the Swedish-Norwegian king Magnusom-Erikssonom, which fixed the actual existing border.

21 This territory included three counties along the river Neiden (Nyavdem) Pasvig (Pazrek) and Pace (Pechenga) with settlements of three groups of East Sami (Skolt).

22 L.M. Poval Russian-Norwegian agreement on the division of the Arctic spaces. Arctic and North. 2012 №6.

23 Dekanozov R.V. 1872 Agreement on Svalbard – Questions of Soviet legal theory. Novosibirsk, 1968. p. 133.

24 From 1974.

the Soviet Union and Norway had fundamentally different standpoints and no willingness to compromise.[25] But on September 15, 2010 Norway and Russia signed a treaty on maritime delimitation and cooperation in the Barents Sea and the Arctic Ocean.[26] By the agreement, the Barents Sea is fully enclosed by the 200-nautical-mile limits of the Spitsbergen Archipelago to the north-west, the Russian Franz Josef Land and Novaya Zemlya to the north and the east, and the adjacent mainland coasts of Norway and Russia to the south.[27]

The 2010 Agreement[28] defines the maritime delimitation line by eight points and splits the disputed area nearly in half. The underlying calculation accounts for the longer Russian coastline, but other factors Russia invoked earlier do not seem to have influenced the boundary line. The northern terminal point of the delimitation line is defined as the intersection of the line drawn through points 7 and 8 and the line connecting the easternmost point and the westernmost point of the still undefined outer limits of the states continental shelves.

Thus, we can state that even the coastal countries agreed to apply to the Arctic spaces 'delimitation tools' of UNCLOS, they are not very much succeeded in overcoming disagreements on its implementation. In retrospect, we can turn to the initial Canadian offer of 1925, followed by the Russian government regulation of 1926, Russian-American delimitation line adopted in 1990, the Norwegian offer dating 1975 which proposed geometrical delimitation based on sector division; so we cannot leave this method to oblivion due the fact that it gives the choice of the delimitation method by the countries having opposite and adjacent coasts.

The Russian-Norwegian agreement is the unique example of compromising which does not offer a widely recognized methodology to resolve all disagreements on delimitation issues which have to turn our attention to the other Arctic states policy and implementation practice.

---

25 Between the two states' preferred borderlines, lay a disputed area of 176'000 square km. with rights to exploit the resources unsettled. The disputed area made up 12 percent of the whole Barents Sea, which is the equivalent of 45 percent of Norway's total land area.

26 Treaty between the Kingdom of Norway and the Russian Federation concerning Maritime Delimitation and Cooperation in the Barents Sea and the Arctic Ocean, Nor.-Russ., Sept. 15, 2010, available at http://www.regjeringen.no/upload/UD/Vedlegg/Folkerett/avtale_engel sk.pdf.

27 From the very beginning of the official negotiations in 1974 the Soviet Union and Norway had fundamentally different standpoints and no willingness to compromise. Between the two states' preferred borderlines, lay a disputed area of 176'000 square kilometres with rights to exploit the resources unsettled. The disputed area made up 12 percent of the whole Barents Sea, which is the equivalent of 45 percent of Norway's total land area.

28 The delimitation line between Norway and Russia in the Barents Sea, agreed in 2010 and effective from 7th July 2011.

3 Respective National Strategy of the Arctic Coastal States in the Arctic

3.1 *U.S. National Strategy in the Arctic*

In the USA there is a good deal of documents forming the national strategy in the Arctic.[32] Under the US legislation, thc US Arctic strategy is supposed to be implemented in a manner consistent with the Constitution and laws of the United States, with the obligations of the United States under the treaties and other international agreements to which the United States is a party, and with customary international law as recognized by the United States, including with respect to the law of the sea. The key term here is 'recognized' and if not, the decision could be based only on the current state interests.

In accordance with the current policy, it is declared by the US pretty much the same objectives as similar documents issued by other countries: protect the Arctic environment and conserve its biological resources; ensure that natural resource management and economic development in the region are environmentally sustainable; strengthen institutions for cooperation among the Arctic nations; involvement the Arctic's indigenous communities in decisions that affect them and enhance scientific monitoring and research into local, regional, and global environmental issues. Nothing we can find in those documents related to the seabed delimitation.

Most of the NSD[29] were adopted in the US by the president G.W. Bush administration (2009). Under President B. Obama, the administration announced its own vision on 'Arctic strategy' (2013)[30] and adopted final rules that require companies working in the offshore Arctic to put in place new safety precautions to prevent and contain oil spills in the remote and forbidding region.[31] Even the US NSD are not very much concrete with regard to the future delimitation policy, we can find some useful provisions:

29 National Security documents.

30 Not many deviations from the previous strategy.

31 The rules are rooted in part in the administration's experiences with Royal Dutch Shell, which encountered a series of setbacks when it pursued exploratory drilling in the Arctic in 2012 and 2015.Under the Bush Administration, roughly 80 million acres in Arctic waters were opened to energy development as part of the 2007–2012 offshore drilling plan. Six sales were planned, but only one occurred – Chukchi Sea Lease Sale 193 – before the DC Circuit Court of Appeals declared the plan illegal. Numerous companies, including Shell, acquired leases for exploration in the Chukchi Sea off the Alaskan Coast. In March 2010, the Obama administration cancelled all future lease sales in the Arctic Ocean, citing a lack of information to support moving forward, but kept the three million acre Chukchi Lease Sale 193 intact, amounting to nearly USD$2.7 billion.

(1) In respect of Extended Continental Shelf and Boundary Issues … "the most effective way to achieve international recognition and legal certainty for our extended continental shelf is through the procedure available to States Parties to the U.N. Convention on the Law of the Sea."
(2) For unresolved boundary in the Beaufort Sea, the United States policy recognizes a boundary in this area based on equidistance."
(3) For the US-Russian boundary the United States expect that the maritime boundary treaty concluded in 1990 will come into force.[32]

From all these we can draw following conclusions: for the US the Arctic region is primarily a maritime domain for which existing policies and authorities relating to maritime areas continue to apply, including those relating to law enforcement and the government will take all actions necessary to establish the outer limit of the continental shelf appertaining to the United States, in the Arctic and in other regions, to the fullest extent permitted under international law, without clear definition what should be understood by this term.

3.2 *Russia's Strategy in the Arctic*

The main goals, strategic priorities and the Russian Federation state policy mechanisms of implementation in the Arctic for the period up to 2020 and further prospects are defined in 'The foundation of the state policy of the Russian Federation in the Arctic'.[33] Both the American doctrine and this document contain pretty much common provisions about strengthening of the bilateral efforts of the region countries and within regional organizations like the Arctic Counsel and Barents/Euroarctic region Counsel; provisions of good-neighborly relations with coastal states, the development of economic, research, cultural cooperation including the field of effective development of natural resources and environment protection in the Arctic; improvement of living conditions of native population and social conditions for economic activity in the Arctic.

Above all these this document states that a strategic priority of Russian policy in the Arctic is the implementation of cooperation between Russia and coastal states in order to delimitate marine spaces on the basis of International Law, mutual agreements taking into account national interests of the Russian Federation and also to decide matters on International legal justification of the outer border of the Arctic zone of the Russian Federation;[34] delimitation of marine spaces in the Arctic Ocean and the provision of mutually beneficial

32 NSC-26, "D".
33 Approved by the President of the Russian Federation of September 18, 2008 N Pr-1969.
34 Paragraph 7(a) Russian Arctic state policy adopted in 2008.

presence of Russia on Spitsbergen archipelago.[35] All mentioned provisions concerning delimitation don't seem exact with the exception of Spitsbergen; although in Article 9 of the document the main state Russian policy goals in the Arctic can be reached on the basis of the preparation of Russian legal acts on the defining of the geographical borders of the Russian Federation in the Arctic.[36]

3.3 *Norwegian Strategy in the Arctic*

On December 1, 2006 the Norwegian Government releases its 'High North Strategy' to clarify long-term Norwegian policy in the Arctic "in order to secure economic resources and its security into the future."[37] It was stated in the document that one of the government's most important priorities in the years ahead will be to take advantage of the opportunities in the High North. The term 'to take an advantage of the opportunities' is not determining, but brief objectives of Norway High North strategy are following:

- continue to build on good neighborly relations with Russia;
- continue to exercise responsibility for combating illegal fishing and managing the renewable fish resources for present and future generations;
- take advantage of the opportunities the Barents Sea presents as a new European energy province in accordance with the principles of sustainable development;
- take environmental and climate considerations into account in all state efforts;
- improve living conditions, opportunities and the quality of life for all those who live in the High North.[38]

Actually, each A-5 country could sign mentioned above provisions as its own and all countries declare that would like to sea not only Barents Sea, but the whole Arctic as a 'ocean of cooperation'.

What Norway consider as the 'conflict of interest' – related to the utilization of fisheries resources and future offshore petroleum resources.[39] Probably, from real practice one should read the Norway Arctic strategy more expanded. One of the issue is the differing views on the geographical scope of the Spitsbergen

35 Paragraph 7(e) Russian Arctic state policy adopted in 2008.
36 Paragraph 9(B) Russian Arctic state policy adopted in 2008.
37 The Norwegian government's High North strategy, p. 1–18 https://www.regjeringen.no/en/dokumenter/strategy-for-the-high-north/id448697/.
38 The Norwegian government's High North strategy, p. 5 https://www.regjeringen.no/en/dokumenter/strategy-for-the-high-north/id448697/.
39 The Norwegian government's High North strategy, p. 16 https://www.regjeringen.no/en/dokumenter/strategy-for-the-high-north/id448697/.

Treaty. Norway's position that the treaty, in accordance with its wording, only applies to the archipelago and the territorial waters. Spain, Iceland and Russia have disputed this right with reference to the provisions of the Spitsbergen Treaty.

It was also stated that Norway's policy towards Russia is based on pragmatism, interests and cooperation. Norway ensure sustainable use of resources and sound environmental management in the Barents Sea with Russia's engagement and Norwegian-Russian cooperation. The Government therefore intends to strengthen cooperation with Russia on ecosystem-based management of the whole Barents Sea, both in the Joint Norwegian-Russian Commission on Environmental Protection and in the Joint Norwegian-Russian Fisheries Commission. This cooperation includes measures to combat illegal, unreported and unregulated fishing.

As it was further written, developing close cooperation with Russia on sound exploitation of the petroleum resources in the Barents Sea should be Norwegian government ambition and both the former and present Norwegian Government accepted President Putin's invitation to forge a strategic partnership between Norway and Russia in the North, but in reality looks rather doubtful. Much more realistic looks Norwegian ambitions of the future role of the Norwegian Armed Forces with obligations to:

- strengthening capacity of the Coast Guard vessels and the Orion aircraft to spend more time at sea and in the air, respectively;
- provide more close cooperation between the civilian and military authorities;
- provide background information for national decision-making through up-to- date surveillance and intelligence;
- maintain the presence of the Norwegian Armed Forces in the High North both to enable Norway to exercise its sovereignty and authority and to ensure that it can maintain its role in resource management.

3.4 *Canada's Strategy in the Arctic*

Canadian Arctic strategy was adopted in 2009 and its main provisions are very similar to all other A-5 countries and focuses on four priority areas: exercising our Arctic sovereignty; promoting social and economic development; protecting the North's environmental heritage; and improving and devolving northern governance, so that Northerners have a greater say in their own destiny.[40] Making more colorful statements like that: the Government of Canada

40 Canada's Northern Strategy: Our North, Our Heritage, Our Future. Published under the authority of the Minister of Indian Affairs and Northern Development and Federal

is firmly asserting its presence in the North, ensuring we have the capability and capacity to protect and patrol the land, sea and sky in our sovereign Arctic territory. We are putting more boots on the Arctic tundra, more ships in the icy water and a better eye-in-the-sky[41] in the same time for our object we can find very little except that Canada dealing with Arctic problems will use cooperation, diplomacy and international law as Canada's preferred approach in the Arctic.[42]

What is more important that Canada claims rights in its continental shelf under the 1997 Oceans Act, where adopted the default length of Canada's continental shelf extends to 200 nautical miles from the straight baselines. On December 6, 2013 Canada made partial submission to the CLCS in respect of Atlantic ocean, continuing the data collection for the Arctic spaces.[43] The dispute in the Beaufort Sea concerns the maritime extension of the land boundary between Yukon and Alaska will certainly influence further delimitation negotiations, which will be under undoubtedly pressureof the assumptionthat the area is considered to be resource-rich. Canada claims that the maritime boundary runs along the 141st meridian as an extension of the territorial boundary agreed with the United States. However, the United States rejects this position, arguing that the boundary must be determined by using the equidistance principle – a recognized mode of maritime delimitation that traces a line at equal distance from the closest land point of each state. This produces a line that reflects more closely the direction of the respective coast lines. Canada and the United States are in effect both promoting the use of a delimitation method that will best serve their respective interests and that will produce, from each of their perspectives, the largest maritime zone possible. Resolution of this dispute is still pending.

Interlocutor for Métis and Non-Status Indians Ottawa, 2009 (http://www.northernstrategy.gc.ca/cns/cns-eng.asp).

41 http://www.northernstrategy.gc.ca/cns/cns-eng.asp.

42 The US and Canada on January 11, 1988 signed the Agreement on Arctic Cooperation between two states. Following the United States Coast Guard Polar Sea's crossing of the Northwest Passage without formal Canadian governmental consent, the Canadian government established straight baselines around Canada's perimeter.

43 Canada is expected to submit a claim to the CLCS, within UNCLOS, to prove the Lomonosov Ridge is a continuation of its own continental shelf.

Canada is expected to claim that the ridge is an underwater extension of Ellesmere Island. If it's proven that the Lomonosov Ridge is an extension of Canada's continental shelf, Canada would obtain unfettered access beyond their EEZ of 200nm, and instead, would gain access to the seabed and its resources across the continental shelf. Canada would be required to submit bathymetry, seismic and gravity data to substantiate its claim.

3.5 *Denmark's Strategy in the Arctic*

Within the 2011 'Kingdom of Denmark: Strategy for the Arctic 2011–2020,' Denmark asserts that it will submit data and other material to the CLCS as a basis for extension of the continental shelf beyond 200nm on three areas near Greenland, including the Lomonosov Ridge. Proof that the Lomonosov Ridge extends from Greenland's continental shelf would give Denmark unfettered access to much of the seabed surrounding the North Pole.

On December 17, 1973 Canada and the Kingdom of Denmark sign the agreement on the delimitation of the continental shelf between Greenland and Canada.[44] The agreement was designed to distinguish territory for the purposes of exploration and exploitation of the natural resources of the continental shelf. Two countries agreed to divide the ocean floor between Canada and Greenland using a median or 'equidistance' line.[45] Both delayed any decision regarding the sovereignty of Hans Island.[46] The treaty also has provisions concerning the possible discovery of hydrocarbons along or near the boundary – neither Party shall issue licenses for exploitation of mineral resources in areas bordering the dividing line without the prior agreement of the other Party as to exact determination of the geographic co-ordinates of points of that part of the dividing line bordering upon the areas in question.[47]

This provision of the agreement makes its legal value in the CS resources exploitation of the void. Article V, of mentioned above agreement, created pretty less critical model of the relations by simple formula: if any single geological petroleum structure or field, or any single geological structure or field of any other mineral deposit extends across the dividing line is exploitable, wholly or in part, from the other side of the dividing line, the Parties shall seek to reach an agreement as to the exploitation of such structure or field. In other words, this formula does not solve problem and postpones its decision to the occurrence of a particular dispute for resource exploitation.

---

44 Agreement between the Government of the Kingdom of Denmark and the Government of Canada relating to the Delimitation of the Continental Shelf between Greenland and Canada (17 December 1973).

45 A line that at every point is an equal distance from the nearest point on each of the two opposing coasts. In this case 109 'turning points' of the 127 were agreed.

46 The agreement distinguished territory between Canada and Greenland. Both delayed any decision regarding the sovereignty of Hans Island – a small 1.3 square kilometer uninhabited island in the Kennedy Channel of the Nares Strait between both Canada's Ellesmere Island and northern Greenland. Both Canada and Denmark (via Greenland) claim the island that exists on both Canadian and Danish sides of the Strait.

47 Michael Byers & James Baker. International Law and the Arctic. Cambridge University Press University Printing House, Cambridge CB2 8BS, United Kingdom, p. 30, 2014.

Both countries agreed in 1984 to extend the work to clarify the computation of the equidistance line[48] south of 75°N which came to an end in 2003. Amendment to the coordinates will be provided in several data and may reduce the number of turning points of the boundary from the present 113 points (south of 75°N).

3.6 *Conclusions*

What conclusions can be drawn on the basis of the Arctic delimitation policy? Despite the fact that the several coastal States have concluded bilateral agreements governing the territorial delimitation they did not completely solve the problem of the seabed delimitation.

Canada, the Kingdom of Denmark (via Greenland), and the Russian Federation each assert that the Lomonosov Ridge is an extension of their own continental shelf. Proof of its continuation, would give the State access to the sea bed and natural resources beyond the current 200 nm limit. The United States claims it to be an oceanic ridge and thus not an extension of any State's continental shelf, and therefore refutes any claim to its ownership.

There is a dispute between Canada and Denmark about delimitation line at Hans Island and in the Lincoln Sea, challenges Lomonosov and claims the USA the direction of the sea border line in the Beaufort Sea.

In 2015, Russia delivered another submission to the CLCS once again claiming the extension of its continental shelf, which stood to potentially acquire nearly one-half of the Arctic Ocean.

The border line with the USA is shown as a direct line coinciding with 'the western limit' in the Article 1 of the Russian-American agreement of 1867 on the Alaska Purchase which comes through the point in the Bering Strait (65 degrees, 30 minutes) in its crossing with the meridian at equal distance of the Kruzenshtern Isle from the Ratmanov Isle, and goes along the unlimited line towards the North until it's fully lost in the Arctic ocean …[49] The Russian-American agreement of 1990[50] 'clarified' the delimitation point and stated that the border in the Bering Strait and further in the Chukchi sea goes along the meridian 168° 58' 37" W up to the limits until it's allowed by International law. It may be admitted that the agreement of 1990 states the delimitation line but it's obvious that it is not enough for the seabed delimitation; if we try to find the outer delimitation point in the Chukchi sea on the

48 The Canadian maps and charts were drawn on the North American Datum (NAD) 1927 and the Danish maps and charts on the Qornoq Datum which uses a different ellipsoid.

49 Full complete Collection of Laws of the Russian Empire. II. Volume XLII. Dep. 1. 1867.

50 The Soviet-American agreement on the maritime border by June 1, 1990.

principle of equidistance,[51] it will be located a bit to the west from the straight line connecting the delimitation point in the strait and the North Pole which used to be drawn on the majority of presentation maps and in some official sources.

Except ostentatious, in our view, pretentiousness coastal states policy, there is another problem which affects the possibility to deal with coastal countries contradictions. Despite the fact that practically all of them tend to stress the necessity to turn the Arctic into the 'peace zone', their military presence in the region is escalated, and there is typical reasoning that polar regions are both vital from economic, ecological point of view, and for the reasons of national security.

On the one hand, all countries' claims are equally categorical in the region; on the other hand, opportunities to provide military presence in the Arctic coastal regions are different.

The USA is the only country which didn't ratify UNCLOS; anyway, it can't be the sign of the drawback of legal policy of the state which has been defining and clarifying its Arctic national interests since 2004. In terms of strategy implementation there are at least 3 main points:

- to make every effort to provide national security;
- to joint efforts with other coastal states within both NATO and the Arctic Counsel keeping the possibility of other unions;
- to maintain the freedom of navigation according to the norms of International Law.

As these are the points from the strategy 'open' part', it may be presumed that the real 'joint effort' is nothing but actions within traditional military alliances, and there is no doubt that the USA wouldn't really tend to cooperate with at least one country among the region states. It is most likely that the last point concerns the efforts to give the Northern sea route the status beyond the laws drawn by the Russian Federation.

The USA does not have big military-industrial bases on the Arctic coast with the exception of temporary base in Barrow which can be regarded as a small base rather than a Navy base. It's obvious that in order to solve problems maintaining the US interests in the Arctic by force which are also connected with the resource extraction activities, will demand the deployment of the Pacific forces command. This fact makes the idea of attracting of the allies owing coastal infrastructure very appealing, what actually are doing the USA getting involved into military cooperation North-European countries within NORDEFCO (Nordic Defense Cooperation).

51 This is not the only method, and is cited as an example.

Canada, as well as Denmark, does not have ground forces in the Arctic zone. Nevertheless, the Canadian government builds so-called 'hubs' there (North Operations Hubs, NOH). Such first hubs were built in Resolute Bay and Iqaluit (the south-east of the Baffin Island), and by 2018 the Canadian Navy ships fuelling point is being planned to be constructed in Nanisivik (North-western part the Baffin Island).

Norway actively supports the NATO 'Arctic vector', and was one of the initiators of 'the NATO Arctic strategy' (January 2009), promotes Scandinavian countries' efforts to create a special security zone, already given the name 'mini-NATO – creation of the defense alliance' of 5 states – Denmark, Iceland, Norway, Finland and Sweden to strengthen their military position in the Arctic.

Russia is the only country which has not only permanent military bases but nuclear forces, military infrastructure objects, including the unique shipbuilding and other industrial enterprises. This complex was formed long ago since the surge of interest to the Arctic.[52]

In response to strengthen the military activity in the Arctic, Russia in 2014 announced about the creation of the Arctic forces within the frames of the North fleet,[53] in 2016 the decision was made to locate special subdivisions in Chukotka. The location of military objects is carried out not only in traditional places of Archangelsk and Murmansk regions, but on the islands – Novaya Zemlya and Franz Josef land, 'polar trainings' have begun more active.

Above all this, Russia has started to use military objects in the peaceful purposes by attracting military personal and staff for the research work. The example is mutual efforts at the station 'Barneo', not far from the North Pole.[54] Moreover, infrastructural objects 'GLONASS' and 'ISS' 'Arctica';[55] communication systems to navigate ships and aircraft on cross-polar routes,[56]

52 The first Naval ships in the North of Russia were built under Peter I, and the first regular Flotilla was formed in 1916 and was named the 'Flotilla of the Arctic Ocean,' which in September 1917 included 89 combat and support ships. In 1937, the Flotilla transformed into the Northern Fleet.

53 "The Northern Fleet – United Strategic Command" (SF-USC) with the of 'military district'.

54 Drifting base 'Barneo' is created in the Arctic every year. Such an object can be attributed to the objects of dual-use, which could be recognized as novelty in the practical development of the Russian Arctic.

55 MSS 'Arctica' is composed of three sub-systems intended for hydrometeorological and climate ('M') monitoring; radar monitoring ice conditions ("P") and provide satellite communication and navigation 'MS'). In the nearest future with the cable laying 'Polarnet System' it will be incorporated to the 'United information space of the Arctic zone of the Russian Federation'.

56 Flights of this type operate 11 airlines, including Russian, Canadian, American and Chinese.

reconstruction of the research objects on the islands of archipelago Severnaya Zemlya, rebuilding of the northern aerodromes (Tiksi, Naryan-Mar, Alykel, Amderma, Nagurskaya, Anadyr and Rogachevo)... are also on 'military shoulders'.

All these facts may serve only as positive circumstances but they cannot change general tendency to militarization of the Arctic, which make chances to compromise pretty low.

4 Addressing and Solving Conflicts regarding Seabed in the High North

So, we may make presentations at international forums, assemblies, sessions of the Arctic counsel assure each other in peaceful intensions, but the fact is – there is no real peace in the Arctic – rivalry is escalated, and all states of the region are not prepared to compromise defending their national interests. Adding the traditional 'NATO spice' on top of that, increase of military trainings in the Arctic zone, already exceeding the level of the cold war, active involvement of Iceland, Finland and Sweden in military games[57] adding to this Russian efforts in strengthening its Northern fleet, which includes the nuclear component the conclusion on unfavorable political and legal environment is quite obvious.

Now, my experienced reader, ask yourself a question – what peaceful methods must be chosen to resolve present delimitation contradictions. Although the conclusion about the need to reduce the intensity of the 'struggle for the Arctic resources' although trivial, but it is the ground of all other efforts. And what delimitation methods would be chosen its implementation in reality without establishing political warming in Arctic seems to be useless. In addition, the principle question whether the Arctic coastal states have to have the same level of responsibilities in the spaces beyond the 200 nm limits as the non-regional states remains unanswered. Formula "apply UNCLOS and enforce freedoms granted by the Convention" without considering the peculiarities of the region,- is completely irresponsible decision.

What could be the principles for the seabed delimitation in Arctic? Theoretically states have three options: to follow the provisions of the UNCLOS, to create specific regime outside UNCLOS frames by bilateral agreement or

57 It is known that the United States, Canada, Norway, Finland, Sweden, Denmark and Iceland propose to create a military group of the rapid response, including ice-breaking ships, amphibious units and air forces.

keep talking maintaining sluggish negotiations or by using the delimitation disputes for periodic outburst of political activity in the region.

A-5 agrees to apply international law provisions for activity in the Arctic Ocean with regard to the delimitation of the outer limits of the continental shelf the coastal states should recognize their obligations and responsibilities under the Law of the Sea in the Arctic, including their commitment to the orderly settlement of any possible overlapping claims.

A-5 also agrees that the provisions of the UNCLOS should be the legal base for the resolving the Arctic claims[58] particular Article 83, which relate to the continental shelf delimitation agreements which are to be based on international law and the uniform geodetic coordinate system[59] to achieve an equitable solution.

In all cases the decisions of the CLCS with adoption of the continental shelf outer limits will need to be delineated on the basis of CLCS recommendations which will be final and binding only after adoption by the coastal states.

The adjacent boundaries were adopted by the three agreements – between Norway and Russia (which is limited by the point 8);[60] between Norway – Denmark[61] and Denmark – Canada. The US and Canada have the dispute in the sea of Beaufort and the Arctic boundary between the US and Russia is uncertain.

According to the decisions established by the International Court of Justice, the starting point is the delimitation line should follow the median line between the two states' coastlines. The starting points of the US-Canadian and the US-Russian boundaries are agreed, but the direction of the first one is in dispute, and the second is uncertain. If even the 1990 US-Russian agreement will be ratified the concept 'as far as permitted under international law'[62]

58 The Ilulissat Declaration was announced on May 28, 2008.

59 As an example – agreement between Canada and Denmark, where from a surveying stand-point, the interesting aspect is the fact that the Canadian maps and charts were drawn on the North American Datum (NAD) 1927 and the Danish maps and charts on the Qornoq Datum which uses a different ellipsoid. The technical experts knew that there was a difference between the geodetic coordinate systems but had no way of knowing the magnitudes. So the practical solution was to set the problem aside for future consideration and to assume that the two coordinate systems were identical.

60 Norway–Russia agreement in the Varangerfjord area 2007 and 2010 Treaty between the Kingdom of Norway and the Russian Federation concerning Maritime Delimitation and Cooperation in the Barents Sea and the Arctic Ocean.

61 On the delimitation of the continental shelf and the fisheries zones in the area between Greenland and Svalbard 2006.

62 USA-USSR 1990 Agreement. Annex: 'From the Initial point, 65° 30' N., 168° 58' 37" W., the maritime boundary extends north along the 168° 58' 37" W. meridian through the Bering Strait and Chukchi Sea into the Arctic Ocean as far as permitted under international law.'

could not be applied for the delimitation of the seabed and subsoil based on the UNCLOS provisions.

In the same time, keeping in mind that the US did not ratified the UNCLOS, based on the customary international law this line could be as far as the agreement between two states will allowed.

In any case, it can be argued that the provisions of the UNCLOS do not prevent the Arctic countries to use bilateral agreements, based on its own method of delimitation, where geographical criteria would greatly simplify the delineation of adjacent spaces up to the Northern Pole. It could also be the ground for an agreement for mutual exploitation of the resources beyond the 200 nm limits in the Arctic seabed by way of a joint development agreement. Such agreement would enable to mutually share the exclusive rights as regards natural resources in the contested areas without abandoning their claims and also without the need for a final resolution of all legal outer limits delimitation issues.

CHAPTER 6

Current Human Impact on Antarctic Seabed Environment and International Law

Yana Evgenyevna Brazovskaya and Gulnara Flurovna Ruchkina

1 Introduction

Antarctica is perhaps the most mysterious and breath-taking continent of our planet by its extreme natural conditions and the absence of a native human population. Located 'opposite the Arctic,' it is the world's largest and coldest desert, the highest continent of the Earth, and, at the same time, a mostly unknown and still under-researched area. Remoteness, permanent glacial cover of the continent, and exceptionally severe weather conditions have apparently also influenced the legal status of the Antarctic territories and the legal regime of their use. Indeed, the 1959 Antarctic Treaty[1] which together with related protocols and conventions form the Antarctic Treaty System (ATS), affirms in its Article I.1 that 'Antarctica shall be used for peaceful purposes only.' Parties to the Treaty have, so to speak, cut the southern polar region[2] off of the possible proliferation of nuclear weapons and conduct of any military activities.[3] The parties also excluded the possibility of regulating the development of industrial development of mineral resources of the Antarctic. However, the Antarctic Treaty aims to facilitate scientific research in Antarctica.[4]

The level of political tensions in the Arctic is much higher than that observed in the Antarctic, with only attempts to raise Antarctic delimitation claims. However, the constant pressure on worldwide natural resources sheds a new light on the more remote areas, including Antarctica, with the subjacent race to claim jurisdiction over those. A first example relates to access to fresh water. A growing number of research shows that water scarcity represents a major threat on vital resources for mankind, threatening the life

1 The 1959 Antarctic Treaty was signed in Washington and entered into force on June 23, 1961, after being signed by the 12 nations that were the Treaty's initial parties. Since then, other nations have acceded the Treaty, which now counts 53 parties.
2 Which covers almost 50 million square kilometres, or 10% of the planet's area, and the mainland is twice the size of Europe. Source: https://ria.ru/spravka/20141130/1035503295.html.
3 Article I.1 and V.1, the 1959 Antarctic Treaty.
4 Articles II, the 1959 Antarctic Treaty.

 | DOI:10.1163/9789004391567_008

conditions of hundreds of millions of people around the world.[5] This has led to increased competition for scarce resources, which can both exacerbate old security dilemmas and create new ones.[6] Access to ice reserves in Antarctica will open almost 80 percent of total fresh water reserves on the planet. A second natural resource is catching attention in the Antarctic. It is highly probable that various kinds of mineral raw materials are present in the interior of Antarctic. Unlike solid minerals, the forecast of hydrocarbon resources in the Antarctic is mainly based on the materials of structural-tectonic zoning, carried out using remote geophysical methods, paleotectonic reconstructions and geohistorical analysis. According to scientists, there may be about 900 significant crude ore deposits in Antarctica. Despite the fragmentation and small size of the overglacial outcrops of the bedrock even in relatively naked mountainous areas of Antarctica, they found many manifestations of ore and non-metallic minerals, representing weak mineralization.[7] According to British scientists, there are coal deposits in the depths of the sixth continent, and American scientists insist that Antarctica has oil and gold.[8] The presence of hydrocarbon reserves[9] in the seabed may therefore reasonably be assumed, with estimated 35–51 billion tons of conditional fuel to be present on the self of Antarctic seas.[10] Those are only two examples in an area which is the least explored of the two circumpolar regions.

2 Jurisdiction Delimitation Claims in the Antarctic

It is believed that Antarctica is the only continent with no history of human habitation,[11] but at present this glacial region is the one most 'densely

5 A. Kushnarenko, 'Lack of Fresh Water: Problems and Solutions,' The W&L, 28.05.2015. Source: <http://thewallmagazine.ru/lack-of-fresh-water/>.

6 Speech of UN Secretary-General Ban Ki-moon at the Security Council session in 2011. Source: http://www.un.org/ru/sections/issues-depth/water/index.html.

7 G.E. Grikurov, G.L. Leichenkov, E.V. Mikhalsky, A.V. Golynsky, V.N. Masolov 'Antarctic mineral resources: geological preconditions and perspectives of development' Source: http://evgengusev.narod.ru/vniio/grikurov-2000.pdf.

8 ITAR-TASS. 'Antarctic: history and problems'. Source: https://tass.ru/spravochnaya-informaciya/628230.

9 The forecast of hydrocarbon resources of Antarctica is based mainly on the materials of structural tectonic zoning by remote geophysical methods, paleotectonic reconstructions, and geo-historical analysis.

10 Antarctica: History and Problems, 03.07.2013. TASS: http://tass.ru/spravochnaya-informaciya/628230.

11 Linda Nowlan. Arctic Legal Regime for Environmental Protection. IUCN Environmental Policy and Law Paper No. 44, p. 41.

populated' by scientists.[12] The lack of native human population has not prevented states from claiming sovereignty over the Antarctic areas. At least seven countries have put forward territorial claims: Argentina, Australia, Chile, France, New Zealand, Norway and the United Kingdom.[13] However, in the view of the present authors, these claims are more of a historical nature. Russia and the United States have reserved their rights; they consider themselves pioneers of Antarctica, entitled to requiring the sovereignty of the territory.[14] It should be noted that the claimant nations did not abandon their territorial claims for Antarctica. For instance, in 2004, Australia sent a request to the UN Commission on the Limits of the Continental Shelf for the Antarctic shelf adjacent to the 'Australian Antarctic Sector,' which, incidentally, was rejected.[15]

The question of the islands and associated continental shelf, located between 40° and 60° south latitude is disputable respectively, whichever the boundary of the Southern Ocean is: the Antarctic Circumpolar current (area of 40° south latitude) or the area regulated by the Antarctic Treaty (area of 60° south latitude).

Both the prospects of accessing in rich natural resources and the effects of climate change have the potential to revive old and new claims. Indeed, the effects of climate change in Antarctica will mainly result in a warmer climate in that part of the Earth, according to scientists. Because of global warming and sea level rise, many of the largest megacities of the planet, as well as small island nations, will disappear. If all glaciers of the mainland melt, Antarctica will become an archipelago (a chain of islands),[16] and the territorial claims issue will appear in much brighter colours.

12 Approximately 29 nations operate seasonal (summer) and year-round stations on the continent and in the coastal zone. The number of scientists engaged in scientific research on the continent and nearby islands varies from around 4,000 in summer, and up to 1,000 in winter; in addition, about 1,000 specialists, including ship crews, are located in the waters adjacent to the continent. As of 2004, the largest number of researches during the summer season was deployed by the USA (about 1,100), Russia, Chile, and Argentina (300), Australia (200) and the United Kingdom (192). During winter, it was Russia, Chile, and Argentina (approximately 200) and the USA (120–130). There are 42 year-round stations in Antarctica (6 belong to Russia and Argentina, 4 to Australia and Chile, 3 to the USA and 2 to the UK).

13 T.B. Mordvinova. A.S. Skaridov, M.A. Skaridova. Polar Law. M., Justitia, 2017, at 200.

14 Who and why is interested in Antarctica? June 07, 2017. Source: http://www.profi-forex.by/news/entry5000037141.html.

15 The United Kingdom, Argentina, New Zealand, and South Africa may submit similar requests for the Southern Ocean shelf. See Recommendations (9 April 2008) Source: http://www.un.org/Depts/los/clcs_new/submissions_files/submission_aus.htm.

16 Antarctic Mainland, 24.06.2017. Source: http://mirplaneta.ru/materik-antarktida-opisanie-relef-klimat.html.

At the present time, the coastal contours of Antarctica are above the Arctic coastline. They are in fact static coastal fast ice, forming ice fronts almost all over, whose height above sea level varies from a few to dozens of meters. Of the total length of the Antarctic coastline (30,000 km), 92 percent are ice banks and only 8 percent are rocky shores. The Southern Ocean is a deep-sea basin. Areas with depths of 3,000 meters occupy about 77 percent of the total area of the ocean. Due to the huge glacial load, the Antarctic shelf is immersed to a depth of up to 500 meters and stretches in a relatively narrow strip, about 280 kilometres on the average.[17]

It should be noted that the international agreement on the delimitation of Antarctic spaces does not exist, as it does for the Arctic. According to A. Skaridov, the international legal doctrine has at least three points of view in this regard, namely that: (i) the areas are outside the jurisdiction of any country, to which the rules of international law apply; (ii) they are areas of joint use of countries (by condominium theory they should be managed by all countries with the assistance of an international organization); (iii) division into sectors to be fully owned by various countries.[18]

While the situation is relatively straightforward in the first two points of view, the theory of sectoral delineation may be less complicated than foreseen to apply in Antarctica. It should be noted that most of the Arctic countries deny the sectoral delineation, but this does not necessarily mean that the situation in Antarctica will cause a similar response,[19] taking into account the previously mentioned claims of the sectoral nature. In order to consolidate rights to a certain sector, justifications such as geographical proximity, right of discovery, or continuous control or occupation of the territory are often referred to.[20] On the one hand, the division of Antarctica into sectors is the simplest and most acceptable way to solve the problem of territorial claims, but on the other hand it is strongly criticized by many lawyers and politicians. For example, US lawyer R. Hayton is of the opinion that geographical proximity, or in other words 'attraction', does not give countries any rights to own

17 Ice Navigation Conditions in the Southern Ocean. A.A. Romanov. Marine Meteorology and Related Oceanographic Activities, Issue No. 35//WMO/TD-No. 783.

18 A. Skaridov, Maritime Law. Moscow, URAIT, 2014, at 167–169. See chapter 5 of this book, A. Skaridov, 'The seabed in the High North – How to address conflicts?'.

19 Although the Arctic and Antarctic regions are similar in many ways, there are also significant differences between them. The Arctic is an ocean surrounded by continents, while Antarctica is a continent surrounded by oceans.

20 N. V Ilyushina, 'Problem of Territorial Division of Antarctica'.Topical Issues of Current International Relations, 2015. Available at <https://cyberleninka.ru/article/n/problema-territorialnogo-razdela-antarktiki>.

Antarctica.[21] He finds it impossible that the rest of the community of nations may be willing to cede all rights to uninhabited lands that may be of strategic importance.

The unilateral establishment of a sectoral division of Antarctica is contrary to the interests of most countries of the world, but assuming that the 1959 Antarctic Treaty ceases to exist and the continent is divided by a sectoral principle, countries will need to measure the breadth of the territorial sea, the Exclusive Economic Zone (EEZ), that is, to resolve the issue of maritime boundaries, and here additional difficulties may arise.

Indeed, in accordance with Article 5 of the United Nations Convention on the Law of the Sea of 1982 (UNCLOS), the normal baseline for measuring the breadth of the territorial sea is the low-water line along the coast.[22] Pursuant to Article 7 of the same Convention, straight baselines should also connect the appropriate points on the shoreline. It is noteworthy that this Convention does not determine the notions of 'shore' and 'shoreline'.

The UN technical expert team determine the shore as a line of contact of the aquatic environment and the land. Since in the establishment of maritime borders, polar borders included, the key is the legality of the definition of the reference base, the definition of what should be understood under the shore, takes a fundamental importance. However, it is not clear what should be understood under 'land' or 'shore' in Antarctic conditions as there is no physically defined concept of 'shore'.[23]

Ice as a physical category may have different origins, but from a legal standpoint, ice mobility is probably its most important hydrographic characteristic. If the outer edge of the glacier, i.e. the line where two ecosystems with different physical characteristics get in contact, is virtually static, that is, not subject to seasonal fluctuations, it can be considered as a coastline. So, the edge of such geomorphological phenomenon as a multi-year glacial fast ice, or the edge of a hummock attached to the shore or fast ice, in case they are naturally connected to the coast and the lack fluctuations due to seasonal cycles may be considered in the polar areas as a coastline in the sense that is applies to the provisions of Articles 5, 7 and 47 of the 1982 UNCLOS to such a line.[24]

Unfortunately, the 1982 UNCLOS did not take into account the peculiarities of establishing borders in constantly ice-covered polar regions. Only in its

21 R.D. Hayton, 'The Antarctic Settlement of 1959', American Journal of International Law (1960: 54), No. 2. at 359–360.

22 For a review of the international process, see chapter 4 of this book, A. Brekke, 'Setting boundaries: Experiences from Norway'.

23 Law of the Sea: a textbook for masters/A.S. Skaridov. – 2nd edition as extended and amended. – M.: URAIT Publishing House, 2012. 167 pages.

24 Ibid. P. 168.

Article 234 it indicates that coastal countries have the right to adopt and enforce non-discriminatory laws and regulations for the prevention, reduction and control of marine pollution from vessels within the limits of the EEZ, where particularly severe climatic conditions and the presence of ice covering such areas for most of the year create obstructions or exceptional hazards to navigation, and pollution of the marine environment could cause major harm to or irreversible disturbance of the ecological balance.

However, the text of the Convention contains provisions which may be considered as an indirect confirmation of the legality of accounting for static ice massifs, which are a continuation of the shore, as a reference point for the outer limits of maritime spaces. For example, natural formations such as reefs have almost the same geophysical dynamics in the south as ice massifs in the north. For the purpose of computing the ratio of water to land, land areas may include waters lying within the fringing reefs of islands and atolls, including that part of a steep-sided oceanic plateau which is enclosed or nearly enclosed by a chain of limestone islands and drying reefs lying on the perimeter of the plateau (paragraph 7, Article 47 of UNCLOS.) The 1982 Convention also reads that in the case of islands situated on atolls or of islands having fringing reefs, the baseline for measuring the breadth of the territorial sea is the seaward low-water line of the reef, as shown by the appropriate symbol on charts officially recognized by the coastal state.

Thus, the placement of points defining the position of the baselines on the outer edge of the foundation of the coastal fast ice may be found not inconsistent with the provisions of UNCLOS, and it may be assumed that the difficulty of delimiting maritime and submarine areas near the Antarctic continent will be related to the uncertainty of the definition of the datum, i.e. baselines.

In addition, with regard to the delimitation of the continental shelf in Antarctica, it should be noted that, under the 1958 Convention on the Continental Shelf, the term 'continental shelf' is used as referring to the seabed and subsoil of the submarine areas adjacent to the coast but outside the area of the territorial sea (Article 1 (a)). But according to UNCLOS, the 'continental shelf of a coastal state' comprises the seabed and subsoil of the submarine areas that extend beyond its territorial sea throughout the natural prolongation of its land territory (Article 76 (1)).

Taking into account the conventional position, it is possible to speak about continental shelf of a coastal state only. The institution of continental shelf is inseparably tied to the institution of territorial sea, hence legally there is no such thing as continental shelf of international territory.

That is, by implication of the 1958 Convention and the 1982 Convention, the continental shelf of a coastal state begins where the bottom of its territorial sea ends. Therefore, due to the fact that no state has its territorial sea washing Antarctica, it is not possible to discuss any continental shelf of Antarctica.

3 Exploitation of Mineral Resources from the Seabed in the Antarctic

To date, the legal regime of Antarctic regions is formed by four basic agreements as part of the ATS:
- The Antarctic Treaty, 1959;
- Protocol to the Antarctic Treaty on Environmental Protection, 1991 (Protocol on Environmental Protection to the Antarctic Treaty (Madrid Protocol);
- Convention for the Conservation of Antarctic Seals, 1972;
- Convention on Conservation of Antarctic Marine Living Resources, 1980 (CCAMLR).

At the time of signing the Antarctic Treaty, the member countries[25] were mainly focused on: security issues (in the Treaty, the parties secured their renunciation of claims to territorial sovereignty (Article IV)); the prohibition of any measures of a military nature, except for the use of military personnel or equipment for scientific research or for any other peaceful purpose (Article I); and the freedom of scientific investigation (Article II). But after the 1998 Madrid Protocol entered into force, some scholars believe that the rules governing the legal regime of Antarctica have essentially acquired an 'ecological vector'.[26]

The analysis of the above rules allows to assert that international law does not regulate the industrial development of mineral resources in Antarctica. In accordance with Article 7 of the 1991 Protocol, 'any activity relating to mineral resources, except for scientific research, shall be prohibited.' However, scientific research in this field are carried out on regular basis, despite the fact that Antarctica is not the best terrain for the placement of drilling rigs. For the purpose of exploration and development of possible mineral deposits, new techniques are being implemented, which will entail the development of

25 The Parties to the Treaty are 50 countries constituting about two thirds of the world population, and 28 countries have become Antarctic Treaty Consultative Parties (ATCP).

26 At the 40th Consultative Meeting, the participants agreed and adopted documents and decisions on the practice of "ecological expeditions" in Antarctica, which implies that research activities in Antarctica is based on the principles of friendly and careful attitude to the environment, i.e. scientists and explorers will strive to minimize the negative impact of their activities on the polar environment.

large deposits in previously undeveloped regions, such as exemplified by the ice Lake Vostok.[27]

Sooner or later, the question of industrial extraction of Antarctic natural resources will be raised, which, given the 'frozen' territorial claims of countries, would require the adoption of a special legal regime for continental Antarctic shelf as an alternative to the continental shelf regime in the sense of the 1982 UNCLOS. Thus, we may state that the established legal regime of the use of Antarctic areas does not solve long-arisen problems of activity on the continent. Recent advances in science and technology, and the overall development of productive forces have made Antarctica more accessible, and the resources of the continent arouse high interest in the region.

The situation in Antarctica is complicated by the fact that the mainland part is covered by an insuperable ice sheet. Consequently, the development of offshore fields in Antarctica will be obstructed by icebergs and the work would take place at a depth of more than 500 meters.[28] The practicability of safe transportation of oil from Antarctica via pipelines is also questionable. Norwegian companies have begun to implement underwater mining complexes in the Arctic. Antarctica in turn has its climatic features, plus it is at a considerable distance from the states interested in mineral resources. According to experts of the Research Institute of Oceanology, currently there are no such technologies that can be effectively used in Antarctica.[29] It is believed that the extraction of Antarctica oil and gas is mostly hampered by super-harsh climatic conditions on the ice continent, which also affects the profitability of the company.

International legal regulation of the Antarctic mineral resources regime was supposed to be governed by the 1988 Convention on the Regulation of Mineral Resources of Antarctica, which was to establish that the search, exploration, and development of Antarctic mineral resources should be exclusively in accordance with the said Convention. The main idea and purpose of the Convention is that the development of resources should not cause any harm to the natural environment. It is controlled by the establishment of conditions and procedures for the development and production of mineral resources.

According to the 1988 Convention, the limit of its distribution is the entire continent of Antarctica and all Antarctic islands, including all ice shelves,

---

27 In Antarctica there are 89 scientific polar stations and bases of various countries, which conduct seasonal (summer) and year-round scientific (including biological, geographical, geological and meteorological) research on the continent and its coastal area.

28 There is oil in Antarctica, but it is impracticable to extract it. Alexander Danilov, Deputy Director (Research), Arctic and Antarctic Research Institute (AANIA), 24.01.2012. Source: RIA Novosti https://ria.ru/eco/20120124/547701814.html.

29 Ibid.

south of 60 degrees south latitude and the seabed and subsoil of adjacent offshore areas up to the deep seabed (up to the limits of the continental shelf).

In view of the special importance of Antarctica as a natural reserve for the development of science, the adoption of the 1988 Convention caused a negative reaction by the international community,[30] which expressed concerns about the possibility of industrial development of Antarctic mineral resources and the related serious threats this may cause to the environment.[31]

4 Environmental Protection of Antarctic Seabed Resources

In order to rectify the situation, as well as to improve the protection of Antarctic ecology and its dependent and related ecosystems, and to increase the guaranteed nature of the use of Antarctic region exclusively for peaceful purposes and to improve the ATS, a special Antarctic Treaty Consultative Meeting at the XI session in 1991 adopted The Protocol on Environmental Protection[32] to the Antarctic Treaty.

It should be noted that the Protocol does not change and does not amend the 1959 Treaty, but only complements it (Section 4, Art. 4).

The protection of the Antarctic environment and its dependent and associated ecosystems should be one of the main factors in the planning and implementation of any activity in the Antarctic Treaty area (Article 3 of the 1998 Protocol).

To achieve these goals, special protective measures should be planned and developed in Antarctica to avoid anthropogenic environmental impact, including through the prohibition of any activity on the industrial development of mineral resources. An exception is provided for scientific research (Article 7).

The protocol prioritized the provision that 'The protection of the Antarctic environment and dependent and associated ecosystems shall be fundamental considerations in the planning and conduct of all activities in the Antarctic Treaty area'. In order to achieve these objectives, special activities should be planned and developed in Antarctica in order to avoid human impact on the environment, in particular by prohibiting any activity for industrial

30 This position was outlined in a resolution adopted by the UN General Assembly at its 43th session in 1988, which expressed "deep regret" in connection with the adoption of the 1988 Convention.

31 Modern International Law of the Sea and Practice of its Application. Monograph. – M.: Nauchnaya Kniga. 2003. – 236 pp.

32 So-called Madrid Protocol.

development of mineral resources, with exception provided for scientific research.[33]

5 Concluding Remarks

Today, the issues of legal regulation of international relations, in particular those directly related to the use of natural resources, take special importance. The solution to these problems is in fact closely linked to the crucial issue of our era, i.e. the struggle for peace. It is in connection with the use of resources that the interests of different countries clash most acutely. The way these interests are reconciled depends largely on whether international cooperation will be established or strengthened; if not, an environment fraught with dangerous aggravation of relations will grow.[34]

A possible solution to the problem of delimitation of maritime areas and therefore of possible claims by different countries for marine resources including the seabed may be the adoption of a special legal regime for the continental shelf of Antarctica and conclusion of bilateral agreements on the delimitation of maritime areas.

It is also likely that Antarctic mineral resources will be considered as a very remote reserve for future generations for a long time. In the near future, there is no reason to expect that the priority of scientific geological and geophysical research will be undermined by unilateral infringement or early lifting of the moratorium on geological prospecting and mining work. Nevertheless, the attention given to the potential resources of Antarctica under the influence of the oil crisis of the early 1970s, leading to the forced elaboration of the International Mineral Exploration Convention, frozen in 1991, indicates the severity of the problem, which for the duration of the moratorium may become more geopolitical than economic.[35]

Most likely, countries with territorial claims in Antarctica, industrialized countries that do not have their own resource base (Japan, Germany) or

33 It is well known that at fluctuations in the number of populations or in biodiversity of communities of living organisms, a negative role can be played not only by man-made impacts, but also by problems of climate and food potential changes, viral diseases of organisms caused by impacts of transboundary transfers in the ocean and atmosphere, or large-scale natural disasters (volcanism, earthquakes, collapse of icebergs, tsunamis etc.).

34 V.A. Avkhadeev Issues of Current Territorial Claims in Antarctica … "Law of the Sea" web magazine, 2008.

35 Mineral Resources of the Arctic: Geological background and development prospects. G.E. Grikurov, G.L. Leichenkov, E.V. Michalski, A.V. Golynsky, V.N. Masolov. // Exploration and protection of subsoil. 2000. No. 12.

whose strategic line is for the conservation of national resources (USA), will be enhancing the scope of research work aimed at revealing the prospects of the region's mineral resources and strengthening their presence on the continent through research, and establishment of stations and bases.

PART 3

Exploitation of Non-Living Resources from the Seabed beyond National Jurisdiction

CHAPTER 7

Commercial Mining Activities in the Deep Seabed beyond National Jurisdiction: the International Legal Framework

Joanna Dingwall

1 Introduction

The deep seabed beyond national jurisdiction covers approximately half of our planet, yet it is the most inaccessible and least explored area on earth.[1] It is home to a wealth of mineral resources, including a variety of valuable metals and rare earth elements.[2] Deep seabed minerals await discovery on volcanic ridges, rocky outposts and amongst the sediment of the ocean floor, typically at depths of up to around 5,000 metres.[3]

The United Nations Convention on the Law of the Sea (UNCLOS or the Convention) established the regime which governs mining in the deep seabed beyond national jurisdiction (referred to in this Chapter as the 'deep seabed' or the 'Area').[4] The Convention defines the Area as comprising 'the seabed and ocean floor and subsoil thereof, beyond the limits of national jurisdiction.'[5] At present, as many coastal States have still to determine the outer boundaries of

1 The author's analysis of the status of the deep seabed mining regime, and the extent of mining activities within it, is current as at 1 November 2018. All web links cited in this Chapter are correct as at that date. The views expressed herein are those of the author alone and do not necessarily represent the views of the Scottish Government or the University of Glasgow. Elements of this Chapter are also addressed by this author in Joanna Dingwall, 'The International Legal Regime Applicable to the Mineral Resources of the Deep Seabed' (2018) 9 Eur YB of Intl Economic L 261–287.

2 International Seabed Authority 'Marine Mineral Resources' (2003) <www.isa.org.jm/files/documents/EN/Brochures/ENG6.pdf>; Jim Hein 'ISA Briefing Paper 02/12 – Prospects for Rare Earth Elements from Marine Minerals' (May 2012) <www.isa.org.jm/files/documents/EN/Pubs/BP2.pdf>.

3 ISA, 'Marine Mineral Resources' (n2).

4 United Nations Convention on the Law of the Sea (adopted and opened for signature 10 December 1982, entered into force 16 November 1994) 1833 UNTS 3, Part XI and Annex III.

5 UNCLOS, art 1(1).

their continental shelves, the precise boundaries of the Area remain 'fluid.'[6] However, to put the size of the Area into perspective, it constitutes approximately 260 million square kilometres, which is around 72% of the total surface area of the oceans.[7]

UNCLOS created the International Seabed Authority (ISA or the Authority), which is the autonomous international organisation charged with overseeing and administering a system for deep seabed mining.[8] The ISA is responsible for regulating the Area and granting contracts (also referred to as licences) to explore for and exploit deep seabed mineral resources.[9] The regime incorporates various key elements, including benefit-sharing aspects and marine environmental protections.[10] Moreover, the ISA is empowered to develop a comprehensive Mining Code, intended to augment the regime and govern the entire lifecycle of deep seabed mining operations.[11]

In relation to deep seabed mining in the Area, the Convention defines 'resources' as 'all solid, liquid or gaseous mineral resources in situ in the Area at or beneath the sea-bed, including polymetallic nodules.'[12] Once extracted from the Area, such resources are referred to as 'minerals.'[13] The Area and its resources constitute 'the common heritage of mankind.'[14] As the UNCLOS

6 Michael W Lodge, ISA Secretary-General, 'Statement to the Open Meeting of the Commission on the Limits of the Continental Shelf' (United Nations, New York, 10 March 2017) 4 <www.isa.org.jm/files/documents/EN/SG-Stats/2017/CLCS-10Mar17.pdf>. For analysis of the boundary between national jurisdiction and the Area, see Chapter 4 of this book, H. Brekke 'Setting maritime limits and boundaries: Experiences from Norway.' See also, e.g., Erik Franckx, 'The 200 Mile Limit: Between Creeping Jurisdiction and Creeping Common Heritage?' (2007) 39 George Washington Intl Rev 467; Erik Franckx, 'The International Seabed Authority and the Common Heritage of Mankind: The Need for States to Establish the Outer Limits of their Continental Shelf' (2010) 25 Intl J Marine and Coastal L 543.

7 Michael W Lodge, 'Satya Nandan's Legacy for the Common Heritage of Mankind' in Michael W Lodge and Myron H Nordquist (eds), *Peaceful Order in the World's Oceans* (Brill 2014) 290, fn22. See also BW Eakins and GF Sharman, 'Volumes of the World's Oceans from ETOPO1' (National Oceanic and Atmospheric Administration (NOAA), National Geophysical Data Center 2010) <www.ngdc.noaa.gov/mgg/global/etopo1_ocean_volumes.html> (the surface area of the oceans is 361.9 million square kilometres).

8 UNCLOS, art 156.

9 UNCLOS, arts 153, 157(1) and see also Annex III.

10 See, e.g., UNCLOS, arts 140, 145, 148.

11 For further consideration of the ISA's Mining Code, see Chapter 25 of this book, K. Svendsen, 'Liability and Compensation for Activities in the Area'.

12 UNCLOS, art 133(a).

13 UNCLOS, art 133(b).

14 UNCLOS, art 136.

deep seabed mining regime is not applicable to living resources, controversy remains as to whether marine genetic resources constitute part of the common heritage.[15]

The focus of this Chapter is to provide an introduction to the Convention's deep seabed mining regime and the extent of commercial activities currently occurring within it. In order to conduct this analysis, Section 2 of this Chapter provides a brief snapshot of the regime's historical development. Section 3 then evaluates the current scope of the UNCLOS deep seabed mining regime and the manner in which commercial actors may participate within it. Thereafter, Section 4 provides a flavour of the status of commercial mining activities in the Area so far. Finally, Section 5 provides concluding remarks on the deep seabed mining regime.

2 The Development of the UNCLOS Deep Seabed Mining Regime

The mineral resources of the deep seabed have considerable economic potential. As early as the 1800s, the HMS Challenger expedition had already discovered the existence of polymetallic nodules on the deep ocean floor.[16] However, the economic prospects of deep seabed resources only began to catch the international community's imagination half a century ago. During the 1960s, the prospective riches of the deep seabed began to seem within humanity's grasp.[17] As technology advanced, for the first time the untold wealth of the seabed appeared tantalisingly close.[18]

15 For analysis of this issue, see: Chapter 10 of this book, T. Scovazzi, 'The Rights to Genetic Resources beyond National Jurisdiction: Challenges for the ongoing Negotiations at the United Nations'; Chapter 11 of this book, M.W. Tvedt, 'Marine genetic resources: A Practical Legal Approach to Stimulate Research, Conservation and Benefit Sharing'. See also, e.g., Konrad Jan Marciniak, 'Marine Genetic Resources: Do They Form Part of the Common Heritage of Mankind Principle?' in Lawrence Martin and others (eds), *Natural Resources and the Law of the Sea: Exploration, Allocation, Exploitation of Natural Resources in Areas under National Jurisdiction and Beyond* (Juris Publishing 2017).

16 See Chapter 3 of this book, Håkon With Andersen, 'A Short Human History of the Ocean Floor'. See also John Murray and Alphonse François Renard, 'Report on Deep-Sea Deposits Based on the Specimens Collected During the Voyage of HMS Challenger in the Years 1872 to 1876' (Neill & Co for HM Stationary Office 1891).

17 See, e.g., John L Mero, *The Mineral Resources of the Sea* (Elsevier 1965).

18 See, e.g., UNGA First Committee (22nd Session) 'Speech by Arvid Pardo' (1 November 1967, 3 pm) UN Doc A/C.1/PV.1516, para 9.

Throughout the 1960s and 1970s, against the backdrop of decolonisation, entitlement to deep seabed resources became an ideological battleground, split into broadly two camps.[19] In one camp were the industrialised States, favouring freedom for all States to exploit deep seabed resources on a first-come, first-served basis. In the other camp were the developing States, fighting for a contrary system whereby deep seabed resources could be exploited only for the benefit of humanity.[20] The latter camp were motivated by the realisation that without careful regulation, to allow freedom for all to exploit deep seabed resources would not be as egalitarian as it appeared.[21] A free-for-all system would in fact preserve deep seabed mining as the domain of the small number of wealthy States possessing the relevant technical expertise, who would be in prime position to carve up the spoils between them.[22]

In a remarkable feat, after decades of wrangling, the developing States achieved their vision in many key respects. The UNCLOS deep seabed regime, which was finally agreed in 1982, is a unique scheme for common resource management. It is underpinned by the principle that the deep seabed constitutes 'the common heritage of mankind' and cannot be alienated unilaterally.[23] This application of the common heritage concept to the deep seabed was a notion first formally introduced before the UN General Assembly by Malta's Ambassador, Arvid Pardo on 1 November 1967.[24] In terms of the content of the common heritage concept, as commentary explains:

19 For further elaboration on the regime's historical development, see, e.g., Martti Koskenniemi and Marja Lehto, 'The Privilege of Universality: International Law, Economic Ideology and Seabed Resources' (1996) 65 Nordic J Intl Law 533, 536–552; RR Churchill and AV Lowe, *The Law of the Sea* (3rd edn, Manchester UP 1999) 224–229; ED Brown, *Sea-Bed Energy and Minerals: The International Legal Regime*, vol 2 (Martin Nijhoff Publishers 2001), Ch 2; Erkki Holmila, 'Common Heritage of Mankind in the Law of the Sea' (2005) 1 Acta Societatis Martensis 187; John E Noyes, 'The Common Heritage of Mankind: Past, Present, and Future' (2012) 40 Denver J Intl L & Policy 447, 459–460; Helmut Tuerk, 'The International Seabed Area' in Malgosia Fitzmaurice and Norman A Martinez Gutierrez (eds), *The IMLI Manual on International Maritime Law*, vol 1 (OUP 2014) 280–282.

20 See, e.g., UNGA Res 2574D (XXIV) (15 December 1969), which called for a moratorium on deep seabed mining in the Area, pending establishment of an international legal regime.

21 Rosalyn Higgins, *Problems & Process – International Law and How We Use It* (Clarendon Press 1994) 130–131.

22 Higgins (n21) 130–131; and Koskenniemi/Lehto 540; Churchill/Lowe 225; Noyes (2012) 459–460 (all n19).

23 UNCLOS, arts 136 and 137, respectively.

24 UNGA First Committee (22nd Session) 'Speech by Arvid Pardo' (1 November 1967, 10.30 am) UN Doc A/C.1/PV.1515. See also UNGA Res 2749 (XXV) (17 December 1970) (the Declaration of Principles Governing the Sea-Bed and the Ocean Floor, and the Subsoil

> The common heritage concept implied that [the deep seabed] was open to use by the international community, but was not owned by it. It required a system of management in which all users had a right to participate as well as an active sharing of benefits and reservation for future generations, and thus also had environmental implications.[25]

As this Chapter addresses further below, all of these elements are captured within the UNCLOS deep seabed mining regime.

Part XI of UNCLOS contains the legal regime applicable to deep seabed mining, and this is elaborated in Annex III, which details the licensing application process and conditions. However, industrialised States viewed Part XI as having swung too far in favour of developing States.[26] The regime's particularly contentious features included requirements for mandatory transfer of technology,[27] and the expansive role envisaged for the ISA's intended mining arm, the Enterprise, together with the subsidisation of the Enterprise by States

---

Thereof, Beyond the Limits of National Jurisdiction) (which captured Pardo's common heritage vision for deep seabed mining). For discussion of precursors to Pardo's speech by nineteenth century jurists, see Myron H Nordquist and others (eds), *United Nations Convention on the Law of the Sea 1982: A Commentary*, vol VI (Martinus Nijhoff Publishers 2002) 6–7; Michael W Lodge, 'International Seabed Authority's Regulations on Prospecting and Exploration for Polymetallic Nodules in the Area' (2002) 20 J Energy and Natural Resources L 270, 271, fn5; Kemal Baslar, *The Concept of the Common Heritage of Mankind in International Law* (Martin Nijhoff Publishers 1998) 80–81.

25 Tuerk (n19) 280. See also LFE Goldie, 'A Note on Some Diverse Meanings of "The Common Heritage of Mankind"' (1983) 10 Syracuse J Intl L and Commerce 69; Michael W Lodge, 'The Common Heritage of Mankind' (2012) 27 Intl J Marine and Coastal L 733. On the concept of the common heritage more broadly, including in relation to the seabed, the moon and Antarctica see, e.g., see Rüdiger Wolfrum, 'The Principle of the Common Heritage of Mankind' (1983) 43 Heidelberg J Intl L 312; Christopher C Joyner, 'Legal Implications of the Concept of the Common Heritage of Mankind' (1986) 35 ICLQ 190; Barbara Ellen Heim, 'Exploring the Last Frontiers for Mineral Resources: A Comparison of International Law Regarding the Deep Seabed, Outer Space, and Antarctica' (1990–91) 23 Vanderbilt J Transnational L 819; Baslar (n24); Graham Nicholson, 'The Common Heritage of Mankind and Mining: An Analysis of the Law as to the High Seas, Outer Space, the Antarctic and World Heritage' (2002) 6 New Zealand J Intl L 177; Jennifer Frakes, 'The Common Heritage of Mankind Principle and the Deep Seabed, Outer Space, and Antarctica: Will Developed and Developing Nations Reach a Compromise?' (2003) 21 Wisconsin Intl L J 409; Edward Guntrip, 'The Common Heritage of Mankind: An Adequate Regime for Managing the Deep Seabed?' (2003) 4 Melbourne J Intl L 376; Rüdiger Wolfrum, 'Common Heritage of Mankind' in R Wolfrum (ed), *The Max Planck Encyclopedia of Public International Law* (OUP 2009); Noyes (2012) (n19) 447.

26 Churchill/Lowe 231; Tuerk 282 (both n19).

27 As originally detailed in UNCLOS, Annex III, art 5.

Parties.[28] Therefore, for over a decade, there was a stalemate. Many industrialised States, including the United States (US), the United Kingdom (UK), Germany, France, Italy, Japan and Russia, refused to ratify UNCLOS, thereby impeding the Convention's entry into force.[29]

Eventually, in 1994, the international community reached a compromise. This took the form of the Agreement on the Implementation of Part XI of UNCLOS (1994 Agreement).[30] Whilst retaining the tenor of the deep seabed regime as envisaged originally by UNCLOS, the 1994 Agreement made significant modifications to its operation, eliminating some of its more controversial aspects.[31] For example, the 1994 Agreement removed from the regime both the subsidisation of the Enterprise and the mandatory transfer of technology requirements.[32] In light of the 1994 Agreement's modification of the deep seabed mining regime, the Convention entered into force on 16 November 1994.

Presently, the UNCLOS deep seabed mining regime (as modified by the 1994 Agreement) enjoys broad support throughout the international community. The overwhelming majority of States are parties to UNCLOS (with perhaps the most notable exception being the US). Currently, UNCLOS has 168

---

28 As mandated previously by art 170(4) of UNCLOS, read in conjunction with arts 171 and 173(2)(b).

29 In numerical terms there were sufficient developing States to bring the Convention into force (only sixty State ratifications were necessary to achieve this, as per UNCLOS, art 308). However, in practice, the Convention system could not operate successfully without political and financial support from industrialised States. See Koskenniemi/Lehto (n19) 534–535, 542–544; Churchill/Lowe (n19) 230–231; Lodge (2002) (n24) 272.

30 Agreement Relating to the Implementation of Part XI of the United Nations Convention on the Law of the Sea of 10 December 1982 (adopted 28 July 1994, entered into force provisionally 16 November 1994 and definitively 28 July 1996) 1836 UNTS 3.

31 On the fundamental changes which the 1994 Agreement introduced, see, e.g., DH Anderson, 'Further Efforts to Ensure Universal Participation in the United Nations Convention on the Law of the Sea' (1994) 43 ICLQ 886; Bernard H Oxman, 'The 1994 Agreement and the Convention' (1994) 88 AJIL 687; Louis B Sohn, 'International Law Implications of the 1994 Agreement' (1994) 88 AJIL 696; ED Brown, 'The 1994 Agreement on the Implementation of Part XI of the UN Convention on the Law of the Sea: Breakthrough to Universality?' (1995) 19 Marine Policy 5; LDM Nelson, 'The New Deep Sea-Bed Mining Regime' (1995) 10 Intl J Marine and Coastal L 189; Koskenniemi/Lehto (n19) 535, 549–551; RP Anand, 'Common Heritage of Mankind: Mutilation of an Idea' (1997) 37 Indian J Intl L 1; Churchill/Lowe (n19) 238, 248–251; Nordquist (n24) 4, 67; Lodge (2002) (n24) 272.

32 See 1994 Agreement, Annex ss2(3) and 5, respectively. In addition, the role of the Enterprise was curtailed significantly; 1994 Agreement, Annex s2. For discussion on all these points, see Nordquist (n24) 4.

States Parties, including the European Union (EU).[33] In addition, the ISA also has some 30 States participating in its activities as observers (including the US), together with various observer international organisations and NGOs.[34] Moreover, many aspects of UNCLOS are recognised as having the status of customary law, binding on all States irrespective of whether or not they are parties to the Convention.[35] Against this backdrop, Section 3 of this Chapter will now address key aspects of the Convention's deep seabed mining regime.

3 The UNCLOS Deep Seabed Mining Regime

The UNCLOS deep seabed mining regime is contained within Part XI of the Convention and elaborated upon in Annex III, as modified by the 1994 Agreement. The provisions of Part XI of UNCLOS and the 1994 Agreement are to be interpreted and applied together as a single instrument; but, in the event of inconsistency between them, the provisions of the latter prevail.[36] In addition, the ISA is augmenting the regime through adoption of rules to regulate deep seabed mining.

The following analysis addresses the key features of the regime (3.1), the means by which commercial operators can participate within the licensing process (3.2) and the position of actors outside of the regime (3.3).

3.1 *Key Elements of the UNCLOS Deep Seabed Mining Regime*

The premise of the UNCLOS deep seabed mining regime is that the deep seabed and its resources are the 'common heritage of mankind'[37] and mining activities

33 The 1994 Agreement has 150 States Parties (all of which are also parties to UNCLOS itself). See United Nations Division for Ocean Affairs and the Law of the Sea, 'Chronological List of Ratifications of, Accessions and Successions to the Convention and the Related Agreements' (3 April 2018) <www.un.org/Depts/los/reference_files/chronological_lists_of_ratifications.htm>.

34 UNCLOS, art 156(3); see also ISA, 'Observers' <www.isa.org.jm/observers> for the current list of ISA observers. See James Harrison, *Making the Law of the Sea* (CUP 2011) 128 on the role of these observers.

35 Churchill/Lowe (n19) 24; Alan Boyle and Christine Chinkin, 'UNCLOS III and the Process of International Law-Making' in Tafsir Malick Ndiaye and Rüdiger Wolfrum (eds), *Law of the Sea, Environmental Law and Settlement of Disputes* (Martinus Nijhoff Publishers 2007) 376, 380; Harrison (n34) 52–56; John E Noyes, 'The Law of the Sea Convention and the United States of America' (2014) 47 Revue Belge de Droit Intl 15, 32.

36 1994 Agreement, art 2(1).

37 UNCLOS, art 136.

must be carried out 'for the benefit of mankind as a whole.'[38] Accordingly, the deep seabed mining system must take the interests of developing States into account and promote their effective participation.[39] Moreover, the ISA is charged with developing mechanisms to guarantee equitable sharing of financial and other economic benefits derived from the Area.[40] Another key element of the regime is the protection of the marine environment.[41] Fundamentally, pursuant to the regime, States and other actors are prohibited from conducting unilateral deep seabed mining activities. No State or entity can unilaterally claim sovereignty over the deep seabed or its resources.[42] Rather, all exploration and exploitation activities in the Area are subject to the permission and oversight of the ISA.[43]

As an international organisation, the ISA has international legal personality, entitling it to the full range of international rights and duties of an international person.[44] The two principal organs which establish the ISA's policies and govern its work are the Assembly, in which all States Parties are represented; and the 36-member Council elected by the Assembly, which functions as the ISA's executive organ.[45] The Secretariat is the ISA's third principal organ and it conducts the ISA's administration.[46] The ISA is also assisted by its Legal and Technical Commission (LTC), which is an organ of the Council,[47]

38 UNCLOS, art 140(1).

39 UNCLOS, arts 148, 152(2).

40 UNCLOS, arts 140(2), 160(2)(f)(i), 160(2)(g) and 162(0)(i).

41 See, e.g., UNCLOS, art 145. For analysis of issues concerning environmental protection in deep seabed mining, see, e.g., Kristina M Gjerde, 'Challenges to Protecting the Marine Environment beyond National Jurisdiction' (2012) 27 Intl J Marine and Coastal L 839; Rüdiger Wolfrum, 'The Contribution of the Regulations of the International Seabed Authority to the Progressive Development of International Environmental Law' in Michael W Lodge and Myron H Nordquist (eds), *Peaceful Order in the World's Oceans* (Brill 2014); Tullio Scovazzi, 'The Exploitation of Resources of the Deep Seabed and the Protection of the Environment' (2014) 57 German YB Intl L 181; Aline L Jaeckel, *The International Seabed Authority and the Precautionary Principle – Balancing Deep Seabed Mineral Mining and Marine Environmental Protection* (Brill 2017) 121–131.

42 UNCLOS, art 137.

43 UNCLOS, art 137(2); see also arts 153 and 157(1) and 1994 Agreement, Annex, s1(1).

44 UNCLOS, art 176. See also *Reparation for Injuries Suffered in the Service of the United Nations Case* (Advisory Opinion) [1949] ICJ Rep 174, 179; Markos Karavias, *Corporate Obligations Under International Law* (OUP 2013) 121.

45 UNCLOS, arts 158(1), 159–162.

46 UNCLOS, arts 158(1), 166.

47 UNCLOS, arts 163(1)(b), 165. The Convention provided also for the Council to be supported by an Economic Planning Commission, but that body is not operational; see UNCLOS, arts 163(1)(a), 164; cf. 1994 Agreement, Annex, s1(4).

and the Finance Committee, a subsidiary organ of the Assembly, composed of representatives of the Council.[48] The Authority's mining arm, the Enterprise, is not yet operational.[49]

The ISA is a unique and unusual body with far-reaching institutional powers of a kind which are arguably unparalleled within the international legal system.[50] It plays the crucial role of custodian of the deep seabed, a role which is enshrined in both UNCLOS and the 1994 Agreement.[51] The ISA's mandate is expressed most clearly in Article 157(1) of UNCLOS, which provides that '[t]he Authority is the Organization through which States Parties shall, in accordance with [Part XI], organize and control activities in the Area, particularly with a view to administering the resources of the Area.' The ISA is entrusted with the role of organising, implementing and controlling activities in the Area 'on behalf of mankind as a whole.'[52] To that end, the ISA has the power to adopt rules and regulations to govern deep seabed mining activities, with the aim of developing a comprehensive Mining Code to regulate exploration and exploitation for all forms of resources in the Area.[53]

The Mining Code adds another layer to the Convention's byzantine deep seabed mining regime and constitutes secondary law by the ISA.[54] The ISA is still in the process of creating a bespoke Mining Code to govern the entire lifespan of deep seabed mining operations. Thus far, the ISA has implemented three sets of regulations, which concern prospecting and exploration activities for three types of minerals (namely, polymetallic nodules, polymetallic

48 1994 Agreement, Annex, s9; see also s3(4), (7).

49 UNCLOS, art 170; 1994 Agreement, Annex, s2.

50 See, e.g., Rüdiger Wolfrum, 'Legitimacy of International Law and the Exercise of Administrative Functions: The Example of the International Seabed Authority, the International Maritime Organization (IMO) and International Fisheries Organizations' in Armin von Bogdandy and others (eds), *The Exercise of Public Authority by International Institutions – Advancing International Institutional Law* (Springer 2010) 917 (citing the ISA is a 'rare' example of an international entity 'exercising functions equivalent to those of States' on the basis that it exercises executive and legislative functions and has jurisdiction to enforce its rules directly via a binding dispute resolution system. As such, the ISA is 'without question, a prime example of what may be referred to as an international administration' (at 934)). See also Jaeckel (n41) 146–148.

51 UNCLOS, art 137(2); see also arts 153 and 157(1) and 1994 Agreement, Annex, s1(1).

52 UNCLOS, art 153(1); see also art 137(2).

53 UNCLOS, arts 140(2), 145, 160(2)(f), 162(o), 165(2)(f), and Annex III, art 17; 1994 Agreement, Annex, s1(1), (5)(f)–(g), (15). For analysis of the ISA's broad powers to progressively develop the deep seabed mining regime, see Harrison (n34) 122–123, 152.

54 Karavias (n44) 121–122.

sulphides and cobalt-rich ferromanganese crusts),[55] together with environmental recommendations.[56] Moreover, the ISA is working at present to develop regulations for the exploitation of all resource types within the Area, through an extensive stakeholder process.[57]

3.2 *Participating within the UNCLOS Licensing Regime for Deep Seabed Mining*

Under the terms of the ISA's licensing regime, only certain actors are entitled to apply to conduct deep seabed mining operations in the Area.[58] The Convention provides that such activities in the Area may be carried out by States Parties to the Convention or by 'state entities or natural or juridical persons which possess the nationality of States Parties or are effectively controlled by them or their nationals, when sponsored by such States.'[59] The ISA's exploration regulations echo this requirement of nationality or control.[60] Therefore, non-State actors, such as private corporations, seeking to participate within the UNCLOS regime must obtain sponsorship from all States of which they are nationals, and '[i]f another State or its nationals exercises effective control, the sponsorship of that State is also necessary.'[61] Only upon receiving appropriate

---

55 ISA, Regulations on Prospecting and Exploration for Polymetallic Nodules in the Area (adopted 13 July 2000) ISBA/6/A/18, (updated 25 July 2013) ISBA/19/C/17 (Nodules Regulations); ISA, Regulations on Prospecting and Exploration for Polymetallic Sulphides in the Area (adopted 7 May 2010) ISBA/16/A/12/Rev.1 (Sulphides Regulations); ISA, Regulations on Prospecting and Exploration for Cobalt-Rich Ferromanganese Crusts in the Area (adopted 27 July 2012) ISBA/18/A/11 (Cobalt Regulations).

56 See, e.g., ISA LTC, 'Recommendations for the Guidance of Contractors for the Assessment of the Possible Environmental Impacts arising from Exploration for Marine Minerals in the Area' (1 March 2013) ISBA/19/LTC/8.

57 For an overview of this process, see ISA, 'Ongoing Development of Regulations on Exploitation of Mineral Resources in the Area' <www.isa.org.jm/legal-instruments/ongoing-development-regulations-exploitation-mineral-resources-area>. The ISA's current draft exploitation regulations are ISA LTC, 'Draft Regulations on Exploitation of Mineral Resources in the Area' (9 July 2018) ISBA/24/LTC/WP.1/Rev.1.

58 For more detailed analysis by this author of the operation of the UNCLOS deep seabed mining regime, see Dingwall (n1).

59 UNCLOS, art 153(2)(b).

60 Nodules Regulations, reg 9(b); Sulphides Regulations, reg 9(b); Cobalt Regulations, reg 9(b).

61 *Responsibility and Obligations of States Sponsoring Persons and Entities with Respect to Activities in the Area* (Advisory Opinion of 1 February 2011) ITLOS Reports 2011, 10 (ITLOS Advisory Opinion) para 77. This is confirmed also in Regulation 11(1)–(2) of the Nodules, Sulphides and Cobalt Regulations. Regulation 11(1) specifies that '[i]f the applicant has more than one nationality (…) each State involved shall issue a certificate of sponsorship.'

sponsorship from a States Party will a non-State actor be entitled to submit an application to the ISA to conduct exploration or exploitation activities.[62]

Indeed, as the Seabed Disputes Chamber (Chamber) of the International Tribunal for the Law of the Sea (ITLOS) has emphasised, the 'notion of "sponsorship" is a key element in the system for the exploration and exploitation of the resources in the Area.'[63] One purpose behind the sponsorship requirement is 'to achieve the result that the obligations set out in the Convention, a treaty under international law which binds only States Parties thereto, are complied with by entities that are subjects of domestic legal systems.'[64] Moreover, by entering into a contract with the ISA, a non-State contractor becomes directly bound to adhere to various international legal obligations concerning deep seabed mining.[65]

To similar effect, Regulation 11(2) requires that '[w]here the applicant has the nationality of one State but is effectively controlled by another State or its nationals, each State involved shall issue a certificate of sponsorship.'

62 The sponsorship scheme does not, however, apply to applications by States Parties, who, by virtue of their sovereign status, are not required to demonstrate sponsorship. See UNCLOS, Annex III, art 4(5): this interpretation is confirmed by ITLOS Advisory Opinion (n61) para 79.

63 ITLOS Advisory Opinion (n61) para 74.

64 ITLOS Advisory Opinion (n61) para 75; as the Chamber acknowledged, this end is also achieved by virtue of the ISA's regulations, which apply to all contractors. Notably, States Parties are responsible for ensuring that any activities that they (or State-sponsored entities) conduct in the Area are carried out in conformity with the Convention regime; UNCLOS, art 139. Any States Party failing to discharge this responsibility will be held responsible at international law. However, as per UNCLOS, art 139(2), a States Party does not incur liability for wrongful activities of its sponsored entity if that States Party has taken 'all necessary and appropriate measures to secure effective compliance' by the sponsored entity; see also arts 153(4) and 138. See further ITLOS Advisory Opinion (n61) para 122 (considering the extent of sponsoring State liability for State-sponsored entities and endorsing a high standard of due diligence, encompassing a legal obligation upon States to apply the precautionary approach and best environmental practices); see also David Freestone, 'Responsibilities and Obligations of States Sponsoring Persons and Entities With Respect To Activities in the Area' (2011) 105 AJIL 755.

65 Karavias (n44) 124 ('[t]he Standard Clauses [to the ISA contract], in turn, transpose the content of the [UNCLOS] provisions regarding activities in the Area and of the [ISA] Regulations into the contractual arrangement, thus providing a nexus between [UNCLOS] and the secondary law enacted by the ISA and the contract for exploration.').

States or State-sponsored entities seeking to explore for and exploit deep seabed resources must submit a plan of work for the ISA's approval. Once approved by the ISA and signed, the plan of work takes the form of a contract between the ISA and the contractor.[66]

The UNCLOS regime was intended to function as a parallel system: States and State-sponsored entities would be permitted to conduct deep seabed mining activities in tandem with the ISA's mining arm, the Enterprise.[67] To that end, it institutes a site-banking system, whereby a contractor's plan of work must specify a total area for exploration or exploitation which is 'sufficiently large and of sufficient estimated commercial value to allow two mining operations' and 'indicate the coordinates dividing the area into two parts of equal estimated commercial value.'[68] On that basis, the ISA designates one part of the submitted area as a 'reserved area.'[69] Before the 1994 Agreement's modifications, the UNCLOS regime envisaged that mining activities in reserved areas would be conducted 'solely' by 'the Authority through the Enterprise or in association with developing States.'[70] However, the 1994 Agreement suspended the role of the Enterprise for the time being.[71]

Under the present regime, the State or entity which contributes a particular area to the ISA as a reserved area now has the right of first refusal to enter into a joint venture agreement with the Enterprise for exploration and exploitation of that reserved area.[72] In the event that that right of first refusal is not exercised, it is open for developing States and entities sponsored by them, including private actors, to apply to explore and exploit the reserved area.[73] In relation to sulphides and cobalt crusts exploration, the ISA modified the

66 UNCLOS, art 153(3) and Annex III, art 3(5); see also 1994 Agreement, Annex, s1(6)(a)(i).

67 UNCLOS, art 153(2).

68 UNCLOS, Annex III, art 8.

69 See UNCLOS, Annex III, art 8 and 1994 Agreement, Annex, s1(10).

70 UNCLOS, Annex III, art 8; see further art 9. See also UNCLOS, Annex IV.

71 1994 Agreement, Annex, s2. In the interim period, the ISA Secretariat performs the Enterprise's functions. The Secretariat is presently conducting a study into the Enterprise's operationalisation; the terms of reference for this study are contained in ISA Secretariat, 'Note on Issues Relating to the Operation of the Enterprise, in Particular, the Legal, Technical and Financial Implications for the Authority and for State Parties' (12 June 2014) ISBA/20/LTC/12, Annex. The Secretary-General anticipates that the Council will consider a full proposal for operationalisation of the Enterprise during 2019; IISD Reporting Services, 'Summary of the Twenty-Fourth Annual Session of the International Seabed Authority (Second Part): 16–26 July 2018' (29 July 2018) 25:168 Earth Negotiations Bulletin 10 <http://enb.iisd.org/download/pdf/enb25168e.pdf>.

72 1994 Agreement, Annex, s2(5).

73 UNCLOS, Annex III, art 9(4).

site-banking system further, permitting an applicant seeking permission to explore for sulphides or cobalt crusts to offer an equity interest in a joint venture arrangement to the Enterprise instead of submitting a reserved area.[74] At present, the LTC is considering whether to amend the Nodules Regulations to align them with the approach to sulphides and cobalt crusts in this regard.[75]

Once exploitation commences, the ISA will oversee redistribution of financial and other economic benefits derived from the resources of the Area. These benefits are to be shared equitably, on a non-discriminatory basis, through a mechanism which the ISA has still to devise.[76] Currently, deep seabed miners contemplating exploitation activities face continued levels of uncertainty, given that levels of fees and royalties and precise terms of exploitation contracts are still to be determined. However, despite this, the ISA's stakeholder consultation process provides a unique opportunity to harness input from a wide range of actors, including commercial operators and environmental experts.[77] This process could facilitate achievement of an exploitation regime which allows commercially viable mining activities while ensuring appropriate safeguards for the environment and the common heritage principles at the heart of the regime.

Indeed, as Section 4 of this Chapter will demonstrate, notwithstanding the developing nature of the existing regime, participation in the Area is increasing. Various States, State actors and private entities are positioning themselves to take advantage of the perceived opportunities afforded by deep seabed resources. However, what is the position of actors which are positioned outside of the UNCLOS regime, such as States which are not parties to the Convention, or their nationals? This following Section will address this point.

3.3 *The Position of Actors outside of the UNCLOS Deep Seabed Mining Regime*

The result of the Convention's system for participation is that non-States Parties to UNCLOS (NSPs) are not eligible to participate within the regime. Equally, non-State actors which have the sole nationality of a NSP cannot

74 Sulphides Regulations, regs 16, 19; Cobalt Regulations, regs 16, 19.

75 ISA LTC, 'Report of the Chair of the Legal and Technical Commission on the Work of the Commission at the First Part of its Twenty-Fourth Session' (26 April 2018) ISBA/24/C/9 para 18; see also ISA Secretariat, 'Issues Related to the Possible Alignment of the Authority's Regulations on Prospecting and Exploration Concerning the Offer of an Equity Interest in a Joint Venture Arrangement' (6 February 2018) ISBA/24/LTC/4.

76 UNCLOS, art 140(2).

77 On this process, see text to n57 above and accompanying citations.

obtain sponsorship and apply to the ISA to explore for or exploit deep seabed resources. Indeed, there is a strong argument that the UNCLOS deep seabed mining regime now constitutes 'the only game in town' and any competing regime would 'surely be condemned as inconsistent with international law.'[78]

Arguably, elements of the Convention's deep seabed mining regime, such as its prohibition on unilateral mining activities in Article 137, may have attained the status of customary law.[79] For example, Koskenniemi and Lehto categorise the UNCLOS deep seabed regime as 'having become part of the normative reality' and maintain that '[u]nilateral mining outside the Convention will be illegal.'[80] Various factors support this conclusion. The UNCLOS deep seabed mining regime is adhered to or recognised by the international community as a whole.[81] This near-universal acceptance of the regime, coupled with the lack of contrary deep seabed mining practice outside of the regime (including by key NSPs),[82] renders it difficult to refute the claim that a State or other actor undertaking unilateral deep seabed mining activities would violate customary law.[83] Egede reaches this conclusion upon an assessment of State practice, concluding that the UNCLOS regime is 'binding on all states (both states parties and non-states parties) and consequently no state, not even non-parties, may unilaterally embark on mining activities in the Area.'[84]

78 Noyes (2012) (n19) 465.

79 Custom is 'evidence of a general practice accepted as law'; Statute for the International Court of Justice (ICJ) (opened for signature 26 June 1945 as annex to UN Charter, entered into force 24 October 1945) UKTS 67 (1946), art 38(1)(b). See also ILC, 'Identification of Customary International Law – Text of the Draft Conclusions as Adopted by the Drafting Committee on Second Reading' (17 May 2018) UN Doc A/CN.4/L.908*, Draft Conclusion 2.

80 Koskenniemi/Lehto (n19) 551–552. Similarly see Jonathan I Charney, 'The United States and the Law of the Sea After UNCLOS III – The Impact of General International Law' (1983) 46(2) L and Contemporary Problems 37, 49; Holmila (n19) 202–205; Edwin Egede, *Africa and the Deep Seabed Regime: Politics and International Law of the Common Heritage of Mankind* (Springer 2011) 66–69; Noyes (2012) (n19) 465; Lodge (2014) (n7) 282–298.

81 As noted at text to n34 above, almost all NSPs to UNCLOS participate within the ISA as observer States. See also Lodge (2012) (n25) 737–738.

82 Cf. *North Sea Continental Shelf Cases* [1969] ICJ Rep 3, paras 72–74 (where the ICJ recognised that treaty provisions may generate new customary law provided that they have a 'fundamentally norm-creating character such as could be regarded as the basis for a general rule of law' (para 72). Moreover, in order for treaty provisions to create new customary rules, 'even without the passage of any considerable period of time, a very widespread and representative participation in the convention might suffice of itself, provided it included that of States whose interests were specially affected' (para 73)).

83 See, e.g., Nelson (n31) 202; Koskenniemi/Lehto (n19) 551–552; Holmila (n19) 205; Egede (n80) 66–69; Noyes (2012) (n19) 465; Lodge (2014) (n7) 282–298.

84 Egede (n80) 69.

Certainly, the only definitive way in which NSPs can render their nationals eligible to undertake mining activities in the Area is through the ratification of UNCLOS. For example, as the US is not a party to UNCLOS, neither the US nor any US nationals are entitled to participate in the ISA's licensing system. Only by ratification of UNCLOS would the US be eligible to apply to the ISA for mining rights or to sponsor US entities who wish to obtain ISA approval to mine. Thus, only by acceding to UNCLOS could the US secure for itself and its nationals legally recognised, internationally enforceable deep seabed mining rights backed by investment protections and binding international dispute resolution options.[85]

Any entity conducting mining operations in the Area without ISA authorisation – and absent the rights of exclusivity and security of tenure that an ISA contract affords[86] – would expose itself to tremendous risk. Given the significant investment required to mount deep seabed mining operations, it does not seem credible that an actor would engage in mining activities without a clear legal basis and enforceable legal title.[87]

This is borne out in practice: NSPs or their nationals are not in fact conducting mining activities in the Area outside of the UNCLOS regime.[88] For example, the US maintains a domestic regime for exploration and exploitation of mineral resources in the Area: the Deep Seabed Hard Mineral Resources Act (US Act), which it enacted in 1980.[89] At present, only two deep seabed exploration licences remain active under the US Act, and both of these are held by US

85 For detailed analysis by this author of the investment protections and dispute resolution options for deep seabed miners within the UNCLOS regime, see Joanna Dingwall, 'International Investment Protection in Deep Seabed Mining Beyond National Jurisdiction' (2018) 19 J World Investment & Trade 890.

86 UNCLOS, art 153(6) and Annex III, arts 3(4)(c), 16; these rights are reflected in Standard Clauses 2.1 and 2.2 of ISA exploration contracts contained within the fourth annex to the Nodules Regulations, Sulphides Regulations and Cobalt Regulations. See also Charney (n80) 50–51; Karavias (n44) 124–125.

87 John Noyes, 'Ocean Resources and US Acceptance of the LOS Convention' (Opinio Juris, 14 June 2012) <http://opiniojuris.org/2012/06/14/ocean-resources-and-u-s-acceptance-of-the-los-convention/>. This sentiment holds equally true for the exploitation phase, once it begins: Brown (2001) (n19) 3 ('the very considerable investment needed to finance such exploitation would become available only if a legal regime could be created under which potential exploiters could acquire secure legal titles').

88 Lodge (2014) (n7) 282–298.

89 Deep Seabed Hard Mineral Resources Act 30 USC §§ 1401–1473 (2002) (USA); see also NOAA, Deep Seabed Mining Regulations for Exploration Licenses 15 Code of Federal Regulations § 970.100 2016.

defence giant, Lockheed Martin.[90] However, as the US has recognised, in order for Lockheed Martin's US licence claims to be afforded 'international recognition' and security of tenure, the US would need to accede to UNCLOS.[91] Indeed, the US Department of State opined that if Lockheed Martin proceeded with exploration activities absent such international recognition, this would violate its licence terms.[92]

In September 2017, the US National Oceanic and Atmospheric Administration (NOAA) addressed the interaction of these US licences with the UNCLOS deep seabed mining regime. Firstly, NOAA acknowledged that the US Act's 'express purpose' is to establish an interim deep seabed mining regime, pending US ratification of UNCLOS.[93] As such, as NOAA explained, a licence under the US Act 'gives the holder the exclusive right to explore a specific area, but only as against other US entities.'[94] On this basis, '[a]ny rights a US company may have domestically are not secured internationally because US companies are not able to go through the internationally recognized process at the [ISA] established for Parties to [UNCLOS].'[95]

In fact, instead of proceeding outside of the UNCLOS system, NSP nationals are structuring their investments in a way that permits participation within UNCLOS. For example, operating within the framework of UNCLOS, the UK has sponsored two applications to the ISA by a British corporation seeking to explore for polymetallic nodules. Both applications were made by UK Seabed Resources Ltd (UKSRL), which was incorporated in May 2012 and is a wholly-owned subsidiary of Lockheed Martin.[96]

---

90 US Department of Commerce, NOAA, 'Deep Seabed Mining – Report to Congress' (December 1995) i <www.gc.noaa.gov/documents/gcil_dsm_1995_report.pdf>.

91 US Department of Commerce, NOAA, 'Extension of Deep Seabed Exploration Licenses: Response to Comments' (30 December 2015) Vol 80, Issue 250 FR 81529, 81530.

92 80 FR 81529 (n91) 81530. To similar effect see also US Department of Commerce, NOAA, 'Deep Seabed Mining: Request for Extension of Exploration Licenses' (29 February 2012) Vol 77, Issue 40 FR 12245, 12246; US Department of Commerce, NOAA, 'Coastal Programs Division' (10 July 2012) Vol 77, Issue 132 FR 40586.

93 US Department of Commerce, NOAA, 'Deep Seabed Mining: Approval of Exploration License Extensions' (7 September 2017) Vol 82, Issue 172 FR 42327, 42328.

94 82 FR 42327 (n93) 42328.

95 82 FR 42327 (n93) 42328.

96 UKSRL is a wholly-owned subsidiary of Lockheed Martin UK Holdings Ltd, which is the UK-based component of Lockheed Martin Corporation. The Lockheed Martin conglomerate is headquartered in Maryland, US.

Provided the test of effective control is met,[97] the deep seabed regime contains no proscriptions to prevent a subsidiary constituted in the territory of a States Party and sponsored by that States Party from applying to engage in activities in the Area in this manner. Indeed, ITLOS has confirmed implicitly the freedom of commercial actors to set up companies in foreign States and acquire the nationality and sponsorship of those States.[98] Although the Chamber cautioned against '[t]he spread of sponsoring States "of convenience"', this was in the context of determining whether developing and developed States were subject to the same responsibilities and liabilities when acting as sponsoring States.[99] Therefore, by pursuing deep seabed mining activities through a UK subsidiary, US national Lockheed Martin managed to circumvent successfully the legal restrictions upon its participation.

In light of the above, deep seabed mining activity outside of the UNCLOS regime appears unlikely to occur at present due to the commercial, financial and legal risks and uncertainty that it would involve for any entity.

4 The Extent of Commercial Activities within the Area

4.1 *Commercial Prospects for Deep Seabed Mining in the Area and Environmental Concerns*

A host of actors are now participating in deep seabed mining activities in the Area. Currently, these activities are still at the exploration phase. Long-term commercial prospects for the deep seabed mining industry remain uncertain given the high costs and technological challenges involved in deep seabed mineral extraction, compounded by other factors such as fluctuating metal prices and environmental concerns.[100]

Notably, deep seabed mining faces major opposition on environmental grounds.[101] On that basis, in January 2018, the European Parliament called

97 See discussion at text to nn59–61 above.

98 ITLOS Advisory Opinion (n61) para 159.

99 ITLOS Advisory Opinion (n61) para 159 (the Chamber found that in order to prevent a race to the bottom, the regulatory burden must be equally applied to developing and developed States, otherwise corporations could choose to route their activities through the former to take advantage of lesser regulation).

100 Rahul Sharma, 'Deep-Sea Mining: Economic, Technical, Technological, and Environmental Considerations for Sustainable Development' (2011) Marine Technology Society J 45(5) 28, 28–31.

101 See overview in Luz Danielle O Bolong, 'Into the Abyss: Rationalizing Commercial Deep Seabed Mining through International Law' (2016) 25 Tulane J Intl & Comparative L 127,

for a moratorium on deep seabed mining until its impacts on the marine environment are more fully understood.[102] Within civil society, a variety of NGOs oppose deep seabed mining, with the Deep Sea Mining Campaign supporting the ban of deep seabed mining activities in both international and national waters, and Greenpeace vehemently protesting future exploitation of deep seabed minerals.[103] Deep seabed exploitation, in particular, presents certain common concerns irrespective of the type of mineral being exploited. These include direct destruction of habitats, together with marine degradation through plumes of seafloor sediments generated by mining activities and other resultant effects, such as underwater noise, vibration and light pollution.[104]

Environmental controversy notwithstanding, the exploitation phase of deep seabed mining may begin in the coming years.[105] As the ISA's Secretary General remarked in February 2018, '[a]t a time when some appear to want to enter into an existential debate about whether deep sea mining should be permitted to go ahead or not, we do well to remember that the international community passed that point already many years ago.'[106] Indeed, Michael Lodge, the current Secretary-General of the ISA, previously described the scale

---

141–146. For a scientific assessment of environmental concerns posed by deep seabed mining, see, e.g., Rahul Sharma, 'Environmental Issues of Deep-Sea Mining' (2015) 11 Procedia Earth & Planetary Science 204. For detailed legal consideration of the balance between mineral exploitation and marine environmental protection, see, e.g., Jaeckel (n41); see also Gjerde (n41).

102 European Parliament, 'International Ocean Governance: An Agenda for the Future of our Oceans in the Context of the 2030 Sustainable Development Goals' (16 January 2018) <www.europarl.europa.eu/sides/getDoc.do?pubRef=-//EP//TEXT+TA+P8-TA-2018-0004+0+DOC+XML+V0//EN&language=EN>.

103 The Deep Sea Mining Campaign is an association of NGOs, including Oxfam Australia and Mining Watch Canada, together with other concerned individuals; Deep Sea Mining Campaign <www.deepseaminingoutofourdepth.org/>. See also Greenpeace International, 'Deep Sea Mining' (Background, 20 March 2014) <www.greenpeace.org/international/en/campaigns/oceans/marine-reserves/deep-sea-mining/>. Greenpeace, alongside over 70 other NGOs, is part of the Deep Sea Conservation Coalition (an umbrella movement seeking to protect deep sea ecosystems): Deep Sea Conservation Coalition <www.savethehighseas.org/>.

104 Sharma (2015) (n101) 205; Jaeckel (n41) 11–14.

105 See, e.g., ISA Press Release, 'Commercialization of Marine Minerals in Deep Seabed Well Within Reach, International Seabed Authority Secretary-General States as He Introduces Annual Report' (19 July 2016) SB/22/11.

106 Michael Lodge, ISA Secretary-General, 'Statement at the Workshop on the Draft Regulations for the Exploitation of Mineral Resources in the Area: Policy, Legal and Institutional Considerations' (London, 12–13 February 2018) 1 <www.isa.org.jm/sites/default/files/documents/EN/SG-Stats/sg-statement_0.pdf>.

of deep seabed mineral deposits as 'staggering.'[107] Based on industry projections, by 2030, deep seabed mining could constitute 10% of all global mining activity and have a value of around $65 billion in 2010 prices.[108] Notably, this level of activity would be only the tip of the iceberg in terms of the potential resources present on the ocean floor. For example, the value of gold deposits lurking on the deep seabed has been calculated at $150 trillion in 2013 prices.[109]

As noted above, thus far, licences for exploration relate to three particular types of minerals: polymetallic nodules, polymetallic sulphides and cobalt-rich crusts.[110] The first of these resource types, polymetallic nodules, are potato-sized rock deposits on the deep ocean floor, containing manganese, iron and other metals, such as copper and nickel. The second resource type, sulphides, generally contain iron, copper and zinc and may contain quantities of gold and silver. Most commonly, such sulphides emanate from hydrothermal vents on the mid-ocean ridge, where tectonic plates meet. Thirdly, cobalt crusts often form on rocky outcrops and ridges in the sea floor and contain elements such as manganese, iron, cobalt, nickel and lead. Each of these seabed deposits comprise relatively high concentrations of valuable metals as compared to equivalent minerals extracted on land.[111] As terrestrial metal stocks become depleted due to rising material consumption, experts anticipate that

107 Michael Lodge, 'Deep Sea Mining: The New Frontier in the Struggle for Resources?' [2014] World Economic Forum <https://agenda.weforum.org/2014/11/deep-sea-mining-the-new-frontier-in-the-struggle-for-resources/>.

108 European Commission, 'Blue Growth – Opportunities From the Marine and Maritime Sustainable Growth' COM (2012) 494 final, 10; Sheila Moorcroft, 'Ocean Mining – a Race to the Bottom' (Shaping Tomorrow, 17 April 2013) <www.shapingtomorrow.com/home/alert/94162-Ocean-mining–a-race-to-the-bottom>.

109 Meghan Miner, 'Will Deep-Sea Mining Yield an Underwater Gold Rush?' *National Geographic News* (3 February 2013) <http://news.nationalgeographic.com/news/2013/13/130201-underwater-mining-gold-precious-metals-oceans-environment/#.UysXr_ldXGA>.

110 For an overview of the geological characteristics of these resource types, see Chapter 1 of this book, Alvar Braathen and Harald Brekke, 'Characterizing the Seabed – A Geoscience Perspective'. See generally Tim Schröder (ed), *World Ocean Review: Marine Resources – Opportunities and Risks*, vol 3 (Maribus 2014).

111 James R Hein and others, 'Deep-Ocean Mineral Deposits as a Source of Critical Metals for High- and Green-Technology Applications: Comparison With Land-Based resources' (2013) 51 Ore Geology Reviews 1; see also Suzanne Goldenberg, 'Marine Mining: Underwater Gold Rush Sparks Fears of Ocean Catastrophe' *The Observer* (2 March 2014) <www.theguardian.com/environment/2014/mar/02/underwater-gold-rush-marine-mining-fears-ocean-threat>.

corresponding costs will continue to rise.[112] In this context, experts predict that metals extracted from the deep seabed will be increasingly sought after.[113]

Moreover, seabed minerals may contain traces of rare earth elements (REES).[114] REES are highly desirable commodities: critical for the manufacturing of green energy technology, such as solar panels, wind turbines and hybrid cars, and in the weapons industry. They are also essential in the electronics industry for production of high-tech gadgets such as smart phones, laptops and flat-screen televisions. The volume of REES available from the deep seabed could potentially exceed global land reserves of REES.[115] In this context, global excitement is growing over potential deep seabed mining for REES in both national and international deep seabed areas.[116] Deep seabed mining for REES could be a viable means by which to secure a reliable supply chain, insulated from potential geopolitical ruptures.

As the following Section will address, in pursuit of deep seabed resources, commercial actors have already made significant investments, and there are indications that commercial mineral extraction could begin in the Area in the years to come.[117]

---

112 UN Environment Programme (UNEP), 'Wealth in the Oceans: Deep Sea Mining on the Horizon?' (Global Environmental Alert Service, May 2014) 1 <https://na.unep.net/geas/archive/pdfs/GEAS_May2014_DeepSeaMining.pdf>; Yves Fouquet and Denis Lacroix, 'Study Summary' in Yves Fouquet and Denis Lacroix (eds), *Deep Marine Mineral Resources* (Springer 2014) 5; European Commission, 'Report on Critical Raw Materials for the EU: Report of the Ad Hoc Working Group on Defining Critical Raw Materials' (May 2014) 9 <http://ec.europa.eu/DocsRoom/documents/10010/attachments/1/translations/en/renditions/native>; Houses of Parliament, Parliamentary Office of Science and Technology, 'Deep-Sea Mining' (POSTnote 508, September 2015) 1–2 <http://researchbriefings.files.parliament.uk/documents/POST-PN-0508/POST-PN-0508.pdf>.

113 European Commission (2012) (n108) 10; UNEP (n112) 1; Houses of Parliament (n112) 1–2.

114 Hein (n2) 1–2; Yves Fouquet and Bruno Martel-Jantin, 'Rare and Strategic Metals' in Yves Fouquet and Denis Lacroix (eds), *Deep Marine Mineral Resources* (Springer 2014) 63ff.

115 Yasuhiro Kato and others, 'Deep-Sea Mud in the Pacific Ocean as a Potential Resource for Rare-Earth Elements' (2011) 4 Nature Geoscience 535, 538.

116 Ian Coles, 'Rare Earth Elements: Deep Sea Mining and the Law of the Sea' (2014) 14 Pratt's Energy L Rep 4, 8–10; Bolong (n101) 134–135.

117 See, e.g., ISA Press Release (n105); see also UNEP (n112); Houses of Parliament (n112). See further David Shukman, 'Deep Sea Mining "Gold Rush" Moves Closer' *BBC News Online* (18 May 2013) <www.bbc.co.uk/news/science-environment-22546875>; Goldenberg (n111); Adam Minter, 'Seafloor Gold Rush Could Have Alarming Impact' *Japan Times* (21 August 2016) <www.japantimes.co.jp/opinion/2016/08/21/commentary/world-commentary/seafloor-gold-rush-alarming-impact/#.W2sBQLllJaQ>; Julie Packard and Chris Scholin, 'The Deep Sea May Soon Be Up for Grabs' *New York Times* (8 June 2018) <www.nytimes.com/2018/06/08/opinion/the-deep-sea-may-soon-be-up-for-grabs.html>.

4.2 *Current Participants in Deep Seabed Mining Activities in the Area*

As of the end of 2018, the ISA has entered into mining contracts with twenty-nine contractors for exploration in relation to the three minerals types currently regulated (nodules, sulphides and cobalt-rich crusts).[118] These figures include a significant increase in the number of contracts granted in recent years. As recently as early 2011, the ISA had approved only eight applications for exploration; by late 2015, that figure had more than tripled.

The total area of deep seabed which the ISA has approved for exploration now surpasses 1.3 million square kilometres: an area approximately equivalent to the land mass of Sweden, Norway, Finland, Iceland and Denmark combined.[119] Although this approved area may seem vast, in fact it represents only half a percent of the entire deep seabed beyond national jurisdiction which is open to licencing by the ISA.

The ISA holds exploration contracts with a wide variety of actors, including States, State enterprises, State institutions, State-controlled corporations and several private corporations. At present, of the existing twenty-nine contracts, one is held by an international consortium of States (exploring for nodules), eight are held by States (exploring for nodules, sulphides or cobalt crusts), and fifteen are held by State enterprises, State institutions or State-controlled corporations (of which nine contracts are for nodules exploration and the remainder concern exploration for sulphides or cobalt crusts). In terms of State or State-controlled activity, the nations which have been most active so far in securing ISA licences are China, Russia, and South Korea. These three States are the only ones to hold exploration contracts relating to all of the three types of mineral which are currently regulated. At present, China has the greatest number of licences of any State or national. China's fourth contract for deep seabed mineral exploration rights (and its second in respect of nodules) was signed in 2017 by the ISA and China's State-owned metals and minerals trading company, China Minmetals Corporation.

The remaining five ISA contractors are private corporations exploring for polymetallic nodules (with two active in reserved areas). The first private

118 Seventeen of the ISA's mining contracts relate to nodules exploration, seven are for sulphides exploration, and five are for exploration of cobalt-rich crusts. For an overview, see ISA, 'Deep Seabed Minerals Contractors' <www.isa.org.jm/deep-seabed-minerals-contractors>.

119 This figure includes all areas approved for exploration, excluding any reserved areas which are not currently subject to exploration. For information on the exploration contracts approved by the ISA to date, including details of size and location of exploration sites, see the ISA website: <www.isa.org.jm/>.

actor to enter a contract with the ISA for nodules exploration was Tonga Mining Offshore Limited (TMOL) in 2012.[120] The contract concerns portions of reserved areas originally contributed by French, German and Japanese State entities and South Korea. A Tongan national, TMOL is a wholly-owned subsidiary of Canadian mining company Nautilus Minerals Inc (Nautilus).

In 2013, Belgian corporation, G-Tec Sea Minerals Resources NV (GSR), and UK corporation, UKSRL, both entered contracts with the ISA.[121] In March 2016, UKSRL entered a second contract with the ISA for nodules exploration.[122] As noted above, UKSRL is a subsidiary of the US defence giant, Lockheed Martin. The remaining private contractor in the Area is Ocean Mineral Singapore Pte Ltd (OMS), a Singaporean corporation with an ISA contract to explore for nodules in a reserved area originally contributed by UKSRL.[123]

In addition to acting directly as contractors, some private corporations also have links to activities in the Area by State-controlled actors. For example, the Cook Islands Investment Corporation (CIIC), a State enterprise, entered a contract with the ISA in July 2016 for nodules exploration.[124] CIIC entered into a joint venture agreement with GSR to jointly explore the area within CIIC's ISA contract, under the proviso that future profits would be shared equally between CIIC and GSR.[125]

---

120 ISA Council, 'Decision of the Council Relating to a Request for Approval of a Plan of Work for Exploration for Polymetallic Nodules Submitted by Tonga Offshore Mining Limited' (19 July 2011) ISBA/17/C/15.

121 ISA Council, 'Decision of the Council of the International Seabed Authority Relating to a Request for Approval of a Plan of Work for Exploration for Polymetallic Nodules Submitted by G-TEC Sea Mineral Resources NV' (26 July 2012) ISBA/18/C/28; ISA Council, 'Decision of the Council of the International Seabed Authority Relating to a Request for Approval of a Plan of Work for Exploration for Polymetallic Nodules Submitted by UK Seabed Resources Ltd' (26 July 2012) ISBA/18/C/27.

122 ISA Council, 'Decision of the Council Relating to an Application for the Approval of a Plan of Work for Exploration for Polymetallic Nodules Submitted by UK Seabed Resources Ltd' (21 July 2014) ISBA/20/C/25.

123 ISA Council, 'Decision of the Council Relating to an Application for the Approval of a Plan of Work for Exploration for Polymetallic Nodules Submitted by Ocean Mineral Singapore Pte Ltd' (21 July 2014) ISBA/20/C/27. OMS is majority owned by Keppel Corporation, a Singaporean corporation and one of the world's largest offshore and marine groups. Minority shares in OMS are held by UKSRL and Lion City Capital Partners Pte Ltd, a Singaporean private investment company.

124 ISA Council, 'Decision of the Council Relating to an Application for the Approval of a Plan of Work for Exploration for Polymetallic Nodules Submitted by the Cook Islands Investment Corporation' (21 July 2014) ISBA/20/C/29.

125 SOPAC, 'Cook Islands Enters New International Seabed Minerals Arrangement in Cooperation with GSR of Belgium' (The Prospect, 3 January 2014); Michael Henry,

In practical terms, joining forces with private actors may provide developing States lacking in technical capabilities and resources, one of the only viable means by which to become active in deep seabed mining.[126] For example, in its application, CIIC stated that its combined approach with GSR would '[enable them] to enhance all relevant synergies in scientific, ecological and economical research and studies to the absolute maximum.'[127]

To sum up, exploration activities in the Area are increasing, and this is especially so for private corporate contractors. Despite remaining uncertainties concerning long-term commercial prospects of deep seabed mining, increasing investment in the industry gives a strong indication that it may soon proceed towards commercialisation.

5 Conclusion

In the coming years, commercial extraction of deep seabed minerals may become feasible. As this Chapter has explored, despite continued uncertainties and challenges within the deep seabed mining industry, commercial investment in deep seabed mining is growing, and the volume of exploration activities in the Area continues to rise. An increasing number of States, State entities and private investors are now participating in deep seabed exploration activities under the ISA's licensing process within the UNCLOS regime. In this context, the ISA Secretary-General anticipates that deep seabed mining beyond national jurisdiction is 'well within reach' and 'attainable in the foreseeable future.'[128]

Under international law, there is a detailed legal framework for deep seabed mining activities within the Convention and associated instruments, including the 1994 Agreement and the ISA's Mining Code. All deep seabed mining activities in the Area are occurring under the auspices of this UNCLOS regime.

'International Seabed Authority Contract Signing' (Cook Islands Investment Corporations Online News, July 2016) <http://ciiconline.com/latestnews/international-seabed-authority-contract-signing/>. In an interesting dynamic, the area which CIIC intends to explore is a reserved area which was originally contributed by GSR pursuant to its 2013 nodules exploration contract with the ISA.

126 Tuerk (n19) 301.

127 ISA LTC, 'Application for Approval of a Plan of Work for Exploration for Polymetallic Nodules by the Cook Islands Investment Corporation' (8 November 2013) ISBA/20/LTC/3 para 13.

128 ISA Press Release (n105).

In practice, one key test for the future success of the regime will be the precise shape of the ISA's expanding regulatory code governing the exploitation phase, including its financial terms and environmental protections. Therefore, the ISA's current efforts to draft workable exploitation regulations, which takes on board the views of stakeholders in the industry, is a crucial movement towards the long-term realisation of a viable deep seabed mining industry in the Area.

CHAPTER 8

Framework Legislation for Commercial Activities in the Area

Erik Røsæg

1 The Problem

When commercial activities, primarily mining,[1] are carried out in the Area,[2] a framework of commercial law is as helpful on sea as on land. For example, workers should be protected against unfair contract terms, and entrepreneurs should be able to mortgage their equipment to ease financing. However, there is no state to provide this legal framework by general legislation. Could it be established in other ways?

The International Seabed Authority (ISA)[3] has a mandate to grant licenses for the exploitation of seabed resources in the Area, including setting conditions to ensure, for example, environmental protection and safety.[4] This could work well to establish some of the desired legal framework, at least to the extent that states are parties to UNCLOS or recognize the Authority.[5]

The ISA possesses considerable power in granting licenses with conditions and revoking them if necessary.[6] In addition, the risk of not getting licenses in the future can have a disciplinary effect.

However, not all kinds of rules can be implemented in this way.[7] For example, conditions for licenses cannot be used to alter the position of third parties to their detriment. An example of framework legislation that cannot be simulated by setting conditions for licenses is arrangements for mortgaging seabed installations.

1 Which activities follows the rules in respect of mining and which other activities there are in the Area are elaborated in section 4 below.
2 The Area is defined in UNCLOS art. 1(1)(1).
3 This is the "Authority" referred to in UNCLOS; see art. 1(1)(2).
4 See below in part 2.1.
5 Notably, the USA is not a party to UNCLOS. However, US firms apparently register subsidiaries in States Parties if they wish to participate in industrial exploitation in the Area.
6 See below in part 2.1.
7 Ibid.

Because of these limitations, there may be a need for regulation that is more comprehensive than what can be achieved by adding conditions to licenses. This paper focuses on rules of this kind. How is the possible need for such legislation taken care of in the Area? How can the necessary commercial framework legislation be implemented?

In the following, the main general jurisdictional bases will be discussed in section 2. These are the powers of ISA under UNCLOS, flag state jurisdiction and the jurisdiction of the sponsoring states and other states over companies. In section 3, some examples of how the jurisdictional problems are resolved or not resolved will be discussed. These examples concern patents, security interests (such as mortgages), labor law, and extra contractual liabilities.

Initially, the discussion will assume that the commercial activity is within the authority of the ISA. In section 4 below, the situation outside the authority of the ISA will be discussed.

2 General Jurisdictional Bases

2.1 *Incidental Jurisdiction of the ISA in Respect of Commercial Framework Legislation*

While coastal states have "sovereign rights" for the purpose of, for example, continental shelf jurisdiction,[8] the International Seabed Authority has more limited powers. In respect of licensing, a number of policies are set out in UNCLOS,[9] and there is a list of conditions (albeit non-exhaustive) in its Annex.[10]

Although the ISA has mainly focused on granting and setting conditions for licenses, its mandate is broader, and can possibly be utilized to create the necessary commercial framework legislation. The mandate for the ISA reads:

8 UNCLOS art. 77.

9 UNCLOS art. 150 ff.

10 UNCLOS, Annex III, in particular art. 17. The conditions are set out in the so-called Mining Code, consisting of a number of regulations. To date, the Authority has issued Regulations on Prospecting and Exploration for Polymetallic Nodules in the Area (ISBA 19/C/17, adopted 13 July 2000) which was later updated and adopted 25 July 2013 (ISBA 19/A/9); the Regulations on Prospecting and Exploration for Polymetallic Sulphides in the Area (ISBA /16/A/12 Rev.1, adopted 7 May 2010) and the Regulations on Prospecting and Exploration for Cobalt-Rich Crusts (ISBA 18/A/11, adopted 27 July 2012). The work is ongoing and, in the following, draft revisions and consolidations will also be referred to.

Article 157 Nature and fundamental principles of the Authority

1. The Authority is the organization through which States Parties shall, in accordance with this Part, organize and control activities in the Area, particularly with a view to administering the resources of the Area.

2. The powers and functions of the Authority shall be those expressly conferred upon it by this Convention. The Authority shall have such incidental powers, consistent with this Convention, as are implicit in and necessary for the exercise of those powers and functions with respect to activities in the Area.

The first part of paragraph 2 apparently excludes implied powers,[11] while the second sentence extends the powers to "incidental powers." This apparent contradiction reflects the power struggle in the negotiations, where the industrialized nations opposed granting wide, implied powers to the ISA.[12] The text was repeated without clarification in the 1994 implementing agreement.[13]

It is difficult to reconcile the text with an absolute ban on necessary, incidental powers relating to organizing and controlling activities in the Area. Such incidental powers could be used to adopt commercial framework legislation if absolutely necessary.

---

11 Without such exclusion, additional powers would be implied. See for example ICJ, Certain Expenses of the United Nations (Article 17, paragraph 2, of the Charter), Advisory Opinion of 20 July 1962, ICJ Reports (1962), 151, 168.

12 V Scatz, art. 157, mn. 7 in Preulss, UNCLOS, 1st ed 2017, ED Brown, Sea-bed energy and minerals: the international legal regime, Vol. 2: Sea-bed mining (2001) p. 193, S Mahmoudi, The law of deep sea-bed mining: a study of the progressive development of international law concerning the management of the polymetallic nodules of the deep sea-bed (1987) p. 281, SN Nandan et al., United Nations Convention on the Law of the Sea, 1982 : a commentary : Vol. 6 : Articles 133 to 191, Annexes III and IV, Final act, Annex I, Resolution II, Agreement relating to the implementation of part XI, Documentary annexes (2002), art. 157 para 157.13(b), FH Paolillo, The institutional arrangements for the international sea-bed and their impact on the evolution of international organizations, Collected Courses of the Hague Academy of International Law, Volume: 188 (1984), p. 273, BH Oxman, The Third United Nations Conference on the Law of the Sea: The Eighth Session (1979), 74 The American Journal of International Law 1 (1980), 15.

13 Agreement relating to the implementation of Part XI of the United Nations Convention on the Law of the Sea of 10 December 1982, Annex to UN Document A/RES/48/263, Annex, s. 1 (1).

Such broad interpretation would be in line with ITLOS'[14] findings that similar powers of a coastal state to regulate fisheries within its EEZ include regulations for offshore bunkering:[15]

> 217. The Tribunal is of the view that the regulation by a coastal State of bunkering of foreign vessels fishing in its exclusive economic zone is among those measures which the coastal State may take in its exclusive economic zone to conserve and manage its living resources under article 56 of the Convention read together with 64 article 62, paragraph 4, of the Convention. This view is also confirmed by State practice which has developed after the adoption of the Convention.

It is not obvious that this case is pertinent. Article 157 relating to the powers of the ISA does not refer to "sovereign rights" as does Article 56 relating to the EEZ. Furthermore, there is as yet no practice – and indeed no state practice – one way or the other relating to the Area. Still, it is likely that a similar broad reading of the powers of the ISA would be permitted, as there is a practical need for it. In particular, this is so when there is no question of creeping jurisdiction on the part of the ISA at the expense of states.

However, if there is a work-around for the problem of lack of commercial framework legislation, it would not be necessary for the ISA to act, and there would be no incidental powers. In the text below, several such workarounds will be discussed.

If the ISA has competence to legislate, final court decisions on the legislation are enforceable in all States Parties to UNCLOS.[16] The ISA may also withdraw licenses[17] and take past performance into consideration when granting new licenses.[18]

The ISA has not yet issued legislation of this kind. There is a reason for this reluctance. Both the secretariat and the decision-making bodies have limited resources, and they are not able to deal with the intricacies of private and procedural law. In addition, the ISA is handicapped because there is no existing

14 The Seabed Dispute Chamber International Tribunal for the Law of the Sea (ITLOS) settles disputes between a state party and the ISA regarding the competence of ISA, UNCLOS art. 187.

15 ITLOS Case No. 19 The M/V "Virginia G" Case (Panama/Guinea-Bissau).

16 UNCLOS, Annex III, art. 21(2).

17 UNCLOS, Annex III, art. 18.

18 UNCLOS, Annex III, art. 10. This article was modified by the 1994 implementing agreement (fn. 13), Annex, s. 1(13).

regime to build on, and each participating state would be likely to insist that its traditions be preserved. Whether or not the ISA has the legal competence to deal with the matters discussed here is thus probably a moot point, as it may not be able to enact change in any event.

2.2 *Flag State Jurisdiction over Ships Involved in Seabed Activities*

The water column above the Area is part of the high seas.[19] In the high seas, the flag state jurisdiction prevails.[20] Flag states therefore have general jurisdiction over ships involved in seabed activities in the Area. There is no indication that the powers of the ISA are exclusive in all matters, so flag state jurisdiction could be a good alternative to the ISA's jurisdiction in private law and other matters not relating to licensing.

In respect of protection of human life, UNCLOS presupposes that the jurisdiction of the ISA should be ancillary to other international provisions, which typically are based on flag state jurisdiction in respect of ships:

> Article 146 Protection of human life
> With respect to activities in the Area, necessary measures shall be taken to ensure effective protection of human life. To this end the Authority shall adopt appropriate rules, regulations and procedures to supplement existing international law as embodied in relevant treaties.

Similarly, flag state jurisdiction supplements the ISA's jurisdiction in respect of pollution:

> Article 209 Pollution from activities in the Area
> 1. International rules, regulations and procedures shall be established in accordance with Part XI to prevent, reduce and control pollution of the marine environment from activities in the Area. Such rules, regulations and procedures shall be re-examined from time to time as necessary.
> 2. Subject to the relevant provisions of this section, States shall adopt laws and regulations to prevent, reduce and control pollution of the marine environment from activities in the Area undertaken by vessels, installations, structures and other devices flying their flag or of their registry or operating under their authority, as the case may be. The requirements of

19 UNCLOS art. 86.

20 UNCLOS art. 87 and 92.

> such laws and regulations shall be no less effective than the international rules, regulations and procedures referred to in paragraph 1.

The coastal states have no special part in the enforcement of such provisions, while flag state enforcement presumably is retained:

> Article 215 Enforcement with respect to pollution from activities in the Area
> Enforcement of international rules, regulations and procedures established in accordance with Part XI to prevent, reduce and control pollution of the marine environment from activities in the Area shall be governed by that Part.

Under flag state jurisdiction, flag states implement a number of convention-based provisions that are impactful in private law matters. In addition to conventions like SOLAS[21] and MARPOL,[22] most flag states are parties to a number of conventions that are concerned with matters other than safety and pollution, such as arresting vessels,[23] labor law,[24] liens and mortgages,[25] and anti-terrorism measures.[26] Also, UNCLOS sets out the responsibility of flag states regarding their vessels.[27] Through these provisions, flag state jurisdiction fills a considerable gap in the legislation concerning activities in the Area.

Not all flag states are party to all conventions, and ship owners are free to choose almost any flag state for their vessels. However, the states with flags of convenience tend to ratify private law conventions, and ship owners do not avoid these flag states, as order and predictability is more important than deregulation in private law. Conventions are often compromises between interested parties and are thus acceptable.

The conventions are not complete codes, despite their broad scope. Therefore, they have to be supplemented by national legislation. Such legislation may vary quite a bit. It could wind up not regulating offshore activities by

21 International Convention for the Safety of Life at Sea (SOLAS), 1974, as amended.

22 International Convention for the Prevention of Pollution from Ships, 1973, as modified by the Protocol of 1978 relating thereto and by the Protocol of 1997 (MARPOL).

23 International Convention Relating to the Arrest of Sea-Going Ships, 1952.

24 Maritime Labour Convention, 2006.

25 International Convention on Maritime Liens and Mortgages, 1993.

26 Convention for the Suppression of Unlawful Acts against the Safety of Maritime Navigation (SUA), 1988, as amended.

27 UNCLOS art. 94.

will, by drafting errors, or simply because no one has considered the need for regulating industrial activity on the high seas, as opposed to ships simply sailing there. Therefore, flag state jurisdiction that is based only on international conventions may leave gaps in the regulation of the activities of ships operating in the Area.

Flag state regulation of ships is common and usually recognized in other states. Even if the legislation may not be given preference under choice of law rules, flag state legislation would prevent a legal vacuum. The ultimate sanction of the flag state against a ship disrespecting its legislation is to withdraw the right to fly its flag, leaving the ship subject to the jurisdiction of any state. In most cases, this would be regarded as a sanction too harsh.

Sometimes, a ship may be governed by the legislation of two cooperating states. The ownership of the vessel is registered in one state, while the vessel is flying the flag of another.[28] For our purposes, this is irrelevant; the sum of the jurisdiction of the two states is neither greater nor smaller than the jurisdiction of a single flag state.

Different vessels involved in the same operation may fly different flags. Thus, different sets of rules may apply to each ship. However, experience from offshore exploitation indicates that this is not a practical problem. If there is a safety aspect to the differences in regulation, the ISA is certainly competent to issue supplementary rules.[29]

The main problem with flag state jurisdiction is that not all units in the Area are ships; for example, they may be platforms and excavators operating on the seabed. This is problematic in two regards.

First, the, the *acquis* of conventions relating to ships may not always apply. There is a need to fill this void with new rules and standards.

Second, the traditional role of the flag state to regulate may be challenged, and it may not even be clear which state is the flag state. A state may invoke personal jurisdiction over the owners, but that may be challenged, for example, by the operator's home state.

Both in respect of flag state jurisdiction of ships and similar jurisdiction of other units, UNCLOS Annex III on Basic Conditions of Prospecting, Exploration and Exploitation restricts the use of flag state type of jurisdiction, etc.:

---

28 This is often referred to as bare boat registration or temporary change of flag.

29 See above in section 2.1.

> Article 21 Applicable law
> 3. No State Party may impose conditions on a contractor that are inconsistent with Part XI. However, the application by a State Party to contractors sponsored by it, or to ships flying its flag, of environmental or other laws and regulations more stringent than those in the rules, regulations and procedures of the Authority adopted pursuant to article 17, paragraph 2(f), of this Annex shall not be deemed inconsistent with Part XI.

This provision only applies in so far as the rules affect the licensee contractor negatively. Labor protection laws may be in this category, while the rule will most likely not affect legislation concerning legal infrastructure providing an option for mortgaging the unit used in the Area. If the licensee is affected, States Parties to UNCLOS cannot legislate, even if there is jurisdiction under the general principles of international law.

However, even when the licensee is affected negatively, flag state regulation of ships in respect of protection of the marine environment[30] is expressly allowed, and so is similar regulation of sponsoring states. Sponsoring states are in a way the "flag state" of the licensee.[31]

The net result of article 21 in respect of commercial framework legislation is that there is a fairly wide discretion of flag states and the like to supplement the regulations of the ISA and UNCLOS if they have jurisdiction on other grounds.

Flag state jurisdiction appears to be dominating in respect of ships, also in the Area.

2.3 *Company Jurisdiction over Companies Involved in Seabed Activities*

In order to obtain a license, a private company must be sponsored by a state:

> Article 153 System of exploration and exploitation
> 2. Activities in the Area shall be carried out as prescribed in paragraph 3: —
> (b) in association with the Authority by States Parties, or state enterprises or natural or juridical persons which possess the nationality of States Parties or are effectively controlled by them or their nationals, when sponsored by such States, or any group of the foregoing which meets the requirements provided in this Part and in Annex III.

30 This follows from the reference to UNCLOS Annex III, art. 17(2)(f).

31 See below in section 2.3.

The licensee must be effectively controlled by a sponsoring state. The sponsoring state (as all states parties) shall assist the ISA,[32] and the sponsoring state has the responsibility to ensure that a licensee shall carry out activities in the Area in conformity with the terms.[33] It is for that reason control is needed. If more than one state has control, they must all sponsor.[34] The state in which a prospective licensee is registered (or in which the licensee is a national) must be a sponsoring state, but that does not always suffice.[35] The involved states may be liable for failure to comply with this part of UNCLOS.[36] The sponsoring state must issue a certificate of sponsorship.[37]

If a state can control a private entity to this extent, perhaps the same mechanism could be utilized to provide supplementary legislation to create commercial framework legislation, subject to the limitations discussed above.[38] But how efficient is this kind of jurisdiction over a licensee or another company in the Area?

A problem could be that jurisdiction over the company is limited to the company. Thus, it is not able to address the relationships between different parties in the Area, such as the relationship between two subcontractors. Two examples of such relationships are provisions for knock-for-knock agreements[39] and channeling of liability.[40] Likewise, the state with jurisdiction over the company cannot regulate mortgages and other third-party relations when the third parties are the general creditors of the company. However, the sponsoring state can exercise its jurisdiction over the licensee or other companies under its control to ensure that it implements certain provisions in subcontracts.

Even when feasible as a regulatory technique, legislation of a sponsoring state may not be easily enforceable. If a company is registered in the

---

32 UCLOS art. 154(4).

33 UNCLOS, Annex III, art. 4(4).

34 UNCLOS, Annex III, art. 4(3).

35 Ibid.

36 UNCLOS art. 139, Responsibilities and obligations of States with respect to activities in the Area, Advisory Opinion, 1 February 2011, ITLOS Reports 2011, p. 10, and Kristoffer Svendsen's contribution to this volume.

37 Se for example Regulations on Prospecting and Exploration for Cobalt-Rich Crusts (fn. 10), reg. 11.

38 UNCLOS Annex III, art. 21; see above in section 2.2.

39 Knock-for-knock means that liability is waived in advance on a reciprocal basis in order to benefit from simpler and more inexpensive insurance arrangements made possible in this way.

40 Channeling of liability means that only one of several potential liability subjects shall be held liable in order to simplify claims handling.

sponsoring state and that state has some jurisdiction thereby, the sponsoring state does not have full control. In particular, this is so if both the assets and the directors of the company are abroad and thus able to escape direct enforcement. If claims are enforced against the licensee to the extent that bankruptcy is declared, that may not take place in the land of registration and may not be subject to the laws of that state.[41] However, winding-up as a sanction is a possibility,[42] as well as withdrawal of the sponsorship of the licensee.[43] These sanctions would in many cases be considered too strong.

In summary, the sponsoring state can contribute to commercial framework legislation in the Area. However, there are such limitations in both the scope and enforcement of such legislation that it is unlikely to be a feasible alternative.

3 Examples

3.1 *Introduction*

There is no doubt that commercial claims may be related to a site in the Area. In most cases, one can find ways to pursue the claims effectively, such as in the state where the damage occurred (in a torts claim) or in the home state of the debtor (in a contract claim). However, some problems are special to the Area in this respect.

First, there is the vacuum problem: Legislation that we take for granted on land or in coastal waters may not extend to the Area. Does a national rule of strict liability for pollution apply in the Area, or only in the territorial waters of the state in question? And do national rules on registration of mortgages in drilling rigs apply in the Area, or only on the continental shelf of the state

---

41 See, e.g., Regulation (EU) 2015/848 of the European Parliament and of the Council of 20 May 2015 on insolvency proceedings (recast), Art. 3, which locates insolvency proceedings to the center of main interest of the debtor (COMI), which may be a state other than the state of registration. This is also the rule in UNCITRAL Model Law on Cross-Border Insolvency with Guide to Enactment and Interpretation (2014), see para 31.

42 In Regulation (EU) 2015/848 (fn. 41), the compulsory winding-up of companies for reasons other than insolvency are not subject to the regulation, and then not necessarily its choice of forum to the COMI (Art. 1(1) of the Regulation).

43 ISA Document ISBA/24/LTC/WP.1/Rev.1, 9 July 2018, Draft Regulations on Exploitation of Mineral Resources in the Area, reg. 22. The sponsoring state may incur liability if it remains passive; see Kristoffer Svendsen's contribution to this volume.

in question? If the rules do not apply,[44] and the ISA has not issued such rules, there is a vacuum problem.

Such a vacuum may hamper commercial activities or leave vital interests without protection.

Second, there is the redundancy problem. It may happen that several states have made some of their legislation or other legal rules applicable worldwide or specifically in the Area. Which set of rules should prevail? This is traditionally governed by conflict of law rules, which vary between jurisdictions. The redundancy problem is likely to be an issue in the Area, as there is no natural presumption for which state's legislation shall apply; all situations will have only weak connections to states, and usually there are connections to many states.

Redundancy may cause a problem of forum shopping – legal action will be taken in the jurisdiction that will apply the rules that most benefit the party bringing the action. It is generally recognized that this is not a desirable system, both because it encourages parties to be the first to bring an action and because the foreseeability for the commercial parties is very limited.

In addition, some of the common choice of law rules are difficult to apply in the Area. One example is the general choice of law principle for property law, wherein the law of the state where the property is situated should be applied (*lex rei sitae*). This is relevant if, for example, there is an ownership dispute on machinery in the Area. Another example is the rule that the form of a document is governed by the law of the state in which the document is issued (*locus regit actum*, relevant if a bill of lading is issued at sea above the Area). New situations call for new private international legislation.

Some clauses in international investment treaties aim at protecting investors against less favorable framework legislation. Such treaties are unlikely to apply in the Area, to the national legislation applicable there and to the voting of states in the ISA. The ISA has not entered into any such agreements.

The following sections will discuss some situations in which the application of private law or private international law may cause problems in the Area in connection with different kinds of commercial framework legislation, namely patens, mortgages, labor law and liability.

44 Or the rules that one state has made applicable are not recognized by the court hearing the case.

3.2 *Patents*

A patent is the exclusive right to manufacture, use, or sell an invention for a certain number of years. There are already a substantial number of patents related to deep seabed mining and other activities in the Area, and it is also likely that many general patents are used there. Is there, and could there be, a similar protection against patent infringements in the Area as on land?

There is no such thing as a patent that grants exclusive rights worldwide. Patent law is national, so each state has to issue a patent. A "world patent" is a collection of such national patents. However, to some extent the states do cooperate and recognize each other's patents.

In line with this recognition, states tend to limit the scope of patents (exclusive rights) issued by them to the territory including the territorial seas.[45] Some states extend the scope of their patents to their own continental shelf.[46] But to my knowledge, no state extends the patents beyond that, to the Area.

The ISA has not issued any rules on patents. Therefore, there is a vacuum problem here; there is no applicable patent law in the Area. This is unfortunate, as patents are likely to be just as beneficial for the commercial life in the Area as elsewhere.

The patented device may be on board or form a part of a ship or another unit operating in the Area. If the ship or unit enters a port, there is still no patent infringement as covered by a special rule in the Paris Convention:[47]

> Article 5ter Patents: Patented Devices Forming Part of Vessels, Aircraft, or Land Vehicles
>
> In any country of the Union the following shall not be considered as infringements of the rights of a patentee:
>
> (i) the use on board vessels of other countries of the Union of devices forming the subject of his patent in the body of the vessel, in the machinery, tackle, gear and other accessories, when such vessels temporarily or accidentally enter the waters of the said country, provided that such devices are used there exclusively for the needs of the vessel;
>
> (ii) the use of devices forming the subject of the patent in the construction or operation of aircraft or land vehicles of other countries of

45 A Stenvik, Patentrett (3rd ed. 2013) p. 282 ff and MCA Kant, Cross-Border Patent Infringement Litigation within the European Union (2015) p. 41 ff.

46 See, for example, Norwegian Petroleum Act, 1996, s. 1–5.

47 Paris Convention for the Protection of Industrial Property, 1883, as amended.

> the Union, or of accessories of such aircraft or land vehicles, when those aircraft or land vehicles temporarily or accidentally enter the said country.

These exceptions are wide, and make it difficult to enforce patents when used in the Area.

When this exception does not apply, a number of activities are defined as patent infringements. An example is the EU law:[48]

> Article 25 Prohibition of direct use of the invention
> A Community patent shall confer on its proprietor the right to prevent all third parties not having his consent:
> (a) from making, offering, putting on the market or using a product which is the subject-matter of the patent, or importing or stocking the product for these purposes;
> (b) from using a process which is the subject-matter of the patent …;
> (c) from offering, putting on the market, using, or importing or stocking for these purposes the product obtained directly by a process which is the subject-matter of the patent.

In some states, acts such as offering the patent can be a patent infringement in the territory even if the patented object is situated outside the territory, for example in the Area.[49]

Despite such rules, patents are not well protected in the Area. States could easily extend their patents legislation to remedy this (which may raise the need for choice of law rules). It is not necessary that the ISA intervenes, and the conditions for the use of its incidental powers are not fulfilled.[50]

3.3 *Mortgages on Property, etc. Used in Seabed Activities*

An activity like deep seabed mining requires a significant amount of capital to invest in ships, rigs, and other equipment. Loan financing is a common way to obtain such capital if the enterprise engaging in the activity can offer the bank a privileged position in its eventual bankruptcy by way of mortgage. In that way, the bank could seize ships, rigs, and other equipment for repayment

48 Convention for the European Patent for the Common Market, annexed to Council Agreement relating to Community Patents, 1989 (89/695/EEC).

49 Stenvik (fn. 45), p. 292–293.

50 See above in section 2.1.

if necessary. A bank is likely to find an enterprise attractive for loans if its security interest in the equipment is legally recognized and protected by a system of law.

Such systems for recognition and protection of mortgagee banks could either be a register similar to land registries or an alternative pledge. The systems vary with the assets offered as collateral. In either case, the question is whether such systems are established or could be established in the Area.

3.3.1 Registrable Items

Recording of mortgages in a registry is well known for ships, and applies equally well to ships used in exploiting the Area. Although such registries are neither uniform nor universal, the systems in different jurisdictions are similar enough to form a single workable system. Ownership and mortgages are recognized based on the ship's registration in a flag state.[51] Forced sales may be carried out if a vessel is present in the jurisdiction, although such sales may not be recognized by all states.[52] Some states have ways to order ships flying their flags to proceed to an appropriate jurisdiction for a forced sale.[53]

Some valuable equipment on the vessel may be considered an appurtenance even if it is not required for navigation.[54] This legal status creates clarity, but may come as a surprise if the owner or mortgagee of the equipment is not the owner or mortgagee of the vessel. In particular, in offshore exploitation, the value of the appurtenances may exceed the value of the vessel.

The same types of rules often apply to floating platforms. Regardless of whether they are considered ships by conventions or customary public international law, such platforms usually have a flag state and they are registered in a particular state, similar to ships.[55] Again, flag state jurisdiction saves the day in respect of mortgages, and coastal state jurisdiction in respect of arrest and enforcement.

Fixed installations could be subject to a similar system. While registration of mortgages is organized by a state when the platform is located on its

---

51 See, for example, International Convention on Maritime Liens and Mortgages (fn. 25), art. 1.

52 Ibid. art. 12.

53 See, for example, the Norwegian Enforcement of Claims Act, 1992, Ch. 11 IV and s. 11–3.

54 See to this E Røsæg, Liens and Mortgages on the Ship – Their Relation to the Charterer's Equipment on Board in O Basurko and JMM Osante, New Trends in Maritime Law : Maritime Liens, Arrest of Ships, Mortgages and Forced Sale (2017) p. 339 ff. and B-E Reinertsen Konow, Løsørepant over landegrenser (2006) p. 176 ff.

55 See, for example, Norwegian Maritime Code, 1994, s. 507.

continental shelf[56] without any particular legal basis in international law,[57] the ISA has not organized a similar registry for the Area. As long as this situation is maintained, platforms could be recorded in a registry similar to a ship registry by a willing flag state. It is likely that that the flag and registered encumbrances would be recognized in the same way as ships and floating platforms. For practical reasons, a forced sale must be carried out while the installation remains onsite, but such a sale would be organized and recognized by the state of registry. However, it is less certain that other states would recognize the sale. In any event, the fixed installation may not be worth much if the associated project is doing poorly.

In all these cases (ships, appurtenances, floating platforms, and fixed installations), unsecured creditors could secure their claims through the legal system of the state of registry. The arrangements for fixed installations may not be widely recognized, and alternatives may be necessary.[58]

3.3.2 Non-Registrable Items

In addition to registrable units and their appurtenances, it is likely that there is other valuable equipment at a site, such as remotely operated vehicles (ROVs). Can these effectively be offered as collateral for financing of, for example, a mining project?

Usually, pledges are governed by the laws of the state in which the equipment is located (*lex rei sitae*).[59] However, in our case, the equipment is not located in a state and thus there are no relevant laws. Therefore, the creditors cannot get the desired certainty that they are secured creditors.

---

56 See, for example, Norwegian Petroleum Act, 1996, ch. 6 and HLJ Roelvink, Het continentaal plat als IPR-aanknopingspunt, SCJJ Kortmann et al (ed.), Op recht : bundel opstellen, aangeboden aan Prof. Mr. A.V.M. Struycken ter gelegenheid van zijn zilveren ambtsjubileum aan de Katholieke Universiteit Nijmegen (1996).

57 UNCLOS art. 77: "The coastal State exercises over the continental shelf sovereign rights for the purpose of exploring it and exploiting its natural resources." There are examples of express powers in this respect, for example Abkommen zwischen Norwegen und Deutschland über den Transport von Gas durch eine Rohrleitung vom norwegischen Festlandsockel und von anderen Gebieten in Deutschland (1993), art. 3.

58 See below in section 3.3.2.

59 See, for example, C Wendehorst art. 43 mn. 83 in HJ Sonnenberger et al., Münchener Kommentar zum Bürgerlichen Gesetzbuch : Bd. 11 : Internationales Privatrecht: Internationales Wirtschaftsrecht, Einführungsgesetz zum Bürgerlichen Gesetzbuche (Art. 25–248) (5. Ed. 2010) and A-R Borner, Comment on the Private Law Aspects of the Nord Stream Pipeline, 52 German Y.B. Int'l L. 355 (2009) p. 360.

It has been suggested that the law of the first jurisdiction where the equipment arrives should be the governing law.[60] However, this will not create foreseeability for the pledgee. A better alternative would be the law of the state of the owner or the law of the last state the equipment was located in before its deployment to the Area. However, these are not widely accepted rules and cannot be counted on to create foreseeability.

Regardless of registration, ownership is likely to be recognized by all courts pursuant to rules that are based on contract and therefore quite similar. An effective pledge on the equipment could therefore be achieved by pledging the shares of the company that owns it, which may be registered in any state. The pledging is then dealt with ashore in a way that creates foreseeability, removed from the legal difficulties of the deep seabed.

Problems may arise if the equipment-owning company goes bankrupt. As mentioned previously, bankruptcy proceedings may occur in a state other than the one in which the equipment-owning company is located.[61] Usually, such proceedings will take place in the state in which the debtor's center of main interest (COMI) is located, which may not recognize the security interest in the equipment or the shares.

Because of this, equipment-owning companies should have a COMI in their states of incorporation. As some bankruptcy courts may look for a common COMI among a group of companies, it may also be wise to ensure that the COMI of the group is located in the state in which the proceedings take place.

With the necessary precautions regarding the location of the bankruptcy, pledging of non-registrable assets in the Area could effectively be arranged by pledging the shares of the company that owns them.

3.3.3 The License

The value of an exploitation license can be used as collateral. Pledging or mortgaging the license does not include the value of the equipment on site.

In the draft regulations, approval of the ISA is necessary to establish such security interests,[62] and the security interest can be registered in a Seabed Mining Register run by the ISA.[63]

60 Ibid., mn. 106. The Uniform Commercial Code of the United States art. 9–301 points to the location of the debtor or the "local law of the jurisdiction in which the wellhead or minehead is located" in these cases.

61 See above in fn. 41.

62 Draft Regulations on Exploitation of Mineral Resources in the Area (fn. 43), reg. 23. This is in line with UNCLOS, Annex III, art. 20.

63 Ibid., art. 78.

The Draft Regulations are intended to safeguard the position of the ISA. There are no provisions of recognition or enforcement of the collateral interest in the license. The license is not to be governed by any specific national law, such as the law of the sponsoring state.[64] The value of a collateral interest in the license in the bankruptcy of the licensee is therefore open to doubt. This point is crucial, as the entire purpose of collateral is to secure the interests of the pledgee or mortgagee in the case of a bankruptcy.

3.3.4 Conclusion

In summary, it seems possible to overcome most problems regarding security interests and property seizure in the Area. Intervention of the ISA is therefore not warranted beyond what has been proposed, perhaps with the exception of a register for fixed installations.

3.4 *Labor Law in the Area*

Deep sea mining is likely to employ a number of people offshore. For onshore mining, employment contracts are subject to mandatory legislation in many states (as well as work safety legislation and the right to form labor unions). How could similar legislation be brought about in the Area?

The ISA previously drafted a provision that would make the relevant legislation of the sponsoring state applicable and require use of ships from flag states that had implemented the key international conventions.[65] Now the approach is to make the contractor responsible for compliance with international and other standards, regardless of the flag of the vessel:[66]

> 2. The Contractor shall ensure compliance with the applicable international rules and standards established by competent international organizations or general diplomatic conferences concerning … the treatment of crewmembers, as well as any rules, regulations and procedures and Standards adopted from time to time by the Council relating to these matters.

64 Ibid., art. 18.

65 Developing a Regulatory Framework for Mineral Exploiting in the Area. Report to Members of the Authority and all stakeholders (ISA 2016), Annex VII (to Annex I) Standard Clauses for Exploitation Contract. This is well in line with UNCLOS art. 146, which defines the task of the ISA in respect of protection of human life to supplement existing international law.

66 Draft Regulations on Exploitation of Mineral Resources in the Area (fn. 43), reg. 32.

3. In addition, Contractors shall:
(a) Comply with the relevant national laws relating to vessel standards and crew safety of their flag State in the case of vessels, or their sponsoring State or States in the case of Installations; and
(b) Comply with the national laws of its sponsoring State or States in relation to any matters that fall outside of the jurisdiction of the flag State, such as worker rights for non-crew members and human health and safety that pertains to the mining process rather than to ship operation.

At a stroke, this provision will make a set of relevant regulations applicable. A diligent sponsoring state or flag state may even issue special regulations for this activity if necessary. The IMO is now preparing safety regulations for crew carriage,[67] which is very relevant for offshore mining.

When enacting the draft provisions on safety, labor, and health standards, the ISA has effectively resolved the matter of labor protection legislation. There is no vacuum. However, in case of redundancy of regulation, there are no choice of law provisions giving the legislation of the flag state or the sponsoring state precedence.

3.5 *Extra Contractual Liability for Damages Caused by Activities in the Area*

Offshore mining will most likely cause damage from time to time. The licensee is responsible for such damage:[68]

> Article 22 Responsibility
> The contractor shall have responsibility or liability for any damage arising out of wrongful acts in the conduct of its operations, account being taken of contributory acts or omissions by the Authority.... Liability in every case shall be for the actual amount of damage.

The basis for liability is not exhaustive,[69] and the licensee may also be held liable under national law, for example where the damage occurs.

---

67 IMO Document MSC 97/22/Add.1, Annex 11, Resolution MSC.418(97) (2016) Interim Recommendations on the Safe Carriage of More Than 12 Industrial Personnel on Board Vessels Engaged on International Voyages.

68 UNCLOS, Annex III.

69 UNCLOS art. 304 and 235. Also the ISA and the sponsoring states may incur liability, see UNCLOS art. 139 and Kristoffer Svendsen's contribution to this volume.

The liability will eventually be backed by insurance[70] and a trust fund.[71] The liability rules will be detailed to some extent.[72] However, the regulation is a long way from becoming a full-fledged liability law.[73] In most states, such a law takes years to develop, and is adapted to local laws and conditions. Yet, this is probably as far as one can get in developing a special torts law for the Area.

4 Commercial Framework Legislation When There Is No ISA Authority

The discussions above have assumed ISA authority. However, this assumption is not always correct. On the one hand, a state that is not a party to the UNCLOS (or companies registered in that state) may be involved. These states are not bound to recognize the decisions of the ISA. The most important example is the USA, but (as already mentioned) American companies tend to operate in the Area via subsidiaries registered in a state party to the UNCLOS.

On the other hand, ISA may not be involved because the activity in question falls outside its mandate. The mandate of the ISA is limited to "activities in the Area", see in particular:

> Article 153 System of exploration and exploitation
> 1. Activities in the Area shall be organized, carried out and controlled by the Authority on behalf of mankind as a whole in accordance with this article as well as other relevant provisions of this Part and the relevant Annexes, and the rules, regulations and procedures of the Authority.

Similar limiting references to the "activities in the Area" are found in other parts of article 153 as well as other key provisions setting out the competence of the ISA.[74]

---

70 Draft Regulations on Exploitation of Mineral Resources in the Area (fn. 43), reg. 38. It is not clear from the wording whether one also has liability insurance in mind.

71 Ibid., reg. 52 ff. This was suggested by ITLOS, see Responsibilities and obligations of States with respect to activities in the Area (fn. 31) para 205.

72 Draft Regulations on Exploitation of Mineral Resources in the Area (fn. 43), Annex X Standard clauses for exploitation contract, section 6–8.

73 See, for example, the issues listed in ISA Discussion Paper No. 4 Enforcement and Liability Challenges for Environmental Regulation of Deep Seabed Mining (1996) p. 24.

74 Articles 140(2), 144, 145, 146, and 147.

The term "activities in the Area" is much narrower than it first appears due to the definitions. Article 1 of UNCLOS defines the term in this way:

> Article 1
> 1. For the purposes of this Convention: —
> (3) "activities in the Area" means all activities of exploration for, and exploitation of, the resources of the Area;

Thus, "activities in the Area" is limited to one specific activity, namely exploration for, and exploitation of, the resources of the Area. The key term "resources" is further narrowed in article 133:

> Article 133
> For the purposes of this Part:
> (a) "resources" means all solid, liquid or gaseous mineral resources in situ in the Area at or beneath the seabed, including polymetallic nodules;

Therefore, the term "activities in the Area" and the mandate of the ISA is limited to all activities of exploration for, and exploitation of, all solid, liquid or gaseous mineral resources in situ in the Area at or beneath the seabed, including polymetallic nodules of the Area. Activities outside the mandate of the ISA include marine scientific research, historical and archeological research, pipelines and cables,[75] catch of sedentary species, thermal energy, CO_2 storage, and utilization of genetic resources at the seabed, if any.

UNCLOS article 157 apparently suggests that the ISA has a wider mandate:

> Article 157 Nature and fundamental principles of the Authority
> 1. The Authority is the organization through which States Parties shall, in accordance with this Part, organize and control activities in the Area, particularly with a view to administering the resources of the Area.

The part starting with "particularly" perhaps suggests that the mandate is wider than "the resources in the Area". However, the first part of the sentence confirms that the mandate is limited to "activities in the Area", in line with the other provisions of UNCLOS. The idea is then presumably to state that when

75 See also UNCLOS art. 87(1)(c) and 112.

organizing and controlling, particular attention should be given to administering the resources.

On this basis, it is submitted that situations exist, in which the ISA is not recognized or that fall outside the authority of the ISA. How would then framework legislation for commercial activities in the Area be dealt with?

The discussions above conclude that regulation in the Area in many cases must rely on flag state jurisdiction and jurisdiction over companies (or individuals). This was so in relation to patens as well as mortgages and other securities.[76] This kind of jurisdiction is available also outside the authority of the ISA. The mechanism requires that all involved states take responsibility, exactly as within the authority of the ISA.

Registering security interests in licenses is possible even if the licenses are issued by a state, and not by the ISA.[77] However, the license is perhaps not much worth when any state can issue a new one with the same right, and the rights under the license might not be recognized by other states.

Labor protection within the authority of the ISA is based on a duty for the licensee to comply with international standards and national legislation of sponsoring states and flag states.[78] Outside the authority of ISA, there are no sponsoring states, but other involved states may enforce rules based on flag state or personal jurisdiction. Such rules may or may not comply with international standards. If they do not, there is no mechanism to enforce compliance. This makes it possible for entrepreneurs to choose to involve states with lenient enforcement or lenient standards.

Extra contractual liabilities for activities subject to ISA authority is based on special, positive law,[79] which is not applicable in other cases. However, there is a long-standing practice that states can seize jurisdiction where the damage has occurred and apply their own substantive torts law or their own choice of law rules.[80] There may be some difficulties enforcing the judgement, and there is no insurance or trust fund to back it.

In sum, the commercial framework legislation is much less well developed outside the scope of the authority of the ISA.

---

76 See section 3.2 and 3.3 above.

77 See section 3.3.3 above.

78 See section 3.4 above.

79 See section 3.5 above.

80 For example PK Mukherdjee, Jurisdictional issues in maritime zones, the high seas and conflict of laws. 24 Journal of International Maritime Law 273 (2018) is apparently based on this assumption.

5 Conclusion

Making a new legal order is not easy, even if only for the needs of the Area. The safest route is to use the legal systems of existing states and in particular the sponsoring states and flag states. This route may create choice of law problems, as the legislation of many states may be relevant. If using local law is not possible, then – and only then – the ISA could develop new rules under its inherent jurisdiction, available only when necessary. Developing such a new law is challenging, and it is not immediately obvious that states will recognize and enforce it.

In patents law, there is a problem that the law is national, and that national patents laws as a rule are not extended to the Area. In respect of security interests such as mortgages, only securities in the license is regulated by the ISA. Securities in the involved ships are subject to flag state jurisdiction, while security interests in other units may have to be arranged by pledging the shares in the company that owns them. Labor law in the area is subject to the applicable international conventions and legislation of flag states and sponsoring states by an obligation put on the contractor to make sure these standards are complied with. Liability rules are still fragmentary, but emerging under the auspices of the ISA.

Outside the scope of the authority of the ISA, the commercial framework legislation is much less reliable than within its authority.

CHAPTER 9

Maritime Security and Deep Seabed beyond National Jurisdiction

Edwin Egede

1 Introduction

The phrase, maritime security, has assumed prominence, mostly due to piratical acts off the coast of Somalia that had a serious impact on international sea trade in the Gulf of Aden, a critical trade corridor linking the Suez Canal and the Indian Ocean.[1] Although maritime security is currently a widely used phrase, its exact scope is not very clear. From a focus on naval sea power, piracy and armed robbery at sea, there is increasingly an expansion of the scope of what falls within maritime security. However, relatively little has been articulated in the discourse on maritime security on its impact on the Deep Seabed beyond national jurisdiction (the Area), which under the United Nations Law of the Sea Convention (UNCLOS) 1982 is declared to be the Common Heritage of Mankind (CHM).

Yet the Area raises certain important maritime security issues such as the emplacement of weapons of mass destruction in the Area, prospects of piracy and terrorism against ships engaged in deep seabed mining (DSM) activities, as well as the challenge that DSM activities could actually raise issues as regard environmental security. Therefore, this chapter aims to investigate possible maritime security issues that may arise in the Area, engaging with what could be considered as a rather traditional state-centric maritime security approach (State/Military implications), as well as certain maritime security implications in the Area from a non-State centric perspective. The chapter begins by exploring the concept of maritime security as an unclear and expanding one (Section 2). Thereafter, it examines the notion of maritime security and the Area from a state-centric perspective (Section 3). Subsequently, it delves into maritime

1 This led to United Nations Security Council involvement. See Edwin Egede, 'Piracy and the East African Region' In Koutrakos, P. & Skordas, A. (eds.) *The Law and Practice of Piracy at Sea: European and International Perspectives.* (Oxford and Portland, Oregon: Hart Publishing, 2014), pp. 249–265.

DOI:10.1163/9789004391567_011

security and the Area, engaging with it from a more non-state centric view-point (Section 4). The chapter ends with some concluding remarks (Section 5).

2 Maritime Security – an Expanding Concept?

Although maritime security has recently become a popular buzz phrase in international relations,[2] the notion of maritime security does underpin a large chunk of the development of the law of the sea.[3] For instance, the whole idea of the territorial sea and the so-called cannon shot rule of measuring this part of the sea was based on the security of the coastal State.[4] In a similar way, the conception of freedom of navigation of the high seas was originally put forward by Grotius based on the economic ground of allowing the Dutch access to the high seas in order to partake in the highly lucrative East Indian trade; it was subsequently utilized to provide justification for big naval powers to patrol freely on the high seas displaying their sea power in their pursuit of security.[5] Furthermore, the age long understanding of pirates as 'enemies of the human race'[6] with all States having universal jurisdiction, depicts that maritime security has always been an intrinsic part of the law of the sea. What, arguably, is novel about the growing notion of maritime security in the law of the sea is the increasing shift from focusing mainly on States actors to an increasing high-profile engagement with non-state actors, including organizations, such as the International Maritime Organization (IMO); private maritime security companies; the perpetrators of maritime security crimes, such as pirates and armed robbers at sea, private fishing trawlers engaged in Illegal, Unreported and Unregulated (IUU) fishing, terrorists groups carrying out their dastardly acts at sea and gangs involved in illegal trafficking of people by sea; as well as human security aspects focused on the rights of victims of maritime crimes, such as seafarers and hostages held by pirates for ransom, as well as vulnerable

2 C. Bueger, 'What is maritime security?' (2015) 53 *Marine Policy*, 159–164 at 159.

3 However, it must concede that the phrase, 'maritime security', was not until recently a vocabulary used in the law of the sea. For instance, the phrase 'maritime security' to the knowledge of this author was not used during the Third United Nations Conference on the Law of the Sea (UNCLOS III) negotiations, neither is it mentioned in the Law of the Sea Convention (LOSC) 1982.

4 W.L. Walker, 'Territorial Waters: the Cannon-Shot Rule,' (1945) 22 *British Yearbook of International Law*, 210–231.

5 T. Scovazzi, 'The Evolution of International Law of the Sea: New Issues, New Challenges', (2001) 289 *Recueil des Cours*, 39–243.

6 Le Louis case (1817) 2 Dods at 210.

migrants at sea. Theoretically, this could be regarded as a swing from a purely realist perspective focusing solely on States to a more liberalist position that recognizes that international relations must necessarily also engage with non-State actors as well.[7] However, it must be pointed out that even in cases when maritime security engages with non-State actors, there still remains a key role for State actors to play, in taking enforcement actions and countering maritime (in)security arising from threats of such non-State actors.[8] They can do so through agencies such as the navy and the coast guards.

Klein points out that '[t]he term "maritime security" has different meanings depending on who is using the term or in what context it is being used.'[9] In her opinion, it may best be understood from two key aspects, namely, traditional security concerns and responses to perceived maritime security threats. The former, she states, primarily refers to border protection, involving preventing incursions into areas that are considered as the sovereign domain of a State. It also refers to power projections, involving a State exercising naval military power in its relationship with other States. The latter, on the other hand, reflects certain steps taken by States to reduce the risk of certain crimes or activities which they believe would prejudice or injure their interests and society.[10] Klein then provides a definition of maritime security as 'the protection of a State's land and maritime territory, infrastructure, economy, environment, and society from certain harmful acts occurring at sea.'[11] This definition is rather restrictive because it appears to over-emphasize more on the interest of the State and deemphasizes engagement with the interests of non-State actors, including international organizations, private companies and individuals, as regards maritime security.

---

7 A. Slaughter, 'A Liberal Theory of International Law' (2000) 94 *American Society of International Law Proceedings*, 240–248.

8 This is not to say that non-State actors, such as private security companies, do not play a role in countering maritime security threats and challenging the monopoly of States in this regard. See: Carolin Liss, 'New Actors and the State: Addressing Maritime Security Threats in Southeast Asia,' (2013) 35(2) *Contemporary Southeast Asia*, pp. 141–162; and Elke Krahmann, 'From State to Non-State Actors: The Emergence of Security Governance' in Elke Krahmann (ed.), *New Threats and New Actors in International Security*, (2005, United States of America, Palgrave Macmillan), pp. 3–19.

9 N. Klein, *Maritime Security and the Law of the Sea* (2011, Oxford: Oxford University Press) at 4.

10 N. Klein, 'Maritime Security' in D.R. Rothwell, A.G., Elferink Oude, K.N.Scott, and T. Stephens (eds), *The Oxford Handbook of the Law of the Sea*, (Oxford University Press, Oxford, 2015) at pp. 582–583.

11 Ibid at 583.

The recently adopted African Charter on Maritime Security and Safety and Development in Africa (Lomé Charter) defines maritime security for the purposes of the Charter as: 'the prevention of and fight against all acts or threats of illicit acts against a ship, its crew and its passengers or against the port facilities, maritime infrastructure, maritime facilities and maritime environment.'[12] Again, this definition appears to be rather limited as it does not focus on maritime security issues that arise from State to State tensions, such as age-long conflicts between the navies of different States and clashes that may arise from maritime delimitation disputes.

Another example of definition of the concept of maritime security, this time in the national context, is the one of the UK National Strategy for Maritime Security, which states that it is: 'The advancement and protection of the UK's national interests, at home and abroad, through the active management of risks and opportunities in and from the maritime domain, in order to strengthen and extend the UK's prosperity, security and resilience and to help shape a stable world.'[13] This rather broad and vague definition is wide-ranging enough to cover virtually every threat arising from the maritime domain. It is also not very clear what 'the active management of risks and opportunities' is in practice and which tools would be used to implement this management.

Bueger, in an interesting article interrogating the meaning of maritime security,[14] concludes that it:

> [...] has no definite meaning. It achieves its meaning by actors relating the concept to others, by attempts to fill it with different issues and by acting in the name of it. If actors agree on the value of maritime security in general terms, its practical meaning will always vary across actors, time and space. Striving for a universally acceptable definition of maritime security is hence an unproductive quest.[15]

---

12 Art.1(1). The Lomé Charter was adopted as a binding treaty by 30 African States on 15 October 2016 at the African Union Extraordinary Summit at Lomé, Togo. The Lomé Charter is available at: http://www.african-union-togo2015.com/en/accueil.

13 UK National Strategy for Maritime Security, May 2014 at 15, https://www.gov.uk/government/uploads/system/uploads/attachment_data/file/310323/National_Strategy_for_Maritime_Security_2014.pdf.

14 Bueger, op.cit. at pp. 160–163 explored the meaning of maritime security using three frameworks: a semiotic viewpoint of seeking to grasp its meaning by exploring its relations with other terms; a constructivist securitization framework and the actual practice of relevant actors.

15 Bueger, op.cit. at 163.

Then again, Germond introduces a geo-political dimension to the exploration of maritime security and points out that: 'Maritime security has to do with (illegal and disruptive) human activities in the maritime milieu, that is to say a certain geographically delimited space. Thus, states are differently impacted by maritime security threats depending on their actual geographical location.'[16] Thus, one could rephrase the renowned statement of the eminent constructivist, Wendt, that: 'maritime security is simply what States make of it!'[17] Although, there is no 'one size fits all' definition of maritime security various intergovernmental organizations and States have adopted strategies or non-binding Codes to deal with maritime security threats.[18] More recently, the African Union has even adopted a binding treaty on maritime security.[19]

In essence, since the exact scope of the concept of maritime security is highly contested, there is nothing that forecloses its application to the Area because there is a real possibility of 'illegal and disruptive human activities',[20] amounting to a threat to security, occurring as regard the Area. This is further supported by the requirement under the 1982 United Nations Convention on Law of the Sea (LOSC) that the Area be used for peaceful purposes only, especially since peace generally in international relations is used along with the

16 B. Germond. 'The geopolitical dimension of maritime security' (2015) 54 *Marine Policy*, 137–142 at 138.

17 A. Wendt, 'Anarchy is what States Make of it: The Social Construction of Power Politics' (1992) 46 *International Organization*, 391–425.

18 See, for example: the 'European Union Maritime Security Strategy: Responding Together to Global Challenges: A Guide for Stakeholders' (EUMSS), Council of the EU, 24 June 2014, and its Action Plan of 16 December 2014, as revised on 26 June 2018; NATO Alliance Maritime Strategy 2011, http://www.nato.int/cps/en/natohq/official_texts_75615.htm; African Union's 2050 Africa's Integrated Maritime Strategy 2014, http://pages.au.int/maritime/documents/2050-aim-strategy-0 ; the US National Strategy for Maritime Security, September 2005, http://www.state.gov/documents/organization/255380.pdf; US Asia-Pacific Maritime Strategy 2015, http://www.defense.gov/Portals/1/Documents/pubs/NDAA%20A-P_Maritime_SecuritY_Strategy-08142015-1300-FINALFORMAT.PDF; and UK National Strategy for Maritime Security 2014. For non-binding Codes, see the following IMO inspired agreements: The Regional Cooperation Agreement on Combating Piracy and Armed Robbery against Ships in Asia (ReCAAP), 2004; the Code of Conduct concerning the Repression of Piracy and Armed Robbery against Ships in the Western Indian Ocean and the Gulf of Aden (Djibouti Code of Conduct), 2009 as revised by the Jeddah Amendment 2017; and the Code of Conduct Concerning the Repression of Piracy, Armed Robbery Against Ships, and Illicit Maritime Activity in West and Central Africa (the Yaoundé Code of Conduct, 2013, http://www.imo.org/en/OurWork/Security/WestAfrica/Documents/code_of_conduct%20signed%20from%20ECOWAS%20site.pdf.

19 See note 12 above.

20 Germond, op.cit at 138.

notion of security.[21] Recently, the International Seabed Authority (ISA) has acknowledged the need to consider maritime security in the Area in the Draft Regulations on Exploitation of the Minerals Resources in the Area, where it requires operators applying for a plan of work to exploit to also include, amongst other things, a health, safety and maritime security plan along with their application.[22]

3 Maritime Security and the Area: a State-Centric Focus

3.1 *Area, Peaceful Purposes and Military Activities: a Matter of Definition*

One of the crucial points raised in Arvid Pardo's renowned address to the United Nations General Assembly in 1967, of the need to establish a regime for the exploitation of the resources of the Area, was that activities in this part of the sea should take place in a 'peaceful atmosphere'. He stressed the rising concern that the Area could '… progressively and competitively [be] appropriated, exploited and used for military purposes by those who possess the

21 See for instance Articles 1(1) and 24(1) of the United Nations Charter which state that the Purposes of the United Nations are: To maintain international peace and security, and to that end: to take effective collective measures for the prevention and removal of threats to the peace, and for the suppression of acts of aggression or other breaches of the peace, and to bring about by peaceful means, and in conformity with the principles of justice and international law, adjustment or settlement of international disputes or situations which might lead to a breach of the peace' and 'In order to ensure prompt and effective action by the United Nations, its Members confer on the Security Council primary responsibility for the maintenance of international peace and security, and agree that in carrying out its duties under this responsibility the Security Council acts on their behalf.

Article 1 of NATO treaty states that: 'The Parties undertake, as set forth in the Charter of the United Nations, to settle any international dispute in which they may be involved by peaceful means in such a manner that international peace and security and justice are not endangered, and to refrain in their international relations from the threat or use of force in any manner inconsistent with the purposes of the United Nations.'

Article 3(f) of the African Union Constitutive Act states: 'The objectives of the Union shall be to … promote peace, security, and stability on the continent.'

22 See Draft Regulations on Exploitation of Mineral Resources in the Area: Prepared by the Legal and Technical Commission, ISBA/24/LTC/WP.1/Rev.1 of 9 July 2018, Draft Regulation 7(3)(f). Annex IV of this Draft Regulations on the nature of the Health, Safety and Maritime Security Plan has not been developed and the annex states as follows: '[To be populated following discussion with the IMO Secretariat, members of the Authority and Stakeholders].'

required technology.'[23] This celebrated address, which called for this part of the sea to be declared as the Common Heritage of Mankind, acted as a trigger for the third United Nations Conference on the Law of the Sea (UNCLOS III) that eventually culminated in the adoption and the eventual coming into force of the LOSC 1982. The Convention requires that: 'The Area shall be open to use exclusively for peaceful purposes by all States, whether coastal or land-locked, without discrimination and without prejudice to the other provisions of this Part.'[24] The phrase 'use exclusively for peaceful purposes' was not defined by the Convention. This therefore raises the issue of whether the peaceful purposes provisions of LOSC completely prohibit all military activities.

The scope of military activities must here be examined. According to Wolfrum, virtually all military activities at sea would fall under at least one of the following activities:

> 1. Navigation on the water surface or in the water column including all military activities connected with navigation.
>
> Navigation and connected activities are performed as routine marine operations or periodic conditioned manoeuvres. They may serve one or more of the following purposes: exercising of ships, co-operation between navy, air force and land forces of one or more nationalities, the latter adding a further co-operation aspect (*e.g.* Ocean Venture 1981), control of the sea, projection of naval presence (e.g. the presence of US units in the vicinity of the Persian Gulf) and deterrence;
>
> 2. Emplacement of sea-based missiles for strategic purposes. This activity is presently fulfilled (mainly) by missile launching nuclear submarines;
>
> 3. The emplacement of sea-based surveillance devices such as fixed acoustic detection systems;
>
> 4. The emplacement of sea-bed based weapons systems for strategic or tactical purposes such as magnetic or acoustic mines against surface ships or submarines. Furthermore, the emplacement of strategic missiles on the seabed has been discussed;
>
> 5. The emplacement of sea-bed based surveillance devices like the fixed acoustic detection array systems which according to some sources have been deployed along the east and west coast of the United States and some strategically important points in the oceans;

23 Para 5 of UN General Assembly 22nd Session, Official Records of Fifth Committee, 1515th Meeting, 1 November 1967 (Arvid Pardo, the Maltese Ambassador to the UN) http://www.un.org/depts/los/convention_agreements/texts/pardo_ga1967.pdf.

24 Art.141, LOSC.

> 6. Military research including the testing of weapons, conducted either on the water surface, in the water column, or the subjacent seabed and subsoil.'[25]

It is doubtful that the list above could be regarded as a completely exhaustive list of what constitutes 'military activities.'

Recently, an arbitral tribunal has had the occasion to give its interpretation of the definition of military activities. The Arbitral Tribunal in the South China Sea Arbitration (Merits) between the Republic of the Philippines and the People's Republic of China,[26] in seeking to determine whether it had jurisdiction over the case, had to ascertain if the Chinese activities in the South China Sea were military in nature. The latter issue was reserved from its previous award on jurisdiction and admissibility.[27] The Tribunal was, however, quick to point out that its remit was merely to consider Article 298(1)(b), which applies to '"disputes concerning military activities" and not to "military activities" as such.' In its view, the relevant question was to consider whether 'the dispute itself concerns military activities, rather than whether a party has employed its military in some manner in relation to the dispute.'[28]

Nonetheless, the decision of the Tribunal in this regard raises some stimulating ideas that could provide some guidance on what are military activities. For instance, the Tribunal points out that what represents a 'quintessentially military situation' is a state-centric conflict that involves the military forces of one State (i.e. Philippines) against a combination of military and paramilitary forces of another State (i.e. China) 'arrayed in opposition to one another.'[29] Furthermore, the Tribunal, in determining that certain Chinese activities

25 Wolfrum, R., "Restricting the Use of the Sea to Peaceful Purposes: Demilitarization in Being?" (1981) 24 *German Yearbook of International Law* p. 200 at 205–6.

26 In the matter of the South China Sea Arbitration before an arbitral tribunal constituted under Annex VII to the 1982 United Nations Convention on the Law of the Sea between the Republic of The Philippines and The People's Republic of China, PCA Case No 2013–19 of 12 July 2016. Available at: <http://www.pcacases.com/pcadocs/PH-CN%20-%20201 60712%20-%20Award.pdf>

27 Art. 298(1)(b) of LOSC allows State Parties to exclude by way of declaration the compulsory jurisdiction under the treaty in 'disputes concerning military activities, including military activities by government vessels and aircraft engaged in non-commercial service, and disputes concerning law enforcement activities in regard to the exercise of sovereign rights or jurisdiction excluded from the jurisdiction of a court or tribunal …' By declaration dated 25 August 2006, China had triggered all of these exceptions listed in Art. 298, including the 'military activities' exception. See Para.161 of PCA Case No. 2013–19 of 12 July 2016 and also the award on jurisdiction and admissibility of 29 October 2015.

28 Para.1158 of PCA Case No. 2013–19 of 12 July 2016.

29 Ibid, para.1161.

were not military in nature, was happy to rely on China's repeated statements and position that the activities were primarily, if not wholly, for civilian use.[30] Ostensibly, this could appear to suggest that a State may solely decide on whether or not their activities at sea is military in nature. However, when this position is read along with the opinion of the Tribunal that it was not determining military activities, as such, but merely seeking to decide on whether Article 298(1)(b) was applicable to the arbitration, it becomes clearer that the Tribunal was merely taking the position that China ought to be estopped from relying on the exclusionary provisions of the Article. China had consistently declared that such activities were civilian and not military.

Clearly military activities would involve military confrontation between two or more states. However, it is clear that it goes beyond this as could be seen from the rather long, but obviously not exhaustive list, of possible military activities at sea provided, as quoted earlier, by Wolfrum.[31]

3.2 *Are All Military Activities in the Area a Security Threat?*

Yet the question arises as to whether all military activities in the Area constitute a breach of the peace thereby raising maritime (in) security concerns. Various States have interpreted this differently to suit their national interests. Certain States, especially developed States with nuclear capacity, interpret this provision as merely prohibiting military activities that has an aggressive purpose. These States argue that this provision would only prohibits military activities that are inconsistent with the United Nations Charter and other obligations under international law, which is what would create maritime (in)security.[32] In support of this viewpoint it has been stated that Article 141, as well as other provisions related to use of the ocean for peaceful purposes,[33] should be interpreted in the light of Article 301 which states: 'In exercising their rights and performing their duties under this Convention, States Parties shall refrain from any threat or use of force against the territorial integrity or political independence of any State, or in any other manner inconsistent with the principles

30 Ibid, paras.925–938, 1024–1028 and 1164.

31 See note 25 above.

32 See M. Nordquist (ed.), *United Nations Convention on the Law of the Sea 1982 – A Commentary*, Vol III (Martinus Nijhoff, Dordrecht, 1995), pp. 90–91; E. Egede *Africa and the Deep Seabed Regime: Politics and International Law of the Common Heritage of Mankind* (Springer, Heidelberg, 2011), pp. 79–80; J Kraska and R. Pedrozo, *International Maritime Security Law*, (Martinus Nijhoff, Leiden, 2013) at pp. 304–309 and M. Lodge, 'The Deep Seabed' and J. Kraska, 'Military Operations' in D.R. Rothwell, A.G. Oude Elferink, K.N. Scott, and T. Stephens, (eds), *The Oxford Handbook of the Law of the Sea*, (Oxford University Press, Oxford, 2015) at 230 and 868–869 respectively.

33 See Articles 88, 143(1), 147(2)(d), 155(2), 240(a), 242(1) and 246(3) of LOSC 1982.

of international law embodied in the Charter of the United Nations.'[34] The United States of America is an example of a State that adopts this viewpoint. In a commentary which accompanied the transmittal by the President of the LOSC 1982, President Clinton stated:

> In furtherance of this principle, article 141 declares the Area to be open to use by all States. Only mining activities are subject to regulation by the International Seabed Authority [...]. Other activities on the deep seabed, including military activities, telecommunications and marine scientific research, may be conducted freely in accordance with principles of the Convention pertaining to the high seas, including the duty to have reasonable regard to other uses.[35]

In addition, the commentary points out that none of the LOSC provisions on 'peaceful purposes' and 'peaceful use' where intended to create new rights or obligations nor impose restraints upon military operations, or impair the inherent right of self-defense, and so long as the military activities are consistent with the principles of international law they were not prohibited by international law generally and the LOSC in particular.[36] Kraska and Pedrozo further cite a 1985 United Nations Secretary-General's Report, which concluded military activities that were consistent with the provisions of Article 2(4) and Article 51 are not prohibited by the LOSC.[37] It would appear that from this standpoint military activities are not prohibited by the LOSC in the Area in the following instances: first, if they do not amount to a threat or use of force against the territorial integrity or political independence of any state, or in any other manner inconsistent with the purposes of the United Nations; second, if the activities are a valid exercise of the right to self-defense or if such activities amounting to the threat or use of force is done with the authorisation of the Security Council; third, if it is not expressly prohibited by a conventional or customary international law norm.

On the other hand, for certain other States, the peaceful purposes provision in Part XI of the LOSC, connotes that all military activities in the Area

---

34 Although Art.301 uses the phrase 'peaceful uses' instead of 'peaceful purposes' it has been pointed out that these two phrases are synonymous. See Kraska and Pedrozo, *International Maritime Security Law*, op.cit. at pp. 305–306 citing the Virginia Commentary.

35 See Letter of Transmittal, White House, October 7, 1994 at p. 61, Senate Treaty Doc.103–39, http://www.foreign.senate.gov/imo/media/doc/treaty_103-39.pdf.

36 Ibid at p. 94.

37 Kraska and Pedrozo, *International Maritime Security Law* at 307.

are strictly prohibited.[38] For instance, the then Organisation for African Unity (OAU) captured this position by stressing that the Area 'should be used exclusively for peaceful purposes [and] [a]ny utilisation of this zone for military purposes is strictly prohibited.'[39]

Francioni, points out that a minority of scholars support the latter view. Using the analogy of certain other treaties it has been argued that there is an absolute prohibition of military activities.[40] For instance, the Antarctica Treaty 1959 in Article 1 titled – Peaceful Purposes – states that:

> 1. Antarctica shall be used for peaceful purposes only. There shall be prohibited, *inter alia*, any measure of a military nature, such as the establishment of military bases and fortifications, the carrying out of military manoeuvres, as well as the testing of any type of weapon.
> 2. The present Treaty shall not prevent the use of military personnel or equipment for scientific research or for any other peaceful purpose.

Francioni points out that early negotiations in the UNCLOS III were based upon a generally prevalent view that the seabed which is the common heritage of mankind was to be completely demilitarized as it was meant to be used for peaceful purposes only.[41] However, a close examination of the relatively detailed provisions of the Antarctica Treaty, in comparison with the provisions of the LOSC on peaceful purposes, would appear to indicate that the LOSC did not intend to prohibit all military activities. While the Antarctica Treaty specifically and explicitly prohibits 'any measure of a military nature', the provision of the LOSC does not explicitly indicate such prohibition.

Conversely, one could criticize those States and scholars that claim that certain military activities may take place in the Area for erroneously conflating the provisions on peaceful purposes in Article 88 (high seas) with that of Article 141(the Area). In doing so, they would wrongly interpret the two

38 See for instance, the Resolution of the Council of Ministers of the Organization of African Unity on the International Zone extending beyond National Jurisdiction, DOC/A/CONF.62/50 of 14 September 1976.

39 Doc.A/CONF.62/50 of 14 September 1976 referred to in UNCLOS III *Official Records* Vol VI, p. 121.

40 See for instance, Antarctic Treaty 1959, Art.1 and Treaty on Principles Governing the Activities of States in the Exploration and Use of Outer Space, including the Moon and Other Celestial Bodies, Art. IV. See Francesco Francioni, 'Peacetime use of Force, Military Activities, and the New Law of the Sea, (1985) 18(2) *Cornell International Law Journal*, p. 203 at 221–225.

41 Ibid at 223–224. See Art.141 of LOSC.

provisions in an identical way.[42] First, the two provisions have different wordings, while Article 88 states that the high seas would be 'reserved for peaceful purposes', Article 141 states that the Area should be used 'exclusively for peaceful purposes.' Second, the provisions apply in essence to two different regimes having two different applicable principles: for the high seas, res communis; and for the Area, common heritage of mankind. It would be recalled that Article 141 points out that the peaceful purposes must be in line with the relevant part of the Convention. Whilst the peaceful purposes as used under Article 88 is compatible with military activities, especially with the long historical connection with such military activities and the whole notion of the freedoms of the sea, it is not so with the common heritage of mankind. The latter principle, which is a relatively recent one, is premised mainly around resource exploitation of this part of the sea, a core activity that is not necessarily reconcilable with military activities.

An interesting practical example, of the conundrums that may arise from these divergent views on the peaceful use of the Area is with regard to manned subsea 'space station' that could be used not only for civilian, but also for military purposes, such as surveillance through, for instance, having sensors to detect the submarines of other States.[43] This raises two interesting issues.

First, whether the use of such mixed-use (or hybrid) installations or structures in the Area would be regarded as military activity. Although, the question of the mixed-use of installations or structures was raised by the Philippines in its arguments in the South China Seas arbitration in relation to whether China's island building activities fell within the ambit of military activities under Article 298(1)(b) of LOSC, the Tribunal did not directly engage with this point.[44] In this author's view, two possible approaches may be adopted to determine this – the initial or original purpose and predominant use approaches. The former focuses on the initial or original purpose for setting up the installation or structure. If such was for civilian purpose then it could be arguably regarded as being used for non-military activities, while if such was originally constructed for a military purpose then it could perhaps be regarded as used for military activities. However, this may raise complexities when such submersible structure or installation though initially set up for civilian purposes is subsequently

42 See Isaak I. Dore, "International Law and the Preservation of the Ocean Space and Outer Space as Zones of Peace: Progress and Problems" (1982)15(1) *Cornell International Law Journal*, pp. 1–61 at 21–22.

43 See 'China is Planning a Massive sea lab 10,000 feet underwater', Bloomberg, June 8, 2016, http://www.bloomberg.com/news/articles/2016-06-07/china-pushes-plan-for-oceanic-space-station-in-south-china-sea.

44 Paras. 893 and 1013–1014, PCA Case No .2013–19 of 12 July 2016.

predominantly used for military purposes. The predominant use approach would probably be a better approach. Here the predominant and prevalent use of the installations or structures would determine whether such installation or structures should be characterized as being used for military activities or not.

Another issue is whether the use of such submersible installations or structures should be regarded as amounting to allowable military activity in the Area. Obviously, as indicated above, States that argue that the peaceful use clause does not prohibit all military activities would take the position that such non-aggressive use of the Area would amount to allowable military activity. According to Treves:

> … listening and other detection or communication devices are considered more acceptable than weapons. Weapons are inherently dangerous, while detection devices are considered more acceptable than weapons. Weapons are inherently dangerous, while detection and communications devices are not. This judgement will probably have some bearings on the solution of conflicts among different uses of the seabed. It seems easier to accommodate the emplacement of detection or communication devices to other uses, such as those related to resources, than the emplacement of weapons.[45]

On the other hand, those who argue that all military activities are prohibited in the Area would obviously take the view that, as long as such activity may be categorised as a military one, it is not allowable in the Area.

3.3 *The Area and the Seabed Arms Control Treaty 1971*

The paramount security concern in the Area is the prospect of emplacement of Nuclear Weapons and other Weapons of Mass Destruction in the seabed, ocean floor and subsoil of the Area. As far back as 1971, prior to LOSC, the Treaty on the Prohibition of the Emplacement of Nuclear Weapons and other Weapons of Mass Destruction on the Seabed and the Ocean Floor and in the Subsoil thereof (the Seabed Arms Control Treaty) was adopted to deal with this issue.[46] This treaty, which has been described as an example of 'superpower

45 Tullio Treves, "Military Installations, Structures, and Devices on the Seabed", (1980) 74(4) American Journal of International Law, pp. 808–857 at 809–810.

46 This treaty of 10 Articles was adopted by a large number of States, (104 in favour, 2 against and 2 abstentions) on February 11, 1971 and came into force on May 18, 1972. So far 94 States have become States Parties. Only France out of the P-5 members of the UNSC is not a Party. See http://www.state.gov/t/isn/5187.htm.

symbiosis',[47] arose from a strictly Statist concern about maritime security concerning the Area. The key concern was the possibility of States, especially the big powers, utilising the Area as another venue for furthering their nuclear arms race. This concern was not far-fetched, especially since during the Cold War there was intense nuclear competition between the bipolar powers the USA and the then Union of Soviet Socialist Republics (USSR), along with their respective allies.

The Convention, which is still in force, prohibits the emplanting or emplacing of any nuclear weapons or other types of weapons of mass destruction, as well as structures, launching installations or any other facilities specifically designed for storing, testing or using such weapons on the seabed, ocean floor and the subsoil thereof beyond the outer limits of the territorial seas (which of course would include the Area).[48] The Treaty makes provision for verification through the observation by State Parties of the activities of other States Parties, provided that observation by the former does not interfere with the lawful activities of the latter.[49] If after such observations there are still doubts as to whether the obligations under the Treaty are being carried out, the State Party carrying out the observations and the other one carrying out the activities giving rise to the doubts are required to consult together with a view to removing the doubts. If, however, the uncertainty still persists, the State Party having the doubts is required to notify the other States Parties to the Treaty, who are to cooperate on further procedures for verification, as they may agree, including the appropriate inspection of objects, structures, installations or other facilities that reasonably may be expected to be in breach of Article 1 of the Treaty. The State Parties located in the region of the activities, including any coastal State, and any other Party so requesting, are also entitled to participate in such consultation and cooperation. After completion of the further procedures for verification, an appropriate report is required to be circulated to other State Parties by the Party that initiated such procedures.[50]

The Treaty also makes provision for situations where the State responsible for the activities that raise doubt cannot be easily identifiable. Here, in the case of the initial step of consultations, the State having doubts is to notify and

---

47 See James Barry Jr., "The Seabed Arms Control Issue 1967–1971 A Superpower Symbiosis" in Richard B. Lillich & John Norton Moore(eds.), *Role of International Law and an Evolving Ocean Law*, (Newport, 1980), pp. 572–585 at 573, where this phrase was used and described as 'a relationship in which advanced States with divergent goals temporarily join forces to achieve a specific end.'

48 See Arts. I and II.

49 Art. III(1).

50 Art. III(2).

make appropriate enquiries with State Parties in the region where the activities has taken place or from any of the other State Parties. If the identity of such State responsible for the activities cannot be ascertained through these inquiries, the inquiring State Party may undertake further verification procedures, including inspection, and shall invite the participation of the State Parties in the region where the activities is taking place, including any coastal State, and any other Party desiring to cooperate.[51] In the event that the consultation and verification process fails to clarify the doubt and there are still serious questions concerning the fulfilment of obligation under the treaty, the State Parties may refer the matter to the United Nations Security Council, who may take action in line with its primary responsibility to maintain peace and security under the UN Charter.[52]

There are, however, some criticisms of the Seabed Arms Treaty. For one, although there are currently a huge number of States Parties to the Treaty,[53] including a number of nuclear weapons States, some known nuclear weapons States, such as France, Pakistan, Israel and North Korea, are yet to become Parties to this important Treaty framework. Another problematic area of the Convention is the escape clause whereby States Parties may withdraw from the Treaty in the following instance:

> Each State Party to this Treaty shall in exercising its national sovereignty have the right to withdraw from this Treaty if it decides that extraordinary events related to the subject matter of this Treaty have jeopardized the supreme interests of its country. It shall give notice of such withdrawal to all other States Parties to the Treaty and to the United Nations Security Council three months in advance. Such notice shall include a statement of the extraordinary events it considers to have jeopardized its supreme interests.[54]

The ground for withdrawal is rather vague and subjective. Furthermore, the requirement that the Party which seeks to withdraw merely gives three months-notice to other States Parties and the Security Council is rather problematic. This appears to be a rather short notice period for such a significant issue as this. The 2003 withdrawal of North Korea from the Nuclear Nonproliferation

51 Art. III(3).
52 Art. III(4).
53 See note 42 above for the number of Parties to the Treaty.
54 Art. IV.

Treaty, with similar provision,[55] is instructive of the challenge with this provision. Obviously, when a State Party withdraws from the Treaty it would be res inter alios acta in relation to such State. However, States who are outside the Seabed Arms Control Treaty framework could only be bound by the obligation prohibiting the emplacement or emplantment of nuclear weapons and WMD in the seabed beyond the territorial sea if it could be argued that this norm has crystallised into Customary International Law.[56]

Another criticism is that the Treaty does not prohibit mobile installations such as a nuclear submarine resting at the bottom of the sea or the deployment of 'crawlers' and also mobile missile platform that may move along the ocean.[57] The ambiguity of the Treaty is also reflected in the reference to structures and installations 'specifically designed for storing, testing or using' WMD, which gives the almost absurd indication that the ban depends on its initial design and not its purpose.[58] Furthermore, the Treaty, for obvious reasons is limited to nuclear weapons, as well as other types of WMDs, but does not really ban nuclear devices to be used for 'peaceful' purposes.[59] It thus does not cover the possibility of nuclear disasters of a scale such as the Chernobyl and Fukushima disasters that may arise from such 'peaceful' use of nuclear materials in the Area, which may have environmental implications and raise issues of environmental (in) security through the actions of States.[60]

What is clear, is that the Seabed Arms Treaty although it seeks to put in place some control over the use of nuclear weapons in the seabed, including the Area, does not effectively denuclearize the seabed, and neither does it create some type of nuclear weapons free zone in the Area.[61] Of course, another major gap of this rather dated Seabed Control Treaty is that it does not cover non-state actors, notably terrorist groups, which this chapter would discuss further in section 4 below.

55 See Art. X of the Nuclear Non-Proliferation Treaty.

56 See North Sea Continental Shelf Cases ICJ Rep.1969, p. 3 at 41–44; Case Concerning Military and Paramilitary Activities in and Against Nicaragua (Merits) ICJ Rep.1986, p. 14 at 98.

57 See Barry Jr., 'The Seabed Arms Control Issue 1967–1971 A Superpower Symbiosis', supra at p. 583 and Dore, 'International Law and the Preservation of the Ocean Space and Outer Space as Zones of Peace: Progress and Problems', supra at 15.

58 See Art. 1 and Dore, Ibid at pp. 15–16.

59 Dore, ibid at p. 13.

60 See below section on non-State centric approach for more discourse on environmental security.

61 For nuclear Free Zones in the World see https://www.un.org/disarmament/wmd/nuclear/nwfz/.

3.4 *Maritime Security and Prospect of Inter-State Tensions over Mining Sites in the Area*

Another key maritime security issue from a State-centric approach is the risk of inter-State tensions over maritime mining sites, which may escalate into conflicts. This threat is reduced in the case of deep seabed mining (DSM) activities due to the crucial role of the International Seabed Authority (ISA) as the intergovernmental organization vested under the LOSC to act on behalf of mankind and who is in charge of allocating mining sites in the Area.[62]

However, a scholar had pointed out that '[a]lthough the International Seabed Authority has shouldered the responsibility for managing the activities on the International seabed beyond national jurisdiction, future exploration is sure to raise many jurisdictional challenges akin to fossil fuel resources, and some claims and counterclaims regarding certain areas can be expected.'[63] Kraska, in an interesting article on Indian Ocean Security,[64] provides the example of India's security concerns, as regards the allocation by the ISA to China, its regional rival, (via the China Ocean Mineral Resources Research and Development Association [COMRA], a State owned entity), of mining site in the Southwest Indian Ocean for the exploration of polymetallic sulphide ore deposit.[65] He points out that this was unsettling for India and states that:

> New Delhi fears it will provide China with "an excuse to operate their warships in [the] area." The Directorate of Naval Intelligence in New Delhi warns that the seabed mining development gives China a reason to "maintain a continuous presence" in the central Indian Ocean. The Ministry of External Affairs called the Chinese seabed mining plan a "worrying development."[66]

This does indicate a potential for State to State tensions over allocation of mining sites for DSM. For instance, India in its recent Maritime Security Strategy 2015 also mentioned its DSM activities in the Indian Ocean as one of its overseas

62 See Arts.137(2) and (3), 153(2)–(6), 157(1) and 162(2)(j)–(l). See Chapter 7 of this book, J. Dingwall, 'Commercial Mining Activities in the Deep Seabed Beyond National Jurisdiction: The International Legal Framework.'

63 Nirmal Verna, 'India and Transnational Maritime Challenges' in Mohan Malik, Maritime Security in Indo-Pacific Perspectives from China, India and the United States of America, (Bowman & Littlefield, London, 2014), pp. 209–214 at 211.

64 James Kraska, 'Indian Ocean Security and the Law of the Sea', (2012) 43 Georgetown Journal of International Law, pp. 434–493, especially at 459–462.

65 Exploration Contract began November 18, 2011 and to end November 17, 2026. See https://www.isa.org.jm/deep-seabed-minerals-contractors.

66 Kraska, op.cit at p. 461.

maritime investments that need to be secured.[67] Furthermore, Japan in its National Security Strategy, under the heading '[r]isks to Global Commons,' which it identifies as one of its national security challenges, points out that '[c]ases of conflicts of interest over the sea are increasing. There is a growing risk of incidents at sea and of possible escalation into unexpected situations.'[68]

4 Maritime Security and the Area: A Non-State-Centric Focus

Contemporary discourse of maritime security seems to be increasingly focused on threats by non-State actors in the maritime domain. This section would explore three possible future security threats by non-State actors with regard to the Area as an important maritime domain, perhaps more so when the actual exploitation in the Area begins, namely: piracy against production support vessels (4.1), maritime terrorism (4.2) and environmental security (4.3).[69]

4.1 *Piracy against Ships and Production Support Vessels (PSVs)*

A critical aspect of deep seabed mining is the use of vessels, ranging from state of the art multipurpose research vessels to Production Support Vessels (PSVs). One of the commonly indicated method of mining operations from available technology and data mining operations in the Area is the use of autonomous underwater vehicles (AUVs), which would be used to survey the seabed prior to extraction, and also remotely operated vehicles (ROVs) that play a role not only in obtaining samples of deposits, but could also be used in mining itself – passing resources extracted from the Area through a steel riser pipe (a riser lift system) to a PSV situated on the high seas.[70] It is important to note that ships

67 Indian Maritime Security Strategy 2015, https://www.indiannavy.nic.in/sites/default/files/Indian_Maritime_Security_Strategy_Document_25Jan16.pdf.

68 National Security Strategy of Japan 2013, p. 5, para.1(4), http://www.cas.go.jp/jp/siryou/131217anzenhoshou/pamphlet_en.pdf.

69 The DSM industry is currently at the exploration stage, however, the ISA has already begun work on drafting the Exploitation Mining Code. See https://www.isa.org.jm/legal-instruments/ongoing-development-regulations-exploitation-mineral-resources-area.

70 See Anon., 'Nautilus sets out its deep sea mining stall,' The Naval Architect, February 2016, http://www.rina.org.uk/Nautilus_sets_out_its_deep_sea_mining_stall.html which mentions a newer innovation of the PSV, the Production Support and Storage Vessel (PSSV). It states that 'the PSSV is an innovative design as the vessel and its mining system are the first of their kind in the world. The ultra-deep water mining process begins at the seafloor where three large mining ROV's cut rock and turn it into slurry for pumping. The slurry is pumped to the PSSV via a Subsea Slurry Lift Pump (SSLP) through a top tensioned riser. At the PSSV the slurry is delivered to a Dewatering Plant (DWP) and then into the vessels' holds for storage. The stoned ore is reclaimed from the ship's holds and offloaded via a

and PSVs play an important role in deep seabed mining in the Area. The ITLOS Seabed Chambers in its first advisory opinion in response to questions formulated by the Council of the ISA in making a distinction between transportation that should be regarded as 'activities in the Area' and that which would not clearly highlights this.[71] According to the Chambers:

> Transportation to points on land from part of the high seas superjacent to the part of the Area in which the contractor operates cannot be included in the notion of "activities in the Area", as it would be incompatible with the exclusion of transportation from "activities in the Area" in Annex IV, article I, paragraph I, of the Convention. However, transportation within that part of the high seas, when directly connected with extraction and lifting, should be included in activities in the Area. In the case of polymetallic nodules, this applies, for instance, to transportation between the ship or installation where the evacuation of water and the preliminary separation and disposal of material to be discarded take place. The inclusion of transportation to points on land could create an unnecessary conflict with provisions of the Convention such as those that concern navigation on the high seas.[72]

With research ships and PSVs having to operate in the high seas to conduct deep seabed mining research or to facilitate seabed mining, as the case may be, there is the risk of piratical attacks on these vessels. The LOSC states that:[73]

> [p]iracy consists of the following acts:
> (a) any illegal acts of violence or detention, or any act of depredation, committed for private ends by the crew or the passengers of a private ship or a private aircraft, and directed:
> (i) on the high seas, against another ship or aircraft, or against persons or property on board such ship or aircraft;
> (ii) against a ship, aircraft, persons or property in a place outside the jurisdiction of any State.

It also includes the act of voluntarily participating in the operation of a ship or aircraft with the knowledge that these are being used for acts of piracy, as

---

cargo handling system into a bulk carrier moored alongside. The bulk carrier then tranships the ore to market.'

71 Responsibilities and Obligations of States with Respect to Activities in the Area, ITLOS Advisory Opinion No. 17 of 2011.

72 Ibid at para.96.

73 Art.101(a) of LOSC 1982(identical to Art.15(1) of the High Seas Convention 1958).

mentioned in the definition above, as well as any 'act of inciting or of intentionally facilitating an act of piracy,' as described above.[74] Due to what is perceived as the heinous nature of piracy it is said that there is universal jurisdiction over the crime.[75] Although, in theory it is easy to discern what piracy is from the definition provided in the LOSC, in reality it may sometimes be contested. There are debatable issues that arise from the definition of piracy that scholars have engaged with.[76] However, it is unlikely, for instance, that acts of violence against a vessel engaged in deep seabed mining activities by a vessel with environmental protesters would be regarded as piracy.[77] Nonetheless, there may be other liability issues that may arise from this, such as criminal or civil damage to property or person.

Currently, there are incidents of piracy in Asia (various seas in Asia), off the coast of Somalia (Indian Ocean) and the Gulf of Guinea, off the West African Coast (Atlantic Ocean), with piratical acts involving petty theft, hijacking of vessels and kidnapping of crew in vessels, as well as well as violent acts of robbery, sometimes accompanied by resource theft.[78] These various regions have adopted instruments clearly indicating that piracy is a major maritime threat and a present maritime security concern.[79] Some of these regions, such as the Indian and Atlantic Oceans, have deep seabed mining sites.[80]

A stimulating issue is whether a PSV or submersibles, such as the AUVs and ROVs, used for deep seabed mining, may be regarded as ships for the purposes of Article 101 of the LOSC? In an article on the application of maritime concepts to seabed mining, Spicer and L'Esperance, evaluated whether sea bed mining vessels and submersibles would qualify as 'ships.'[81] They indicated that

---

74 Art.101(b) and (c) of LOSC.

75 Art.105 of the LOSC. Also see E. Kontorovich and S. Art, "An Empirical Examination of Universal Jurisdiction for Piracy," *American Journal of International Law* (2010:104), 436–453.

76 See R.Churchill, "The Piracy Provisions of the UN Convention on the Law of the Sea – Fit for Purpose?" In *P.* Koutrakos & A. Skordas (eds.) *The Law and Practice of Piracy at Sea: European and International Perspectives.* (Oxford and Portland, Oregon: Hart Publishing, 2014), pp. 9–32 for interesting and insightful analysis of the problems with defining piracy.

77 See In the Matter of the Artic Sunrise Arbitration (The Kingdom of the Netherlands v. The Russian Federation, (Award on Merits), PCA Case Nº 2014–02, https://pcacases.com/web/sendAttach/1438.

78 See Egede, 'Piracy and the East African Region' in P. Koutrakos, & A. Skordas (eds.), op.cit. 249 at 255–257.

79 See note 18 above.

80 See ISA Contractor areas, https://www.isa.org.jm/deep-seabed-minerals-contractors/overview.

81 W. Spicer and P. L'Esperance, 'Seabed Mining and the application of Maritime Law Concepts', *LawyersIssue*, July 11, 2016, http://www.lawyerissue.com/seabed-mining-and-the-application-of-maritime-law-concepts/.

a quick survey of the rather voluminous legislation and case law of certain States would appear to suggest that for an object to be a ship it must satisfy the following requirements: partial navigational use; navigational capabilities; navigation through or above water; vessel under construction and that the mode of propulsion is irrelevant.[82] They then suggested that the PSV may satisfy many of the common elements of ships, especially when they are independently navigating between extraction sites. They were, however, rather skeptical on whether PSVs permanently moored or positioned to engage in DSM activities for extended period of time would be regarded as a ship. Further, they take the view that the status of AUVs and ROVs as ships is rather ambiguous.[83] This would need more clarity. If we are to accept the position of Spicer and L'Esperance on PSVs, an attack on a PSV would only be piracy when such PSV qualifies as 'ships.'

4.2 *Maritime Terrorism*

Another risk to ships and PSVs engaged in DSM activities is maritime terrorism. The United Nations, as far back as 1988, had adopted The Convention for the Suppression of Unlawful Acts against the Safety of Maritime Navigation 1988, and the subsequent Protocols, as one of its sectoral Conventions to counter terrorism.[84] This Convention criminalises attacks against vessels beyond the outer limits of the territorial sea of States which may involve acts such as, a person or group of persons seizing or exercising control over a ship by force or threat of such force; the committing of acts of violence against individuals on a ship or destroys such ship or causes damage to such ship or its cargo.[85] In recent times, a series of incidents occurred, such as the failed attack on USS The Sullivans in January 2000, or the successful attacks on the USS Cole in October 2000, and the French supertanker, MV Limburg in October, 2002, Superferry 14 just outside Manilla Bay in the Philippines in 2004. As a result, efforts have focused on addressing the issue of terrorist attacks on ships and installations at sea, by upgrading maritime terrorism to a major maritime security threat.[86] The possibility of such attacks against vessels engaged in DSM could therefore not be ignored.

82 Ibid at p. 4.

83 Ibid.

84 Preamble 3 states that the Convention was a response to 'world-wide escalation of acts of terrorism in all of its forms.'

85 Art.3.

86 See for instance, Art. V(d) of the EU Maritime Security Strategy 2014; Indian Maritime Security Strategy 2015 at pp. 38–39 and Part X of the 2050 Africa's Integrated Maritime Strategy, 2014.

Furthermore, especially since the 9/11 terrorist attack against the USA, there have been concerns about terrorist groups having access to nuclear weapons and other types of WMDs.[87] The Seabed Control Treaty, as mentioned above, as a State-centric treaty, does not in any way cover the possibility of non-State actors emplacing nuclear weapons or other WMD in the seabed, including the Area. Article 1(1) of the Seabed Control Treaty declares that '[t]he States Parties to this Treaty undertake not to emplant or emplace …', while Article 1(2) says that '[t]he States Parties to this Treaty undertake not to assist, encourage or induce any State to carry out activities referred to in paragraph 1 …', thereby effectively excluding non-State actors, such as terrorist groups from the ambit of the Convention. As a response, the United Nations Security Council resolutions, such as resolutions 1540 and 1887,[88] as well as the Convention for the Suppression of Acts of Nuclear Terrorism 2005, have sought to fill this gap.

4.3 *Environmental Security*

One of the key risks of deep seabed mining is the chance of it having serious adverse impacts on the marine environment. According to Greenpeace, deep seabed mining is likely to cause serious environmental damage that may destroy vital habitats and could lead to the killing of marine life, including the extinction of certain unique species, as well as potentially generating toxic sediments that could contaminate the food chain.[89] These environmental concerns have been described at various times as a 'new Ocean threat,'[90] 'an emerging threat to our Oceans',[91] 'a global threat to our Oceans',[92] 'a serious

87 See the testimony of George Tenet before the US Senate Select Committee on Intelligence, 'The Worldwide Threat 2004: Challenges in a Changing Global Context,' 24 February 2004, http://www.nti.org/media/pdfs/10.pdf?_=1316466791.

88 UNSC Resolution 1540(2004) and Resolution 1887(2009).

89 See Indian Ocean Observatory, 'Deep seabed mining will cause serious environmental damage' - Greenpeace', March 10, 2016, http://www.theioo.com/index.php/en/insight/item/486-deep-seabed-mining-will-cause-serious-environmental-damage-greenpeace See also, Michelle Allsopp, Clare Miller, Rebecca Atkins, Steve Rocliffe, Imogen Tabor, David Santillo & Paul Johnson, 'Review of the Current State of Development and Potential for Environment Impacts of Seabed Mining Operations,' http://www.greenpeace.to/greenpeace/wp-content/uploads/2013/07/seabed-mining-tech-review-2013.pdf.

90 Richard Steiner, 'Deep Sea Mining a New Ocean Threat' The Huffington Post, 20 October, 2016, http://www.huffingtonpost.com/richard-steiner/deep-sea-mining-new-threa_b_8334428.html.

91 Greenpeace, 'Deep seabed mining: an emerging threat to our oceans', 31 July 2013, http://www.greenpeace.org/international/en/news/Blogs/makingwaves/Deep-seabed-mining/blog/46110/.

92 WWF South Africa, 'Seabed Mining a global threat to our Oceans: International Contexts', http://awsassets.wwf.org.za/downloads/sosc_factsheet3_web.pdf.

threat to the stability of Oceans systems and processes'[93] and 'next frontier of maritime insecurity.'[94]

With the protection of the environment increasingly becoming an issue foremost on the agenda of the international community, there is a growing interest in environmental security since the end of the Cold War. Although, the concept of environmental security is essentially a contested one, the academic literature is steadily engaging with the notion that there is some kind of linkage between the environment and security, and the extent to which environmental degradation may be regarded as a security threat.[95]

As far back as 1990, the United Nations Secretary-General stated that: 'ecological threats to the marine environment are also increasingly seen as a more serious threat to national security.'[96] Also, the United Nations Secretary-General High Level Panel 2004 Report on Threats, Challenge and Change identified some crucial linkages between environmental degradation and security.[97] While the linkage between environmental degradation and security may be easier to discern when such degradation arises from military activities of State actors, for instance, in the case of the Area, the possibility of a nuclear disaster as a result of military activities by nuclear submarines in the seabed, it is sometimes difficult to fathom such linkages when the degradation is caused by the activities of non-State actors. For instance, NATO in recognition of the possibility of environmental security threats arising from military activities states that: 'the Alliance is working to reduce the environmental effects of military activities and to respond to security challenges emanating from the environment.'[98] But, even at that, the NATO, in explaining environmental security, appears to recognise that its remit, in this regard, may actually go beyond the direct impact of military activities on the environment, to engaging with certain environmental issues that may have an indirect effect on security by stating:

93 Sylvia Earle, 'Deep Seabed Mining: An Invisible Land Grab', July 20, 2016, https://www.mission-blue.org/2016/07/deep-sea-mining-an-invisible-land-grab/.

94 Maurice Beseng, 'Will deep sea mining be the next frontier of maritime insecurity in Africa?' July 12, 2016, http://www.maritimesecurity.global/2016/07/12/deep-sea-mining/.

95 Nina Græger, "Environmental Security?" (1996)33(1) *Journal of Peace Research*, pp. 109–116 and Karen Hulme, "Environmental Security: Implications for International Law,"(2009)19(1) *Yearbook of International Environmental Law*, pp. 3–26.

96 Para.36, Law of the Sea, Report of the Secretary-General, Doc. A/45/721 & Coor.1 of 19 November 1990. See also NILOS, International Organizations and the Law of the Sea, Documentary Yearbook, Vol.6, (London, Martinus Nijhoff, 1990), pp. 70–114 at 79.

97 Section III, Paras.53–55, UNGA Doc.A/59/565 of 2 December 2004, https://documents-dds-ny.un.org/doc/UNDOC/GEN/N04/602/31/PDF/N0460231.pdf?OpenElement.

98 ENVIRONMENT – NATO'S STAKE, http://www.nato.int/cps/en/natohq/topics_91048.htm.

> Based on a broad definition of security that recognizes the importance of political, economic, social and environmental factors, NATO is addressing security challenges emanating from the environment. This includes extreme weather conditions, depletion of natural resources, pollution and so on – factors that can ultimately lead to disaster, regional tensions and violence. The Alliance is looking closely at how to best address environmental risks to security in general as well as those that directly impact military activities.[99]

However, more and more there has been a push to see environmental security as going beyond the limited prism of its linkage with military activities. Some scholars, such as Barnett, adopt a broader notion of environmental (in)security by linking environmental degradation with human security.[100] He explains environmental (in)security as follows:

> Environmental insecurity is defined here as the vulnerability of people to the effects of environmental degradation. So environmental insecurity is more than the physical processes of environmental degradation; it includes the way this degradation affects the welfare of human beings … Environmental insecurity is very much about risk. In the first instance, a risk to biosphere integrity entails risks to human health.[101]

Hulme points out that the broadest definition of environmental security would include 'man's ability to impact the stability and viability of the biosphere.'[102]

Even recently, we see that maritime security strategy instruments include the environmental issues, although the latter do not necessarily link directly to military activities, but become indirectly a vital maritime security challenge.[103] Some of these strategy instruments provisions are broadly framed to also regard environmental degradation by non-State actors in the relevant maritime zone as a maritime security threat. For instance, the European Union Maritime Strategy 2014, includes as part of maritime security risks and threats, environmental risks, which it relates to: the 'unsustainable and unauthorized exploitation of natural and marine resources, threats to biodiversity, IUU fishing, environmental degradation due to illegal or accidental discharge,

---

99 Environmental Security, Ibid.

100 J. Barnett, *The Meaning of Environmental Security: Ecological Politics and Policy in the New Security Era*, (London, Zed Publishers, 2001).

101 Ibid at p. 17.

102 Hulme, 'Environmental Security: Implications for International Law,' op.cit. at p. 9.

103 See for instance, Art. V(g) of the EU Maritime Security Strategy 2014 and Part XI of the 2050 Africa's Integrated Maritime Strategy 2014.

chemical, biological and nuclear pollution, in particular sea-dumped chemical munitions and unexploded ordnance.'[104]

As a result of the possibility of serious environmental impacts due to deep seabed mining, the protection of the environment is high on the agenda of the ISA and it is taking steps to minimize and manage the environmental risks of DSM.[105] As an illustration, the first advisory opinion of the ITLOS Seabed Disputes Chambers on the Responsibilities and obligations of States sponsoring persons and entities with respect to activities in the Area was mainly devoted to engaging with the extent of sponsoring States' obligation under LOSC 1982 for the protection of the environment as regard DSM activities by contractors sponsored by such States.[106] These contractors, including not only State entities, but also private commercial entities, such as Multinational Corporations.[107] As a matter of fact the Chambers in this advisory opinion sets the highest standards of due diligence and goes ahead to endorse a legal obligation to apply the precautionary approach, best environmental practices, and Environmental Impact Assessment (EIA).[108] Thus environmental security is an issue of vital concern as regard the Area.

104 Ibid. This has revised by a 2018 update and action plan which stresses that environmental degradation due to illegal or accidental discharge is a key maritime risks and threat. See https://www.consilium.europa.eu/en/press/press-releases/2018/06/26/maritime-security-eu-revises-its-action-plan/.

105 See Arts.139, 145, 209 and 215 of LOSC. Also, the ISA in its recent draft strategic plan for the period 2019–2023 gave the following as its mission statement: 'The mission of the International Seabed Authority is to be the organization through which States Parties organize and control activities in the Area, which is the common heritage of mankind, to promote the orderly, safe and responsible management and development of the resources of the Area for the benefit of mankind as a whole, including through the effective protection of the marine environment and contributing to agreed international objectives and principles, including the Sustainable Development Goals. This will be accomplished by developing and maintaining a comprehensive regulatory mechanism for commercial deep seabed mining that incorporates effective protection of the marine environment and of human health and safety, the equitable sharing of financial and other economic benefits from activities in the Area and allows for fully integrated participation of developing States through knowledge and best practice exchange consistent with the principle of the common heritage of mankind.' See ISBA/24/A/4 of 21 May 2018.

106 ITLOS Advisory Opinion of 1 February 2011, Case No. 17, https://www.itlos.org/cases/list-of-cases/case-no-17/.

107 See list of contractors, https://www.isa.org.jm/deep-seabed-minerals-contractors.

108 ITLOS Advisory Opinion of 1 February 2011. For analysis of this decision see David Freestone, 'Advisory Opinion of the Seabed Disputes Chamber of International Tribunal for the Law of the Sea on "Responsibilities and Obligations of States Sponsoring Persons and Entities With Respect To Activities in the Area"', (2011)15(7) *ASIL Insights*, https://www.asil.org/insights/volume/15/issue/7/advisory-opinion-seabed-disputes-chamber-international-tribunal-law-sea-

5 Conclusion

Maritime security, though a popular buzzword in contemporary times, is a nebulous concept that may mean different things to different people. In practical terms it is a matter that has relevance both in maritime zones within national jurisdiction and that beyond. This chapter has sought to explore key maritime security issues that are of relevance to the Area. It explored maritime security from the State-centric perspective of using the Area for exclusively peaceful purposes, and the variance in interpretation of what this actually means, as well as issues arising from the Seabed Arms Control Treaty framework, and its limitations. In addition, it explored the possibility of sponsoring States tensions with regard to mining sites allocated by the ISA. It further engaged with certain possible non-State centric maritime security future scenarios that the international community may need to take on as time goes on, especially when exploitation actually begins in the Area, such as piratical and terrorist acts against vessels involved in DSM, as well as the issue of environmental security arising from the risk of environment degradation due to DSM activities.

The Area, which has been described as 'a new frontier' for future mineral development and contains extremely valuable mineral resources, is a vital maritime zone that should be kept secure so users of this maritime domain would be protected. This is in line with the Common Heritage of Mankind which emphasizes the peaceful use of this maritime space and the importance of the effective protection of human lives.[109] ISA, as the international organization charged with the responsibility to 'organize and control activities in the Area,'[110] need to start thinking seriously about developing some sort of maritime security strategy as regard the Area to avoid irreparable damages to security interests of States, interested parties and the environment.

109 Arts.141 and 146 of LOSC.

110 Art.157(1) of LOSC.

PART 4

Exploitation of Marine Biodiversity and Living Resources on the Seabed beyond National Jurisdiction

CHAPTER 10

The Rights to Genetic Resources beyond National Jurisdiction: Challenges for the Ongoing Negotiations at the United Nations

Tullio Scovazzi

1 A New Negotiation

On 19 June 2015, following the recommendations of the Ad Hoc Open-ended Informal Working Group to Study Issues Relating to the Conservation and Sustainable Use of Marine Biological Diversity beyond Areas of National Jurisdiction, the United Nations General Assembly adopted by consensus Resolution 69/292, relating to the development of an international legally-binding instrument under the United Nations Convention on the Law of the Sea (UNCLOS) on the conservation and sustainable use of marine biological diversity of areas beyond national jurisdiction.

A Preparatory Committee (Prepcom), chaired by Mr. Eden Charles, was established by Resolution 69/292 with the mandate to make substantive recommendations to the General Assembly on the elements of a draft text of an international legally-binding instrument under the UNCLOS. The negotiations addressed four main topics, intended as a "package", in the sense that none of them can be separated from the others, namely:

- marine genetic resources, including questions on the sharing of benefits;
- measures such as area-based management tools, including marine protected areas;
- environmental impact assessments;
- capacity building and transfer of marine technology.

On the basis of the recommendations of the Prepcom, the General Assembly, by Resolution 72/249 of 24 December 2017, decided to convene an intergovernmental conference, with a view to developing the above mentioned instrument as soon as possible. The first session of the conference, chaired by Ms. Rena Lee, was held in September 2018.

Because of its scope and objectives, the present negotiation is likely to become a turning point in the progressive development of international law of the sea. A number of issues that are not fully covered by the UNCLOS are being addressed and hopefully will be regulated under an internationally

DOI:10.1163/9789004391567_012

agreed regime. However, the path towards the new legal instrument is far from being an easy one. While many States feel that there is a need for a new agreement to fill gaps in the UNCLOS, a minority of States express scepticism about such an instrument. They think that a better implementation of existing instruments would be sufficient to address the questions posed by conservation and sustainable use of marine biological diversity beyond national jurisdiction.

The topic of marine genetic resources, including questions on the sharing of benefits, presents a number of conceptual, political and legal difficulties that probably make it the most challenging aspect of the negotiation. After the second session of the Prepcom (2016), the chairperson published a document with his understanding of possible areas of convergence of views and possible issues for further discussion emanating from the discussion.[1] After the first session of the conference (2018), the chairperson issued a paper, called "President's aid to negotiations".[2] It presents options of provisions, including the "no text" option, that reflect the different positions taken by the negotiating States. Both documents show that fundamental differences persist as regards a number of matters on which the future regime should be built. This chapter reviews the main pending questions relating to the genetic resources regime.

2 The Concept of Common Heritage of Mankind

Under Art. 136 of the UNCLOS, the "Area", that is the seabed and ocean floor and subsoil thereof beyond the limits of national jurisdiction, and its resources, are the common heritage of mankind. This is the main innovating aspect of the UNCLOS with respect to the previous law of the sea regime. It is based on a new concept which is completely different from both the traditional concepts of sovereignty, which applies in the territorial sea and, to a certain extent, the Exclusive Economic Zone (EEZ), or freedom, which applies on the high seas. The common heritage of mankind is a third option (*tertium genus*), referring to a particular kind of resources located in the seabed beyond national jurisdiction.

The principle of common heritage of mankind was launched in a memorable speech made on 1 November 1967 at the United Nations (U.N.) General Assembly by the representative of Malta, Mr. Arvid Pardo. The opportunity for

1 See the document *Preparatory Committee Established by General Assembly Resolution 69/292: Development of an International Legally Binding Instrument under the United Nations Convention on the Law of the Sea on the Conservation and Sustainable Use of Marine Biological Diversity of Areas beyond National Jurisdiction – Chair's Overview of the Second Session of the Preparatory Committee*, available on the website of the United Nations.

2 U.N. Doc. A/CONF.232/2019/1 of 3 December 2018.

proposing it came from the technological developments which were expected to lead in a relatively short time to the commercial exploitation of the polymetallic nodules located on the surface of the deep seabed and containing some valuable minerals, such as manganese, nickel, cobalt and copper. The application of a regime of sovereignty was likely to lead to a series of competitive extensions by coastal States of the limits of national jurisdiction on the sea bed. The application of the regime of freedom was likely to lead to a rush towards the exploitation of economically and strategically valuable minerals under a first-come-first-served criterion. According to Mr. Pardo, the consequences of both possible scenarios would have been equally undesirable. They would have encompassed political tension, economic injustice and risks of pollution. In a few words, "the strong would get stronger, the rich richer".

The basic elements of the regime of common heritage of mankind, applying to the seabed beyond the limits of national jurisdiction, are the prohibition of national appropriation, the destination of the Area for peaceful purposes, the use of the Area and its resources for the benefit of mankind as a whole, with particular consideration for the interests and needs of developing countries, as well as the establishment of an international organization entitled to act on behalf of mankind in the exercise of rights over the resources. The assumption that sovereign States are bound to share the profits resulting from the exploitation of some natural resources could be considered as the second most revolutionary idea ever conceived in the framework of international law (of course, the first is the prohibition to make war, as embodied in the U.N. Charter).

The proposal by Malta led to Resolution 2749 (XXV), adopted on 17 December 1970, whereby the U.N. General Assembly solemnly declared that "the sea-bed and the ocean floor, and the subsoil thereof, beyond the limits of national jurisdiction (...), as well as the resources of the area, are the common heritage of mankind" (Art. 1). It should be noted that, according to this resolution, all the resources of the seabed beyond national jurisdiction, fall under the common heritage of mankind regime.

Today the basic elements of the concept of common heritage of mankind can be found in Part XI of the UNCLOS. The Area and its resources are the common heritage of mankind (Art. 136). No State can claim or exercise sovereignty over any part of the Area, nor can any State or natural or juridical person appropriate any part thereof (Art. 137, para. 1). The Area can be used exclusively for peaceful purposes (Art. 141). All rights over the resources of the Area are vested in mankind as a whole, on whose behalf an international organization, that is the International Sea-Bed Authority (ISBA), is entitled to act. Activities in the Area are carried out for the benefit of mankind as a whole, irrespective of the geographical location of States, whether coastal or land-locked, and taking into particular consideration the interests and needs of developing States (Art. 140, para. 1). The ISBA provides for the equitable sharing of financial and other economic benefits derived from activities in the Area through an appropriate mechanism (Art. 140, para. 2).[3]

As it is well known, in 1982 the text of the UNCLOS was submitted to vote after all efforts to reach consensus had been exhausted. It received 130 votes in favour, 4 against and 17 abstentions. Many developed States were among those which cast a negative vote or abstained. The main criticism was addressed to the regime of the Area. According to the developed States, it would have discouraged mining activities by individual States and private concerns, would have unduly favoured the monopoly of activities by the ISBA, would have burdened the contractors with excessive financial and other obligations relating also to the field of transfer of technology and would have disregarded the interests of industrialized countries in the decision-making procedures of the Council, the executive organ of the ISBA.

In 1994 it was clear that the UNCLOS was expected to formally enter into force without the participation of many developed countries, that is without the participation of the only States having the command of the high technological and financial capability required to engage in deep seabed mining activities. To avoid this danger, a new negotiation began on Part XI of the UNCLOS. It resulted in the Agreement Relating to the Implementation of Part XI of the UNCLOS, which was annexed to Resolution 48/263, adopted by the General Assembly on 17 August 1994. This resolution, while reaffirming that the Area and its resources are the common heritage of mankind, recognizes that "political and economic changes, including in particular a growing reliance on market principles, have necessitated the re-evaluation of some aspects of the regime for the Area and its resources".

The provisions of the Part XI Implementation Agreement and those of Part XI of the UNCLOS "shall be interpreted and applied together as a single instrument" (Art. 2). However, in the event of any inconsistency between the Part XI Implementation Agreement and Part XI itself, the provisions of the former shall prevail. In fact, the label of "implementation agreement" is a diplomatic device – a fig leaf, in non-diplomatic language – that covers the evident reality that the UNCLOS was amended and several provisions were changed within the original system for exploitation of the resources of the Area.

Following the adoption of the Implementation Agreement, the UNCLOS has today achieved a broad participation (with some notable exceptions). Although modified, the original spirit of the UNCLOS is not betrayed. The

3 See chapter 7 of this book, J. Dingwall, 'Commercial Mining Activities in the Deep Seabed Beyond National Jurisdiction: The International Legal Framework'.

principle of common heritage of mankind is still there and remains a major source of inspiration for a treaty that aims not only at the codification, but also the progressive development of international law.

Since several years the ISBA has been working on its mandate. In 2000, the ISBA Assembly approved the Regulations on prospecting and exploration for polymetallic nodules and in 2010 the Regulations on prospecting and exploration for polymetallic sulphides. The approval of a third set of regulations, relating to prospecting and exploration for cobalt-rich ferromanganese crusts has taken place in 2012. Several plans of work for exploration for polymetallic nodules and polymetallic sulphides have so far been approved by the ISBA Council. On 1 February 2011 the Seabed Disputes Chamber of the International Tribunal for the Law of the Sea rendered an advisory opinion on *Responsibilities and Obligations of States Sponsoring Persons and Entities with Respect to Activities in the Area*, as requested by the ISBA Council, which provides important clarifications on a number of aspects of the mining regime.[4]

However, the prospects coming from the mineral resources in the Area remain uncertain, as some factors have a negative impact on progress towards their commercial exploitation. They include the great depths at which deposits occur, the high costs involved in research and development of mining technology and the fact that, under current economic conditions, deep seabed mining may remain uncompetitive if compared to land-based mining.

3 The Conflicting Views on the General Principles Applicable to Genetic Resources beyond National Jurisdiction

In the meantime, the exploitation of genetic resources found beyond the limits of national jurisdiction has become a commercially promising activity. The deep seabed is not a desert, despite extreme conditions of cold, complete darkness and high pressure. It is the habitat of diverse forms of life associated with typical features, such as hydrothermal vents, cold water seeps, seamounts or deep water coral reefs. In particular, it supports forms of life that present unique genetic characteristics. For instance, animal communities of microorganisms, fish, crustaceans, polychaetes, echinoderms, coelenterates and molluscs live in the complete absence of sunlight where warm water springs from tectonically active areas (so-called hydrothermal vents). These communities, which do not depend on plant photosynthesis for their survival, rely on specially adapted micro-organisms able to synthesize organic compounds

4 For further consideration of the ISBA's Mining Code, see Ibid, chapter 7 of this book, J. Dingwall.

from the hydrothermal fluid of the vents (chemosyntesis). The ability of some deep seabed organisms to survive extreme temperatures (thermophiles and hyperthermofiles), high pressure (barophiles) and other extreme conditions (extremophiles) makes their genes of great interest to science and industry.[5]

So far, only few States and private entities have access to the financial means and sophisticated technologies needed to reach the deep seabed and to take samples of organisms found there, in order to study and isolating in laboratories genetic material deriving from such organisms. The result of this kind of activity could be the patenting of commercially valuable products to put them on the market.

Neither the UNCLOS, nor the 1992 Convention on Biological Diversity (CBD) provide any specific regulatory framework for the genetic resources of the seabed beyond national jurisdiction. This could be considered as an evident gap in international law of the sea.[6]

5 See chapter 2 of this book, E. Ramirez-Llodra, 'Deep-Sea ecosystems: biodiversity and anthropogenic impacts'.

6 On this question see L. Glowka, 'The Deepest of Ironies: Genetic Resources, Marine Scientific Research, and the Area', in (eds.) *Ocean Yearbook* (Brill, 1996) at p. 156; T. Scovazzi, 'Mining, Protection of the Environment, Scientific Research and Bioprospecting: Some Considerations on the Role of the International Sea-Bed Authority' (2004) *International Journal of Marine and Coastal Law*, 383; S. Arico and C. Salpin, *Bioprospecting of Genetic Resources in the Deep Seabed: Scientific, Legal and Policy Aspects* (United Nations University, 2005); D.K. Leary, *International Law and the Genetic Resources of the Deep Sea* (Brill Nijhoff, 2006); A.G. Oude Elferink, 'The Regime of the Area: Delineating the Scope of Application of the Common Heritage Principle and Freedom of the High Seas' (2007) *International Journal of Marine and Coastal Law* at p. 143; F. Millicay, 'A Legal Regime for the Biodiversity of the Area', in Nordquist, Long, Heider and Moore (eds.), *Law, Science and Ocean Management* (Brill, 2007), p. 739; de La Fayette, 'A New Regime for the Conservation and Sustainable Use of Marine Biodiversity and Genetic Resources Beyond the Limits of National Jurisdiction' Vol. 24:2 (2009) *International Journal of Marine and Coastal Law*, p. 221; F. Armas-Pfirter, *How Can Life in the Deep Seabed Be Protected?* (2009) *ibidem*, p. 281; L. Ridweway, 'Marine Genetic Resources: Outcomes of the United Nations Informal Consultative Process', (2009) *ibidem*, p. 309; R. Barnes, 'Entitlement to Marine Living Resources in Areas Beyond National Jurisdiction', in A.G. Oude Elferink& E.J. Molenaar (eds.), *The International Legal Regime of Areas beyond National Jurisdiction: Current and Future Developments*(Brill, Leiden, 2010), p. 83; T. Scovazzi, 'The Seabed Beyond the Limits of National Jurisdiction: General and Institutional Aspects', *ibidem*, p. 43; A. Jørem and M. Walløe Tvedt, 'Bioprospecting in the High Seas: Existing Rights and Obligations in View of a New Legal Regime for Marine Areas beyond National Jurisdiction'29 (2014:2) *International Journal of Marine and Coastal Law*, p. 321; L. Glowka, 'Marine Genetic Resources within and beyond the Limits of National Jurisdiction: Challenges and Opportunities Posed by Existing and Emerging International Legal Framework and Processes', in M.C. Ribeiro (ed.), *30 Years after the Signature of the United Nations Convention on the Law of the Sea* (Coimbra, 2014) p. 251; J. Wehrli and T. Cottier, 'Towards a Treaty Instrument on Marine Genetic Resources', *ibidem*, p. 517.

In 2006, the subject of the international regime for such resources began to be discussed within the already mentioned Working Group. Opposite views were put forward by the States concerned. Several States, especially within the group of developing countries, took the position that the UNCLOS principle of common heritage of mankind applies also to marine genetic resources and that the mandate of the ISBA should cover also such resources. Other States, in particular some developed countries, relied on the principle of freedom of the high seas, which would imply freedom of access to, and unrestricted exploitation of, deep seabed genetic resources.

This basic difference of views persists. On 18 December 2015 the chairperson of the Prepcom invited delegations who wished to do so to submit papers with their views on the elements of a future legally binding instrument.[7]

On the one hand, according to the views expressed on 9 September 2016 by the United States,

> there is no legal gap in regard to marine genetic resources in areas beyond national jurisdiction. Rather, these resources fall under the high seas regime of international law as reflected in the Law of the Sea Convention (LOSC). Marine genetic resources (MGR) in areas beyond national jurisdiction are not covered by the provisions pertaining to the International Seabed Authority or the Area (Part XI), except as part of the marine environment that must be protected in connection with 'activities in the Area' (which are defined as activities of exploration for and exploitation of the resources of the Area; in the context of the Area, 'resources' are expressly defined to include only mineral resources).
>
> We support application of the concept of the common heritage of mankind to mineral resources in the Area, as is clearly articulated in the Law of the Sea Convention. However, we do not support the application of this concept beyond that, and in particular, we oppose any application of the concept of 'common heritage of mankind' to marine genetic resources in areas beyond national jurisdiction. [...]
>
> In the high seas regime under international law, no State nor any other entity has sovereign rights over MGR in areas beyond national jurisdiction. Anyone can freely access such MGR in accordance with international law. [...]

7 The papers are available on the website of the United Nations.

> MGR in areas beyond national jurisdiction fall under the high seas regime of the law of the sea, and we do not want to see restrictions placed on those resources.[8]

On the other hand, according to the paper by Costa Rica, the principle of common heritage of mankind has a broad content that goes beyond the mineral resources of the Area:

> Common heritage of mankind is a principle of international law that states that the cosmos, defined territorial areas and elements of humanity's common heritage (cultural and natural) are common to humankind. The principle states that areas of Antarctica, the sea bed, and outer space cannot be monopolized for the benefit of one state or group of states alone, for they are to be used for the benefit of all mankind.[9]

Besides relying on the above mentioned General Assembly Resolution 2749 (XXV),[10] Costa Rica also points out that Art. 137 UNCLOS

> contemplates a general provision that encompasses all resources of the Area ('All rights in the resources of the Area are vested in mankind as a whole'), and a specific one for mineral resources ('The minerals recovered from the Area, however, may only be alienated in accordance with this Part and the rules, regulations and procedures of the Authority').
>
> In this regard all other resources from 'the Area' are also common heritage of mankind.
>
> This is the case of the marine genetic resources of the Area and any other resources that may be discovered in the future.[11]

Also in the view of the Federated States of Micronesia (paper of 14 March 2016),

> the living resources of ABNJs [= areas beyond national jurisdiction] are the common heritage of humankind, deserving of coordinated conservation and sustainable use by the international community lest the resources are forever depleted. This designation extends to marine genetic resources of ABNJs, in light of their potential for providing important benefits for

8 Paper by the United States (*supra*, n 7), p. 1.

9 Paper by Costa Rica (*supra*, n 7), p. 5.

10 *Supra*, para. 2.

11 Paper by Costa Rica (*supra*, n 7), p. 6.

> the health and livelihoods of all humankind for generations to come if properly studied and sustainably exploited. The sole exception to this designation among the living resources of ABNJs is fish, but only to the extent that existing international, regional, and subregional instruments, institutions, and other regulatory entities do not currently allow for such a designation to attach to the fish stocks they regulate.[12]

In the already mentioned document containing his overview of the second session of the Prepcom,[13] the chairperson remarked that

> discussions will need to intensify to identify ways to bridge the divergent views of delegations regarding the application of the high seas freedom and the common heritage of mankind in relation to marine genetic resources of areas beyond national jurisdiction, including questions on the sharing of benefits.

It may be asked whether the common heritage of mankind and the freedom of the high seas are mutually exclusive or could apply concurrently in an international instrument. In the already mentioned "President's aid to negotiations" paper[14] the following alternatives are provided:

> "Common heritage of mankind;
> *No text*;
> Freedom of the high seas;
> The freedom of the high seas shall govern the provisions for access to marine genetic resources of areas beyond national jurisdiction, while the common heritage of mankind shall govern their exploitation".

4 Some Considerations on the Conflicting Views

Both the conflicting views move from the frequently repeated assumption that the UNCLOS is the legal framework for all activities taking place in marine spaces. For instance, General Assembly Resolution 73/124 of 11 December 2018 on "Oceans and the Law of the Sea", following several previous resolutions on the same subject, emphasizes in the preamble "the universal and unified

12 Paper by Micronesia (*supra*, n 7), para. 5.
13 *Supra*, n 1.
14 *Supra*, n 2.

character" of the UNCLOS and reaffirms that it "sets out the legal framework within which all activities in the oceans and seas must be carried out".

However, such an assumption is not completely true.[15] There is no doubt that the UNCLOS is a cornerstone in the field of codification of international law. It has been rightly qualified as a "constitution for the oceans," "a monumental achievement in the international community," "the first comprehensive treaty dealing with practically every aspect of the uses and resources of the seas and the oceans," an instrument which "has successfully accommodated the competing interests of all nations."[16] Nevertheless, the UNCLOS, as any legal text, is linked to the time when it was negotiated and adopted (from 1973 to 1982). Being itself a product of time, the UNCLOS cannot stop the passing of time. While it provides a solid basis for the regulation of many matters, it would be illusory to think that the UNCLOS is the end of legal regulation. International law of the sea is subject to a process of natural evolution and progressive development which involves also the UNCLOS.

In particular, the UNCLOS cannot work the miracle of regulating those activities that were not foreseeable in the period when it was being negotiated. At this time, very little was known about the uses of the genetic properties of marine organisms. For evident chronological reasons, the economic value of this kind of uses was not taken in consideration by the UNCLOS negotiators. When dealing with the special regime of the Area and its resources, they had only mineral resources in mind. This is evident from the plain text of the UNCLOS, in which the expressions "genetic resources" and "bioprospecting" do not appear anywhere. By regulating today genetic resources no pre-existing balance could be altered for the simple reason that genetic resources were not included in any UNCLOS balance.

It is a matter of fact that the term "activities" in the Area is defined as "all activities of exploration for, and exploitation of the resources of the Area" (Art. 1, para. 1) and that the term "resources" of the Area is defined as "all solid, liquid or gaseous mineral resources *in-situ* in the Area at or beneath the seabed, including polymetallic nodules" (Art. 133, *a*). This means that the present UNCLOS regime of common heritage of mankind does not include the non-mineral resources of the Area and that the rules conceived for the exploration

15 See T. Scovazzi, 'Is the UN Convention on the Law of the Sea the Legal Framework for All Activities in the Sea? The Case of Bioprospecting', in D. Vidas (ed.), *Law, Technology and Science for Oceans in Globalisation*, (Brill Leiden, 2010), p. 309.

16 T. Koh, *A Constitution for the Oceans*, in U.N., *The Law of the Sea – Official Text of the United Nations Convention on the Law of the Sea with Annexes and Index*, New York, 1983, p. xxiii.

and exploitation of mineral resources cannot be extended to other resources located therein.

However, this does not prevent States from extending the most innovating principle of common heritage of mankind to newly discovered resources not covered by the UNCLOS, if they wished to do so. Yet, such an extension would be a natural evolution within the spirit of the UNCLOS, a treaty based on General Assembly Resolution 2749 (XXV)[17] and aiming at contributing "to the realization of a just and equitable international economic order which takes into account the interests and needs of mankind as a whole and, in particular the special interests and needs of developing countries, whether coastal or land-locked" (UNCLOS preamble).

The scope of the regime of the Area is already today broader than it may be believed at first sight. Under the UNCLOS, the legal condition of the Area as common heritage of mankind has an influence also on the regulation of activities that, although different from mining activities, are located in this space. The regime of the Area encompasses subjects which are more or less directly related to mining activities, such as marine scientific research (see Art. 143, para. 1), the preservation of the marine environment (see Art. 145) and the protection of underwater cultural heritage (see Art. 149). As far as the first two subjects are concerned, it is difficult to draw a clear-cut distinction between what takes place on the seabed and what in the super-jacent waters.

An interesting remark is that bioprospecting, an activity that is currently understood as the search for commercially valuable genetic resources, can already be considered as falling under the UNCLOS regime of marine scientific research. While the UNCLOS does not provide any definition of "marine scientific research", Art. 246, which applies to the EEZ and the continental shelf, makes a distinction between two kinds of marine scientific research projects, namely those carried out "to increase scientific knowledge of the marine environment for the benefit of all mankind" (Art. 246, para. 3) and those "of direct significance for the exploration and exploitation of natural resources, whether living or non-living" (Art. 246, para. 5, *a*). This distinction supports the view that, under the UNCLOS logic, research activities of direct significance for the purpose of exploration and exploitation of natural resources also fall under the general label of "marine scientific research". It follows that bioprospecting is also covered by Art. 143, para. 1, UNCLOS, which sets forth the rule that "marine scientific research in the Area shall be carried out exclusively

17 *Supra*, para. 2. Under this resolution, all the resources of the seabed beyond national jurisdiction fall under the common heritage of mankind regime.

for peaceful purposes and for the benefit of mankind as a whole."[18] This provision refers to any kind of marine scientific research and is not limited to research on mineral resources. Yet, the reading of Art. 143 UNCLOS in combination with Art. 246 UNLOS contradicts the assumption that there is an absolute freedom to carry out bioprospecting in the Area.[19] States which are active in bioprospecting in this space are already bound to contribute to the benefit of mankind as a whole.[20] There is an inextricable link between marine scientific research and bioprospecting. A research endeavour organised with the intent to increase human knowledge may well result in the discovery of commercially valuable information on genetic resources.

While a specific regime for genetic resources is lacking, the aim of sharing the benefits among all States, which was the main aspect of the seminal proposal made by Mr. Pardo, can still be seen as a basic objective embodied in the UNCLOS spirit. Also in the field of genetic resources, the application of the principle of freedom of the sea, that is based on a "first-come-first-served" or "freedom-of-fishing-for-resources" approach, leads to inequitable consequences. New cooperative schemes, based on provisions on access and sharing of benefits, should be envisaged as the basis of a future regime for marine genetic resources beyond national jurisdiction. This would also be in full conformity with the principle of fair and equitable sharing of the benefits arising out of the utilization of genetic resources, set forth by Art. 1 of the CBD and

18 Art. 241 UNCLOS is also relevant in the discussion on the legal condition of the genetic resources of the deep seabed. It provides that "marine scientific research activities shall not constitute the legal basis for any claim to any part of the marine environment or its resources".

19 Art. 143, para. 3, UNCLOS grants the States the right to carry out scientific research in the Area, but obliges them to cooperate with other States and the ISBA in various fields, including the dissemination of results. Also this provision refers to any kind of marine scientific research in the Area.

20 As stated by F. Francioni, 'Genetic Resources, Biotechnology and Human Rights: The International Legal Framework', in F. Francioni (ed.), *Biotechnologies and International Human Rights*, (Hart Publishing, 2007), p. 14, "the principle of common heritage in its substantive aspect is, like any norm of international law, capable of being applied in a decentralised manner by states. Even in the absence of *ad hoc* institutions every state is under an obligation to respect and fulfil the principle of the common heritage by ensuring that subjects within its jurisdiction do not act contrary to its object and purpose. This would be the case if a state authorised or negligently failed to prevent biotechnological activities in common spaces that had the effect of causing severe and irreversible damage to the unique biodiversity of that space. Similarly, a state would fail the common heritage if it authorised exclusive appropriation of genetic resources without requiring equitable sharing of pertinent scientific knowledge and without ensuring that a fair portion of economic benefits accruing from their exploitation be devoted to the conservation and sustainable development of such common resources".

by Art. 10 of the Protocol on Access to Genetic Resources and the Fair and Equitable Sharing of Benefits Arising from their Utilization (Nagoya, 2010).[21]

The major conceptual, but also practical, obstacle to be addressed by the ongoing negotiations on marine genetic resources beyond national jurisdiction is to be found in the ghost of a contradiction. On the one hand, it may not seem politically expedient to rely on the concept of common heritage of mankind, because this could since the beginning prevent many hopes to reach consensus among all the States involved in the negotiations. On the other, to meet evident requirements of equity, any future regime on genetic resources located in a "common" space could hardly result in the granting of all the benefits therefrom only to a few developed States. This brings into the picture the need for a regulation of access and sharing of benefits, which are the two typical elements of the principle of common heritage of mankind.

5 Other Relevant Questions

Besides the basic dilemma between a heritage-based or a freedom-based approach, progress on the way for the drafting of a new regime for marine genetic resources is also linked to the capacity of States to address and solve a number of specific issues which are far from being trivial.

In the already mentioned document containing his overview of the second session of the Prepcom,[22] the chairperson listed the following possible areas of convergence of views on marine genetic resources:

- "Usefulness of agreeing on working definitions of marine genetic resources and other key concepts at the preliminary stage;

---

21 Hereinafter: Nagoya Protocol. According to Art. 10, "Parties shall consider the need for and the modalities of a global multilateral benefit-sharing mechanism to address the fair and equitable sharing of benefits derived from the utilization of genetic resources and traditional knowledge associated with genetic resources that occur in transboundary situations or for which it is not possible to grant or obtain prior informed consent. The benefits shared by users of genetic resources and traditional knowledge associated with genetic resources through this mechanism shall be used to support the conservation of biological diversity and the sustainable use of its components globally". While the Nagoya Protocol does not apply to areas beyond national jurisdiction, it could become a source of inspiration for a future regime applying to resources located in such areas. Another source of inspiration could be the International Treaty on Plant Genetic Resources for Food and Agriculture (Rome, 2001), concluded under the auspices of the Food and Agriculture Organization of the United Nations (FAO). On this treaty, see C. Chiarolla, *Intellectual Property, Agriculture and Global Food Security* (Edward Elgar, Cheltenham, 2011).

22 *Supra*, n 1.

- Usefulness of drawing on definitions contained in existing instruments;
- Guiding principles and approaches constitute a cross-cutting issue;
- Benefit-sharing for non-monetary benefits;
- The rights of coastal States over their continental shelf should be respected;
- Benefit-sharing should/should also/could contribute to conservation and sustainable use of marine biodiversity of areas beyond national jurisdiction;
- Benefit-sharing should be beneficial to current and future generations, build capacity to access marine genetic resources of areas beyond national jurisdiction, and not be detrimental to research and development".

This list does not seem to reflect much convergence. A convergence to define key concepts at a preliminary stage does not mean that agreement on the content of the definitions has already been reached.[23] Moreover, any convergence on specific aspects of benefit-sharing seems conditional on an agreement on "whether to have a benefit-sharing mechanism" itself, which instead is listed among the issues requiring further discussions. Notable is that convergence exists on the assumption that the rights of coastal States over their continental shelf should be respected. This should be understood in the sense that the new regime will not affect coastal States' rights over the genetic resources found in the so-called extended continental shelf, that is in the seabed beyond the 200 n.m. and up to the limit of the continental margin (see the definition of continental shelf given by Art. 76 UNCLOS).[24]

Among the still pending issues related to genetic resources, the following seem particularly important.

5.1 *Access and Benefit Sharing Regimes*

In the already mentioned document containing the Prepcom chairperson's overview of the second session of the Prepcom,[25] the issues on "whether to regulate access to marine genetic resources of areas beyond national jurisdiction or not", "whether to have a benefit-sharing mechanism", "whether to include monetary benefits or not" and "whether access to resources *ex situ* / resources *in silico* / genetic sequence data should be included in an access and

23 According to the restrictive position taken by the United States, "marine genetic resources should be limited to material from living organisms containing functional genetic units of heredity. The definition should not include material such as enzymes or other proteins or information generated from MGR [= marine genetic resources] such as genetic sequence data": paper by the United States (*supra*, n 7), p. 1.

24 Under Art. 82 UNCLOS, the coastal States' obligation to make payments or contributions in kind in respect of the exploitation of the continental shelf beyond 200 nautical miles is limited to the exploitation of the "non-living resources" of the extended continental shelf.

25 *Supra*, n 1.

benefit-sharing regime" are listed among those requiring further discussions. The four issues are closely connected.

According to the United States, no access regime should be put in place at the international level, as it would be contrary to the principle of freedom of the high seas:

> In the high seas regime under international law, no State nor any other entity has sovereign rights over MGR [= marine genetic resources] in areas beyond national jurisdiction. Anyone can freely access such MGR in accordance with international law.[26]

In the United States' view, any future regime should not put in question the right of ownership over marine genetic resources. Access itself to genetic resources, in the sense of capacity building for those States that need it, should be considered as the main, if not the only, benefit that could be shared:

> As we do not have to discuss issues of ownership of MGR, we are instead free to share ideas on how sharing benefits might allow us to best achieve our overarching conservation objectives, and how such benefit sharing arrangements might work. Benefit sharing must be considered in the context of how any benefit sharing might allow us to achieve our conservation objectives. We do not want to advance any benefit sharing conditions that might create operational inefficiencies or otherwise obstruct beneficial research or development activities. In our view, especially given the difficulty for many to even access MGR, access to MGR in areas beyond national jurisdiction could itself be considered a benefit, and it is important for this group to discuss how we might be able to advance access as a benefit.[27]
>
> MGR in areas beyond national jurisdiction fall under the high seas regime of the law of the sea, and we do not want to see restrictions placed on those resources. If, however, a new instrument were to include a benefit-sharing regime, the benefits should focus on capacity building and conservation. At the last Prepcom session we heard compelling descriptions of the difficulties some scientists, particularly from developing countries, face in terms of having access to BBNJ [= biodiversity

26 Paper by the United States (*supra*, n 7), p. 2.

27 *Ibidem*, p. 2.

> beyond national jurisdiction]. Increased access to BBNJ, in ways acceptable to States, could be an example of positive benefit-sharing.[28]

A different view, based on strong access and benefit sharing regimes that presuppose an international machinery, is put forward by those States that are in favour of the principle of common heritage of mankind. As stated in the paper by Micronesia,

> it is the FSM's [Federated States of Micronesia's] view that marine genetic resources (MGRs) – being the common heritage of humankind and a key component of the marine biological diversity to be regulated by the BBNJ instrument – must be subject to an access and benefit sharing (ABS) regime that is robust, equitable, and properly attuned to the needs and interests of developing countries. The sharing of benefits accruing from access to and exploitation of MGRs of ABNJs is essential to ensuring that the BBNJ instrument will be acceptable for SIDS [Small Island Developing States] like the FSM with longstanding historical and cultural connections to and reliance on the Ocean's bounty.[29]

In Micronesia's view, the ABS machinery should play in favour of developing States and could include the establishment of an international fund and the prohibition of activities not complying with the future international instrument:

> For example, the BBNJ instrument can allow developed country Parties to sponsor exploration and exploitation of MGRs of ABNJs by private contractors – similar to the exploration and exploitation of non-living resources in the Area – in exchange for financial contributions by those Parties and/or contractors to an ABS fund, which will then be disbursed in an equitable manner to BBNJ States Parties. The BBNJ instrument can also encourage developing country Parties to sponsor private contractors as well, in exchange for granting those developing country Parties preferential access to the ABS fund. The sizes of financial contributions from developed country Parties may be based on a flat rate, or be proportionate to the geographical scope of the ABNJs to be explored and/or exploited by the private entities they contract. The BBNJ instrument can prohibit access to MGRs – including any benefits that accrue from

28 *Ibidem*, p. 3.

29 Paper by Micronesia (*supra*, n 7), para. 6.

their exploration and exploitation – that does not comply with this ABS regime.[30]

What kinds of benefits, if any, should be shared is another difficult question linked to the previous one. The Nagoya Protocol provides in an annex a detailed list of monetary and non-monetary benefits.[31] But it is open to discussion, especially in the case of monetary benefits, which among them could be relevant, *mutatis mutandis*, also for the future instrument being discussed in the present negotiations on marine biological diversity of areas beyond national jurisdiction.

According to the paper submitted on 29 August 2016 by Nauru on behalf of the group of twelve Pacific Small Island Developing States, the following forms of benefit sharing could be considered in the future implementing agreement:

30 *Ibidem.*

31 Pursuant to the Annex: "1. Monetary benefits may include, but not be limited to: (a) Access fees/fee per sample collected or otherwise acquired; (b) Up-front payments; (c) Milestone payments; (d) Payment of royalties; (e) Licence fees in case of commercialization; (f) Special fees to be paid to trust funds supporting conservation and sustainable use of biodiversity; (g) Salaries and preferential terms where mutually agreed; (h) Research funding; (i) Joint ventures; (j) Joint ownership of relevant intellectual property rights. 2. Non-monetary benefits may include, but not be limited to: (a) Sharing of research and development results; (b) Collaboration, cooperation and contribution in scientific research and development programmes, particularly biotechnological research activities, where possible in the Party providing genetic resources; (c) Participation in product development; (d) Collaboration, cooperation and contribution in education and training; (e) Admittance to ex situ facilities of genetic resources and to databases; (f) Transfer to the provider of the genetic resources of knowledge and technology under fair and most favourable terms, including on concessional and preferential terms where agreed, in particular, knowledge and technology that make use of genetic resources, including biotechnology, or that are relevant to the conservation and sustainable utilization of biological diversity; (g) Strengthening capacities for technology transfer; (h) Institutional capacity-building; (i) Human and material resources to strengthen the capacities for the administration and enforcement of access regulations; (j) Training related to genetic resources with the full participation of countries providing genetic resources, and where possible, in such countries; (k) Access to scientific information relevant to conservation and sustainable use of biological diversity, including biological inventories and taxonomic studies; (l) Contributions to the local economy; (m) Research directed towards priority needs, such as health and food security, taking into account domestic uses of genetic resources in the Party providing genetic resources; (n) Institutional and professional relationships that can arise from an access and benefit-sharing agreement and subsequent collaborative activities; (o) Food and livelihood security benefits; (p) Social recognition; (q) Joint ownership of relevant intellectual property rights".

Monetary:
A trust fund could be created to fund capacity building initiatives for developing countries, in particular SIDS [= Small Island Developing States]. Given our special circumstances, this trust fund could provide a special allocation to SIDS.

This trust fund could be funded by both:

Royalties or milestones payments from the exploitation of MGRs could be transferred to a trust fund. Not all MGR-related research lead to lucrative outcomes. Therefore, a system focusing solely on royalties or milestones payments will not suffice.

Mandatory fees: Proponents of MGR related activities could be required to financially contribute to a trust fund.

Non-monetary:
Technology transfer refers to instruments, equipment, vessels, processes and methodologies to produce and use knowledge to improve study and understanding of ocean/coastal nature/resources:

Proponents of MGR related activities could be required to transfer specific technology.

In the SAMOA Pathway it was agreed that Marine technology transfer should consist of appropriate, reliable, affordable, modern and environmentally sound technologies (including software and equipment) and know-how (based on SAMOA Pathway, para. 111).

Knowledge sharing and access to information: Possible consideration for a clearinghouse mechanism.

Capacity building: Proponents of MGR related activities could be required to provide capacity building to SIDS. Elements of capacity building could include as an initial matter:

– the provision of education/training in science and technologies, policy and governance, including through joint research efforts supported through the establishment of a global scholarship fund, and enhanced through collaboration in research and development on marine genetic resources;

– support for and development of regional centres of excellence (such as the University of South Pacific) to address regional needs and provide long-term education and training.[32]

32 Paper by Nauru (*supra*, note 7), p. 6.

The definition to be chosen for marine genetic resources has also an influence on the question of the access and benefit sharing regimes. According to the United States,

> it is best to limit the definition of marine genetic resources to *in situ* collection. Including *ex situ* samples and procedures in the definition of MGR would introduce a range of complex variables, such as how materials are collected, transported, and stored. These would dramatically complicate the operation of BBNJ benefit-sharing and move us farther away from achieving our objectives.[33]

However, the Pacific Small Island Developing States stress the importance of being provided access also to *ex situ* marine genetic resources:

> (...) special consideration for developing countries, in particular SIDS, in the access to MGRs activities is important. This requires capacity to undertake such research and prospecting activity or research *ex-situ*. It also requires appropriate technology. Furthermore, access to the data gathered from accessing these resources should also be provided. This requires capacity to understand and use the data. Therefore, the concepts of transparency and traceability of MGRs are important.[34]

The already mentioned "President's aid to negotiations" paper[35] shows that the alternative between both kinds of benefits (monetary and non-monetary), on the one hand, and only non-monetary benefits, on the other, is a crucial question that still needs to be settled:

> For the purposes of this Part, the term 'benefits' shall mean monetary and non-monetary benefits;

---

33 Paper by the United States (*supra*, n 7), p. 2. As regards the so-called *in silico* genetic resources, the United States makes the following remarks: "Indeed, for purposes of clarity, we should refer to information taken from MGR by its proper name: genetic sequence data, or GSD, and not use the term *in silico*. GSD is information and its sharing can promote uses of GSD in research and development. If GSD is included, and a decision were made to attempt to trace the downloading and use of such information, how would that work? We struggle to envision a scenario that could be workable. How could we manage benefit-sharing (and promote compliance) if data, something that is freely and openly shared as part of research best-practices, were included in it?" (*ibidem*).

34 Paper by Nauru (*supra*, n 7), p. 6.

35 *Supra*, n 2.

> For the purposes of this Part, the term 'benefits' shall mean non-monetary benefits. It includes but is not limited to capacity-building, the exchange and public availability of information and scientific knowledge, access to samples and sample collections, access to technology and technical knowledge and transfer of technology.
>
> Benefits arising from the use of marine genetic resources of areas beyond national jurisdiction may be shared on a voluntary basis.

The last option could be seen as very disappointing for all those who believe in the principle of common heritage of mankind.

If a strong access and benefit sharing mechanism is put in place, the question should be asked about the role of the ISBA within the future regime for marine genetic resources. It would be logical to involve an international organization that has already acquired knowledge and experience in activities on the seabed beyond the limits of national jurisdiction and in the protection of its unique environment. Yet, the mandate of the ISBA deserves close scrutiny, especially if it is to be understood not only as an entity involved in marine mining activities in competition with others, but as the international organisation which bears the main responsibility to realize a just and equitable economic order of the oceans and seas. Nothing prevents States from expanding the responsibility of the ISBA and granting it some management competences also in the field of genetic resources.

5.2 *Intellectual Property Rights*

Another issue requiring further discussions is whether to address intellectual property rights (IPRS) in the future international instrument.

It is a matter of fact that only few States and private entities have the financial means and sophisticated technologies needed not only to reach the genetic resources of the deep seabed, but also to develop commercially valuable products therefrom. Already in 2005, a document issued by the Subsidiary Body on Scientific, Technical and Technological Advice (SBSTTA) of the CBD pointed out that

> reaching deep seabed extreme environments and maintaining alive the sampled organisms, as well as culturing them, requires sophisticated and expensive technologies. (…) Typically, the technology associated with research on deep seabed genetic resources involves: oceanographic vessels equipped with sonar technology, manned or unmanned submersible vehicles; *in situ* sampling tools; technology related to culture methods;

> molecular biology technology and techniques; and technology associated with the different steps of the commercialization process of derivatives of deep seabed genetic resources. With the exception of basic molecular biology techniques, most of the technology necessary for accessing the deep seabed and studying and isolating its organisms is owned by research institutions, both public and private. To date, only very few countries have access to these technologies.[36]

The prospects for commercial application of marine bioprospecting activities are quite promising, especially in the pharmaceutical sector. However, the IPRs legislation of several States does not compel the applicant for a patent to disclose the origin of the genetic materials used. This may prevent the establishment of an international regime which would provide for the joint ownership over, or other form of benefit sharing from, IPRs related to genetic resources found in the seabed beyond national jurisdiction.

It is not surprising that those States that support the application of the principle of common heritage of mankind also envisage substantive changes in the present regime of IPRs. As stated by Micronesia,

> the study of MGRs is a cutting edge field, and as such, it is subject to competing and complex intellectual property (IP) considerations. It is the FSM's view that the Ocean must be an open source for research innovation and solutions for the ills and challenges of humankind, particularly the living resources of ABNJs. Although discoveries and commercial innovations are encouraged with regard to MGRs of ABNJs, such innovation should not cater exclusively to profit motives, but should instead respect the common interest of humankind in benefitting from the sustainable use of such MGRs. Discoveries among living resources of ABNJs that are useful for the further preservation of biodiversity and MGRs should not be beholden to exclusive rights of discrete private entities or individuals. IP rights should not be applied with such rigidity in pursuit of exclusively commercial ends that they preclude the use of the targeted living resources to ameliorate some of the common concerns of humankind. Thus, it is the FSM's view that any IP rights that attach to MGRs of ABNJs under the BBNJ instrument must be secondary to the entitlements of the

36 *Status and Trends of, and Threats to, Deep Seabed Genetic Resources beyond National Jurisdiction, and Identification of Technical Options for their Conservation and Sustainable Use*, doc. UNEP/CBD/SBSTTA/11/11 of 22 July 2005, paras. 12 and 13.

> international community that flow from the designation of MGRs as the common heritage of humankind. An ABS regime for MGRs of ABNJs must reflect this treatment of IP rights.[37]

Any proposals of innovations in rules regarding IPRs have proved to be a major subject of discussion in the ongoing negotiations. The already mentioned "President's aid to negotiations" paper[38] under the heading "intellectual property rights" lists options ranging from compulsory disclosure of the origin of marine genetic resources involved in patents to "no text".

5.3 *Genetic Resources of the Water Column*

The United States puts forward that MGRs found in the high seas water column above the seabed should not be included in any benefit sharing arrangement. They rely on the precedent established by Art. 77, para. 4, UNCLOS, where species on the seabed (sedentary species) are treated differently from species found in the super-jacent waters.[39]

Other States take the position that the future regime should be the same and should apply to all marine spaces beyond national jurisdiction, irrespective of whether the genetic resources are found in the waters or on the seabed.

In the Prepcom chairperson's overview, the issue of "whether to include marine genetic resources of the water column beyond areas of national jurisdiction in a benefit-sharing regime" is listed among the issues requiring further discussions. Conflicting options appear also in the "President's aid to negotiations" paper.

5.4 *Role of Traditional Knowledge*

The view has been expressed that those indigenous coastal communities who possess special knowledge about living resources in marine areas beyond national jurisdiction should be afforded special consideration under the future regime. According to Micronesia,

> the ancestors of the current indigenous inhabitants of the islands and atolls of the FSM navigated the wide expanse of the Pacific Ocean using, among other things, the creatures of the Ocean as guides and sources of sustenance, based on ancient knowledge about their behavior patterns and nutritional values. In ABS regimes under the Nagoya Protocol

37 Paper by Micronesia (*supra*, n 7), p. 2.

38 *Supra*, n 2.

39 Paper by the United States (*supra*, n 7), p. 3.

> [...]to the Convention on Biological Diversity (CBD), entities wishing to access biological diversity under the jurisdiction of a CBD Contracting Party must first secure the informed consent of indigenous communities in the jurisdiction who are considered custodians of that biological diversity. Similarly, for the BBNJ instrument, it is the FSM's view that an entity wishing to study, explore, and/or exploit MGRs in a particular ABNJ must at the very least consult the indigenous communities of the coastal States whose EEZs abut that ABNJ prior to commencing their engagement with the MGRs. The ABS regime for the MGRs can also prioritize those indigenous communities in terms of receiving benefits from the regime, including benefits from the aforementioned ABS fund.[40]

In the Prepcom Chairperson's overview, the issue of the "role of traditional knowledge in the conservation and sustainable use of marine biodiversity of areas beyond national jurisdiction" is listed among the issues requiring further discussions. The question is still open, at least insofar it is linked to agreement on the access and benefit sharing regimes.

5.5 *The Double Nature of Fish*

While listed in the Prepcom chairperson's overview as an issue requiring further discussions, "whether to take into account the distinction between fish used for its genetic properties and fish used as commodity when developing a definition" should not become a major point of contention. Even though different options are listed, the "President's aid to negotiations" paper goes in the direction that the future instrument will not apply to fish used as a commodity.

A distinction between fishing and the exploitation of marine genetic resources is needed and is conceptually clear enough. Both fishing and the exploitation of marine genetic resources presuppose the taking of living resources from the sea. However, what is important for the fishermen is the whole body of the exploited resource or tangible parts of it, in order to harvest it for consumption. Fishing activities aim at exploiting large quantities of given living resources to produce the maximum yield from the target species. The purpose of those who look for genetic resources is different. Here the bodies are considered mainly as receptacles of genes. The objective is to seize the functional units of heredity to determine whether and how they can be used or stored waiting for a future use. For this kind of activity there is no need of large quantities of living resources, as quality and difference are much more significant than quantity and similarity. Unlike the case of fisheries, the

40 Paper by Micronesia (*supra*, note 7), para. 8.

added value of the use of genetic material is tremendous and questions of patents and protection of intellectual property can easily arise. Although genetic resources could be extracted also from fish or other marine living resources normally used as commodities, it would be completely illogical to apply to activities directed at marine genetic resources the rules of the UNCLOS relating to conservation and management of the living resources of the high seas (for example, the determination of the total allowable catch). It would equally be illogical to transform the future implementation agreement into a treaty devoted to the management and allocation of fish resources.

As remarked by the United States,

> (…) if marine genetic resources from a fish are used for their genetic properties, they should be treated as other MGR under any new instrument. There would be no reason to treat a gene from a fish differently than a gene from any other marine organism. If, however, fish are used as a commodity, then many would fall under existing regimes, including regional fisheries management organizations, and should not be addressed here.[41]

5.6 *The "Straddling" Genetic Resources*

The already mentioned Nagoya Protocol, which applies to the components of biological diversity found in areas within the limits of its national jurisdiction (see Art. 4, *b*, CBD) grants a number of rights to the country of origin of genetic resources, intended as "the country which possesses those genetic resources in *in-situ* conditions" (Art. 2 CBD). As areas of deep waters can also be located within the EEZ, depending on the broad or reduced extension of the continental margin of a given coastal State, it may happen that the same genetic resource is found both within and beyond areas of national jurisdiction. It would potentially fall under two different regimes, namely those of the Nagoya Protocol and the future implementation agreement. The problem would get worse if the implementation agreement were to cover also the genetic resources of high seas waters.

The "President's aid to negotiations" paper includes, as an option, a detailed provision addressing the thorny question of "straddling" genetic resources:

> Activities with respect to marine genetic resources of areas beyond national jurisdiction that are also found in areas within national jurisdiction shall be conducted with due regard to the rights and legitimate interests of any coastal State under the jurisdiction of which such

41 Paper by the United States (*supra*, n 7), p. 2.

> resources are found. Consultations, including a system of prior notification, shall be undertaken with the State concerned, with a view to avoiding infringement of such rights and interests. In cases where activities with respect to marine genetic resources of areas beyond national jurisdiction may result in the exploitation of marine genetic resources which are found in areas both within and beyond national jurisdiction, the prior consent of the coastal State concerned shall be required.

6 Conclusive Remark

The work for a third UNCLOS implementation agreement is ongoing.[42] If the agreement will ever be adopted, it could lead to a major improvement in international law of the sea and to a more equitable system of exploitation of marine resources. Given the difficulties and intricacies of many among the issues under discussion, the work needs to be pursued by the States involved in a spirit of moderation and through an effort constructive imagination. Such qualities have not yet sufficiently materialized. Time will tell about the future meetings.

42 See T. Scovazzi, 'The Negotiations for a Binding Instrument on the Conservation and Sustainable Use of Marine Biological Diversity beyond National Jurisdiction', Vol. 70 (2016) *Marine Policy*, p. 188.

CHAPTER 11

Marine Genetic Resources: a Practical Legal Approach to Stimulate Research, Conservation and Benefit Sharing

Morten Walløe Tvedt

1 Introduction*

In September 2018, the negotiation of a new international legally binding instrument (ILBI) for the Area Beyond National Jurisdiction (ABNJ) took the first really concrete steps under the United Nations Convention on the Law of the Sea (UNCLOS[1]).[2] Prior to the General Assembly establishing this Intergovernmental Committee there have been meetings in a Preparatory Committee and Ad Hoc Working group, dating back more than a decade.[3] The second meeting of the Intergovernmental Committee spent, at its meeting in March–April 2019, two and a half days discussing the role of marine genetic resources (MGR). In the working document 'Chair's streamlined non-paper on elements of a draft text of an international legally-binding instrument' that might be the first textual step towards what might be a protocol to UNCLOS, MGR are allocated six pages drawing on experiences of other regulations of genetic resources.[4]

There are a number of special regulatory and factual features for marine genetic resources that cannot be copy-pasted from any of the existing regimes

* The research behind this chapter is conducted under the project *BiosPolar* funded by the Norwegian Research Council "Legal frameworks for bioprospecting and bio-innovation in Polar Regions (BiosPolar)" (Project number 257631/E10).

1 United Nations Convention on the Law of the Sea, 1833 UNTS 3, adopted on 10 December 1982.

2 According to the General Assembly resolution 72/249 of 24 December 2017.

3 The Ad Hoc was established by resolution of the United Nations General Assembly, *Oceans and the Law of the Sea* (A/RES/59/24, 2004). For more information about the early steps, see https://www.un.org/Depts/los/biodiversityworkinggroup/biodiversityworkinggroup.htm.

4 https://www.un.org/depts/los/biodiversity/prepcom_files/Chairs_streamlined_non-paper_to_delegations.pdf Leary focuses on the many topics on which there is yet no consensus, in Leary, D. (2019) Agreeing to disagree on what we have or have not agreed on: The current state of play of the BBNJ negotiations on the status of marine genetic resources in areas beyond national jurisdiction, Marine Policy. Volume 99, pp. 21–29.

on Access and Benefit Sharing (ABS).[5] The ABS system of Convention of Biological Diversity (CBD) or its Nagoya Protocol (NP)[6] does not apply directly to ABNJ.[7] One obvious feature is the lack of a state with sovereign rights to and jurisdiction over marine genetic resources. Consequently, there are no existing institutions to grant access, provide a prior informed consent and be the counterpart in a contract.

Even if the 'Chair's non-paper' from 2019 is built on definitions and concepts from existing ABS regimes, they are not legally binding in the ABNJ. Thus, a new instrument in the ABNJ can choose other approaches. The UNCLOS itself takes both a *resource* approach and an *activity* approach in its regulation.[8] The resource approach appears in the regulation of 'living resources'. The concept of 'living resources' focuses rather on bulk harvest and is not geared towards 'marine genetic resources'. Rules concerning 'marine scientific research' constitute the most relevant activity for the topic here. There is some overlap between MRS and bioprospecting as both cover types of scientific research, whereas bioprospecting is broader.[9] Since neither the marine genetic resources nor the activity bioprospecting are regulated for the high seas and deep seabed, it must be concluded that there is regulatory room for the negotiations.

5 The EU suggested in 2008 taking the multilateral mechanism under the International Treaty on Plant Genetic Resources for Food and Agriculture (ITPGRFA, International Treaty on Plant Genetic Resources for Food and Agriculture, 2400 UNTS 303, adopted on 3 November 2001) as a reference point for the discussions. Though most Parties welcomed the proposal (Broggiato, A. (2008) 'Marine Biological Diversity in Areas Beyond National Jurisdiction', *Environmental Law and Policy*, 38(4), 182–188, at p. 186) the Plant Treaty is made for facilitating plant breeding which a different technical field. Differences are also highlighted by Leary, explaining that regime is tailored for mineral resource exploitation and does not answer how benefit sharing would be done when inventions are patent protected (Leary, D. (2009) 'International Law and the Genetic Resources of the Deep Sea', in Vidas, D. (ed.) *Law, Technology and Science for Oceans in Globalisation: IUU Fishing, Oil Pollution, Bioprospecting, Outer Continental shelf.* Martinus Nijhoff, Leiden, pp. 353–69, at p. 366).

6 Convention on Biological Diversity, 1760 UNTS 79, adopted on 5 June 1992, article 15. Nagoya Protocol on Access to Genetic Resources and the Fair and Equitable Sharing of Benefits Arising from their Utilization to the Convention on Biological Diversity, adopted on 29 October 2010.

7 Its provisions apply in relation to each contracting party to areas within national jurisdiction and may only be extended to ABNJ as a result of flag state jurisdiction. See *ibid.*, article 4(a) and (b). See as well Chapter 11 of this book, Tullio Scovazzi, 'The Rights to Genetic Resources beyond National Jurisdiction: Challenges for the ongoing Negotiations at the United Nations'.

8 Jørem, A. and M.W. Tvedt (2014). "Bioprospecting in the High Seas: Existing Rights and Obligations in View of a New Legal Regime for Marine Areas beyond National Jurisdiction." The International Journal of Marine and Coastal Law 29: 321–343, at p. 324 *sig.*

9 Jørem and Tvedt conducts a thorough study of these concepts, see *ibid.*, pp. 327–333.

The basic idea of introducing regulation of any aspects of marine genetic resources is to change the behaviour of researchers and commercial firms using the resources or conducting the activities. Meanwhile, Parties to UNCLOS are states. If an international legally binding instrument is to have any potential to bind private parties, it is not enough to address the questions in international law; the content of a new instrument must be designed in a manner possible to be implemented by and enforced on private legal persons. One goal of this chapter is to explore options for how to make an ABS system for ABNJ legally binding on users. This means that all suggestions are tested against whether it would be possible to oblige private parties to a system of both access and benefit sharing.

The ultimate objective of this chapter is to identify the legal elements required for a functional model of a binding access and benefit sharing system that also promotes global research and development based on marine genetic resources.

This chapter contributes to the debate by identifying the legal solutions to the system being proposed (Section 2). After having looked at the practical system, Section 3 reviews the legal tools that are already available without amending the UNCLOS, and, based on a review of the bioprospecting activities envisaged (Section 4), review the set of key obligations related to registering of samples and depository in a repository (Section 5). The chapter ends with a discussion of the necessary institutional framework (Section 6), before summarizing the solution proposed (Section 7).

2 Short Description of the Proposed 'Realistic and Functional' System for ABS

There seems to be a growing consensus among legal scholars that an ABS regime needs to include States without capacity to conduct cruises themselves by making samples and information available to a broader audience than to the collector alone.[10] This approach to a regulatory system also draws moral legitimacy from distributive theories, as the *Theory of Justice*.[11] The essence of

10 For one of the most recent proposals: Broggiato, A. (2018), *Mare Geneticum: Balancing Governance of Marine Genetic Resources in International Waters*, The International Journal of Marine and Coastal Law 33, doi 10.1163/15718085-13310030, pp. 3–33, about the rationale at pp. 10–12 and 14–16.

11 This could be based on a morally based fairness argument; however, it has been conceptualized in different ways. Fairness, a core perspective, is that of Rawls's, as described in *A Theory of Justice* (Rawls, J. (1999) *A Theory of Justice: Revised Edition*, Belknapp Press,

this conception of fairness is that any new legal tool must contribute at least so that the worse off be put in a slightly better position.

Increasingly, it is suggested making samples available by storing them in *ex situ* conservation conditions sites.[12] Accessible collections can ensure sustainable use,[13] encourage research,[14] and lead to the development of commercially interesting products or processes.[15] The establishment of just one global repository would not be a practical solution since it will by necessity be very far from at least some of the oceans where collections take place. Therefore, a distributed global system repositories with branches bound together in a global network is probably the only practical solution.

One practical idea for governing MGR including both monetary and non-monetary benefit-sharing was developed by Tvedt and Jørem in 2013.[16] In

---

Cambridge (MA)). Although Rawls did not himself apply his theory to international issues without modification (Rawls, J. (1993) 'The Law of the Peoples', *Critical Inquiry,* 20(1), 36–68, pp. 36–68; Rawls, 1999, pp. 331–5). UNCLOS, preamble, paragraph 5. On the justice of using genetic resources, see also D. Schroeder, and T. Pogge, 'Justice and the Convention on Biological Diversity', *Ethics and International Affairs,* (2009) 23(3), 267–80. The link to fairness has been contested, see R.D. Simpson, 'Biodiversity Prospecting: Shopping the Wild Is Not the Key to Conservation', *Resources,* (1997) 126, 12–5.

12 CBD, Article 2, paragraph 13 defines *in situ* conservation as "the conservation of ecosystems and natural habitats and the maintenance and recovery of viable populations of species in their natural surroundings [...]". *Ibid.*, Article 2, paragraph 8: *Ex situ* conservation is the "conservation of components of biological diversity outside their natural habitats".

13 Drawing the lines back to sustainable development understood as "limitations imposed by the state of technology and social organization on the environment's ability to meet present and future needs." in World Commission on Environment and Development, *Report of the World Commission on Environment and Development: Our Common Future* (Annex to document A/42/427, 1987), part IV, paragraph 1. See also P. Sands, *Principles of International Environmental Law* (Cambridge University Press, Cambridge, 2003), p. 253.

14 This is a utilitarian argument of taking advantage of biological material. See e.g. Ad Hoc Open-Ended Informal Working Group, *Letter dated 15 May 2008 from the Co-Chairpersons of the Ad-Hoc Open-Ended Informal Working Group to Study Issues Relating to the Conservation of Marine Biological Diversity in Areas Beyond National Jurisdiction to the President of the General Assembly* (A/63/79), paragraphs 6 and 10. See also Harden-Davies, H. (2017) *Deep-sea genetic resources: New frontiers for science and stewardship in areas beyond national jurisdiction,* Deep Sea Research Part II: Topical Studies in Oceanography. Volume 137. March 2017, pp. 504–513. On the effects of the economic downturn for biotechnology, see G. Giovannetti, and G. Jaggi, *Beyond Borders: Global Biotechnology Report 2012,* (Ernst & Young, 2012), n.p., pp. 25–39.

15 Tiller, R. et al., (2019), *The once and future treaty: Towards a new regime for biodiversity in areas beyond national jurisdiction,* Marine Policy. Volume 99, pp. 239–242.

16 Tvedt, Morten Walløe and Ane E. Jørem. "Bioprospecting in the High Seas: Regulatory Options for Benefit Sharing" in Journal of World Intellectual Property 16 (2013) 3–4.

the 2018 article *Mare Geneticum*, similar ideas were described by Broggiato and Vanagt et al.[17] The remainder of this chapter will scrutinize the legal requirements and possibilities of such a factual system that strike the balance between open access and commercial interests in marine genetic resources.

Such an ABS system builds on the successive steps of access, deposit, utilisation and then creation of benefit sharing. Broggiato et al. elaborate further on the manner to address those successive steps and suggest a list of eight actions, as described below. Their analysis is very normative in arguing how the system should be set up. Although this list has the benefit of providing an outline of how an ABS system could set up in the ABNJ, it remains vague in terms of legally enforceable elements. They merely mention that 'these conditions will need to be encoded into legal obligations' suggesting a clickwrap.[18] Given that each of these steps entails different legal challenges, for each of the actions identified, the present author has indicated in the second column the related legal aspects that will need to be addressed.

The elements in the right column will be further developed in the next sections, with the aim of transposing the different actions on the list into legally

TABLE 11.1

Proposed steps in Broggiato et al.	Required legal action
Submit and register plans for research or bioprospecting (at pp. 17–18)	To be binding on the users, this needs to be an obligation in national law based on flag State jurisdiction
Register the material sampled in the "OPEN" system	Can be imposed either in national law or in a contract entered into at an earlier point of time (otherwise voluntary)
Deposit samples of what has been collected (at p. 22)	The flag State can set this as a criterion before a cruise, for this to be binding after the cruise/collection has ended, it would have to have been translated into a contractual obligation.

pp. 150–167 and presentation by Tvedt at FNI side-event 23th August 2013.

17 Broggiato, A. (2018), *Mare Geneticum: Balancing Governance of Marine Genetic Resources in International Waters*, The International Journal of Marine and Coastal Law 33, doi 10.1163/15718085-13310030, pp. 3–33.

18 Ibid. at p. 22.

TABLE 11.1 (*cont.*)

Proposed steps in Broggiato et al.	Required legal action
Research according to the plans submitted initially (at pp. 17–18)	Any submitted plans are non-binding unless that are made binding by a contract.
Update information in the data base when transferring to subsequent user	Theoretically possible to regulate in a contract, but there are considerable enforcement challenges.
Embargo period for others or the payment of exclusivity fee (at p. 22)	The collection could regulate the criteria for when others can use the material in its criteria for receiving material.
Commercialisation (at pp. 22–24)	Other types of legislation than that of bioprospecting, like patent law or that of product approval, come into play. A contract must set these rules before bioprospecting starts.
Benefit sharing as a percentage of the gross sales of a product (at pp. 22–24 and 28–29)	A national law of users imposing an obligation of payment or a contractual obligation.

binding obligations. Indeed, in their discussion, Broggiato et al. presuppose legally binding criteria, without discussing the legal tools for making the respective steps binding on the user. If these steps are not turned into binding obligations on the users (and not only in international law) then the system will end up becoming voluntary on the users. To ensure users change their behaviour, however, the question of how to make obligations legally binding and enforceable must be answered. Any lack of means to enforce obligations renders the system to be voluntary and therefore not legally functional[19] and enforceable.[20] This chapter takes the body of proposals as a point of departure and discusses what would be required to make their proposals into a legally workable system.

A central contribution with the present chapter is to reflect on the legal tools that could introduce regulation on marine genetic resources in the ABNJ,

19 Functionality of ABS is discussed by Tvedt (2017) in Beyond Nagoya: Towards a Legally Functional System of Access and Benefit-sharing in Global Governance of Genetic Resources Access and Benefit Sharing after the Nagoya Protocol. Edited by Sebastian Oberthür and G. Kristin Rosendal. New York, Routledge, 2014. pp. 158–178.

20 Young, T., & Tvedt, M. (2017). "Preface". In Drafting Successful Access and Benefit-sharing Contracts. Leiden, The Netherlands: Brill | Nijhoff.

without creating heavy regulatory burdens or bureaucratic procedures for the users. Broggiato et al. set out as an overall virtues: "[t]o reduce the transaction costs and to maximize predictability, which are necessary to attract investments from the private sector, a fixed percentage would be preferable over a case-by-case negotiation."[21] It is a general criticism against CBD and NP that the ABS system does not meet these criteria. The concept of benefit sharing is inspired by the CBD and NP. MGR-issues in the ABNJ do not necessarily need to be resolved in an identical manner, since there is no sovereign rights of states beyond the national jurisdiction. A major challenge is to embed a system in existing rules for the ABNJ while avoiding general ABS challenges.

3 Legal Tools Available without Amending the UNCLOS

Since amending an existing international treaty or convention is a challenging task, this chapter is taking a pragmatic approach by rather proposing legally binding tools and elements that do not require the existing text of UNCLOS to be changed. This chapter takes the existing international law for granted and rather identifies how the Intergovernmental Committee can adapt new tools supplementing the existing ones.

There is an underlying tension between the freedoms of the high seas and the competence of the flag state to apply regulations. One approach to impose obligation on private persons is by using the flag state jurisdiction combined with a contractual obligation on the users of MGR at the point of time of accessing the material.

There are some articles in UNCLOS that are particularly relevant to bear in mind in this discussion: Article 118 of UNCLOS deals with the Cooperation of States in the conservation and management of living resources. It reads as follows:

> States shall cooperate with each other in the conservation and management of living resources in the areas of the high seas. States whose nationals exploit identical living resources, or different living resources in the same area, shall enter into negotiations with a view to taking the

21 Broggiato, A. et al. (2018), *Mare Geneticum*, at p. 10. It is interesting to observe that their proposal and the general guideline for their discussion was proposed by some countries in Ad Hoc Open-ended Informal Working Group, *Letter Dated 30 June 2011 from the Co-Chairs of the Ad Hoc Open-ended Informal Working Group to the President of the General Assembly* (A/66/119, 2011), paragraph 6.

> measures necessary for the conservation of the living resources concerned. [...]

Article 118 establishes an obligation to collaborate for conservation and management of living resources. Flag State jurisdiction can be used as a legal tool in the UNCLOS in order to implement obligations on the users. The system for making samples available can be seen as a way different countries can work together.

Another relevant provisions is UNCLOS Article 263 on "Responsibility and liability", which establishes responsibility on States for their nationals. Article 263 provides as follows:

> 1. States and competent international organizations shall be responsible for ensuring that marine scientific research, whether undertaken by them or on their behalf, is conducted in accordance with this Convention.
> 2. States and competent international organizations shall be responsible and liable for the easures they take in contravention of this Convention in respect of marine scientific research conducted by other States, their natural or juridical persons or by competent international organizations, and shall provide compensation for damage resulting from such measures.
> 3. States and competent international organizations shall be responsible and liable pursuant to article 235 for damage caused by pollution of the marine environment arising out of marine scientific research undertaken by them or on their behalf.

The responsibility of States to control 'marine scientific research' can be used as a model for them also to control their national when conducting bioprospecting. Drawing on this obligation, the possibility of regulating in greater detail how governments must apply their flag state jurisdiction, is not a new idea when it comes to changing the behaviour of their citizens.

4 Bioprospecting Activities on the High Seas and in the Area

Bioprospecting is defined as the search for new and useful biological and genetic resources. Activities are taking place both on the high seas and potentially in the deep seabed. The research for and exploitation of MGR in the Area[22] have generated more debate and a more substantial body of literature[23] than those on the high seas. Expeditions for bioprospecting are often referred to as "cruises", fulfilling a range of different purposes, including population studies, taxonomic mapping of species occurrence, commercially oriented exploration and systematic collection of samples. One first step in a potential ABS system is to monitor and register the activities. In that context, the central question is to establish how a legally binding regime could make a register mandatory.

In that respect, and because the concept of "genetic resources" is not defined in UNCLOS, it could be useful to look at how the Convention deals with other related activities. Two main activities in the area which are declared as freedoms are "fishing" (Art. 87.1.e, UNCLOS) and "marine scientific research" (Art. 256 and 257, UNCLOS). Fishing has a very different purpose than bioprospecting and the rules regulating fishing are not suited to be applied directly; especially since fishing has large quantities of one or specific species as a goal, whereas bioprospecting generally requires small quantities.[24]

To answer the question make a register mandatory under the Convention, it is necessary to look at the link between an obligation to register and the existing right in the High Seas or the deep seabed. According to respectively Articles 256 and 257 of UNCLOS, which both deal with marine scientific research, "[a]ll States, irrespective of their geographical location, and competent international organizations have the right, in conformity with this Convention, to conduct marine scientific research in the water column beyond the limits of the exclusive economic zone." Article 257 refers instead to Part XI of the Convention and to the Area.

---

22 The Area is the "seabed and the ocean floor and the subsoil thereof beyond national jurisdiction". UNCLOS, article 1(1).

23 See e.g. L. Glowka, 'The Deepest of Ironies: Genetic Resources, Marine Scientific Research, and the Area' (1996) *Ocean Yearbook*, 12, 154–78; A.G. Oude Elferink, 'The Regime of the Area: Delineating the Scope of Application of the Common Heritage Principle and Freedom of the High Seas', (2007) *The International Journal of Marine and Coastal Law*, 22(1), 143–75; N. Matz-Lück, 'The Concept of the Common Heritage of Mankind', in E.J. Molenaar, and A.G. Oude Elferink, (eds) *The International Legal Regime of Areas Beyond National Jurisdiction: Current and Future Developments* (Martinus Nijhoff, Leiden, 2010), pp. 61–75; T. Scovazzi, 'The Seabed Beyond the Limits of National Jurisdiction: General and Institutional Aspects', in Oude Elferink, A.G. and Molenaar, E.J. (eds) *The International Legal Regime of Areas Beyond National Jurisdiction: Current and Future Developments* (Martinus Nijhoff, Leiden, 2010), pp. 43–60.

24 See Jørem, A. and M.W. Tvedt (2014). "Bioprospecting in the High Seas: Existing Rights and Obligations in View of a New Legal Regime for Marine Areas beyond National Jurisdiction." The International Journal of Marine and Coastal Law 29: 321–343, at pp. 325–327.

The detailed meaning of 'marine scientific research' as referred to in those articles is not defined in the Convention.[25] Because it is unlikely that the wording of Articles 256 and 257 will be modified and because this chapter intends to avoid the need for amendment, a possible solution would be to subsume the obligation to register under these rules. These articles allocate rights to States and not directly to private parties. The wording here does not include any obligation to register, nor does it include an obligation to register 'marine scientific research' in a clearing-house register for control purposes.

To compare, the argument has often been raised in CBD-related discussions that 'non-commercial' uses shall be left unregulated and that the regulatory obligation only kicks in when use turns 'commercial'. This dichotomy depends on the *intention* of the user at the point of time of initiating a cruise, which may be difficult to assess. The *objectively manifest activity*, however, is "collecting samples" both for non-commercial and commercial purposes. It is not possible to establish externally verifiable criteria to assess whether the intention is non-commercial or commercial, and therefore this dichotomy cannot be used as a legal criterion triggering different sets of obligations.

Regulating bioprospecting in the ABNJ will be less bureaucratic and create more legal certainty if the activity of 'collecting samples' is regulated regardless of the subjective intentions of the collector. The proposal here is that the actors at this point in time sign a standard contract. Such a contract needs to regulate scenarios for both non-commercial activities along with commercial ones. From the user-perspective, it is less bureaucratic to sign one contract regardless of what he has in mind when carrying out the collection. The counter-argument would be that such a contract would include a number of obligations that do not kick in since the trigger-point is never reached. This is not a unique situation in bioprospecting, since e.g. software contracts set out a large number of clauses that are not even observed. The rest of this article will show solutions for a one-size-fits-all standard contract and outline different aspects to be included in it to safeguard the different elements from collection, through research and utilisation to the creation of economic gains, identifying needs from the legislative level and need for preparing institutions.

To implement a rule requiring registration by users would require flag States to impose an obligation on their research vessels (and vessels in general since they could be potential bioprospectors) to register their activities. Since it is not unusual for researchers from different nations to participate on a cruise

25 See Jørem, A. and M.W. Tvedt (2014). "Bioprospecting in the High Seas: Existing Rights and Obligations in View of a New Legal Regime for Marine Areas beyond National Jurisdiction." The International Journal of Marine and Coastal Law 29: 321–343., pp. 327.

and in one vessel, the regulatory requirement must be put on the flag State responsible for all activities on board regardless of the nationality of the individual researchers on board. To make such an obligation binding the international regime will have to establish an obligation on the flag States targeting how they regulate the participants in the cruises on vessels sailing their flag. Thus, an international regime needs to oblige the contracting states to require the user to enter into the standard contract, foreseeing and providing solutions for all most relevant potential scenarios.

5 Obligation to Register the Samples and Depositing in a Repository

During and after the collecting activities have ended, two actions should be required: first, to register the sample taken and, second, deposit a duplicate or copy-sample of the biological material in a global repository system.

Grieber suggested as recently as 2011 that "One approach to non-monetary benefit sharing would be to establish a common pool of biological material retrieved from the high seas."[26] Further, Tvedt and Jørem noted that: "Such a common pool collection model for samples from the high seas would entail a continuum of marine resources being open to all."[27] The basic idea behind this option is to spur innovation by maintaining the material available for research and development for a greater scientific and perhaps also commercial audience than those with the financial resources to bioprospect in the high seas.[28] An example of *ex situ* conservation is keeping reproductive material in collections as a resource for research, or for reintroduction of endangered species

26 Greiber, 2011, pp. 36, 46–7. On the development of the open source movement, see e.g. Mandrusiak, L. (2010) 'Balancing Open Source Paradigms and Traditional Intellectual Property Models to Optimize Innovation', *Maine Law Review,* 63(1), pp. 313–6. See e.g. E.C. Kamau, and G. Winter, (eds) *Common Pools of Genetic Resources: Equity and Innovation in International Biodiversity Law,* (Routledge, London, 2013). See e.g. A. Broggiato, 'Marine Genetic Resources Beyond National Jurisdiction – Coordination and Harmonisation of Governance Regimes', *Environmental Policy and Law,* (2011) 41(1), p. 36; also D. Leary, 'International Law and the Genetic Resources of the Deep Sea', in Vidas, D. (ed.) *Law, Technology and Science for Oceans in Globalisation: IUU Fishing, Oil Pollution, Bioprospecting, Outer Continental shelf.* (Martinus Nijhoff, Leiden, 2009) p. 362.

27 Jørem, A. and M.W. Tvedt (2014). "Bioprospecting in the High Seas: Existing Rights and Obligations in View of a New Legal Regime for Marine Areas beyond National Jurisdiction." The International Journal of Marine and Coastal Law 29: 321–343.

28 Jørem, A. and M.W. Tvedt (2014). "Bioprospecting in the High Seas: Existing Rights and Obligations in View of a New Legal Regime for Marine Areas beyond National Jurisdiction." The International Journal of Marine and Coastal Law 29: 321–343.

into the wild. Other rational for repositories is to make the samples available for research and product-development by a broader audience.[29]

Establishing a common pool of MGR makes it necessary to discuss and clarify legal questions related to the pool itself and the degree of commonness of the material therein. Tvedt and Schei have identified five legal and practical topics to clarify for repositories.[30] The following sections develop three topics from a practical-legal perspective: the subject matter to be included in such a common pool (Section 5.1); the conditions for exclusivity for the depository (Section 5.2); and, the definition of the access rights to the material in the respository and limitations to such rights by others (Section 5.3). A final issue deals with the possibility of imposing obligations related to transfer (Section 5.4).

5.1 *The Subject Matter Deposited in the Repository*

Concerning the subject matter to be included at the repository there is a limited number of possibilities. Examples of biological material that can be kept in a duplicate collection are living specimens from micro-organisms to higher organisms; biological material that is not living but still intact as an organism; biological material in alternate form; or, any type of dried material. This ranges from aquarium-style living samples to highly prepared laboratory samples of biological material. These different biological conditions or formats of the material will require different forms for storage capacity and will open for different types of uses.

The most relevant point of time to ask for these samples/accessions to the repository is when the cruise comes to an end. At that moment, a legal regulation is likely to have less effect since a private person has the material in hand. If then no strings were attached to the material when it was collected, it is difficult to attach new legal regulations to the material. A contract entered into at the moment of entering upon the cruise could require duplicates in any of these conditions along with passport data to be left with the repository.

---

29 For a comparison of impacts of seabed mining and fossil fuel extraction compared to bioprospecting, see E., Ramirez-Llodra, P.A., Tyler, M.C., Baker, O.A. Bergstad, and M.A. Clark, 'Man and the Last Great Wilderness: Human Impact on the Deep Sea', (2011) *PLoS One,* 6(7), pp. 11–5.

30 Tvedt, M and Schei, P.J., 2009, The legal status for MARBANK and other marine biobanks in Norway, FNI Report 6/2009 identify the following legal questions: defining the material or information included in the pool; regulate the legal position of the depositor; the management of the pool itself; the legal position of the user of material taken out of the pool; and finally questions regarding ownership of the collection itself.

From any of these biological samples, different new expressions of the biological material can be produced: taxonomic information; ready-made assays; biochemical composition; DNA sequencing; digital sequence data; screening of the entire genome; developing metagenome and synthesized new molecules copying those found in nature. Typically, these new abstractions from the original samples are steps in research and development activities. A contract agreed to at the point of leaving for a cruise of collection needs also to regulate aspects relating to any information, knowledge or use of these biological subject matters.[31] Later than prior to collection, a user has no incentive to enter into a contract or undertake any obligations restricting the right to the sampled material. Any contractual obligation that is not included into the initial contract becomes voluntary. This situation is parallel to the Plant Treaty that establishes a system for access to plant genetic resources on the terms and conditions set out in its Standard Material Transfer Agreement. (There are weaknesses with the SMTA and it is currently undergoing a renegotiation.) Common for the two situations is that any obligation that is not included in the contract tends to be voluntary for the user. A first lesson to be drawn here is that all aspects must be regulated in the first contract.

A next important lesson is that since the UNCLOS is not using the terminology 'genetic resources', the new international legally binding instrument can be regulating the subject matter (res) of the obligations in a more precise manner. The term 'genetic resources' as it is defined in the CBD entails considerable difficulties for a contract.[32] If the ILBI copies those definitions rather than being more specific in what a contract regulates, the same lack of clarity as we see in the CBD/NP today will be reproduced for marine resources. The ILBI needs to ensure it is does not become outdated even before its signing, and could be more technology-neutral is it leaves flexibility to the standard contract to define the subject matters regulated by the contractual obligations. The important lesson to learn from contract law concerning the obligation to share any sample, information or knowledge is that the contract needs to be precise and that the language in the CBD/NP is not necessarily suited for the purpose of contract language.

31 I.e. "describing biological processes or experiments that are simulated by a computer program", *A Dictionary of Biology* Oxford University Press, Oxford, 2012.

32 Young, T., & Tvedt, M. (2017). "Preface". In Drafting Successful Access and Benefit-sharing Contracts. Leiden, The Netherlands: Brill | Nijhoff, in Chapter 5, at p. 113 sig.

5.2 *Degree of Exclusivity for the Collector*

Making samples available is a trade-off versus the rights of the collector. Tvedt and Jørem noted "three main alternatives [to exclusivity]: samples are held in trust on behalf of the depositor; samples are given without restriction; samples are pooled under certain specific conditions."[33] Broggiato et al. call this an "embargo period" where the access by others is restricted for a longer or shorter period of time.[34]

Seemingly the most attractive option from the perspective of incentives to invention is to provide the depositor a period of exclusivity as a main trade-off in accepting an obligation to deposit the samples and other subject matter in the global marine repository. This is a crucial point where the contract with the depositor could open for individually adapted solutions; either by leaving parts open for negotiations or leaving the contract with two or three standard options. Practically, the contract could stipulate three binding options for the depositor to choose from at a later stage. The most important observation is that if these questions are not resolved in the contract prior to collection, it is virtually impossible to impose any obligation on the user when having the material in hand, except from those imposed by national acts. CBD implementation has shown a reluctance by countries to impose a general benefit-sharing obligation in the acts.

In setting up the information-sharing parts of the repositories the system must be designed in a manner that is compatible to other legal systems handling information. The system must reserve publication to a point of time agreed by the depositors. Publication of academic articles is based on a principle of contributing to the state of knowledge. Material or information must be made available to the public in a manner and at a time securing publication possibilities. Patenting a new invention based on the material or information is also assessed from the perspective of the technical patent-law definition of 'prior art' – which means that state of publications that the patent is measured against. Prior art is the baseline against which novelty and inventiveness are assessed. Making a sequence available or publishing sequence data can be regarded as 'prior art' and probably prevent a patent from being granted. This relationship between the system of repositories and other legal systems needs to be clarified.[35] If the system fails to secure the secrecy until any rights

---

33 Tvedt, M. and Jørem, A. "Bioprospecting in the High Seas: Regulatory Options for Benefit Sharing" in Journal of World Intellectual Property 16 (2013) 3–4. pp. 150–167, at p. 157.

34 Broggiato et al. 2018, p. 21–22.

35 Agreement on Trade-Related Aspects of Intellectual Property Rights, adopted on 15 April 1994, article 27(1). There are different understandings of the "novelty" requirement, ranging from an absolute novelty requirement to a local novelty requirement, see

are secured it would undermine the incentive to deposit material, reduce the incentive to do research or innovation based on the material or information.

5.3 *Conditions for the Use by Others*

The objective of deposits in repositories is to make samples available for other. This does not mean that the material shall be without restrictions on the second users. From the perspective of the collector the conditions for the second user needs to be predictable and take his interests into account. Technically legally, it is not as urgent to set these criteria and conditions in the first contract with the collector before the cruise. However, to strike a balance and ensure the rights of the collector, at least the core topics should preferably be outlined in the first contract.

To compare again with the plant sector, the whole idea of the International Agricultural Research Centres (IARCS) of the Consultative Group on International Agricultural Research (CGIAR) for plants and the whole rationale for the Plant Treaty is that parties other than the depositor shall have access. In this system, the interest of the collector is seldom reflected in the conditions set on the second user. The collections of plant have been instrumental in promoting plant breeding.[36] The importance to make material available and conserved is the main rationals – less attention has been given to make monetary benefits back to the provider. For the global society to get a similar effect on innovation based on marine biological material, repositories can play a similarly important role. Therefore, the contract needs to clarify and specify the conditions whereby others may use the material. Unlike the plant sector, the collector needs to have incentives in making the samples available through this system. The manner in which the conditions on the second user is formulated has potential to create positive incentives.

The rights of others to apply for patent protection also needs to be specified. Other users need to know the conditions on which they invest in research on these samples. In the case where the second user arrives at a patentable

---

P.W. Grubb, *Patents for Chemicals, Pharmaceuticals and Biotechnology: Fundamentals of Global Law, Practice and Strategy*, (Oxford University Press, Oxford, 1999), pp. 54–8.

36 On the history and functioning of the CGIAR, see C. Fowler, and P. Mooney, *Shattering: Food, Policies and the Loss of Genetic Diversity*, (University of Arizona Press, Tucson, 1990) pp. 150–1; FAO (1997) *The State of the World's Plant Genetic Resources for Food and Agriculture*, FAO, Rome, p. 253; S. Louafi, 'Collective Action Challenges in the Implementation of the Multilateral System of the International Treaty: What Roles for the CGIAR Centres?', in Halewood, M., Noriega, I.L. and Louafi, S. (eds) *Crop Plant Genetic Resources As a Global Commons: Challenges in International Law and Governance.* (Earthscan, Abington, 2013) pp. 310–28.

invention, the principle of sharing part benefits of the invention with the original pool would be imperative for an open access system to work. To create a positive incentive for the collector to deposit samples, one could consider to impose a condition on the second user that the collector has a right to a licence of the patented invention when the new invention is based on "his" material. An alternative could be that a small portion of the turnover created by the patent finds it ways back to the collector.

Both mechanisms could be specified in the collection contract as well in the later contract between the collection and the new user. Since these conditions potentially only gives rights to the collector, it is not as urgent that all details are spelled out in the original contract.

5.4 *On the Possibility of Imposing Obligations on Transfer*

A central topic for ABS contracts is the transfer to third parties of any of the material, information or knowledge.[37] Solutions to this complex contract law question must be resolved in the original contract. Ideally, each of the subject matter elements described in the section above should have its tailor-made transfer solution. In first generation ABS contract, this question was set on hold since it was perceived too complex and the initial agreement postpones how this question shall be regulated till later. However, this approach is not recommendable as at this later stage the negotiation positions are different at the point of time of leaving for a cruise and the situation when a final product has been developed and is ready to be transferred.

6 Institutional Options

In this model for access and benefit sharing two institutions are needed: (i) the repositories where accessions and other subject matter are kept; and (ii) the institute with competence to represent the common interest in the negotiation of the contract.

As mentioned above the repositories needs to be a global system, but also to be geographically distributed. Since the infrastructure is expensive and much more advanced than for plants building on the existing repositories is probably the only viable solution. For plants a freezer with stable electricity is sufficient to maintain a good part of seed samples. For living organisms, often collected

37 Young, T., & Tvedt, M. (2017). "Preface". In Drafting Successful Access and Benefit-sharing Contracts. Leiden, The Netherlands: Brill | Nijhoff., Chapter 6 is devoted to third party transfer in ABS contracts.

in deep seas, it is far more difficult and costly to conserve the genetic base. Seeds can also be stored for a long time with high percentage of germination. The regeneration of higher organisms is a much more technically complex process. Also bacteria and viri raises their particular challenges.

Here again the flag State jurisdiction can be triggered. The flag State could be required to enter into the contract on behalf of the public and common interest.[38] Since the nation-state has limited interest in invoking obligations on its own citizens, there must be an obligation in the implementing agreement for them to take the common interest into the negotiations. A contract-based system has its benefits in the flexibility. The same flexibility there is a risk of lessening the burdens for the flag State's own collectors, so there is a clear role for a global body.

Perhaps regional fisheries management organisations or frameworks like the OSPAR Convention[39] may serve as contracting partners. Nevertheless, the body responsible for the contracts needs standardised templates with a level of discretion to adapt to the individual situation.

7 Summary of Suggested Solution to the ABS-System

This chapter suggests the use of two main regulatory instruments as tools for making a marine genetic resources pool possible: first, regulating the flag State principle and, second, drafting a comprehensive and clear contract that should be entered into before collection starts. In developing a contract, the terms and conditions should learn from contract law analysis and observe that the SMTA for the Plant Treaty has not succeeded. The contract (or SMTA) must be a valid, binding, implementable and enforceable document. The details in how such a contract could look like is a too comprehensive task for this chapter.

38 UNCLOS, article 92(1), and Pursuant to UNCLOS Article 91 there must nonetheless be a "genuine link between the State and the ship".

39 Convention for the Protection of the Marine Environment of the North-East Atlantic, 2354 UNTS 67, adopted on 22 September 1992.

CHAPTER 12

Deep-Sea Bottom Fisheries and the Protection of Seabed Ecosystems: Problems, Progress and Prospects

Richard Caddell

1 Introduction

Of the myriad industrial activities currently undertaken in the marine environment, fisheries have progressively emerged as 'the most widespread source of anthropogenic physical disturbance to global seabed habitats'.[1] Demersal fisheries (i.e., those targeting benthic and benthopelagic species, whose core habitats comprise the seabed or areas in very close proximity to it) have been conducted for centuries and are of enduring commercial and nutritional importance to many States. However, as with numerous other fishing practices, technological advances have dramatically transformed demersal fisheries over the past half-century.[2] While seabed fisheries were historically concentrated in coastal and inshore locations, the steady depletion of shallow-water stocks has subsequently forced many fishers to pursue offshore and deep-sea alternatives.[3] Consequently, the mean depth of global fisheries has expanded four-fold since the mid-1960s,[4] as industrial fleets have increasingly targeted new deep-water opportunities, notably those located within areas beyond national jurisdiction (ABNJ). Although fishing is by no means a new use of the seabed, the poorly regulated proliferation of fisheries into unprecedented depths has nevertheless generated novel environmental and management concerns, with worrying implications for the enduring health and integrity of benthic ecosystems.

1 J.G. Hiddink *et al.*, 'Global Analysis of Depletion and Recovery of Seabed Biota After Bottom Trawling Disturbance' (2017) 114 PNAS 8301, 8301. Even in regions noted for an extensive offshore industrial presence, the impact of fisheries on seabed ecosystems often outweighs that of all other anthropogenic activities combined: A.R. Benn *et al.*, 'Human Activities on the Deep Seafloor in the North East Atlantic: An Assessment of Spatial Extent' (2010) 5 *PLoS One* e12730.

2 W. Swartz *et al.*, 'The Spatial Expansion and Ecological Footprint of Fisheries (1950 to Present)' (2010) 5 *PLoS One* e15143.

3 T. Morato *et al.*, 'Fishing Down The Deep' (2006) 7 *Fish and Fisheries* 24, 31.

4 E.A. Norse *et al.*, 'Sustainability of Deep-Sea Fisheries' (2012) 36 *Marine Policy*, 307, 308.

Demersal fisheries deploy an array of gear to catch their respective target species, all of which involve at least some contact with the seabed, thereby posing varying degrees of collateral risk to the benthic environment.[5] Some equipment – such as traps and pots – ultimately exert a superficial impact upon marine ecosystems.[6] Others – notably gillnets and longlines – inflict little physical damage to the seabed itself, but have provoked rather different regulatory anxieties due to their propensity for incidental catches of non-target species, particularly marine mammals.[7] Instead, bottom fisheries[8] – especially those engaged in bottom trawling,[9] involving the dragging of weighted nets over the seabed – are by some distance the most ecologically injurious forms of demersal fishing. They also remain among the most prevalent, with approximately 20–25% of all current global seafood landings attributed to bottom trawling.[10]

The advent of bottom trawling can be traced back to the late fourteenth century,[11] although historical objections to its impacts were often confined to overfishing and the displacement of small-scale fishers by industrial fleets.[12] Nevertheless, even by the 1860s, regular complaints were raised over the 'scouring' of the seabed by trawl nets.[13] More recent ecological vexations over bottom trawling have concerned the scraping of vulnerable benthic sediments,[14]

5 M.J. Kaiser *et al.*, 'Prioritization of Knowledge-Needs to Achieve Best Practices for Bottom Trawling in Relation to Seabed Habitats' (2016) 17 *Fish and Fisheries* 637, 639.

6 F. Stephenson *et al.*, 'Experimental Potting Impacts on Common UK Reef Habitats in Areas of High and Low Fishing Pressure' (2017) 74 *ICES Journal of Marine Science* 1648, 1657.

7 A.J. Reid, 'The Looming Crisis: Interactions between Marine Mammals and Fisheries' (2008) 89 *Journal of Mammalogy* 541, 542–543.

8 Bottom fisheries are defined by the UN Food and Agriculture Organization (FAO) as any fishery that uses gear 'that either contact or are likely to contact the sea floor during the course of the fishing operation': A. Bensch *et al.*, *Worldwide Review of Bottom Fisheries in the High Seas* (FAO: Rome, 2009), 2.

9 Bottom trawling is an umbrella term for the use of a variety of gear, each of which uses variations that encapsulate weighted features with the capacity to impact both hard- and soft-bottom benthic ecosystems: see further J.W. Valdermansen, T. Jørgensen and A. Engås, *Options to Mitigate Bottom Habitat Impact of Dragged Gears* (FAO: Rome, 2007), 5–18.

10 Hiddink (n.1), 8301 (extrapolating global catch data collated by the FAO since 2009).

11 C. Roberts, *The Unnatural History of the Sea* (Island Press: Washington DC, 2007), 131–2.

12 T.K. Kerby, W.W.L. Cheung and G.H. Engelhard, 'The United Kingdom's Role in North Sea Demersal Fisheries: A Hundred Year Perspective' (2012) 22 *Reviews in Fish Biology and Fisheries* 621, 629.

13 R.H. Thurstan, J.P. Hawkins and C.M. Roberts, 'Origins of the Bottom Trawling Controversy in the British Isles: 19th Century Witness Testimonies Reveal Evidence of Early Fishery Declines' (2014) 15 *Fish and Fisheries* 506, 515.

14 F.G. O'Neill and A. Ivanović, 'The Physical Impact of Towed Demersal Fishing Gears on Soft Sediments' (2016) 73(Supplement) *ICES Journal of Marine Science* 5, 12.

damage to submarine features,[15] the removal of habitat-forming species[16] and the disturbance of complex benthic ecosystems which may in turn further compromise fish productivity.[17] Moreover, while bottom trawling may inflict specific localised impacts upon the seabed, these fisheries also represent a significant contribution to the cumulative footprint of anthropogenic activities in the global oceans, hence their effects may be exacerbated in tandem with those of other industries.[18] Seabed ecosystems are slow to recover from such impacts: although data are currently limited, full regeneration appears likely to be a multi-decadal process and one fraught with scientific uncertainty.[19]

The prospective impact of bottom trawling is amplified in the case of deep-sea fisheries. In order to facilitate fishing at advanced depths, larger and heavier equipment is required to ensure the effective operation of the trawl, hence these activities are often significantly more destructive to the seabed than trawling undertaken in shallower waters.[20] Likewise, trawling is often conducted repeatedly and intensively in particular locations, notably in the vicinity of seamounts and other fragile submarine features due to habitual aggregation of fish in many such areas,[21] thereby representing a serious threat to the integrity of seabed ecosystems.[22] Such fisheries also pose considerable risks to long-term sustainability of their target species. Indeed, in order to survive at advanced depths, in an environment characterised by cold temperatures, little light and limited productivity, deep-sea fish typically

15 M.R. Clark *et al.*, 'The Impacts of Deep-Sea Fisheries on Benthic Communities: A Review' (2016) 73(Supplement) *ICES Journal of Marine Science* 51, 52.

16 S.F. Thrush, K.E. Ellingsen and K. Davis, 'Implications of Fisheries Impacts to Seabed Biodiversity and Ecosystem-Based Management' (2016) 73(Supplement) *ICES Journal of Marine Science* 44, 45.

17 J. Collie *et al.*, 'Indirect Effects of Bottom Fishing on the Productivity of Marine Fish' (2017) 18 *Fish and Fisheries* 619, 634.

18 Thrush (n.16), 45–46. Nor are benthic ecosystems the only fragile features at risk – concerns have also been raised over the impacts of trawling upon underwater cultural heritage: M.L. Brennan *et al.*, 'Quantification of Bottom Trawl Fishing Damage to Ancient Shipwreck Sites' (2016) 371 *Marine Geology* 82.

19 M.R. Clark et al., 'Little Evidence of Benthic Community Resilience to Bottom Trawling on Seamounts After 15 Years' (2019) 6 *Frontiers in Marine Science* 1, 13.

20 Clark (n.15), 52.

21 Although a definitive accounting remains elusive for all fish species for which seamounts form an essential habitat, current projections indicate that approximately 800 separate species of fish may be considered seamount species, while a considerable array of additional species aggregate around such features at some stage during their life cycles: T. Morato, W.W.L. Cheung and T.J. Pitcher, 'Vulnerability of Seamount Fish to Fishing: Fuzzy Analysis of Life History Attributes' (2006) 68 *Journal of Fish Biology* 209.

22 A. Pusceddu *et al.*, 'Chronic and Intensive Bottom Trawling Impairs Deep-Sea Biodiversity and Ecosystem Functioning' (2014) 111 PNAS 8861, 8861.

exhibit delayed sexual maturity, slow growth and a high maximum age, hence stocks may become rapidly depleted with little scope to regenerate swiftly.[23] Compounding these concerns, catches from deep-sea trawling provide a negligible contribution to current nutritional needs, accounting for less than 0.5% of global landings,[24] while causing significant long-term damage to the seabed in the process. Moreover, deep-sea fishing in ABNJ is conducted by relatively few national fleets, thereby inflicting a strikingly disproportionate degree of harm upon the global commons relative to its numerical participants.

Despite these concerns, until the turn of the present century, many bottom fisheries were subject to surprisingly minimal regulation, as regional fisheries management organisations (RFMOs) either lacked an express mandate to address benthic ecosystems or had yet to be established in respect of significant portions of the global oceans, while little provision had been made for seabed fishing within the overarching legal framework provided by the United Nations Convention on the Law of the Sea 1982.[25] This has since been addressed through the graduated development of a distinct tier of regulation for seabed ecosystems, which has emerged largely *ab initio* over the course of the past two decades. The contemporary oversight of deep-sea bottom fisheries accordingly represents an intriguing case-study of the development of regulatory standards concerning the seabed. In this regard, regulatory momentum has been primarily channelled through the United Nations General Assembly (UNGA), which has adopted a series of highly influential Resolutions calling for the protection of vulnerable marine ecosystems (VMEs) on the seabed. The political impetus provided by multiple UNGA Resolutions has in turn prompted the adoption of complementary conservation and management measures by RFMOs, flag states and other pertinent actors. As this Chapter will demonstrate, considerable progress has since been made towards the development of uniform standards for bottom fishing, although the implementation of this framework remains far from complete. Such shortcomings will accordingly need to be further addressed in order to operationalise the commitments established under this collective of UNGA Resolutions, and thus to fully protect benthic ecosystems from the adverse impacts of fishing activities.

To this end, this Chapter first outlines the emergence of global standards towards the protection of VMEs, examining the limits of global instruments and tracing the development of the pertinent UNGA commitments (Section 2).

23 Norse (n.4), 317.

24 L. Victorero *et al.*, 'Out of Sight, but Within Reach: A Global History of Bottom-Trawled Deep-Sea Fisheries from >400m Depth' (2018) 5 *Frontiers in Marine Ecology* 1, 4.

25 1883 UNTS 396 [hereinafter 'LOSC'].

The Chapter then examines the extent to which these commitments have been implemented by RFMOs and other pertinent actors (Section 3), before advancing a series of conclusions as to the future regulatory landscape concerning bottom fisheries (Section 4).

2 The Legal Framework for the Regulation of Deep-Sea Bottom Fisheries

2.1 *Fisheries, the Seabed and the Limitations of the LOSC*

Until the early twenty-first century, deep-sea bottom fisheries were largely overlooked by international law. Even as deep-sea trawling began to expand on an unprecedented scale from the mid-1970s, such practices remained largely inured from regulatory scrutiny as the 'out of sight, out of mind' nature of bottom fishing in remote locations ensured that there was little popular appreciation of its deleterious impacts upon the seabed. Instead, multilateral fisheries regulation continued to focus primarily on the management of staple pelagic stocks, such as salmon or tuna. This is perhaps unsurprising, since deep-sea fish have not traditionally constituted an attractive cohort of target species, either physically or economically. However, by the 1980s a number of deep-sea species had undergone something of a gastronomic renaissance. Having initially been taken – and often discarded – as undesirable by-catch, prominent seafood chefs began to appreciate that their versatile flavour and texture allowed for considerable culinary creativity.[26] This subsequently created an unprecedented demand for particular fish that had rarely been pursued – or regulated – as a target stock. Coupled with the tactical rebranding of many species to render their previous sobriquets rather more palatable to discerning diners – for instance, 'slimehead' and 'toothfish' have been rechristened as 'orange roughy' and 'Chilean seabass' respectively[27] – distant water deep-sea fishing was swiftly transformed into a highly lucrative undertaking. At the same time, however, this unexpectedly elevated demand starkly exposed acute governance gaps in the international framework for the regulation of fisheries and seabed ecosystems.[28]

---

26 G.B. Knecht, *Hooked: A True Story of Pirates, Poaching and the Perfect Fish* (Allen & Unwin: Sydney, 2006), 81–86.

27 For a sobering account of the extent of this practice see J.L. Jacquet and D. Pauly, 'Trade Secrets: Renaming and Mislabeling of Seafood' (2008) 32 *Marine Policy* 309, 311–313.

28 K.M. Gjerde and D. Freestone, 'Unfinished Business: Deep-Sea Fisheries and the Conservation of Marine Biodiversity Beyond National Jurisdiction' (2004) 19 *International Journal of Marine and Coastal Law* 209, 209.

A 'perfect storm' of three regulatory lacunae can be seen to have engendered a rapacious frontier mentality towards deep-sea demersal fish. Firstly, RFMOs had yet to be established for substantial geographical portions of the global oceans, meaning that there were few pre-existing governance structures through which such fisheries could be regulated. Secondly, of the comparatively small number of RFMOs that were in existence at the material time, most had been established to regulate a single species and had no express competence over deep-sea stocks. And thirdly, of the four regulators that in principle held a mandate over deep-sea fish,[29] these powers were seemingly limited to setting stock allocations rather than addressing the wider ecosystem impacts of bottom fisheries. Although (as considered further below) these bodies have tended to interpret their powers proactively in order to advance protective measures for deep-sea ecosystems, arguably only CCAMLR possessed undisputed competence over both the stocks and their surrounding environment.[30] This effective regulatory vacuum duly facilitated a 'gold rush' among certain fleets, which invested heavily in new technology in order to exploit seabed fisheries resources before meaningful management actions could be elaborated.[31] Consequently, by the late 1990s leading estimates suggested that an area of up to half the size of the global continental shelf was being trawled annually,[32] facilitated by a dearth of regulation that ensured that 'people trawl almost anywhere they want, and the sea's equivalents of ancient forests are becoming cattle pastures by default, not by design'.[33]

Little overt guidance towards the responsible pursuit of demersal fisheries was forthcoming from the LOSC. The LOSC elaborates a broad framework for fisheries competences, under which states may utilise fisheries resources on the high seas[34] and in their respective exclusive economic zones (EEZs),[35] subject to obligations concerning the conservation and management of the stocks

29 Namely the North-East Atlantic Fisheries Commission (NEAFC), the Northwest Atlantic Fisheries Organisation (NAFO), the General Fisheries Commission for the Mediterranean (GFCM) and the Commission for the Conservation of Atlantic Marine Living Resources (CCAMLR).

30 E.J. Molenaar, 'Addressing Regulatory Gaps in High Seas Fisheries' (2005) 20 *International Journal of Marine and Coastal Law* 533, 538.

31 A. Merrie *et al.*, 'An Ocean of Surprises – Trends in Human Use, Unexpected Dynamics and Governance Challenges in Areas beyond National Jurisdiction' (2014) 27 *Global Environmental Change* 19, 26.

32 L. Wattling and E.A. Norse, 'Disturbance of the Seabed by Mobile Fishing Gear: A Comparison to Forest Clearcutting' (1998) 12 *Conservation Biology* 1180, 1180.

33 *Ibid.*, 1193.

34 Articles 87(1)(a) and 116.

35 Article 62.

in question.[36] Nevertheless, these provisions of the Convention merely establish generalised requirements for the conduct of fisheries: within the EEZ, stocks must be managed in a way that consistently allows for harvest at maximum sustainable yield and that overfishing does not occur as a result, while also taking into consideration the inter-dependence of fish stocks,[37] with high seas fishing conducted subject to the more ambiguous obligation to exercise 'due regard' towards the interests of others.[38]

Nor were demersal fisheries considered in Part VI of the LOSC, which addresses the continental shelf. Indeed, proposals to specifically regulate benthic and benthopelagic species within the ambit of these provisions – given that they are 'intimately associated with the seabed' – were expressly rejected during the negotiation of the Convention.[39] Instead, fisheries entitlements on the continental shelf under the LOSC are restricted to the exercise of sovereign rights over sedentary species.[40] Such species are defined as 'organisms which, at the harvestable stage, either are immobile on or under the seabed or are unable to move except in constant physical contact with the seabed or the subsoil'.[41] This has created sporadic controversy over the status of particular stocks of crustaceans and molluscs, whose biological attributes and behaviour do not always neatly align with the legal understanding of a 'sedentary' species, while also raising questions over the limits of RFMO competences towards certain seabed species.[42] In any event, the demersal fish targeted by trawling clearly do not meet the Article 77 criteria and thereby lie outside the ambit of Part VI. Such fisheries instead remain subject to the general position

36 Articles 61 (EEZ) and 117–119 (high seas).

37 Article 61.

38 Article 87(2).

39 S. Borg, *Conservation on the High Seas: Harmonizing International Regimes for the Sustainable Use of Marine Resources* (Edward Elgar: Cheltenham: 2012) 151.

40 Article 77(1).

41 Article 77(4).

42 As recently considered by the Norwegian courts where, in two separate prosecutions against Latvian-flagged crabbing vessels, Norway has sought to enforce its sovereign rights over snow crabs as a lucrative sedentary species: see further I. Dahl and E. Johansen, *The Norwegian Snow Crab Regime and Foreign Vessels – A Commentary on the* Juras Vilkas *Decision of the Øst-Finnmark District Court*; available on-line at http://site.uit.no/jclos/2017/03/29/the-norwegian-snow-crab-regime-and-foreign-vessels-a-commentary-on-the-juras-vilkas-decision-of-the-ost-finnmark-district-court/ and T. Henriksen, *The* Senator *Case – A New Turn in Norway's Dealings with Foreign Vessels Operating in the Waters off Svalbard*, available on-line at https://site.uit.no/jclos/2019/02/28/the-senator-case-a-new-turn-in-norways-dealings-with-foreign-vessels-operating-in-the-waters-off-svalbard/.

prescribed in relation to the EEZ, since '[t]he rights of the coastal State over the continental shelf do not affect the legal status of the superjacent waters'.[43]

Similarly, the LOSC made no direct provision for the regulation of fishing in those areas of the seabed that lie outside the confines of national jurisdiction (the 'Area'). The provisions of Part XI, which pertain to the Area, clarify that the concept of 'resources' in this context is restricted to minerals.[44] As with the continental shelf regime, trawl fishing in the Area is thus subject to the provisions governing the high seas.[45]

A more intriguing position arises where the coastal state has exercised its right under Article 76 of the LOSC to extend its continental shelf beyond the orthodox 200 nautical mile limit. This may lead to conflicts between the interests of the coastal state in protecting the seabed environment in this area – either for its own intrinsic value or to safeguard a lucrative aggregation of sedentary species – and those of another state intending to undertake bottom trawling by exercising its long-established rights to fish on the high seas, as guaranteed under Article 87(2). A coastal state cannot unilaterally impose a blanket ban on trawling across the entirety of its outer continental shelf, since Article 78(2) specifies that '[t]he exercise of the rights of the coastal State over the continental shelf must not infringe or result in any unjustifiable interference with navigation and other rights and freedoms of other States as provided for in this Convention'. Instead, this raises the possibility that more localised restrictions could represent a *justifiable* interference with the exercise of high seas freedoms. Indeed, a sense of coexistence between states has long been considered necessary where there are bifurcated entitlements applicable to overlapping areas of maritime jurisdiction: as confirmed by the International Tribunal for the Law of the Sea (ITLOS) in the *Bangladesh/Myanmar* case, 'the legal regime of the continental shelf has always coexisted with another legal regime in the same area ... each coastal State must exercise its rights and perform its duties with due regard to the rights and duties of the other'.[46] By analogy to practice in other industries, Mossop suggests that a coastal state could prospectively introduce limited restrictions on bottom trawling in these

43 Article 78(1).

44 Article 133.

45 Article 135.

46 *Dispute concerning delimitation of the maritime boundary between Bangladesh and Myanmar in the Bay of Bengal (Bangladesh/Myanmar)*; Judgment of 14 March 2012: para 475.

areas, provided that interference with legitimate high seas freedoms is minimal and appropriate consultations have been undertaken.[47]

Nevertheless, given the limited ability of the coastal state to control activities in the superjacent water column of the high seas, such restrictions are likely to be markedly strengthened where they operate in tandem with those of a multilateral body, such as an RFMO or a Regional Seas Organisation (RSO) – and are arguably weakened where such collaboration is not forthcoming. For instance, an RSO may implement a marine protected area (MPA) or an RFMO may establish concurrent restrictions on fishing in areas of the high seas that overlay locations of the outer continental shelf for which a coastal state has sought to curtail bottom fisheries. As noted below, given that ABNJ are primarily regulated on a sectoral basis, the optimal solution for seabed locations of particular ecological sensitivity would be for each of these designations to operate in tandem. In a limited number of cases, such designations have been successfully aligned: most notably in the context of the Altair and Antialtair Seamounts located on the outer continental shelf of Portugal, which have been protected from trawling by the Portuguese authorities, with parallel fishing restrictions imposed in the superjacent water column by the pertinent RFMO, NEAFC, alongside an overlapping MPA designation under the Convention for the Protection of the Marine Environment of the North East Atlantic,[48] which has thus secured the highest possible degree of protection for these features.[49] This is not the case for all such features purportedly protected by states on their outer continental shelves, however, as other Portuguese designations attest.[50]

This is not to suggest that delicate seabed features are devoid of legal protection under the LOSC. Indeed, benthic ecosystems are more directly regulated under Part XII of the LOSC, addressing the protection and preservation of the marine environment, with Articles 192 and 194 of particular relevance. Both provisions have been subject to extensive judicial consideration in recent years to clarify their application and scope. Article 192 rather concisely provides that '[s]tates have the obligation to protect and preserve the marine environment'. In a fisheries context it has been recently acknowledged that Article 192 'extends to the prevention of harms that would affect depleted, threatened, or

47 J. Mossop, *The Continental Shelf Beyond 200 Nautical Miles: Rights and Responsibilities* (Oxford University Press: Oxford, 2016), 195.

48 2354 UNTS 67 [hereinafter 'OSPAR Convention'].

49 See further Mossop (n.47), 218–220.

50 *Ibid.*, 220.

endangered species indirectly through the destruction of their habitat',[51] which not only provides a basis to protect seabed ecosystems, but also imposes an obligation of due diligence upon flag states to ensure that nationally-registered vessels adhere to relevant conservation and management measures in the conduct of fishing activities.[52]

Similarly, Article 194(5) provides that measures undertaken pursuant to Part XII of the LOSC 'shall include those necessary to protect and preserve rare or fragile ecosystems as well as the habitat of depleted, threatened or endangered species and other forms of marine life'. In the *South China Sea* Arbitration, it was considered that the general obligation to protect the marine environment advanced in Article 192 is 'given particular shape in the context of fragile ecosystems by Article 194(5)'.[53] Although this did not concern bottom fishing *per se*, the Tribunal nonetheless declared that if a state is deemed to have knowledge that its vessels are conducting destructive fishing activities, a failure to enforce such measures may constitute a breach of the obligation of due diligence.[54] Like Article 192, recent jurisprudence has clarified that Article 194 is also to be viewed expansively, and is 'not limited to measures aimed strictly at controlling pollution and extends to measures focussed primarily on conservation and the preservation of ecosystems'.[55] As with Article 78(2), when exercising the obligations prescribed under Article 194, states 'shall refrain from unjustifiable interference with activities carried out by other States in the exercise of their rights and in pursuance of their duties in conformity with this Convention'.[56] The concept of unjustifiable interference has been recently considered to be 'functionally equivalent' to the obligations to give 'due regard' under Article 56(2) and of good faith under Article 2(3) of the LOSC.[57] The extent of the obligation under Article 194(4) remains essentially context-dependant, although the Arbitral Panel considered that due regard as applied under Article 56(2) will depend upon the nature of the rights held, their importance, the extent of the anticipated impairment, the nature and importance of the activities and the availability of alternative approaches

51 *South China Sea Arbitration (Republic of the Philippines* v. *Peoples' Republic of China)*, Award of 12 July 2016; para 959.

52 *Request for an Advisory Opinion Submitted by the Sub-Regional Fisheries Commission (SRFC)*; paras 118–140.

53 *South China Sea Arbitration* (n.51), para 959.

54 *Ibid.*, paras 964–966.

55 *Chagos Marine Protected Area Arbitration* (*Mauritius v. UK*), Award of 18 March 2015; para 538.

56 Article 194(4).

57 *Chagos Marine Protected Area Arbitration* (n.55), para 540.

and will involve 'at least some consultation'.[58] Provided this broad balancing act is undertaken, the possibility for the legitimate interference with fishing rights otherwise guaranteed under the LOSC was pointedly left open in the *Chagos Marine Protected Area Arbitration*, although this would require 'significant engagement ... to explain the need for the measure and to explore less restrictive alternatives'. Accordingly, Article 194 provides a prospective basis for restricting bottom fishing in areas of known sensitivity, albeit based on meaningful consultations and with the possibility that intended activities may still proceed to a lesser extent in the intended location or be displaced to another area of the seabed.

2.2 *The Evolving UNGA Commitments and the Emergence of the VME Concept*

Notwithstanding the ongoing elaboration of obligations under Part XII of the LOSC, the protection of benthic ecosystems from the damaging impacts of bottom fisheries has been primarily driven by a series of key UNGA Resolutions adopted over the course of the past fifteen years, which have subsequently proved influential in the adoption of unified standards and policies by RFMOs, flag states and other actors. By the late 1990s, concerns began to mount over the proliferation of bottom fisheries and the effective lack of regulation for many demersal species. In some cases, regulation was largely localised and bilateral, as exemplified by the pioneering arrangement between Australia and New Zealand to set allocations for orange roughy in the South Tasman rise.[59] While such arrangements were predominantly focused on catch limits, allocations and stock analysis, they also introduced localised prohibitions on trawling and demersal fishing in particular locations, albeit that bottom fishing could still proceed with the express authorisation of the parties.[60] Such initiatives were however few and far between, leading to increased calls for a more globalised tier of regulation over deep-sea bottom fisheries.

58 *Ibid.*, para 518.

59 See further E.J. Molenaar, 'The South Tasman Rise Arrangement and Other Initiatives on Management and Conservation of Orange Roughy' (2001) 16 *International Journal of Marine and Coastal Law* 77. As Molenaar observes, a series of orange roughy fisheries in the Southern Hemisphere provided a strong test for international fisheries regulation in the interim period between the adoption and entry into force of the UN Fish Stocks Agreement.

60 Arrangement Between the Government of Australia and the Government of New Zealand for the Conservation and Management of Orange Roughy on the South Tasman Rise 2000; reproduced in Molenaar, *ibid.*

Despite an emerging degree of support for the improved governance of demersal fisheries, there were strong divergences of opinion over which institution would be best placed to institute effective regulation and whether these activities were best addressed through binding or non-binding means.[61] In this respect, the UNGA offers particular advantages in providing both a suitably global platform for such discussions, while also avoiding the more loaded issue of whether such standards should be legally binding and thereby trigger substantive consequences if they are not fully implemented. Although UNGA Resolutions are not technically binding,[62] in recent years they have nevertheless had considerable traction as a regulatory spur to address destructive fishing practices.[63] This proved to be particularly true in the context of large-scale driftnet fishing on the high seas, with a series of UNGA Resolutions in the early 1990s providing a compelling stimulus for specific regulatory activity by RFMOs and other actors,[64] to the extent to which commentators have suggested that these standards now represent customary international law.[65]

The UNGA driftnet Resolutions provided a helpful regulatory template to promote policies to address the impacts of bottom fishing on seabed ecosystems. In 2002, the UNGA recognised for the first time the need to consider 'the management of risks to marine biodiversity of seamounts and certain other underwater features'.[66] This precipitated a protracted and circular series of discussions within a variety of different multilateral bodies as to the most

61 See D.A. Balton and D.C. Zbicz, 'Managing Deep-Sea Fisheries: Some Threshold Questions' (2004) 19 *International Journal of Marine and Coastal Law* 247, 252–255.

62 As Harrison observes, UNGA Resolutions are not law-making tools but have had a more nuanced influence on state practice as 'a means of drawing attention to the current threats to fish stocks and encouraging international efforts taking place in other institutions to address them': J. Harrison, *Making the Law of the Sea: A Study in the Development of International Law* (Cambridge University Press: Cambridge, 2013), 204.

63 See further R. Caddell, 'International Fisheries Law and Interactions with Global Regimes and Processes' in R. Caddell and E.J. Molenaar, *Strengthening International Fisheries Law in an Era of Changing Oceans* (Hart: Oxford, 2019) 133, 135–137.

64 On the elaboration of these provisions see D.R. Rothwell, 'The General Assembly Ban on Driftnet Fishing' in D. Shelton (ed.) *Commitment and Compliance: The Role of Non-Binding Norms in the International Legal System* (Oxford University Press: Oxford, 2003), 121,126–131, W.T. Burke, M. Freeberg and E.L. Miles, 'United Nations Resolutions on Driftnet Fishing: An Unsustainable Precedent for High Seas and Coastal Fisheries Management' (1994) 25 *Ocean Development and International Law* 127, 137–144 and R. Caddell, 'Caught in the Net: Driftnet Fishing Restrictions and the European Court of Justice' (2010) 22 *Journal of Environmental Law* 301, 301–304.

65 G.J. Hewison, 'The Legally Binding Nature of the Moratorium on Large-Scale High Seas Driftnet Fishing' (1994) 25 *Journal of Maritime Law and Commerce* 557, 578–580.

66 Resolution 57/141 of 12 December 2002, para 56.

appropriate forum through which bottom fishing concerns might be directed.[67] Since 2003, the UNGA has amalgamated its previously fragmented pronouncements on fisheries matters into a specific – and often hard-fought – annual Resolution on Sustainable Fisheries. In 2004 the protection of the seabed was addressed for the first time in the UNGA Sustainable Fisheries Resolution.[68] In four core paragraphs, Resolution 59/25 lamented a general lack of regulatory competence over particular marine ecosystems, calling upon States to apply the precautionary approach to prohibit 'destructive fishing practices, including bottom trawling that has adverse impacts on vulnerable marine ecosystems', until appropriate conservation and management measures have been adopted.[69]

This represents the first – albeit undefined – use of the term 'vulnerable marine ecosystem' which has subsequently become a notable addition to the lexicon of global marine governance. The commitment to seek the interim prohibition of 'destructive fishing practices' on a case-by-case basis nonetheless fell short of the full moratorium on trawl fisheries in ABNJ that had been sought by activists – and had indeed been initially present in earlier drafts of the Resolution.[70] Resolution 59/25 called upon those RFMOs with competence over bottom fisheries to 'urgently' adopt conservation and management measures in accordance with international law to address destructive fishing practices, those inflicting adverse impacts on VMEs and to ensure compliance with such measures. In particular, RFMOs that lacked these competences were requested to extend their pre-existing mandates to regulate bottom fisheries,[71] while the international community was urged to cooperate in the creation of new RFMOs for unregulated areas of the global oceans, which would be endowed with these powers *ab initio* and thereby be able to address deep-sea fishing as an immediate operative priority upon their inception.

In 2006, a further and arguably more influential Resolution was adopted by the UNGA, expressing dissatisfaction with the rate of progress since 2004 and calling for steps to be taken 'immediately' to sustainably manage deep-sea fish

---

67 See further L.A. Kimball, 'Deep-Sea Fisheries of the High Seas: The Management Impasse' (2004) 19 *International Journal of Marine and Coastal Law* 259, 263–272.

68 Resolution 59/25 of 17 November 2004.

69 Para 66.

70 Y. Takei, *Filling Regulatory Gaps in High Seas Fisheries: Discrete High Seas Fish Stocks, Deep-Sea Fisheries and Vulnerable Marine Ecosystems* (Martinus Nijhoff: Leiden/Boston: 2013) 112. The rather vague wording of the Resolution also suggests that, while politically prompted by the excesses of trawl fisheries, its application was not confined to such activities.

71 Para 68.

stocks and VMEs.[72] In a notable departure, paragraph 83 of Resolution 61/105 listed for the first time a series of targeted action points for RFMOs to complete, notably an assessment of whether individual bottom fishing activities have significant adverse impacts on VMEs, and to ensure that if so they are managed to prevent such impacts, or not authorized to proceed; to identify VMEs and determine whether bottom fishing activities would cause significant adverse impacts to such ecosystems; to close such areas to bottom fishing and ensure that such activities do not proceed unless conservation and management measures have been established to prevent significant adverse impacts; and to require RFMOs to cease bottom fishing where VMEs are encountered.

Resolution 61/105 therefore represents a more nuanced series of restrictions upon bottom fisheries, whereby such commitments are triggered only where there is a threat of a *significant adverse impact* (SAI) on VMEs. The onus is therefore placed upon identifying locations within which VMEs are present and in establishing a precautionary requirement for vessels to cease fishing upon encountering such features, or to ensure that these areas are pre-emptively closed to bottom fishing until it may be established that no such encounters are likely to result. In principle, this allows for the co-existence of fisheries and environmental restrictions in locations that have been identified as susceptible to the adverse impacts of deep-sea bottom fishing if not managed proactively.

However, Resolution 61/105 also raised immediate interpretive difficulties, since the threshold by which a marine ecosystem may be considered 'vulnerable' and an adverse impact deemed 'significant' – and, moreover, the precise circumstances under which an 'encounter' may be considered to have occurred in the first place – was not defined, hence the FAO was called upon to elaborate detailed practical guidance on these issues.[73] As with previous pronouncements, Resolution 61/105 also recommended the expedited development of interim measures to address locations without operational RFMO coverage.[74] In a novel departure, Resolution 61/105 also directed specific action points to states in areas for which no competent authority was in existence, with flag states requested to cease the national authorisation of fishing vessels in ABNJ without a competent RFMO or to unilaterally introduce measures applicable to nationally-registered ships to implement the broad commitments advanced therein.[75]

72 UNGA Resolution 61/105 of 8 December 2006, para 80.

73 Para 89.

74 Para 85.

75 Para 86.

In 2008, following an extensive technical consultation process, the FAO adopted an influential set of international Guidelines to frame the practical implementation of the relevant UNGA Resolutions for fisheries exploiting deep-sea species 'in a targeted or incidental manner'.[76] The Guidelines have an express application to the high seas – although states are also encouraged to adopt these approaches where appropriate within their national waters – and are applicable to fisheries for which the total catch includes species that can only sustain low exploitation rates and the fishing gear used in this process is likely to contact the seafloor during the course of fishing operations.[77] The overarching objectives of the Guidelines are therefore to ensure the long-term and sustainable use of marine living resources in the deep-sea and to prevent significant adverse impacts upon VMEs in the process.[78] Reinforcing the earlier UNGA commitments, states and RFMOs are also requested to adopt and implement measures consistent with the precautionary and ecosystem approaches to fisheries management to identify areas in which VMEs are known or likely to occur and to take action using the best available information.[79] Nevertheless, the Guidelines are expressly non-binding in nature,[80] and their regulatory effect instead is considered to have been promoted through the due diligence obligation incumbent upon flag states.[81]

The Guidelines were swiftly endorsed in turn by the UNGA, which called upon states to act to secure their implementation 'immediately, individually and through regional fisheries management organizations and arrangements'.[82] Nevertheless it was observed that the operative paragraphs of the previous UNGA Resolutions had 'not been sufficiently implemented in all cases'.[83] Accordingly, Resolution 64/72 reiterated the action points called for in Resolution 61/105, adding a further commitment to promote the adoption of conservation and management measures to ensure the long-term sustainability of deep-sea stocks and associated species, particularly through setting

76 *International Guidelines for the Management of Deep-Sea Fisheries in the High Seas* (FAO, Rome: 2008), para 5. For an extensive discussion of the formulation of the Guidelines and their prospective legal effect see L. Korseberg, 'The Law-Making Effects of the FAO Deep-Sea Fisheries Guidelines' (2018) 67 *International and Comparative Law Quarterly* 801.

77 Para 8. The total catch is defined herein as 'everything brought up by the gear', reinforcing the notion that the Guidelines ought to be applied even where deep-sea species are taken incidentally.

78 Para 11.

79 Para 12.

80 Abstract to the Guidelines.

81 Korseberg (n.76), 830.

82 Resolution 64/72 of 4 December 2009, para 113.

83 Para 118.

appropriate levels for fishing effort, capacity and catch limits.[84] This was reinforced in 2011, wherein the UNGA observed that 'despite the progress made, the urgent actions called for in the relevant paragraphs of resolutions 61/105 and 64/72 have not been fully implemented in all cases'.[85] In a notable departure to the previous instruments, Resolution 66/68 further called for the strengthening of assessment procedures so as 'to take into account individual, collective and cumulative impacts, and for making the assessments publicly available, recognizing that doing so can support transparency and capacity-building globally'.[86]

Since 2011, these commitments have been regularly affirmed and endorsed, most notably in 2016, where the UNGA called upon states and RFMOs to use the full set of criteria in the FAO Guidelines to identify VMEs and to assess SAIs, to ensure that impact assessments are reviewed periodically and updated where there is a 'substantial change in the fishery' or where there is relevant new information and that conservation and management measures are based upon best available science, especially with regard to improving the effective implementation of thresholds and move-on rules.[87]

3 Bottom Fisheries and Benthic Ecosystems: Assessing the Impact of the UNGA VME Commitments

Thus far, eight bodies have established competence over bottom fishing of deep-sea stocks, each of which has adopted a series of measures for the regulation of these fisheries. These are the North-East Atlantic Fisheries Commission (NEAFC), the Northwest Atlantic Fisheries Organisation (NAFO), the Southeast Atlantic Fisheries Organisation (SEAFO), the Southern Indian Ocean Fisheries Agreement (SIOFA), the North Pacific Fisheries Commission (NPFC), the General Fisheries Commission for the Mediterranean (GFCM), the South Pacific Regional Fisheries Management Organisation (SPRFMO) and the Commission for the Conservation of Atlantic Marine Living Resources (CCAMLR). This cohort is also complemented by the European Union (EU), which has also developed standards for bottom fishing in respect of its Member States fishing in ABNJ.

Despite an expanding degree of regulation for deep-sea bottom fisheries in ABNJ, the work of these bodies rather defies neat comparative analysis. Each

84 Para 119(d).
85 Resolution 66/68 of 6 December 2011, para 129.
86 Para 129(a).
87 Resolution 71/123 of 13 February 2017, para 180.

RFMO presides over a unique set of ecological conditions (and thereby differing volumes of VMEs to manage), different fishing industries and varying political, material and financial resources and priorities. Accordingly, the following section examines progress towards three key cumulative requirements of the various UNGA Resolutions, namely the need to identify VMEs and to regulate encounters, the establishment of fishing footprints and the elaboration of area closures, and the rectification of problematic governance gaps in respect of bottom fisheries and sensitive seabed ecosystems.

3.1 *VMEs Designations and Encounter Management*

The primary commitment established across the various UNGA Resolutions is the need to identify VMEs and to institute appropriate precautionary management measures in order to protect such areas from significant adverse impacts of bottom fishing. As noted above, these commitments have been given practical voice through the FAO's Deep-sea Fisheries Guidelines. A primary stage in protecting an individual seabed site is therefore to identify it as 'vulnerable', a notion considered by the Guidelines to be

> related to the likelihood that a population, community, or habitat will experience substantial alteration from short-term or chronic disturbance, and the likelihood that it would recover and in what time frame … The most vulnerable ecosystems are those that are both easily disturbed and very slow to recover, or may never recover.[88]

Accordingly, the Guidelines acknowledge that this is itself a variable concept. Features that are 'physically fragile or inherently rare' may be vulnerable to most impacts, while locations may be more resilient and therefore not necessarily 'vulnerable' in particular contexts.[89] At the heart of this distinction appears to lie an inherent tension between advocates of blanket preservation, and those states with a vested socio-economic interest in bottom fishing. Indeed, as Korseberg observes, there was a relatively limited attendance at the FAO's Technical Consultation in 2008 at which the Guidelines were adopted, but those states that did attend were prominent fishing nations.[90] A number of states have also strenuously observed that bottom fishing does not necessarily result in catastrophic benthic damage, especially in largely featureless areas of the global seabed, and (perhaps less convincingly in the ABNJ context) that

88 Para 14.
89 Para 15.
90 Korseberg (n.76), 824.

such fisheries continue represent a significant source of global food security.[91] In this regard, fishing states have largely managed to stave off the threat of blanket closures: the Guidelines expressly provide that the risks to a marine ecosystem are to be measured 'by its vulnerability, the probability of a threat occurring and the mitigation means applied to the threat'.[92]

The Guidelines also elaborate a series of representative characteristics that should be used as criteria in identifying VMEs, namely the uniqueness or rarity of an area or ecosystem; the functional significance of the habitat; the fragility of the area; the life history traits of component species that would make recovery difficult (as exhibited in many species of deep-sea fish); and the structural complexity of an ecosystem.[93] Although a further, non-exhaustive, series of examples are listed in an Annex to the Guidelines, the individual geological conditions of each region mean that the ultimate decision as to whether a particular site or ecosystem constitutes a VME is made by the RFMO in question. The practice of RFMOs has thus been to develop extensive individual lists of indicator species and ecosystems in order to assess its unique regulatory priorities.

In this respect, there has been a strong emphasis upon sharing examples of best practice between RFMOs, as mandated both by successive UNGA Resolutions and the Guidelines themselves.[94] This was initially problematic, leading to complaints that support tools were initially limited[95] and that there were few opportunities for institutional learning between RFMOs.[96] With

---

91 *Impacts of fishing on vulnerable marine ecosystems: Actions taken by States and regional fisheries management organizations and arrangements to give effect to paragraphs 66 to 69 of General Assembly resolution 59/25 on sustainable fisheries, regarding the impacts of fishing on vulnerable marine ecosystems. Report of the Secretary-General*; UN Document A/61/154, para 59.

92 Para 16.

93 Para 42.

94 *Guidelines*, para 29.

95 *Actions taken by States and regional fisheries management organizations and arrangements to give effect to paragraphs 83 to 90 of General Assembly resolution 61/105 on sustainable fisheries, including through the 1995 Agreement for the Implementation of the Provisions of the United Nations Convention on the Law of the Sea of 10 December 1982 relating to the Conservation and Management of Straddling Fish Stocks and Highly Migratory Fish Stocks, and related instruments: Report of the Secretary*-General; UN Document A/64/305, para 202.

96 *Actions taken by States and regional fisheries management organizations and arrangements in response to paragraphs 80 and 83 to 87 of General Assembly resolution 61/105 and paragraphs 113 to 117 and 119 to 127 of General Assembly resolution 64/72 on sustainable fisheries, addressing the impacts of bottom fishing on vulnerable marine ecosystems and the long-term sustainability of deep-sea fish stocks: Report of the Secretary-General;* UN Document A/66/307, para 207.

regard to the former concern, considerable progress has been made by the FAO in developing a distinct VME database, which has consolidated a map of areas that are both restricted and open for fishing, as well as maintaining a central repository of information concerning indicator lists for VMEs by individual RFMOs.[97] Attempts to improve coordination between the various RFMOs with competences over the deep-sea environment have, however, proved arguably less productive. Notwithstanding some initial optimism that these RFMOs could develop collaborative practices akin to the Kobe process of Cooperation to streamline regulatory and administrative activities on issues of common concern, which has proved to be of significant utility to streamlining the work of tuna RFMOs,[98] only one such meeting has been convened to date.[99]

Where VMEs have been identified, the Guidelines also outline the circumstances under which a SAI may be deemed to have occurred. Such an encounter involves ecosystem integrity being compromised in a manner that '(i) impairs the ability of affected populations to replace themselves; (ii) degrades the long-term natural productivity of habitats; or (iii) causes, on more than a temporary basis, significant loss of species richness, habitat or community types. Impacts should be evaluated individually, in combination and cumulatively'.[100] The risk of a significant adverse impact therefore triggers further mitigation strategies, primarily in the form of the 'move-on' rule that requires a vessel to retreat to a particular distance away from the VME in question. To this end, a number of RFMOs have developed encounter protocols, which are predominantly based upon the incidental catch of indicator species above threshold levels.[101]

Nevertheless, concerns have been raised that the thresholds established by a number of RFMOs are contingent upon a certain volume of live by-catch. This, for instance, has been the approach of NAFO, NEAFC and SEAFO, yet it is considered a poor tool to identify an encounter since the equipment used is designed exclusively to catch fish and is therefore not conducive to the effective sampling of benthic areas.[102] Similarly, given that cold-water reefs (which represent a substantial proportion of current VMEs) habitually comprise a

97 For the most recent consolidated version, see http://www.fao.org/in-action/vulnerable-marine-ecosystems/vme-indicators/en/.

98 See J. Harrison, 'Key Challenges Relating to the Governance of Regional Fisheries' in R. Caddell and E.J. Molenaar (eds), *Strengthening International Fisheries Law in an Era of Changing Oceans* (Hart: Oxford, 2019) 79, 99.

99 *Record of the Meeting of the Deep-Sea Secretariats Contact Group, June 2016*, 4.

100 Para 16.

101 See for example, SEAFO CM 30/5 and CCAMLR CM 22–07.

102 P.J. Auster *et al.*, 'Definition and Detection of Vulnerable Marine Ecosystems on the High Seas: Problems with the "Move-On" Rule' (2011) 68 *ICES Journal of Marine Science* 254, 258.

framework of primarily dead coral, significant damage may be legitimately inflicted upon a VME without triggering the 'move-on rule'.[103] There is also a risk that evidence of an encounter may be lost during the retrieval of a net, hence otherwise responsible fishers may be oblivious to an impact, while a longstanding objection to 'move-on' approaches remains the tacit toleration of a documented degree of environmental harm incumbent in the process.

Ultimately, definitively establishing the existence of a VME or whether a SAI has occurred is a complex task, and one that will be subject to constant revision in the light of on-going data-collection requirements. This is especially true given the very limited baseline knowledge of remote seabed areas. The capacity of parties to RFMOs to constantly monitor such ecosystems – even within the areas in which significant research activities have been conducted, let alone those that are currently unfished and largely unexplored – is highly variable, while assessing whether such areas might be sufficiently resilient to prospectively support a degree of fishing is also an exceptionally difficult undertaking. Consequently, while the Guidelines have provided a pathway through which prospective VMEs may be identified and made subject to avoidance techniques, calls for institutional restraint are prevalent and there remains a vocal constituency of advocates for a moratorium on bottom fishing, especially in ABNJ.[104]

3.2 *Fishing Footprints and Protected Areas*

Once VMEs have been identified and appropriate encounter protocols and mitigation strategies have been advanced by RFMOs, a further commitment established within the multiple UNGA Resolutions is to prevent fishing activities where they may have a SAI upon such sites. In this respect, RFMOs can be seen to have advanced two core strategies towards addressing the impacts of fishing in the areas under their jurisdiction. In the first instance, participants have been required to identify their existing fishing footprints – i.e. those locations in which some fishing activity has previously been conducted – wherein fishing is intended to be focused for the foreseeable future. In this manner, the unfished area is to be largely left alone, unless a participant wishes to expand its footprint by initiating an exploratory fishery. If so, any such endeavours are to be subject to prior approval and conducted under strict controls. The second response is the institution of a more traditional area closure, in which fishing is prohibited in order to protect VMEs.

103 UN Document A/66/307 (n.96), para 46.

104 Norse (n.4), 317.

Since the adoption of the UNGA Resolutions, RFMOs have engaged in a process of mapping the current extent of fished areas with their jurisdictional purviews. In this manner, one key policy has been to seek to confine fisheries within their current locations, while managing VMEs within these areas through encounter protocols and, if necessary, fisheries closures. Within the residue of these unfished areas, bottom fishing is *prima facie* constrained, although parties may apply to extend their fishing footprints and establish spatially and temporally restricted exploratory fisheries in these locations. Exploratory fisheries are fundamentally different to commercial fisheries and are primarily characterised by an emphasis on low-effort fishing, stringent observer coverage, extensive data-collection commitments and a requirement that activities may not proceed without the express prior approval of the management body.

The process is exemplified by the practice of NEAFC, which in 2008 adopted an Interim Exploratory Bottom Fishing Protocol for New Bottom Fishing Areas.[105] Under these arrangements NEAFC, in consultation with the International Council on Exploration of the Sea (ICES), maintains a full itinerary of such areas in an Annex to the Recommendation which establishes a definitive, yet adjustable, list of existing fishing footprints. These footprints can be extended, but any activities conducted therein remain classed as 'exploratory' and subject to prior approval, based on the submission of a Notice of Intent to fish, alongside a harvesting plan, mitigation plan and a 'sufficient system' to record data.[106] Similar systems have been established by SEAFO,[107] the NFPC,[108] SPRFMO,[109] and, especially, CCAMLR,[110] on whose pioneering system of exploratory fisheries such procedures have been largely modelled.[111]

A more traditional policy response to the problem of the degradation of particular seabed ecosystems on the part of RFMOs has been the designation of restricted areas for bottom fishing, or the outright prohibition of this equipment. In this respect, significant area closures have been instituted by NEAFC,

105 Recommendation XVI:2008.

106 Articles 6 and 7 of Recommendation 9:2015; see Consolidated Text of all NEAFC Recommendations on Regulating Bottom Fishing, available on-line at www.neafc.org.

107 CM 30/15.

108 CMM 2016–05: Bottom Fisheries and Protection of Vulnerable Marine Ecosystems in the Northeast Pacific Ocean and CMM 2016–06: Bottom Fisheries and Protection of Vulnerable Marine Ecosystems in the North-western Pacific Ocean.

109 CMM 03–2014 (Management of Bottom Fishing in the SPRFMO Convention Area).

110 CM 22–06: Bottom Fishing in the Convention Area.

111 See further R. Caddell, 'Precautionary Management and the Development of Future Fishing Opportunities: The International Regulation of New and Exploratory Fisheries (2018) 33 *International Journal of Marine and Coastal Law* 199, 212–234.

NAFO, SEAFO and CCAMLR.[112] Moreover, the GFCM has established three separate Fisheries Restricted Areas (FRAs) for a more modest volume of territory. Most emphatically, bottom trawling has been prohibited by the GFCM in any area below 1000 metres,[113] while CCAMLR has banned the use of all commercial bottom trawling in the high seas areas under its purview.[114] Nevertheless, concerns have been raised that mid-water trawling, where nets are piloted through the water column as opposed to being dragged across the seabed, retains considerable scope to damage fragile submarine features, especially seamounts. One notable regulatory departure in this regard is the approach adopted by SPRFMO, which has defined bottom-trawling as including mid-water trawling on seamounts, given the propensity for contact with seabed features even at this comparatively more elevated depth.[115] This remains a minority view, however, although similar policies were considered by NAFO in 2015, which failed to find consensus on the issue but instead imposed particular restrictions on the design and deployment of mid-water trawl gear.[116]

Beyond the closures initiated under the auspices of RFMOs, an intriguing recent development has been the elaboration of a new sub-species of area closure led by the fishing industry itself. Such closures have been most closely associated on an international level within the Southern Indian Ocean, while on a national level, extensive seabed areas have also been closed to dredging and bottom trawling within the EEZ of New Zealand. In the context of the Indian Ocean, SIOFA entered into effect in June 2012, although early meetings failed to engender confidence that this new organisation would be able to address VMEs swiftly and effectively. Despite a strong recognition of the need to implement the UNGA VME commitments, the parties were unable to agree on substantive conservation and management measures to address bottom fisheries and instead came to the bland agreement that individual participants should 'endeavour' to limit deep-sea trawling.[117] Prior to the inauguration of SIOFA, in 2006 the main deep-sea fishing operators established the Southern

112 For a full inventory, see http://www.fao.org/in-action/vulnerable-marine-ecosystems/vme-indicators/en/.

113 REC29/2005/1.

114 CM 22-05: Restrictions on the use of bottom trawling gear in high-seas areas of the Convention Area. Under this provision, bottom trawling could prospectively be deployed for scientific purposes.

115 *Report of the Second Scientific Committee Meeting*, at p. 15.

116 D Diz, 'The Seamounts of the Sargasso Sea: Adequately Protected?' (2016) 31(2) *International Journal of Marine and Coastal Law* 359, 366–67.

117 G. Wright *et al.*, 'Advancing Marine Biodiversity Protection through Regional Fisheries Management: A Review of Bottom Fisheries Closures in Areas Beyond National Jurisdiction' (2015) 61 *Marine Policy* 134, 144.

Indian Ocean Deepsea Fishers Association (SIODFA) and voluntarily designated eleven individual sites as Benthic Protected Areas, albeit with some criticism that these locations were arguably too deep for a viable commercial fishery.[118] Additional sites within more orthodox fishing grounds were designated as Benthic Protected Areas in 2013. Although these sites remain voluntary in nature – and as an industry initiative they do not carry the prohibitive value of RFMO authority – they have nevertheless had considerable traction within the region, since the benefits of membership of SIODFA are contingent upon compliance with these measures, while individual states have also made this a formal condition for vessel licensing,[119] suggesting that commercial incentives can be at least as effective as the traditional command-and-control approach adopted by RFMOs.

Similarly, on a national level, an extensive series of Benthic Protection Areas covering over 1.1 million km^2 was designated throughout the EEZ of New Zealand in 2007 in which trawling and dredging is precluded, a measure that was also, to the surprise of many domestic commentators, proposed by the national fishing industry.[120] In a manner reminiscent of the designation of fishing footprints collated by RFMOs, these sites are currently located in areas in which no fishing activities occur, with their prospective value considered to be in curtailing the future expansion of trawling. This approach has also been adopted in neighbouring Australia, were approximately 58% of the national EEZ is also closed to bottom trawling – albeit encompassing significant locations in which no trawling occurs.[121] Nevertheless, in New Zealand the proponents of these measures were not entirely motivated by ecological altruism and the initial proposal came with heavy strings attached, with the Deepwater Group seeking an indefinite moratorium on future fisheries closures in the national EEZ.[122] Eventually, this *quid pro quo* was restricted to an agreement not to create further marine protected areas until 2013.[123] Moreover, the sites are confined to the seabed itself, with mid-water trawls permitted in the water

118 *Ibid.* (noting however that this could be a useful designation if fishing activities seek to move deeper in future years).

119 G. Wright and J. Rochette, 'Regional Management of Areas beyond National Jurisdiction in the Western Indian Ocean: State of Play and Possible Ways Forward' (2017) 32 *International Journal of Marine and Coastal Law* 765, 777–778.

120 J. Helson *et al.*, 'Private Rights, Public Benefits: Industry-Driven Seabed Protection' (2010) 34 *Marine Policy* 557, 559–563.

121 T.K. Mazor *et al.*, 'Trawl Exposure and Protection of Seabed Fauna at Large Spatial Scales' (2017) 23 *Diversity and Distributions* 1280, 1280.

122 Helson (n120), 560.

123 T.D. Eddy, 'On the Need for Meaningful Marine Protected Area (MPA) Standards' (2013) 23 *Aquatic Conservation: Marine and Freshwater Ecosystems* 481, 482.

column, although considerable monitoring restrictions were placed upon any fishing activities taking place in these areas.[124] Critics of this approach have argued that these areas are not true MPAs as they merely restrain specific activities yet leave open the possibility of further industrial activities in superjacent waters.[125] Indeed, the national authorities have long been accused of using these designations as a means of inflating the volume of national MPA coverage, culminating in a rather contrite declaration in April 2019 that Benthic Protection Areas would no longer be counted within the domestic inventory of MPAs.[126]

A further issue of considerable significance in seeking to ensure that holistic protection is applied to sensitive seabed ecosystems is the extent to which fisheries closures are able to dovetail effectively with the management policies of other sectoral regulators. As noted above, seabed features are best protected where both the bethos and the overlaying water column are subject to complementary conservation measures. One striking example of this approach is the pioneering arrangements between NEAFC and OSPAR, whose respective jurisdictional areas intersect within the north-east Atlantic region. The legal mandate of NEAFC is confined to fisheries management, while OSPAR's competence to address "non-polluting human activities" strictly excludes any consideration of fisheries.[127] In discharging this mandate, OSPAR has placed considerable emphasis upon establishing a network of MPAs and has a comparatively lengthy history of promoting interactions with other organisations in this capacity.[128]

In 2008, a MOU was concluded between NEAFC and OSPAR[129] to explore areas of mutual interest and formalise a basis for potential future collaboration, for which the most significant outcome was the adoption in 2014 of a Collective Arrangement on cooperation on MPAs in ABNJ. In 2009, NEAFC had closed a series of areas to bottom fishing that broadly corresponded to the designations within the OSPAR MPA network, notably within the Charlie

124 Helson (n.120) 564.

125 T.D. Eddy, 'One Hundred-Fold Difference between Perceived and Actual Levels of Marine Protection in New Zealand' (2014) 46 *Marine Policy* 61, 64–66.

126 'Conservationists Win Battle Over Government's "Fishing Whoppers"' https://www.stuff.co.nz/national/politics/111766497/conservationists-win-battle-over-governments-fishing-whoppers.

127 Article 4 of Annex V to the OSPAR Convention, added in 1998.

128 See further E.J. Molenaar and A.G. Oude Elferink, 'Marine Protected Areas in Areas beyond National Jurisdiction: The Pioneering Efforts under the OSPAR Convention' (2009) 5 *Utrecht Law Review* 5, 16.

129 Reproduced on-line at http://www.ospar.org/about/international-cooperation/memoranda-of-understanding.

Gibbs and Mid-Atlantic Ridge MPAs. The Collaborative Arrangement therefore addresses specific locations of mutual interest within the region, which are outlined in Annex I and are jointly maintained by both organisations. While not exclusively focused on area-based management – promising lines of cooperation have also emerged for marine litter and shark conservation – the Collaborative Arrangement provides a platform for data exchange and updates on amendments to the respective restricted areas, with annual meetings having been convened since 2015 to promote these objectives further.

The OSPAR/NEAFC arrangements illustrate both the opportunities and the complexities facing purported collaborative exchanges of this nature. The Collaborative Arrangement seeks to include other pertinent global and regional actors to minimise potential interference with Annex I areas.[130] OSPAR and NEAFC have thus encouraged the International Maritime Organisation (IMO) and International Seabed Authority (ISA) – which both have the capacity to advance sectoral management tools – to participate in this process, albeit with little success. As with many synergistic endeavours between multilateral bodies, incompatible meeting schedules have inhibited interactions with the IMO.[131] More significantly, however, such initiatives have met with internal resistance from IMO participants that are geographically and economically removed from shipping activities in the region opposed to devoting time and resources on matters of more localised concern.[132] Meanwhile, the ISA has considered its participation to be 'premature' in the absence of a clearly defined project-based role in the region.[133] Wariness about open-ended collaborative demands has also been expressed internally within the OSPAR Commission, with some participants concerned that cross-sectoral management represents a significant but small aspect of an extensive portfolio of activities that could impede the pursuit of more immediate regulatory priorities.[134]

130 The International Commission for the Conservation of Atlantic Tunas (ICCAT) is also identified as a potential partner, given its application to tuna fisheries in the region. Similarly, OSPAR has developed a rather more concise MOU with the North Atlantic Salmon Conservation Organisation (NASCO), although it is more ambiguous and prescribes few action points.

131 *Aide Memoire and Key Actions Resulting from the First Meeting under the Collective Agreement*, para 2.7.

132 *Aide Memoire and Key Actions Resulting from the Second Meeting under the Collective Agreement*, para 3.6.

133 *Ibid.*

134 D. Freestone *et al.*, 'Can Existing Institutions Protect Biodiversity in Areas beyond National Jurisdiction? Experiences from Two On-Going Processes' (2014) 49 *Marine Policy* 167, 173.

3.3 *Governance Gaps*

A further issue of strong concern within the various UNGA Resolutions has been the need to close the significant governance gaps that were exposed by initial attempts to regulate deep-sea fisheries. As observed above, bottom fisheries have long been deceptively under-regulated on an international basis. Many RFMO apply only to single species, while until relatively recently large swathes of the global oceans were not subject to any management coverage. Moreover, as observed above, those RFMOs that did technically exercise a mandate over deep-sea stocks generally lacked the legal competence to fully protect seabed ecosystems. In this respect, closing gaps in regulatory coverage has been an important, and largely successful, element of the commitments established under the UNGA Resolution.

Since the turn of the current century, a suite of new RFMOs – notably SIOFA, SEAFO, SPRFMO and the NPFC – have been inaugurated and have expressly established the protection of VMEs as a central tenet of the mandates of their constituent bodies. This is most clearly illustrated by the NPFC, for which the preamble to its constituent treaty expressly references the relevant UNGA Resolutions, while the prevention of significant adverse impacts from fisheries upon VMEs is established as a 'general principle' for this body.[135] Similarly, consideration of VMEs is established as a specific aspect of the mandate of the scientific fora of SPRFMO.[136] More significantly, perhaps, the conclusion of these instruments was preceded by a series of interim arrangements focused on the regulation of bottom fishing in these areas.

Allied to this, more longstanding RFMOs have been prepared to interpret their mandates – and, indeed, reformulate their constituent provisions – in a broad manner so as to establish a clear degree of management control over bottom fisheries. This is exemplified by the practice of NEAFC, which closed three deep-sea sites to bottom trawling and fishing with static gear, including gillnets and bottom longlines effective from 2005 onwards.[137] This was a very far-sighted development at the material time, preceding the seminal UNGA Resolution 59/25 by some months, with NEAFC therefore operating in uncharted waters with no global guidance in place. Moreover, it might be questioned whether *stricto sensu* NEAFC possessed the requisite regulatory

---

135 Convention on the Conservation and Management of High Seas Fisheries Resources in the North Pacific Ocean (Seoul, 1 April 2012, in force 19 July 2015); preamble and Article 3(e).

136 Convention on the Conservation and Management of High Seas Fishery Resources in the South Pacific Ocean (Auckland, 14 November 2009, in force 24 August 2012) [2012] ATS 28; Article 11.

137 Recommendation 05–2005: Recommendation for the Protection of Vulnerable Deep-Water Habitats.

competence to do so, which necessitated a very broad interpretation of the Commission's powers on the part of its constituent Members in approving these closures.[138]

More fundamentally, older structures have undertaken a process of considerable reform in recent years to allow them to more centrally address the environmental issues associated with deep-sea fisheries. In this respect, a series of provisions have been 'retro-fitted' into the NAFO Convention, through extensive textual revisions adopted in 2007 in order to promote an ecosystem approach to fisheries management, which eventually entered into effect on 18 May 2017.[139] Notwithstanding the recent formalisation of these arrangements, many of these obligations had been applied provisionally throughout this interim period,[140] which enabled NAFO to adopt a series of measures to promote the protection of VMEs. A similar approach has been adopted by the International Commission on the Conservation of Atlantic Tunas (ICCAT), again largely motivated to advance additional protection for VMEs as promoted through the relevant UNGA commitments.[141]

A further important factor has been the role of flag states, which had also been requested to take action to ensure that their vessels fish in a manner that is sympathetic to the benthic environment. In this respect, particular states and entities have been prepared to adopt unilateral standards where their vessels operate in areas lacking RFMO coverage. A particular example is the South-West Atlantic, whereby political complications have precluded the likely establishment of a RFMO for these waters for the foreseeable future. A volume of deep-sea fishing has been conducted in the region, predominantly by Spain, for which a series of voluntary closures have been instituted. More broadly, in 2008, in seeking to implement the pertinent UNGA Resolutions, the EU adopted a Regulation specifically addressing the actions of its Member States in ABNJ for which no RFMO has been established or interim measures have not yet been agreed for the protection of VMEs.[142] Under this provision, such

138 Molenaar (n.30), 538–39.

139 Convention on Cooperation in the Northwest Atlantic fisheries (Ottawa, 24 October 1978, in force 1 January 1979) 1135 UNTS 369 (Article 1(h)). The consolidated version of this instrument, incorporating the 2017 amendments, is available on-line at https://www.nafo.int/Portals/0/PDFs/key-publications/NAFOConvention-2017.pdf.

140 Resolution 1/08 of 26 September 2008 on the Interpretation and Implementation of the Convention on Future Multilateral Cooperation in the Northwest Atlantic Fisheries.

141 See further Z. Scanlon, 'The Art of "Not Undermining": Possibilities within Existing Architecture to Improve Environmental Protections in Areas beyond National Jurisdiction' (2018) 75 *ICES Journal of Marine Science* 405–416, 410.

142 Council Regulation (EC) No 735/2008 of 15 July 2008 on the protection of vulnerable marine ecosystems in the high seas from the adverse impacts of bottom fishing gears [2008] *Official Journal* L201/8.

activities may only be conducted pursuant to a special permit[143] and the use of bottom gear is prohibited in areas 'where no proper scientific assessment has been carried out and made available'.[144] A permit may only be issued upon submission of a detailed fishing plan involving an assessment of the potential impacts of fishing in the area,[145] for which any breach is considered a 'serious infringement' of the Common Fisheries Policy.[146] In response, Spain has thus far closed nine separate areas to deep-sea bottom fishing by its vessels. In 2017, a further Regulation entered into force banning deep-sea fishing by EU vessels in the North-east Atlantic at depths of 800 metres,[147] although the unilateral policies introduced by the EU for this region has generated some disquiet that such measures might exercise an undue influence over the future trajectory of the regulation of VMEs within NEAFC.[148]

4 Conclusions

While little appreciated in the popular perception of risks to the benthic environment, bottom fisheries have quietly emerged as the most pressing current threat to seabed ecosystems. Although trawl fisheries have constituted a source of regulatory discontent for over six hundred years, meaningful standards to address the increasingly insidious impacts of such activities on the global seabed have only systematically emerged over the course of the present century. Bottom fisheries have proved to be an ecologically expensive food source, inflicting damage with a multi-decadal recovery timescale, while also decimating stocks of deep-sea fish that have often proved manifestly ill-suited to the scale and impact of commercial fishing. Such activities have also, for the most part, been conducted in an effective regulatory vacuum, with deep-sea fisheries representing not only a particularly striking example of the 'tragedy

143 Article 1(1).

144 Article 6(1).

145 Article 3.

146 Article 10.

147 Regulation (EU) No. 2016/2336 of the European Parliament and of the Council of 14 December 2016 establishing specific conditions for fishing for deep-sea stocks in the north-east Atlantic and provisions for fishing in international waters of the north-east Atlantic and repealing Council Regulation (EC) No. 2347/2002 [2016] *Official Journal* 354/1; On the development of this provision see further G.A. Oanta, 'The European Union's Reform of Deep-Sea Fisheries in the North East Atlantic' (2017) 32 *International Journal of Marine and Coastal Law* 589.

148 G.A. Oanta, 'International Organisations and Deep-Sea Fisheries: Current Status and Future Prospects' (2018) 87 Marine Policy 51, 57.

of the commons', but also serving as a stark metaphor for the failure of international legal frameworks to respond effectively to the challenges of depleted fisheries and damaged oceanic ecosystems.[149]

The response to the challenges presented by deep-sea bottom fishing represents an intriguing case study of the development of regulatory standards for the seabed. In this respect, conservation and management policies have been largely driven through the UNGA, which has in turn designated RFMOs as the primary vehicles through which this problem is to be addressed. Consequently, since 2004 a growing suite of commitments has quietly emerged, which have been implemented by a variety of actors over the past decade. To this end, a series of positive outcomes have been achieved. Governance gaps have been noticeably closed, with new RFMOs having emerged in recent years, many of which feature a specific and unprecedented commitment towards protecting VMEs. Other structures have used the regulatory impetus provided by the UNGA to undertake significant reforms, placing ecosystem considerations more centrally within their revised mandates. Where regulatory lacunae persist, flag states have demonstrated a willingness to step into the breach and promote standards that are rapidly becoming recognised as universal examples of best practice towards threatened seabed environments.

Nevertheless, particular challenges remain and, despite this laudable progress, the commitments elaborated by the relevant UNGA Resolutions have still not been fully realised. Notwithstanding the value of the FAO Guidelines, bottom fisheries may still legitimately wreak considerable damage to the seabed before environmental obligations are engaged. The relevant instruments remain replete with opaque trigger-points for action, the identification of VMEs on a global basis is very much a work in progress and the 'move-on' rule remains a blunt instrument for addressing seabed conservation. Meanwhile the knowledge base concerning seabed ecology – both within, but especially outside – current areas of fishing activity remains extremely patchy. Considerable efforts have been made to elaborate area-based management tools, although a strong emphasis has been placed on developing closures in locations in which fishing does not occur and is unlikely to advance for many years. Regulatory innovations have emerged, particularly those led by industry, although in key instances a primary motivation has been to stave off stronger protective standards. Similarly, despite some notable successes in the North Atlantic region, it has proved difficult to align area-based management

149 K.M. Gjerde, 'High Seas Fisheries Management under the Convention on the Law of the Sea' in D. Freestone, R. Barnes and D. Ong (eds), *The Law of the Sea: Progress and Prospects* (Oxford University Press: Oxford, 2006) 281, 295.

tools so as to provide holistic protection to the seabed from multiple sources of anthropogenic activity. With the prevailing legal framework having pointedly bifurcated the regulation of the seabed and that of the water column, the experience of deep-sea bottom fisheries reinforces the need for joined-up management of benthic locations and the superjacent water column in a manner that has proceeded to a regrettably limited degree to date.

PART 5

Principles Applicable to Sovereign States When Exploiting Seabed Resources within National Jurisdiction

CHAPTER 13

Review of National Legislations Applicable to Seabed Mineral Resources Exploitation

Saul Roux and Catherine Horsfield

1 Introduction

In parallel to growing interest for seabed exploration and mining in the high seas, there has been increased interest for seabed mining activities within the national jurisdictions of coastal states. Seabed mining in the high seas and within national jurisdictions is not unconnected. Seabed mining in the high seas provides a good indication of the nature, status quo and future of seabed mining in national jurisdictions and vice versa. Notably, the direction of seabed mining in the high seas will influence the policy directions of nation states.[1]

This chapter seeks to provide a broad overview of seabed mining within the national jurisdictions of sovereign coastal states. Section 2 provides a brief outline of international legal and governance instruments that are intended to provide a framework for national policy and decision-making on seabed mining.

Section 3 seeks to provide a comparative analysis of legal and policy approaches adopted by various countries in relation to seabed exploration and exploitation. This aims to offer insights into legal principles and mechanisms that have been used in response to seabed mining. This comparative analysis focuses on seabed mining, in respect of both exploration and exploitation, within the exclusive economic zones of five coastal states namely: New Zealand, Australia, Namibia, Mexico and Papua New Guinea. These cases are assessed primarily due to the fact that seabed mining proposals have been concentrated in these coastal states. Furthermore, they represent a diversity of policy approaches and regulatory regimes. These approaches have been diverse, due

1 Lawson, T (2015). Global Opposition Is Mounting Against the Latest Environmental Abuse – Deep Sea Mining. Truthout <http://www.truth-out.org/news/item/34206-global-opposition-is-mounting-against-the-latest-environmental-abuse-deep-sea-mining?

 | DOI:10.1163/9789004391567_015

to different legal and governance regimes operating in these countries, ranging from permanent bans, moratoria, strategic environmental assessments, assessments by environmental agencies and preparation for exploitation.

Section 4 provides a more in-depth case study of South Africa, focusing on marine phosphate prospecting. The South African case provides strong insights into the potential risks of seabed mining applications in jurisdictions with limited regulatory and governance capacity. Issues such as socio-economic and environmental risks related to marine phosphate extraction and legal and governance challenges for South Africa are discussed in this section.

Section 5 provides a brief synthesis of approaches adopted by other countries and the legal and governance principles that underpinned these approaches. Overall, this seeks to provide a structure for decision-making on seabed mining in the national jurisdictions of coastal States.

2 Seabed Mining in the High-Seas or National Jurisdictions

It is important to differentiate between seabed mining within the exclusive economic zones of coastal States and seabed mining in the high seas or 'the Area' that lies beyond national jurisdictions. In the case of the high seas, seabed mining is governed by the United Nations Convention on the Law of the Sea (UNCLOS) and associated Mining Codes, and regulated by the International Seabed Authority.[2] Seabed mining applications that fall within a country's exclusive economic zone (within national jurisdiction) are regulated by that country's domestic law. The international deep seabed area (the Area), regulated by the International Seabed Authority (ISA) comprises about 260-million square kilometres. A total of 85-million square kilometres of ocean falls within exclusive economic zones.[3]

National jurisdictions usually extend to 200 nautical miles (370 km) seaward from baselines running along the shore. However, countries may submit Continental Shelf Extension claims (See Chapter Brekke). The International Seabed Authority has no role in determining these claims. This is decided by the Commission on the Limits of the Continental Shelf, also established by UNCLOS. Maritime boundaries between states are generally decided by bilateral negotiation, sometimes with the aid of independent judicial bodies.

2 See Chapter 7 of this book, J. Dingwall, 'Commercial Mining Activities in the Deep Seabed Beyond National Jurisdiction: The International Legal Framework.'

3 www.isa.org.jm.

2.1 *The United Nations Convention on the Law of Seas*

The United Nations Convention on the Law of the Sea (UNCLOS) confers coastal states with a broad range of sovereign rights and duties in relation to their exclusive economic zones. Such rights and duties relate to utilisation of living resources, fisheries management, species management, enforcement and compliance, exploitation of non-living resource, marine scientific research, the protection and preservation of the marine environment and any other activities related to economic exploitation and exploration, such as the production of energy from the water, currents and winds.[4] This includes the seabed, subsoil and water column.

Articles 76–85[5] outline the rights and duties of coastal States over the continental shelf. These rights relate primarily for the purpose of exploring and exploiting natural resources. These natural resources consist of mineral and other non-living resources of the seabed and subsoil together with living organisms belonging to sedentary species (immobile). In brief, the powers vested in coastal states over their continental shelves only relate to exploitation of specific resources.

In terms of Article 192 of UNCLOS, signatory States have a general obligation to protect and preserve the marine environment within and outside their jurisdiction. Article 194(1) directs signatory States to take all measures necessary 'to prevent, reduce and control pollution of the marine environment from any source, using for this purpose the best practicable means at their disposal'. States are also required to take all measures necessary to ensure activities within their jurisdiction do not cause damage to other States and their environment. This duty is important in the case of seabed mining, where the impacts may extend to the exclusive economic zones (EEZ) of neighbouring states. In particular, the impact on fishery resources and migratory fish stocks are important.

In terms of Article 194(2) measures should be put in place to minimise, to the fullest possible extent 'pollution from installations and devices used in exploration or exploitation of the natural resources of the seabed and subsoil'. These should, in particular, include measures to regulate the design, construction and operation of such installations or devices. These measures must 'protect and preserve rare or fragile ecosystems as well as the habitat of depleted, threatened or endangered species and other forms of marine life'.

The legal obligation of signatory States in respect of seabed mining has been determined and outlined by the Seabed Disputes Chamber of the International

4 Article 55–Article 75.

5 Part VI of UNCLOS.

Tribunal for the Law of the Sea.[6] This determination requires State laws and regulations governing seabed mining to be 'no less effective than international rules, regulations and procedures'. Thus State laws cannot be less effective than the International Seabed Authority Mining Code. The Seabed Disputes Chamber further determines that:

> States have a direct obligation under international law to ensure that seabed mining activities are governed in accordance with the precautionary approach, employing best environmental practice and conducting prior environmental impact assessment.

In other words 'an effective state response to these obligations ultimately requires an appropriate national legislative framework' to regulate seabed mining.[7] Granting seabed mining rights in the absence of such a legislative framework would accordingly be contrary to UNCLOS.

2.2 *Other International and Regional Agreements*

There is a wide range of other international agreements that are of relevance to seabed mining in national jurisdictions, if the nation in questions is signatory to these. These include the Convention on the Prevention of Marine Pollution by Dumping of Wastes and Other Matter (London Convention), Convention on the Conservation of Migratory Species of Wild Animals and the Convention on Biological Diversity.

Furthermore, there are a number of regional treaties that govern marine activities between signatory nations.[8] For example, Angola, Namibia and South Africa are signatories to the Benguela Current Convention (BCC).

---

6 Seabed Disputes Chamber of the International Tribunal for the Law of the Sea, 'Responsibilities and Obligations of States Sponsoring Persons and Entities With Respect to Activities in the Area – Advisory Opinion', 1 February 2011, list of cases: no. 17 para. 214: UNCLOS Articles 208(3) and 209(2) 01.

7 Benkenstein, A. (2014). Seabed Mining: Lessons from the Namibian Experience. SAIIA.

8 Including Convention for Co-operation in the Protection and Development of the Marine and Coastal Environment of: West and Central African Region (1981); Mediterranean Sea (1976); Wider Caribbean Region, (1983); Eastern African Region (1985); South-east Pacific (1981); North-east Atlantic OSPAR Convention (1992); South Pacific Region (1986); Baltic Sea Area (1992); Framework Convention for the Protection of the Marine Environment of the Caspian Sea; Kuwait Regional Convention for Co-operation on the Protection of the Marine Environment from Pollution (1978); Regional Convention for the Conservation of the Red Sea and the Gulf of Aden Environment (1982).

The BCC provides a good example of the interaction between seabed mining and regional agreements, as Namibia and South Africa have existing proposals for marine phosphate mining and prospecting respectively.

The BCC aims to facilitate cooperation and mutual responsibility of the three signatory nations in respect the Benguela Current Large Marine Ecosystem. The BCC seeks to ensure environmentally responsible industrial development and cooperation between the three nations in order to protect biodiversity, maintain ecological integrity and minimise risk of long-term or irreversible impacts caused by human activities. The Preamble states that a priority of the BCC is to minimise pollution from marine mining. A further priority is to ensure that policies, laws and regulations of the three signatory nations are harmonised so that 'industrial activities in one country [do] not impact the coastal or marine environment of another country' (ibid).

Relevant Articles of the BCC related to seabed mining include Article 4 which outlines 'General Principles' to be followed by signatory nations. These include:

a) The cooperation, collaboration and sovereign equality principle;
b) sustainable use and management of the marine resources;
c) the precautionary principle;
d) prevention, avoidance and mitigation of pollution;
e) the polluter pays principle; and
f) protection of biodiversity in the marine environment and conservation of the marine ecosystem.

In order to give effect to the objective of the BCC and the principles above, signatory parties are required to:

1) take all possible steps to prevent, abate and minimise pollution and take the necessary measures to protect the marine ecosystem against any adverse impacts;
2) undertake environmental impact assessments for proposed activities that are likely to cause adverse impacts on marine and coastal environments;
3) apply management measures based on the best scientific evidence available;
4) where possible, reverse and prevent habitat alteration and destruction; and
5) protect vulnerable species and biological diversity.

The BCC requires signatory states to 'agree on, where necessary, measures to prevent, abate and minimise pollution caused by or resulting from exploration and exploitation of the continental shelf and the seabed and its subsoil'. Article 8 further requires the Commission agree on conservation and management measures concerning transboundary marine resources and the environment'.

The BCC thus would require a cooperative approach to seabed mining to ensure that 'activities in one country do not impact on the marine environment of another country'. Such regional treaties are particularly important where signatory nations share marine and fishery resources where a seabed mining activity of one nation would negatively impact on the marine environment, fishery resources and ocean economy activities of a neighbouring country.[9]

3 Seabed Mining in National Jurisdictions

This section aims to provide a comparative analysis of legal and policy approaches adopted by other countries in relation to seabed exploration and exploitation. It offers insights into legal mechanisms and principles that have been used in response to seabed mining. Proposals for seabed mining, for both exploration and exploitation, within the exclusive economic zones of coastal States have been concentrated on six nations or areas, namely New Zealand, Australia, Namibia, South Africa, Mexico and the Pacific Islands (including Fiji, Papua New Guinea, Tonga, Vanuatu, Solomon Islands and Cook Islands).[10] This section will review each of these in turn.

In brief, there has been considerable resistance to seabed extraction proposals in all of these cases and in most cases governments have opted for a cautious approach to seabed mining in the form of moratoria, permanent bans or refusal of project proposals. Whereas Namibia and the Northern Territory of Australia placed a moratorium on phosphate mining and bulk seabed mining respectively, New Zealand's Environmental Protection Agency refused its first application for consent to undertake marine phosphate mining and Mexico's Secretary of Environment and Natural Resources (SEMARNAT) denied environmental authorisation for its first marine phosphate mine proposal. Several other jurisdictions have also established moratoria on oil and gas exploration and production in their exclusive economic zones. Papua New Guinea has approved an exploitation right for seabed mining. However, actual seabed mining operations have not yet commenced.

3.1 *Namibia: Moratorium on Marine Phosphate Mining*

Namibia's environmental management framework was established by the Environmental Management Act, 2007 (EMA). The EMA sets out a range of

9 Ministerial meetings of the Benguela Current Commission (BCC, 2013).

10 REPRISK (2015). Deep Sea Extractive Activities: Seabed Mining and Deep Sea Drilling.

environmental principles intended to mitigate and avoid adverse impacts of activities. Regulations developed in terms of EMA list certain activities that may not be undertaken without an environmental clearance certificate.

The Act lists 'resource removal, including natural living resources' as an activity that requires an environmental clearance. The Regulations moreover specifically refer to 'mining and quarrying' activities including 'resource extraction, manipulation and forms of mining or extraction of any natural resources whether regulated by law or not.'

Mining licenses in Namibia are granted under the Minerals (Prospecting and Mining) Act, 1992 with the application process set out in s.91. The Minerals Act provides for attaching environmental conditions in issuing mining licenses and requires

> An estimate of the effect which the proposed prospecting operations and mining operations may have on the environment and the proposed steps to be taken in order to minimize or prevent any such effect.

Furthermore, before mining may commence, the environmental commissioner must decide on whether the activity requires an environmental impact assessment.

In 2011 the Ministry of Mines and Energy granted mining licences to two companies, Namibian Marine Phosphate and LL Namibia Phosphates, for the exploitation of phosphate deposits off the coast of Namibia.[11] The mining licences were granted subject to the approval of an Environmental Impact Assessment. In January 2012 Namibian Marine Phosphate submitted a draft Environmental Impact Assessment and Environmental Management Plan for the proposed Sandpiper Phosphate Mining Project (The Sandpiper Project).

The Sandpiper Project would involve dredging the seabed in Namibia's exclusive economic zone at depths of 3m and removing large volumes of sediment, up to 5.5 million tonnes annually. The operations expected an annual production of up to 3 million tonnes of phosphate rocks. Sediment would be stored on the vessels and thereafter transferred to land for processing which involves separation of phosphate from other marine sediment.[12]

---

11 Benkenstein, A. (2014). Seabed Mining: Lessons from the Namibian Experience. SAIIA.

12 Midgley, J., 2012. Environmental Impact Assessment Report for the Marine Component. Sandpiper Project. Proposed recovery of phosphate enriched sediment from the marine Mining Liscence Area No. 170 off Walvis Bay Namibia – Environmental Scoping Report (Final Report, April 2012).

As part of the Environmental Impact Assessment, specialist studies were undertaken that included potential impacts on seabirds and marine mammals, fish and fisheries, water column dynamics, macrobenthos and jellyfish.[13]

There was considerable opposition to the Sandpiper Project. A number of civil society organisations (both local and international)[14] and fishery industry associations[15] strongly opposed the granting of rights and campaigned for a review of the decision.[16] Furthermore, independent scientists raised strong objections against the Sandpiper Project and the EIA based on environmental impacts. These included the direct destruction to benthic habitats; release of hydrogen sulphide concentrations; release of heavy metals and bioaccumulation in the food chain in turn impacting on fisheries; reduced phosphorous content in the seabed which results in oxygen depletion and toxicity of processing plants on shore near Walvis Bay.[17]

In response to that opposition and the potential impacts on Namibia's fishing industry, the Minister of Fisheries and Marine Resources took steps to establish a moratorium on marine phosphate mining. This led to a cabinet decision in September 2013 to establish an 18 month moratorium on marine phosphate mining. The decision outlined a number of concerns from the Namibian Government:

> Government is concerned that the removal of soft sediment from the seabed along with living organisms and the suspension of fine sediment in the seawater may affect the functioning of the marine ecosystem negatively. The ministry is further concerned that any contamination or suspended particles from mining activities may have adverse effects on the fish larvae and their development and thus [we] need to investigate the breeding and nursing areas, breeding patterns of the commercially

13 Benkenstein, A. (2014). Seabed Mining: Lessons from the Namibian Experience. SAIIA.

14 Including Earth Organisation Namibia, the Deep Sea Mining Campaign and Swartkopmund Matters.

15 Including the Namibian Hake Fishing Industry Association and the Confederation of Namibian Fishing Associations.

16 Komnenic, A. (2013). Namibia imposes moratorium on coastal phosphate mining. Mining.com<http://www.mining.com/namibia-imposes-moratorium-on-coastal-phosphate-mining-41520/> Accessed April 2016.

17 Currie, B. (2014). Investigation of cumulative impacts on the marine ecosystem from mining of phosphorites off the Namibian coast; Solbakken, et al. (2014). Preparing for an Assessment of Environmental Impacts from Onshore and Offshore Phosphate Mining Activities in Namibia.

> important species in relation to the mining activities, as well as the current dynamics.[18]

A major rationale for the moratorium was lack of knowledge on the potential impacts of marine phosphate mining on the marine environment and fisheries resources. The Minister of Fisheries and Marine Resources announced that 'information collected during the moratorium would provide an indication as to whether marine phosphate mining and the fishing industry can co-exist.' The Minister further stated that:

> The Namibian government is committed to ensure that proper scientific investigations are carried out that will allow it to make an informed decision about the future of phosphate mining in Namibia.[19]

Accordingly, the Namibian government subsequently commissioned an independent study and environmental impact assessment of the impacts of marine phosphate mining aligned to broader strategic environmental assessments on the Namibian marine environment. The Fisheries and Aquaculture section of the Norwegian-based Foundation for Scientific and Industrial Research (SINTEF) and the Institute for Marine Research were commissioned to conduct these studies.[20]

In 2017 Namibia's Environmental Commissioner issued an environmental clearance certificate for the Sandpiper Project. This led to a public outcry resulting in the Minister of Environment and Tourism cancelling the environmental clearance certificate. Uncertainty and contestation related to the Sandpiper Project is on-going and the issue is highly politicised.

3.2 *New Zealand: Refusal of Marine Phosphate Mining Application*

On 14 May 2014 Chatham Rock Phosphate Limited (Chatham Rock) applied to the Environment Protection Authority for marine consent[21] to mine phosphate in Chatham Rise, located 400 km east of Christchurch. The application

---

18 Namibian Government Cabinet Moratorium, 2013.

19 Haufiku, M. (2014). Moratorium on phosphate mining remains. New Era https://www.newera.com.na/2014/04/11/moratorium-phosphate-mining-remains/ Accessed May 2016.

20 Draft reports related to the overall plan for the Strategic Environmental Assessments have been completed by SINTEF (Solbakken, et al. (2014). Preparing for an Assessment of Environmental Impacts from Onshore and Offshore Phosphate Mining Activities in Namibia).

21 In terms of the Exclusive Economic Zone and Continental Shelf (Environmental Effects) Act 2012 (the EEZ Act).

sought consent to commence operations over a 10 192 km² area for a 35 year period. Chatham Rock Phosphate Limited later withdrew the application for operations in an Eastern mining block on 1 August 2014 reducing the total area to 5 207 km². In its application Chatham Rock sought a mining permit (MPL 55549) to mine a total area of 1 052 km² over 35 years.[22] The application sought to mine three 10km² each year at depths between 250m and 450m with an expected annual production of up to 1.5 million tonnes.[23] In the first five years of operations a mining area restricted to 820 km² was proposed. Chatham Rock provided a description of the proposed mining operation to be undertaken:

> The mining was to be carried out by a specially built or modified vessel using a mining system designed by Royal Boskalis Westminster nv (Boskalis). Phosphorite-bearing material was to be retrieved from the seabed by means of a trailing suction drag-head and mechanically processed on board the vessel. Phosphorite nodules greater than 2 mm would be separated from other material using sieves and logwashers and stored on the vessel. Waste material would then be released close to the seabed, using a discharge (sinker) pipe with a diffuser.[24]

A Decision-Making Committee (DMC) was appointed by the Environment Protection Authority to decide on the application. The application by Chatham Rock drew widespread public attention, with over 294 submissions and large public hearings held by the DMC in Wellington, Hamilton and the Chatham Islands.[25] On 10 February 2015 the DMC completed deliberations and refused consent.

New Zealand has a comprehensive legal regime on governing seabed mining, which deserves attention. The legislative framework on determining marine consents for marine mining is provided for in its Exclusive Economic Zone and Continental Shelf (Environmental Effects) Act 2012 (the EEZ Act). A large number of provisions in the EEZ Act were considered in the Chatham Rock Phosphate application.[26]

22 Currie, D. (2015). Decision on marine consent application: Chatham Rock Phosphate Limited to mine phosphorite nodules on the Chatham Rise.

23 Ibid.

24 Chatham Rock Phosphate, 2014.

25 Currie, D. (2015). Decision on marine consent application: Chatham Rock Phosphate Limited to mine phosphorite nodules on the Chatham Rise.

26 Environmental Protection Authority, 2015.

Section 10(1) outlines the purpose of the Act, which is 'to promote the sustainable management of the natural resources of the exclusive economic zone and continental shelf.'[27]

Section 10(2) defines sustainable management as 'managing the use, development and protection of natural resources in a way, or at a rate, that enables people to provide for their economic wellbeing while—

(a) sustaining the potential of natural resources (excluding minerals) to meet the reasonable needs of future generations; and
(b) safeguarding the life-supporting capacity of the environment; and
(c) avoiding, remedying or mitigating any adverse effects of activities on the environment.

Section 10(3) states that to fulfil the purposes of the Act, decision-makers must consider decision-making criteria relevant to a particular application[28] and apply information principles[29] set out in the Act.

Section 20 lists activities that may only be undertaken in the Exclusive Economic Zone with marine consent. Activities requiring marine consent in terms of Section 20 include:

> the removal of non-living natural material from the seabed or subsoil; the disturbance of the seabed or subsoil in a manner likely to have an adverse effect on the seabed or subsoil; the deposit of anything or organism in, on or under the seabed; and the destruction, damage or disturbance of the seabed or subsoil in a manner that is likely to have an adverse effect on marine species or their habitat.

Section 59(2) sets out the criteria that need to be considered in the determination of an application for a marine consent. Relevant criteria that the EPA must consider include:

(a) any effects on the environment or existing interests of allowing the activity, including—
    (i) cumulative effects; and
    (ii) effects that may occur in New Zealand or in the waters above or beyond the continental shelf beyond the outer limits of the exclusive economic zone;

27 Section 10(1).
28 In this case s59.
29 s61.

(b) the effects on the environment or existing interests of other activities undertaken in the area covered by the application or in its vicinity, including—
(ii) effects that may occur in New Zealand or in the waters above or beyond the continental shelf beyond the outer limits of the exclusive economic zone; and
(c) the effects on human health;
(d) the importance of protecting the biological diversity and integrity of marine species, ecosystems, and processes;
(e) the importance of protecting rare and vulnerable ecosystems and the habitats of threatened species;
(f) the economic benefit to New Zealand of allowing the application;
(g) the efficient use and development of natural resources;
(h) the nature and effect of other marine management regimes; and
(i) best practice in relation to an industry or activity; and
(j) the extent to which imposing conditions under section 63 might avoid, remedy, or mitigate the adverse effects of the activity; and

In addition, the EPA must have regard to any advice received from the Māori Advisory Committee[30] and regard to the value of the investment in the activity of the existing consent holder.[31] In terms of the Act, the decision-maker must also consider the effects of the proposed activities on existing rights. Existing rights may include:

> an interest a person has in any lawfully established existing activity such as rights of access, navigation and fishing; the settlement of historical claims under the Treaty of Waitangi Act 1975; the settlement of a contemporary claim under the Treaty of Waitangi as provided for in an Act, including the Treaty of Waitangi (Fisheries Claim) Settlement Act 1992; and a protected customary right or customary marine title recognised under the Marine and Coastal Area (Takutai Moana) Act 2011[32]

Section 61 outlines information principles that a decision-maker must apply to the determination of a marine consent application. Section 61 requires decision-makers to make decisions on the best available information and must consider inadequacy and uncertainty of information in its decision. Section 61 directs that 'when the information available in relation to an application

30 S.59(2)(d).
31 S59(3).
32 Definition of existing interest, Section 4 of the EEZ Act.

is uncertain or inadequate, decision-makers must favour caution and environmental protection.' The Act further requires decision-makers to use their powers to gain as much information as possible by requesting information from the applicant, obtaining advice or commissioning reviews or reports.

Section 62(2) states:

> For the avoidance of doubt, the application may be refused if the decision-makers consider that they do not have adequate information to determine the application.

In reaching its decision the Decision Making Committee considered the following effects on the environment:

a) The significant and permanent adverse effects on the benthic environment.
b) The effects of the return of waste material to the seabed following processing aboard the vessel.
c) Effects on the trophic web (including primary production, microbes and zooplankton), fish and other pelagic fauna, rock lobsters, water quality and seabirds.
d) The effects of mining related noise, including on marine mammals; and the risks to biosecurity and human health.

The Decision Making Committee further considered the following threats to existing interests:

a) Interests arising from Treaty of Waitangi settlements.
b) Commercial fishing; marine eco-tourism; customary fishing; and other vessels traversing the area.
c) Effects of the proposal on other marine management regimes, including the Mid Chatham Rise Benthic Protection Area.
d) Interests of Chatham Islanders and the effects of the proposal on Māori and Moriori cultural interests (Currie, 2015).

In summary, the Decision Making Committee refused the application on four grounds:

1. Environmental impact: The Decision Making Committee firstly found that the impacts of the drag head on benthic fauna would be highly destructive. These effects could not be 'avoided, remedied or mitigated'. The mining would take place in the Mid Chatham Rise Benthic Protection Area, an area protected from trawling and dredging. The mining operations would destroy stony coral communities which were regarded by the Decision Making Committee as rare and vulnerable ecosystems. The

damage would be irreversible' altering the habitat permanently to soft sediment habitats. Furthermore, waste material and effluent returned to the seabed would have considerable effects on benthic habitats both within the mining blocks and the wider environment.[33]

2. Lack of knowledge: There was a lack of information to enable certainty on the impacts on the environment and existing interests. This was attributed to a lack of scientific knowledge of the receiving environment and of the proposed mining operations, regarded as 'the first seabed mining project ever undertaken at such depths anywhere in the world'. With a lack of available knowledge, the Decision Making Committee was accordingly required to 'favour caution and environmental protection'.[34]
3. Socio-economic impacts: The Decision Making Committee considered economic factors and the 'likely economic benefit to New Zealand of the proposal'. In this consideration the Decision Making Committee determined the following:

   'It was not persuaded that the proposal's economic benefit to New Zealand would be of the significance argued by the applicant, or that reliance could be placed on economic benefits as a potential offsetting factor'.

   It found that 'seabed mining would be unlikely to generate more than a modest economic benefit to New Zealand and the quantum and distribution of that benefit remained uncertain. This had to be weighed against the significant and permanent adverse effects on the benthic environment' and on other existing interests, particularly the fishing industry.
4. Mitigating impacts: The Decision Making Committee considered whether specific conditions that could be applied to the mining proposal could 'avoid, remedy or mitigate the adverse effects of the activity.' In this consideration it found that impacts 'could not be mitigated by any set of conditions or adaptive management regime that might reasonably be imposed'.

After weighing up factors prescribed to make a decision the Decision Making Committee found that the 'application could not be approved either in part or in whole'. The application for marine consent was accordingly refused.[35]

33 Currie, D. (2015). Decision on marine consent application: Chatham Rock Phosphate Limited to mine phosphorite nodules on the Chatham Rise.

34 Ibid.

35 Ibid.

3.3 *Northern Territory Australia: Moratorium on Seabed Mining*

In Northern Territory Australia a number of seabed mining proposals and preliminary exploration rights were granted for manganese exploitation in the shallow waters of Groote Eylandt. Mining companies involved included Northern Manganese Limited, Groote Resources Limited and Groote Eylandt Mining Company.[36]

Northern Manganese Limited sought environmental approval to conduct exploratory activities in 2012. It applied for 7 mineral exploration licenses, covering an approximately of 3 856km^2 in the shallow offshore Northern Territory. This was known as the Blue Mud Bay Project. Northern Manganese Limited also holds exploration rights for eight tenements (1 723 km^2) of shallow marine terrain and two islands near Groote Eylandt. This project is known as the Groote Eylandt Project.[37]

The granting of exploration rights generated widespread public concern. In response, the Northern Territory government placed a moratorium on seabed mining.[38] The Policy Statement of the Moratorium stated that seabed mining is a new and evolving worldwide industry with a minimum number of generally accepted practice standards. The methods applied in seabed mining operations are rapidly changing and there is limited information available on:

a) actual or potential impacts on environment and other resource industries; and
b) methods for managing impacts

This lack of available information decreases the ability of decision-makers to:

a) accurately assess appropriate methodology for management of industry, its development, and sustainability;
b) develop appropriate conditions and authorisations.

The moratorium covered both granting titles permitting exploration for minerals or mining and on issuing authorisations within the coastal waters of the Northern Territory until a review of actual or potential impacts is undertaken.[39] The review was intended to form the basis for an assessment of possible future development and sustainability of industry. The review was required to:

a) examine practices adopted internationally and in Australia to identify environmental best practices;

---

36 EPA Northern Territory (2012). Interim Report: Seabed Mining in the Northern Territory.

37 Ibid.

38 Northern Territory Government. (2012). Moratorium on exploration and mining in coastal waters of the Northern Territory until 2015.

39 Ibid.

b) identify likely impacts (direct and associated) of seabed mining on the environment and other resources, including commercial and recreational fishing;
c) examine mitigation strategies to manage impacts of exploration and mining;
d) consider advice from Aboriginal Areas Protection Authority – measures to avoid and protect sacred sites.

The overall aim of the review was to:

a) Identify appropriate standards for the Territory which adequately address needs of the community in respect of industry "best practice", protection of the environment, protection of social and cultural impacts, mitigation strategies and community involvement.
b) permit development of regulations and guidelines for assessment of applications to ensure consistency.

The Environmental Protection Agency of Australia produced an Interim Report (2012) which recommended a highly cautious approach to seabed mining and the extension of the moratorium until better knowledge of impacts is available. The moratorium was initially put in place for a period of three years, from March 2012 to March 2015. However, following the lapse of the moratorium period, the Northern Territory extended the moratorium period for a further three years and placed a complete ban on seabed mining in Groote Eylandt.

3.4 *Mexico: Refusal of Environmental Authorisation of a Marine Phosphate Mine*

In September 2014, Odyssey Marine Explorations (an American company) submitted an environmental impact assessment for its planned 'Don Diego' marine phosphate mine in Baja, California's San Ignacio lagoon, near Ulloa Bay.[40]

The project intended to mine 225,000 acres of seabed in five work sites. It was planned that each site would be exploited for 10 years, resulting in a 50-year long project. Overall, the project planned to extract 350 million tons of phosphate sand from the seabed.[41]

The project was opposed by many stakeholders including the Interamerican Association for Environmental Defense (AIDA), San Juanico residents, the

40 Exploraciones Oceanicas, S. de R.L de C.V. (2014) 'Environmental Impact Assessment Non-Technical Executive Summary'. Available online: http://www.dondiego.mx/wp-content/uploads/2014/06/DD_EIA-Non-Technical-Summary.pdf. Accessed 25 May 2016.

41 Moguel, S. (2016). Mexico protects loggerhead turtles from the Don Diego mine. AIDA. http://www.aida-americas.org/release/mexico-protects-loggerhead-turtles-don-diego-mine Accessed June 2016.

Centro Mexicano de Derecho Ambiental (CEMDA), local fishing cooperatives, BCS Noticias, WildCoast and Save the Waves Coalition. Civil society raised several concerns including potential impacts on Gray whales, Blue wales, Humpback wales and Loggerhead turtles as a result of noise, disturbance and radioactive releases from marine phosphate mining.[42]

In 2016, the Secretary of Environment and Natural Resources (SEMARNAT) denied environmental authorisation for the Don Diego marine phosphate mining project. The environmental authority found that measures presented by the company for protecting Loggerhead turtles were based on inconsistent information. Furthermore, the authority found that the economic benefits of the project could not prevail over the protection of the natural resources of Ulloa Bay, particularly in relation to threatened species subject to strict standards of protection.[43]

3.5 *Pacific Islands: Seabed Mining Applications*

A number of Pacific Island Nations, namely; Papua New Guinea, Fiji, Tonga, Vanuatu, Solomon Islands and the Cook Islands have granted permits for deep sea mining exploration. The Solwara 1 Project was the most advanced. In 2009 Nautilus Minerals (a Canadian seabed mining company) was granted a mining right for the Solwara 1 Project, for seabed mining in the Bismarck Sea.[44] It appears that Papua New Guinea government's rationale for granting these rights relates to the potential revenue that could stem from royalties.[45]

There has been significant opposition to the Solwara 1 Project from local communities and not for profit organisations related to the impacts on marine and coastal ecosystems and communities. Recently, the World Bank (2016) published a report titled 'Precautionary Management of Deep Sea Mining Potential in Pacific Island Countries'.[46] The report recommended that Pacific Island countries that are supporting or considering deep sea mining activities should 'proceed with a high degree of caution to avoid irreversible damage to

42 Ibid.

43 Ibid.

44 Steiner, R. (2009). Independent Review of the Environmental Impact Statement for the proposed Nautilus Minerals Solwara 1 Seabed Mining Project, Papua New Guinea. Conducted for the Bismarck-Solomon Seas Indigenous Peoples Council, Madang, Papua New Guinea.

45 Rosenbaum, H. (2011). Out of our depth! Mining the Ocean Floor in Papua New Guinea.

46 World Bank (2016). Pacific Possible: Precautionary Management of Deep Sea Mining Potential in Pacific Island Countries.

their ecosystems'. The report further stresses the need for strong governance arrangements to ensure that appropriate social and environmental safeguards are in place.

More recently, Nautilus Minerals has run into financial difficulties, and on 21 February 2019, the company filed for protection from creditors under the Canadian Companies' Creditors Arrangement Act (CCAA).

3.6 *Bristol Bay: Moratorium on Oil and Gas Activities*[47]

Oil and gas leasing was first allowed in Bristol Bay in 1986 when the U.S. federal government set aside 5.6 million acres of the southern region of Bristol Bay for oil and gas leasing.[48] The decision was opposed widely by fishing industry, indigenous communities (Alaska Native Tribes), environmental organisations, the State of Alaska, conservation organisations, recreational users and local seafood and tourism businesses, organised in a coalition called the Fish Basket Coalition.[49] The Trustees for Alaska and other conservation groups, indigenous communities and the State of Alaska initiated court proceedings to stop the lease sale. The court case was unsuccessful and oil and gas leases were issued.[50]

Potential impacts to fishery resources and marine habitats, cited in independent impact assessments included 'seismic testing, air pollution from ship traffic and the ever-present risk of oil spills'. In particular, impact studies stated that 'the area's harsh weather, rough seas, ice and strong currents would make cleanup and containment of an oil spill difficult, if not impossible'.[51]

In 2010 the Obama Administration cancelled the lease sale and provided temporary protection to Bristol Bay by withdrawing the area for consideration for oil and gas development until 2017. This was followed by steps to permanently protect Bristol Bay. On 16 December 2014, the Obama Administration announced an indefinite Presidential Moratorium on oil and gas drilling in Bristol Bay using powers conferred under the Outer Continental Shelf Lands Act (1953).[52]

---

47 A further example of a moratorium on offshore oil and gas is Antarctica. Since 1961, the continent has been administered under the Antarctic Treaty, an international agreement to preserve the continent for peaceful scientific study. In 1991 the Madrid Protocol placed a moratorium on mining in Antarctica for fifty years. [Please give details.] This was ratified by all treaty nations by 1998. There is a strong international campaign aimed at establishing a moratorium on oil and gas drilling in the Arctic Sea.

48 WWF International (2011). The value of commercial fisheries near Bristol Bay, Alaska.

49 Ibid.

50 Ibid.

51 Ibid.

52 The White House (2014). President Obama Protects Alaska's Bristol Bay From Future Oil and Gas Drilling. Press release. https://www.whitehouse.gov/the-press-office/2014/12/16/

The basis for the moratorium was partly socio-economic considerations. Bristol Bay supports a $2 billion annual fishing industry, regarded as one of the world's most valuable fisheries and provides up to 40% of America's wild caught seafood. It supports indigenous fishing communities and subsistence fishers throughout Alaska and Pacific Northwest. It is home to Native American sacred lands. The area provides considerable tourism opportunities and revenue, up to $100 million in annual recreational fishing and tourism activities.[53]

4 Marine Phosphate Mining in South Africa

This section provides an in-depth case study into seabed mining in South Africa. The South African case provides strong insights into the potential risks of seabed mining applications in jurisdictions with limited regulatory and governance capacity.

In 2012 and 2014, South Africa's Department of Mineral Resources granted three prospecting rights for marine phosphate to Green Flash Trading 251 (Pty) Ltd (GFT 251), Green Flash Trading 257 (Pty) Ltd (GFT 257) and Diamond Fields International Ltd (DFI Ltd) respectively.

The rights extend over a considerable portion of South Africa's marine environment, together covering more than 150 000 km^2 or 10% of its exclusive economic zone. The GFT 251 and GFT 257 prospecting rights cover an area approximately 63 637 km^2 and 44 389 km^2 respectively, located off South Africa's West Coast and Southwest Coast. The DFI Ltd prospecting right extends over 47 468 km^2 within South Africa's Outeniqua Basin.[54]

These prospecting rights have lapsed, and representatives of GFT 251 and GFT 257 have publicly announced, in November 2016, that marine phosphate mining in South Africa would not be financially feasible based on their assessments. Nevertheless, these areas have been earmarked by South Africa's government for seabed mining, despite the fact that they coincide with several existing fishery footprints, critically endangered ecosystems and a number

president-obama-protects-alaska-s-bristol-bay-future-oil-and-gas-drilling.

53 WWF International (2011). The value of commercial fisheries near Bristol Bay, Alaska.

54 Diamond Fields International. (2013). Environmental Management Plan for the proposed Marine Phosphate Prospecting by Diamond Fields International Ltd in the Outeniqua West Licence Area on the Eastern Agulhas, Offshore Mossel Bay.

of recently proclaimed offshore marine protected areas. Some of the benthic habitat types that coincide with the proposed seabed mining areas don't exist anywhere else in the world.[55]

The next section details issues that need consideration in regulating and governing seabed mining in South Africa. Regulatory approaches adopted in other countries may shed light into some of the regulatory and governance challenges seabed mining poses to South Africa.

4.1 *Environmental Impacts*

There is relatively limited knowledge of the impacts of seabed mining on marine environments. However, preliminary assessments outline considerable and irreversible impacts on marine ecosystems and fishery resources, signifying the need for longer term investigations.[56] These studies recommend that potentially considerable impacts coupled with the lack of knowledge available on seabed mining, warrants regulatory caution.

WWF-South Africa commissioned a report on potential impacts of marine sediment mining in the west and south coasts of the Western Cape, South Africa.[57] The report found that seabed mining in existing prospecting areas would have severe and potentially irreversible impacts on seabed habitats. The destruction and permanent alteration of marine habitats would be inevitable in light of the type of mining proposed.[58]

The report found that seabed mining could cause significant and irreversible impacts to biodiversity, ecosystem functioning and fishery resources. In particular, the prospecting rights for marine phosphate mining directly overlap with critically endangered ecosystems, ecologically and biologically significant areas and areas earmarked for protection in terms of an Operation

55 Green Flash Trading. (2012a). Environmental Management Plan in the ocean off Cape Columbine and Cape Infanta, Western Cape Province. Green Flash Trading 257 (Pty) Ltd; Green Flash Trading. (2012b). Environmental Management Plan in the ocean off Adam Se Baai and the area to the south towards Table Bay, Western Cape Province. Green Flash Trading 251 (Pty) Ltd.

56 See Currie, D. 2015; NT Australia EPA, 2012; Rosenbaum, 2011.

57 Currie, J. (2013). Brief Overview of Potential Ecosystem Impacts of Marine Phosphate Mining in the Western Cape, South Africa.

58 Ibid.

Phakisa[59] initiative to establish a network of marine protected areas.[60] Some of the benthic habitat types that coincide with the prospecting areas don't exist anywhere else in the world.

There is also limited knowledge of the ecological importance of seabed ecosystems. It is known that benthic habitats provide habitat, feeding, spawning and breeding areas for a staggering variety of marine species, many of which are commercially important.[61] However, there are considerable gaps in knowledge on the relationship between the seabed and critical ecosystem services.

4.2 *Socio-Economic Impacts*

As seabed mining would have negative impacts on fish stocks it would also impact on revenue, jobs and livelihoods associated with the fishing industry. It is thus important that the socio-economic contributions of fishing are understood and considered in decision-making on seabed mining.

South Africa's commercial fishing industry employs approximately 27 000 people directly and 100 000 people indirectly.[62] In the Western Cape, where the prospecting rights are concentrated, the fishing industry contributes approximately 2% to Gross Geographic Product.[63]

As mentioned, the marine phosphate prospecting areas directly overlap with fishing footprints of several fisheries including hake longline, tuna pole, west coast rock lobster, small pelagics and chokka squid sector. Together, these five fishing sectors provide up to 23 000 jobs.[64] Notably, the prospecting rights coincide with South Africa's only Marine Stewardship Council accredited fishery, which employs 12 000 people and generates approximately R4 billion in revenue annually.[65]

---

59 Operation Phakisa is a South African government initiative that seeks to rapidly unlock the economic potential of South Africa's oceans, through fast tracking development of five growth areas, namely: Marine Transport and Manufacturing; Offshore Oil and Gas; Aquaculture; Marine Protection Services and Ocean Governance; and Coastal Tourism.

60 Operation Phakisa. (2014). Marine Protection Services and Governance Final Lab Report.

61 Sink, K. and Attwood, C. (2008). Guidelines for Offshore Marine Protected Areas in South Africa. SANBI Biodiversity Series 9. South African National Biodiversity Institute, Pretoria.

62 Department of Agriculture, Forestry and Fisheries, (2010). Agriculture, Forestry and Fisheries Integrated Growth and Development Plan (GDP), DAFF Sector GDP.

63 Ibid.

64 Brick, K. and Hasson, R. (2016). Valuing the socio-economic contribution of fisheries and other marine uses in South Africa: A socio-economic assessment in the context of marine phosphate mining.

65 Blaine, S. (2012). Miner's marine phosphate plans could harm MSC-accredited fishery. <http://www.bdlive.co.za/articles/2012/07/20/miner-s-marine-phosphate-plans-could-harm-msc-accredited-fishery> Business Day Live Accessed July 2016; Lallemand, P.,

Furthermore, latest estimates suggest that there are approximately 8 078 small-scale fishers in South Africa, with the Western Cape home to approximately 1 667 small scale fishers.[66] Small scale fishers directly rely on heathy marine ecosystems for livelihood, nutrition, food security and income. Small scale fishing plays a critical role in providing employment and access to protein in coastal communities. There are approximately 44 coastal communities dotted along the coastline of the Western Cape.[67]

Seabed mining would likely have detrimental impacts on tourism, particularly coastal and marine tourism. Bulk marine sediment mining requires additional coastal developments that have associated impacts. These include desalination plants, beneficiations facilities and processing plants as well as increased shipping. In relation to the Sandpiper Project, residents of Walvis Bay were highly 'concerned about the potential release of toxic elements in the land-based phosphate processing phase'.[68]

Thus, decision-making, policy and regulation related to seabed mining would need to consider potential negative impacts on existing users and if impacts cannot be avoided, regulatory mechanisms for compensation would need to be devised. With regards to the latter, there is potential difficulty in assessing appropriate compensation for users from renewable industries where their economic benefit is potentially indefinite?

4.3 *South Africa's Legal and Governance Framework*

South Africa's legal and governance framework was not designed to deal with seabed mining. Without specific regulation relevant to seabed mining, applications would tend to be dealt with through existing terrestrial mineral and environmental management legislation. The Mineral and Petroleum Resources Development Act, 2002 (MPRDA) and National Environmental Management Act, 1998 (NEMA), South Africa's existing legislation, are inadequate to govern seabed mining.

Bergh, M., Hansen, M., Purves, M., (2016). Estimating the economic benefits of MSC certification for the South African hake trawl fishery.

66 Isaacs, M. and Hara, M. (2015). Backing small-scale fishers: Opportunities and challenges in transforming the fish sector. Published by the Institute for Poverty, Land and Agrarian Studies.

67 Ibid.

68 Benkenstein, A. (2014). Seabed Mining: Lessons from the Namibian Experience. SAIIA.

4.3.1 The Mineral and Petroleum Resources Development Act, 2002 (MPRDA)

The Mineral and Petroleum Resources Development Act does not deal explicitly with seabed prospecting and mining. Many provisions (particularly related to consultation and environmental impact) in the MPRDA are difficult to apply or translate to seabed mining. For instance, legal requirements to consult with landowners, neighbours, community members and other interested and affected parties act as a check and balance against inappropriate prospecting and mining licensing decisions. The importance of a lively civil society has repeatedly been acknowledged by South Africa's Constitutional Court. This check and balance will not be as strong for marine environments where there are no landowners, neighbours or community members.

Marine prospecting applications are accepted or refused by Regional Managers. However, South Africa's exclusive economic zone has not been divided along regional Department of Mineral Resources' administrative lines and functions. Furthermore, the Department of Mineral Resources has no overarching strategy or policy on seabed prospecting and mining.[69] There are no specific regulatory instruments or guidelines that provide assistance to decision-makers in respect of seabed mining. Nor is there any national and regional database of seabed prospecting and mining or other resource compiled to facilitate the Department of Mineral Resources' consideration of applications.

The Department of Mineral Resources has limited capacity for compliance monitoring and enforcement (CME) of terrestrial operations. In some cases, environmental CME is entirely absent. Accordingly, South Africa's department responsible for environmental management of mineral resource extraction has limited capacity for, or experience, guidance and knowledge of monitoring and enforcing compliance with environmental management plans or programmes and conditions of environmental authorisations in respect of seabed prospecting and mining.

Similarly, mineral authorities across the globe face significant governance and regulatory challenges with terrestrial prospecting and mining related to, amongst other challenges; environmental and social impact; consultation; access to information; rehabilitation and an inability to close operations. These challenges would, in all probability, be magnified in respect of seabed prospecting and mining.

69 In terms of s.3(3) of the MPRDA 'the Minister must ensure the sustainable development of South Africa's mineral and petroleum resources within a framework of national environmental policy, norms and standards while promoting economic and social development'.

Notably, seabed mining poses significant challenges related to rehabilitation. For example, in terms of s.24(2N)(f) an environmental management programme must as 'far as reasonably practicable' include 'measures to rehabilitate the environment affected to its natural or predetermined state'. Under s.24(P)(1) 'an applicant for an environmental authorisation, relating to prospecting, exploration, mining or production' must provide financial provision for rehabilitation. Based on environmental assessments in other countries, rehabilitation in respect of seabed mining seems unlikely, as the 'resultant ecosystem changes in such mined areas could be considered permanent, as recovery to pre-disturbed sediment structures (and hence similar habitat) would occur on geological time scales'.[70]

4.3.2 National Environmental Management Act, 1998 (NEMA)

Seabed mining would potentially fail to comply with a number of provisions in NEMA and similar environmental management legislative frameworks in other countries. Firstly, there is a lack of knowledge of impacts of seabed mining. In such circumstances, NEMA requires a precautionary approach which takes into account the 'limits of current knowledge about the consequences of decisions and actions'.[71] Seabed mining thus potentially conflicts with the precautionary principle.

Secondly, NEMA requires development to avoid, minimise or remedy disturbance of ecosystems and loss of biological diversity. In view of independent impact assessments and strategic environmental assessments undertaken in relation to seabed mining, it seems that seabed mining would fail to comply with principles of avoiding, minimising or remedying impact.

Thirdly, NEMA and environmental management legislative frameworks in other countries are usually underpinned by a sustainable development principle, which requires a balancing of social, economic and environmental factors. In South Africa, NEMA requires that development applications should be considered, assessed and evaluated with due regard to the 'social, economic and environmental impacts of activities, including disadvantages and benefits'. Fulfilment of this principle is threatened by the potential socio-economic and environmental impacts of seabed mining.

---

70 Currie, B. (2014). Investigation of cumulative impacts on the marine ecosystem from mining of phosphorites off the Namibian coast.; Levin, L.; Le, J.T; and Carson, R. (2016). 'Incorporating ecosystem services into environmental management of deep-seabed mining'. Deep-Sea Research.

71 Section 2(4)(vii).

In sum, there are concerns in South Africa's legal, governance and institutional frameworks in relation to seabed mining. Whilst there are gaps in the MPRDA, NEMA has principles that cannot be upheld if seabed mining were to proceed. This would see a seabed mining industry that is unregulated and not subject to state monitoring or enforcement of its compliance with licences and environmental laws. This in turn could facilitate severe and irreversible damage to marine environments and fisheries resources.

5 Conclusion: Legal Principles and Frameworks for Seabed Mining

The approach, policy positions or decisions adopted by coastal States in relation to seabed mining within their jurisdictions has generally been cautious, with due regard to the precautionary principle. These decisions and approaches have been underpinned by a number of principles, including: lack of knowledge on impacts of seabed mining alongside lack of knowledge of benthic ecosystems (information principle); considerable environmental and socio-economic risks (precautionary principle and sustainability principle); insufficient regulatory, legal, policy, institutional frameworks to cope with seabed mining (governance principle); potential impacts on communities particularly small-scale, traditional and subsistence fishing (fairness principle).

In Namibia and Northern Territory Australia's case, a time-based moratorium has been established, alongside commitments to undertake long-term research and reviews on issues such as: potential environmental impacts; potential impacts on existing marine users, particularly fishing industry (in the form of socio-economic impact assessments or cost-benefit analysis); reviews of potential management regimes to mitigate or reduce impacts; review of potential conditions, permissions and authorisations to regulate seabed mining; and reviews in view of developing potential legislative, regulatory, policy and institutional frameworks in relation to seabed mining.

The rationale for these moratoria and decisions and the cases above can be distilled into a number of prerequisites that these national jurisdictions have committed to before allowing seabed mining to proceed. These are:

1. **The undertaking of a strategic environmental assessment**: This was a central criteria and tool in Namibia's decision to establish a moratorium on seabed mining. A Strategic Environmental Assessment allows for a long-term and detailed study on potential and actual environmental

impacts[72] and on the socio-economic benefits and drawbacks of developing this industry. An environmental impact assessment (EIA) appears to be an inadequate mechanism to deal with risks posed by seabed mining. An EIA only investigates the environmental implications of a single development, at a project level. It does not consider cumulative impacts and assumes that a broader policy decision has already been made, to allow a specific type of development to occur. An activity such as seabed mining, on the other hand, requires a proactive, higher level policy decision on whether it is appropriate or not to allow seabed mining in a specific context in the first place. Such a decision can be guided by a Strategic Environmental Assessment.

2. **Socio-economic assessments and cost-benefits analysis:** Decision-making on seabed mining further requires proper analysis of potential socio-economic impacts of seabed mining on existing industries and associated jobs and livelihoods – many of which are renewable, unlike seabed mining.[73] This requires that costs and benefits are weighed up, so that trade-offs are properly informed. In particular, it requires considerations of impacts on commercial and small-scale fishing and associated jobs and livelihoods. Socio-economic assessments and cost-benefits analysis may be part of a strategic environmental assessment, however, in the case of New Zealand and Australia Northern Territory, socio – economic considerations were assessed separately and guided decisions accordingly.
3. Open deliberation on **alternatives to seabed mining** in respect of resource requirements: An argument asserted in applications for marine phosphate mining in New Zealand, South Africa, Mexico and Namibia is that phosphate is critical for food security and terrestrial phosphate supplies are dwindling whilst demand is increasing[74] . These justifications require further interrogation in decision-making and policy related to seabed mining. Decision-making should be informed by assessments on on the overall desirability of allowing seabed mining in view of

72 Such an SEA can also provide for the collection and management of adequate baseline information in areas where seabed mining is proposed.

73 See Brick and Hasson (2016) for a valuation of South African fishing industry in the context of marine phosphate mining.

74 Diamond Fields International. (2013). Environmental Management Plan for the proposed Marine Phosphate Prospecting by Diamond Fields International Ltd in the Outeniqua West Licence Area on the Eastern Agulhas, Offshore Mossel Bay; Green Flash Trading. (2012a). Environmental Management Plan in the ocean off Cape Columbine and Cape Infanta, Western Cape Province. Green Flash Trading 257 (Pty) Ltd.

alternative sustainable supply strategies, based on an understanding of market dynamics, use profiles, options for resource recovery and improved management.

4. The development and implementation of **appropriate policies, strategies, laws and regulations** in relation to seabed mining:[75] This entails developing a comprehensive framework to guide decision-making, environmental management and compliance monitoring and enforcement. It requires decision-makers to proactively determine the suitability of seabed mining, before development proposals are invested in and considered. This moreover requires considerable institutional and human resource capacity development.[76] This was an approach adopted by Northern Territory Australia, which provided for a review to 'identify the appropriate standards' and the 'development of regulations, guidelines, or both, for the assessment of applications to ensure consistency of assessment procedures and appropriate determinations'. Developing a regulatory framework for seabed mining should include proper consideration and alignment of regional commitments in respect of potential impacts of seabed mining on neighbouring coastal countries.
5. Establishment of **networks of marine protected areas, no-go-areas and fisheries management areas** which prohibit and restrict seabed mining, in order to ensure benthic habitats and marine ecosystems receive adequate protection and fishery resources are secured. In many jurisdictions marine protected areas prohibit or provide restrictions on seabed mining in line with provisions on the purposes of marine protection. For example, South Africa's National Environmental Management: Protected Areas Amendment Act, 2014 (NEMPAAA) explicitly provides for the restriction or prohibition of any activities in a marine protected area that may have adverse effects on the purposes of declaring such a marine protected area in the first place.[77] Section 48(1) explicitly states:

   > Despite other legislation, no person may conduct commercial prospecting or mining, exploration, production or related activities (in a marine protected area).

75 This should include the promulgation of proper technical standards and industry guidelines to minimise damage of operations.

76 This should include proper regulatory provision for seabed mining in national marine spatial planning processes and ocean economic development strategies.

77 s.22A(g).

In particular, where the purpose of declaring an MPA relates to the conservation and protection of benthic ecosystems, species and habitats, the argument for restricting mining, prospecting, exploration and production activities is heightened.

6. The establishment of legal provisions to ensure **liability for environmental damage** is assigned to mining operators through appropriate assessment, strict collection and ring-fenced. This is in line with the 'polluter pays' principle entrenched in international law and environmental management regimes of many countries, whereby 'national authorities should endeavour to promote the internalisation of environmental costs and the use of economic instruments, taking into the account that the polluter should, in principle, bear the cost of pollution'.[78] This should include the establishment of legal mechanisms for compensation to existing marine users for negative impacts stemming from seabed mining.
7. Establishment of legal mechanisms to facilitate **proper consultation and stakeholder engagement.** This is particularly necessary in view of the fact that checks and balances in the form of legal obligations to consult with landowners, neighbours, community members and other interested and affected parties, are not as robust in relation to marine environments.

These provide a robust set of guidelines or recommendations, based on comparative approaches and processes adopted by other countries, for policy positions and regulatory frameworks for sovereign coastal states to ensure that seabed mining does not disproportionately impact on marine and coastal ecosystems and marine based industries that depend on functioning ecosystems, and associated jobs and livelihoods.

78 Principle 16, Rio Declaration, 1992.

CHAPTER 14

European Union Law and the Seabed

Finn Arnesen, Rosa Greaves, and Alla Pozdnakova

1 Introduction

The European Union (EU) is a supranational organization established under international law. Nevertheless, over several decades, it has evolved into a suis generis international entity with its own legal order, significant competences and an effective enforcement mechanism. The Court of Justice of the European Union (CJEU), which under the Treaty has exclusive jurisdiction to interpret EU law and to review the legality of any EU secondary legislation,[1] has emphasised the unique nature of the EU.[2] The EU Member States did not delegate but transferred some sovereign rights to the EU meaning that only the EU has competence to act, either internally or externally, in a number of fields of activity. Furthermore, the transfer of sovereignty has not only occurred by expressed provision in the EU Treaties but also by the EU adopting extensive secondary legislation and thus occupying the field.[3] This latter process means that for some activities, the initial shared competence to act has been replaced by exclusive EU competence.

In this chapter the focus is on the extent to which the EU has competence to regulate the seabed under the sovereign jurisdiction of the EU Member States and the manner in which the competence has been exercised. The chapter will explain first the extent of the EU's competence in respect of the relevant seabed (Section 2). Then, the chapter will, by way of illustration of the level of policy and legislative activity, consider two areas affecting the seabed: the EU's policy in respect of seabed mining and the application of EU environmental protection and liability legislation to the seabed (Section 3). In neither area has there been EU legislation adopted specifically in respect of the seabed. However, several policy papers have been published by the European Commission on seabed mining which demonstrates the Commission's intention for the EU to be an

1 Article 263 TFEU.

2 E.g. *Van Gen den Loos*, 26/62 [1963] ECR 1.

3 When the EU and the EU Member States share a competence, the latter lose their 'competence' (power to take decisions) when the EU decides to regulate in that area. The EU is said to 'occupy the field' meaning that the EU Member States no longer have the right to legislate in the covered area. Case 22/70 *Commission v Council* (ERTA Case) ECLI:EU:C:1971:32.

DOI:10.1163/9789004391567_016

active player in exploiting the seabed such as deep seabed mining. Similarly, current environmental protection legislative measures can be interpreted so as to extend their scope to the seabed. The last part of the chapter will provide some concluding observations (Section 4).

2 EU Competences in Respect of the Seabed

According to Article 5 of the Treaty on European Union (TEU), the EU competences are governed by the principle of conferral. This has two implications worth noticing in the context of the issues discussed in this chapter. First, the EU may only act within the limits conferred upon it by the TEU or the Treaty on the Functioning of the European Union (TFEU). Secondly, these competences may only be used to achieve the objectives set out in those treaties. Thus, competences not transferred to the EU, remain with the Member States. Competences conferred, may either become an exclusive EU competence, or a competence shared with the Member States.[4] In matters concerning the use of the seabed, the competence is shared – with a possible caveat for measures affecting the use of the seabed taken under the common fisheries policy. However, the EU does not have competence relating to the continental shelf and the seabed unless the Member States have such competence under public international law.

The TEU and TFEU are – for the time being – a last step in the development of what is now known as the European Union. Today the competence conferred on the EU to regulate matters pertaining to the market place is supplemented by TFEU part Three, Titles XX and XXI – on environment and energy respectively.

In order to comprehend fully the impact of EU law on matters pertaining to the use, and non-use, of the seabed, one has to acknowledge the effects of the general provisions of the TFEU on the four freedoms, i.e. free movement of goods, services people and capital, competition between undertakings and state aid on activities relating to the seabed. These effects are, to the extent EU law is applicable, independent of any legislative acts adopted by the EU addressing seabed use.

Article 191 TFEU provides that EU policy on the environment shall contribute to the pursuit of four objectives. Among these we find traditional environmental aims such as the preservation, protection and improvement of

4 Articles 2 to 6 TFEU.

the quality of the environment, but also prudent and rational utilization of natural resources. Article 192 TFEU provides the legal basis for legislative acts implementing EU's policy on the environment. These measures are to be adopted by majority voting according to the ordinary legislative procedure provided for in Article 114 TFEU. However, according to Article 192 (2) TFEU, some decisions nevertheless require unanimity. Among these are measures affecting 'land use' and measures 'significantly affecting a Member State's choice between different energy sources'. Whether 'land use' also cover use of the seabed, may be open to debate.

Article 194 provides for an EU policy on energy, and a legal basis for measures implementing this policy. The measures taken according to Article 194 TFEU may, 'without prejudice to Article 192(2), however not affect a Member State's right to determine the conditions for exploiting its energy resources, its choice between different energy resources and its general structure of energy supply.

As a legal topic, the EU and the use of the seabed have two dimensions. One dimension being the external, i.e. issues pertaining to the powers of EU institutions in interaction with third states – both within the framework of conventions to which EU is a party, like the UN Convention on the Law of the Sea (UNCLOS), and in relation to conventions where only (some) of the Member States are parties. As a matter of EU law, the EU has exclusive competence in certain areas covered by the Convention, and shared competence in others. Pursuant to Article 5(1) of Annex IX to the Convention, a declaration specifying these areas has been given.

The other dimension is the internal dimension, i.e. issues pertaining to EU competence to regulate activities taking place outside the territories of the Member States. This internal dimension gives rise to two discussions. The first being to what extent EU law applies to measures taken by Member States that relate to activities taking place outside their territory but under their jurisdiction. The other being to what extent the EU has competence to oblige the Member States to undertake certain activities, for instance explore for natural resources, make certain uses of the seabed, lay (or allow) cables and pipelines, etc.

In public international law, terms like 'territory', 'continental shelf', and 'economic zone' are used to define the sovereignty and jurisdiction of states. Turning to Article 52 TEU, we see that this provision proclaims that the EU Treaties apply to the Member States, and that the territorial scope of the treaties is specified in Article 355 TFEU. Article 355 TFEU does however not shed much light on the application of EU law outside the territories of the Member

States. It is nevertheless established through CJEU case-law that EU law applies where Member States have sovereign powers. This can be illustrated by *Salemink*,[5] where the CJEU held:

> Since a Member State has sovereignty over the continental shelf adjacent to it – albeit functional and limited sovereignty (see, to that effect, Case C-111/05 *Aktiebolaget* NN [2007] ECR I-2697, paragraph 59) – work carried out on fixed or floating installations positioned on the continental shelf, in the context of the prospecting and/or exploitation of natural resources, is to be regarded as work carried out in the territory of that State for the purposes of applying EU law (see, to that effect, Case C-37/00 *Weber* [2002] ECR I-2013, paragraph 36, and Case C-6/04 *Commission v United Kingdom* [2005] ECR I-9017, paragraph 117).
>
> A Member State which takes advantage of the economic rights to prospect and/or exploit natural resources on that part of the continental shelf which is adjacent to it cannot avoid the application of the EU law provisions designed to ensure the freedom of movement of persons working on such installations.[6]

The case concerned the application of EU law on the free movement of persons, but the same applies to other fields of EU law, national sovereignty carries EU law on its back, not unlike the snail carries its shell. Thus, the applicability of EU law to issues relating to geographical areas outside the territories of the Member States has to be decided on the basis of an interpretation of the relevant EU legal act and the Member States' jurisdiction over the issue and the area, and then made subject to possible functional caveats.

As a matter of comparison, under the Agreement on the European Economic Area (EEA Agreement), this may be different, as Article 126 EEA uses the term 'territory', and expressly proclaims that the agreement is applicable to territories of the EFTA States and the territories to which EU law applies. Thus, the position of Norway is that the EEA Agreement, and thus EEA law, does not apply to matters concerning the Norwegian continental shelf, as these do not take place on Norwegian territory. The EFTA Surveillance Authority (ESA) has voiced a different opinion.[7]

---

5 Case C-347/10, ECLI:EU:C:2012:17.

6 Case C-347/10 *Salemink*, paras 35 and 36.

7 Finn Arnesen, comments on Art. 126 EEA, in F. Arnesen, H.H. Fredriksen, H.P. Graver, O. Mestad and C. Vedder (eds.) Agreement on the European Economic Area – A Commentary, Nomos Verlagsgesellschaft, Baden-Baden 2018. The EEA Agreement and activities off-shore, SIMPLY [2010] p. 17.

Whether the EU may adopt measures obliging the Member States to undertake certain activities, depends on the powers conferred on the EU by the Member States, and the prerequisites for executing those powers. As far as environmental issues are concerned, we have seen that measures affecting land use require unanimity, as do measures affecting a Member State's choice between energy resources,[8] while EU measures shall not affect a Member State's right to determine the conditions for exploiting its energy resources.[9]

3 The Exercise of EU Competences with Respect of the Seabed

Two areas in respect of the seabed have been selected to illustrate the extent of the EU's competence to regulate the seabed and the manner in which it seeks to play a significant role internationally. Seabed mining, meaning the extraction of minerals[10] from the seabed from a depth of more than 200 metres, is an economic activity at an embryonic stage but with immense potential to develop into a major industrial activity. It is therefore not surprising that the EU has expressed interest in playing a major role in shaping its governance.[11] It is therefore appropriate to explore how the EU will achieve its objective (Section 3.1). Environmental protection and liability, on the other hand, is an area where the EU has already adopted a number of significant legal instruments, some of which have had impact on marine and maritime activities. In this section of the chapter, the extent to which the legislation on environmental protection and liability applies to the seabed will also be considered (Section 3.2).

3.1 *Seabed Mining (Minerals)*

The first thing to be remembered about seabed mining is that it has not yet taken place anywhere in the world on a commercial basis. However, the idea of mining the seabed for the extraction of minerals and rare earth elements is not

8 Article 192 TFEU.

9 Article 194(2) second paragraph.

10 Extraction of marine aggregates such as sand and gravel are excluded. Minerals are to be understood as raw materials found on or under the seabed.

11 Joint Communication to the European Parliament, the Council, the European Economic and Social Committee and the Committee of the Regions, entitled 'International ocean governance: an agenda for the future of our oceans' JOIN(2016)49 final of 10 November 2016.

new. The main reasons that have prevented the growth of this new industry are environmental issues and costs.[12]

As far as the EU is concerned seabed mining is an economic activity with huge potential for EU undertakings engaged in seabed mining activities such as technology providers who are rapidly expanding the capabilities of underwater technology.[13] Thus the EU, within the limits of its competence as discussed in Section 2, is certainly ensuring that it has a role to play in shaping this industry as part of its general industrial policy. The European Commission is thus engaged in commissioning a variety of studies and providing funding for projects[14] that will reduce knowledge gaps associated with seabed mining as well as identifying the benefits and drawbacks of seabed mining operations. In addition, the Commission has issued a number of policy documents, 'communications' with relevance to seabed mining.[15]

UNCLOS, to which both the EU and the EU Member States are parties, provides that each coastal State may regulate seabed mining in maritime areas under its national jurisdiction but it does not provide any legal framework as to how the task should be exercised. As far as seabed mining is concerned beyond the maritime areas under the jurisdiction of States, the so-called 'Area', UNCLOS, supplemented by Part XI Implementation Agreement, provides an incomplete legal framework for seabed mining.[16]

Although UNCLOS governs the use of the oceans and their resources, sectoral activities such as seabed mining are the responsibility of an international institution, the International Seabed Authority (ISA).[17] In its Joint Communication

12 For example, a Canadian company, Nautilus Minerals, was granted a 20 year lease by the Papua New Guinean government to mine offshore but soon terminated their project due mainly to costs.

13 E.g. vacuum pumps and remotely operated vehicles.

14 E.g. Managing Impacts of Deep-seA reSource exploitation (MIDAS) project – a multidisciplinary research programme investigating the environmental impacts of extracting mineral and energy resources from the deep-sea environment. It is funded under the European Commission's Framework 7 programme and was started in 2013 for a period of 3 years <http://www.eu-midas.net/>; Study on deep sea mining <https://webgate.ec.europa.eu/maritimeforum/en/node/3617>; Blue Mining project also funded under the Framework 7 programme <http://www.bluemining.eu>.

15 E.g. Communication of the Commission: Blue Growth opportunities for marine and maritime sustainable growth COM(2012)49 final.

16 For a discussion of the international legal framework, see Chapter 7 of this book, J. Dingwall, 'Commercial Mining Activities in the Deep Seabed Beyond National Jurisdiction: The International Legal Framework'.

17 Ibid, for details of competences, powers and actions.

on international ocean governance published in November 2016,[18] the European Commission stresses that the current international framework is not adequate to make sure that management of the oceans is sustained. The Communication highlights that ISA has not yet adopted a mining code setting out rules and procedures to govern seabed mining.[19] In addition, the Commission expresses concern at the lack of coordination between the various international organisations with sector specific responsibilities for the oceans.[20]

The importance of this Commission document should not be underestimated. The strategy, namely the publication of policy papers identifying inefficiencies in international frameworks and setting out a number of proposed actions is not an uncommon practice before the Commission goes on, at a future date, to make legislative proposals for action and promote their adoption by the European Council and the European Parliament. Indeed in this Joint Communication, the Commission emphasises its experience of ocean management[21] to claim that the EU is 'well placed to shape international ocean governance'.[22]

The Joint Communication specifies a number of actions that are to be initiated. In the context of seabed mining the relevant action (Action 1) relates to filing the gaps in the international ocean governance framework. The Commission commits to producing guidance 'on the exploration and exploitation of natural resources on the seabed areas under national jurisdiction, [in order] to assist coastal Member States to respect their duty under UNCLOS to protect and preserve the marine environment.'[23]

As to the internal dimension, it is clear from CJEU case law[24] that EU law applies to maritime areas over which EU Member States have jurisdiction. Given that UNCLOS provides that each coastal State may regulate seabed mining in maritime areas under its national jurisdiction then it follows that the national law of the relevant EU Member States and EU law apply to seabed mining operations within those maritime areas. It is also clear that no EU

18 Joint Communication to the European Parliament, the Council, the European Economic and Social Committee and the Committee of the Regions entitled 'International ocean governance: an agenda for the future of our oceans, JOIN(2016)49 final.

19 Such a code is being prepared.

20 (n18), p. 3.

21 Marine Strategy Framework Directive; Maritime Spatial Planning Directive; reformed Common Fisheries Policy; and its maritime transport policy.

22 (n18), p. 4.

23 (n18), p. 6.

24 Case 61/77 *Commission v Ireland* [1978] 417, paras 45 to 51.

legislation has been adopted specifically to govern exploration or extraction and exploitation of seabed mining either within maritime areas under the national jurisdiction of EU Member States[25] or beyond. Thus, until the EU occupies this field by adopting specific legislation, the EU Member States have competence to act.

There is no doubt that the EU legislative measures, which are the most likely to have a potential significant impact on seabed mining activities, are the directives that have been adopted to protect the environment.[26] A discussion of the application of these directives to seabed mining will be provided in the second section of this part of the chapter. However, given the fact that seabed mining is an economic activity, it is highly likely that a whole range of other secondary legislation, which has been adopted to regulate market operators in their economic activities, will also be applicable. There are, for example, EU directives which impose reporting obligations on undertakings, such as the obligation on large and listed companies to publish information on payments made to governments.[27]

One major concern that is worth noting is the following. There are no international standards for seabed mining at present so the coastal states may adopt their own legislation accordingly which may give rise to a risk of different standards being applied. If this were to happen in respect of EU Member States it is highly likely that the EU would act under the competence granted under Article 114 TFEU[28] and adopt legislation on the basis that harmonisation of national laws in respect of seabed mining was necessary to safeguard the EU's internal market objective. An alternative softer approach may be to promote the adoption of EU standards for seabed mining operations. This could, of course, also be problematic as you could have a different standard for the Area (ISA's responsibility) from that being applied to maritime areas under the jurisdiction of EU Member States for the same activity.

By way of summary the EU's position as far as its competence and policy on seabed mining is concerned is reasonably clear. EU law applies to maritime areas under the jurisdiction of EU Member States. Seabed mining is an

---

25 These include not only 'the territorial sea' (i.e. 12 nautical miles from the baseline) but also the exclusive economic zone, the 'EEZ' (up to 200 nautical miles) and adjacent 'continental shelf' which comprises the seabed and subsoil of the submarine areas beyond the territorial sea.

26 See the 2015 Study to investigate the state of knowledge of deep sea mining by ECORYS and MRAS – <https://webgate.ec.europa.eu/maritimeforum/en/node/373>.

27 Accounting Directive, Directive 2013/34, OJ 2013 L182, on annual financial statements.

28 This Treaty provision provides the legal base for the adoption of EU approximation measures whose objective is the establishment and functioning of the internal market.

economic activity and will be treated like any other economic activity within the internal market. Where necessary, the European Commission is likely to propose legislation to ensure that the seabed mining activity, within the jurisdiction of the EU Member States, is regulated and thus to ensure that the objectives of the internal market are achieved: no discrimination; free movement of goods, services people and capital, competition between undertakings and state aid.

Given that the EU and the EU Member States are party to UNCLOS and UNCLOS is an integral part of the EU legal order, the EU will respect international norms adopted under its auspices. Thus, as far as seabed mining activities in the Area are concerned, any activity will be governed by international law and the EU Member States will be encouraged by the European Commission to adopt the necessary laws under UNCLOS to protect and preserve the marine environment.

Nevertheless, in its latest Communication,[29] the European Commission has made it clear that it will not wait for ISA's mining code before adopting its own guidelines to regulate seabed mining activities in maritime areas under the national jurisdiction of the EU Member States.

3.2 *Environmental Regulation of Seabed Activities*

In the discussion above we have examined EU competence to regulate activities on the seabed and concluded that this competence is, generally, shared but to-date has not been exercised. However, by contrast to measures related to the seabed mining, several environmental measures relevant to the offshore sector have been adopted by the EU.

This section focuses on how the EU has exercised its competence to protect the environment from the pollution caused by seabed activities of the Member States and considers whether EU measures are adequate to address the environmental issues posed by the new uses of the seabed.

It is debatable whether environmental protection in the offshore sector should be achieved by the EU rather than by the Member States. Even in the absence of EU action some steps have been taken by EU Member States at national level to counteract risks of environmental damage which may result from offshore operations. In particular, this is achieved through a system of authorization of offshore activities, which require undertakings, amongst other obligations, to obtain licences and meet other requirements (as the case may be) prior to starting offshore operations.

29 Above (n18).

Moreover, the State's duty to protect the environment from the pollution caused by offshore activities is set out in the international instruments. First, a general duty is imposed in UNCLOS[30] on States to prevent, reduce and control pollution of the marine environment arising from or in connection with seabed activities subject to their jurisdiction. Secondly, two regional agreements, the Convention for the Protection of the Marine Environment of the North-East Atlantic (the OSPAR Convention)[31] and the Offshore Protocol of Barcelona Convention,[32] do regulate important environmental aspects of offshore activities in the North East Atlantic and in the Mediterranean Sea. By virtue of EU's accession to these agreements, their provisions are part of the EU's legal order and, therefore, binding on the EU and its Member States.

It is outside the scope of this chapter to give a detailed account of the national and international environmental safety rules applicable to the activities on the seabed within Member States' jurisdiction. It is sufficient to submit briefly that, in spite of the existence of some international and national safety regulations in this sector, important gaps and inconsistencies still exist. Importantly, the national approaches to risk management and safety culture in the offshore sector have been found to vary significantly among the EU Member States.[33] Given the gaps and diversity of approach to risk management, the EU may play an important role in contributing to the improvement in the level of protection of the seabed environment across the EU by adopting secondary legislation harmonizing the national approaches and thus ensure that a minimum standard of safety applies in all EU Member States active in the offshore sector.

As mentioned previously, Article 191(2) TFEU sets out the following principles for EU environmental action which are also highly relevant to the offshore sector: the precautionary principle; the principle that preventive action should

30 Article 208 UNCLOS. Council Decision 98/392/EC, OJ 1998 L179/1, of 23 March 1998 concerning the conclusion by the European Community of the United Nations Convention of 10 December 1982 on the Law of the Sea; the Agreement of 28 July 1994 relating to the implementation of Part XI thereof.

31 Council Decision 98/249/EC, OJ 1998 L104/1, of 7 October 1997 on the conclusion of the Convention for the protection of the marine environment of the north-east Atlantic.

32 Council Decision 2013/5/EU, OJ 2013 L4/13, of 17 December 2012 on the accession of the European Union to the Protocol for the Protection of the Mediterranean Sea against pollution resulting from exploration and exploitation of the continental shelf and the seabed and its subsoil.

33 SEC(2011) 1293 final, Commission Staff Working Paper. Impact Assessment *Accompanying the document* Proposal for the Regulation of the European Parliament and of the Council on safety of offshore oil and gas prospection, exploration and production activities, COM(2011) 688 final.

be taken; the principle that rectification of environmental damage at source is a priority; and the principle that the polluter should pay. If not all, then at least some of these principles have counterparts in international law, and therefore are relevant (although not necessarily binding) on EU Member States.[34] However, the scope and contents of the principles (and their legal effect) is unclear in international law and EU Member States' approaches to them may differ significantly.[35] EU measures may, therefore, provide precise content and effect to the principles.

A Common EU Action must be able to raise to an acceptable level the safety legislation of those EU Member States which fall below and to overcome divergences in the national approaches to the offshore safety culture and pollution preparedness among Member States. It is also important when drawing up a Common Action to take into consideration the probable transboundary impact of an offshore pollution incident in Member States' waters.

These ambitious aims can only be achieved if EU measures are adopted in a timely and effective manner which may not necessarily be possible due to the legislative procedures described in Section II above.[36] Furthermore, the obligations set out in the EU legislative measure should be formulated precisely and in sufficient detail.[37] Thus, in making the choice between a regulation or a directive the European Commission is likely to prefer a regulation which is directly applicable[38] and does not require (or allow) any national implementing measures, while at the same time giving a possibility for EU legislators to envisage adoption of implementing legal acts where necessary.

However, the objectives of Article 191 TFEU are to be achieved by the adoption of harmonisation measures which means directives. Directives, by contrast to regulations, do not have direct effect and are only binding upon the EU Member States as to the result to be achieved, leaving to the national authorities the choice of form and methods of implementation.[39] Accordingly,

---

34 See generally Nicolas de Sadeleer. *Environmental Principles: From Political Slogans to Legal Rules*, OUP, 2002, p. 91 et seq. Note that Article 191 is only binding on EU, not Member States.

35 See, e.g., Daniel Bodansky, 'Deconstructing the Precautionary Principle,' at pp. 381–391 in D.D. Caron and H.N. Scheiber (eds), *Bringing New Law to Ocean Waters*, 2004, Koninklijke Brill N.V.

36 Kramer (2015), p. 55.

37 Ibid.

38 Direct applicability means that the EU regulation is incorporated into of the national legal orders of the EU Member States on publication in the EU's Official Journal. Regulations also have 'direct effect' meaning that individuals can rely on provisions of EU regulations before national courts.

39 Article 288(3) TFEU.

the obligations and responsibilities of the offshore actors provided in the relevant EU directives and examined in more detail below must be properly transposed in the national laws of Member States before they become applicable to these actors.[40]

Over time, the EU adopted several secondary legislative measures which give effect to the principles laid down in Article 191 TFEU in respect of the protection of the marine environment. It is important to note that EU's provisions regulating environmental issues in the offshore sector are found in a range of directives. Some of these measures have general application and regulate all sectors, including the offshore sector, unless expressly excluded from their scope. The relevant directives[41] are: the Marine Strategy Framework Directive;[42] the Environmental Liability Directive;[43] the Environmental Impact Assessment Directive;[44] Directive on the conservation of natural habitats and of wild fauna and flora (Habitats Directive).[45]

Initially, EU legislation adopted specifically for the offshore sector addressed only access to the markets of the EU Member States. For example, the Prospection, Exploration and Production of Hydrocarbon Directive (1994)[46] mentions briefly the protection of the environment and biological resources as one of the grounds for a Member State to impose conditions and requirements on the activities regulated by the Directive.[47]

This Directive does not set out any specific criteria for Member States' decision to rely on the protection of the environment. However, the authorization

40 Above (n36), p. 55.

41 It is outside the scope of this section to address all these general directives in detail.

42 Directive 2008/56/EC, OJ 2008 L164/19, of the European Parliament and of the Council of 17 June 2008 establishing a framework for community action in the field of marine environmental policy (Marine Strategy Framework Directive).

43 Directive 2004/35/CE, OJ 2004 L143/56, of the European Parliament and of the Council of 21 April 2004 on environmental liability with regard to the prevention and remedying of environmental damage (Environmental Liability Directive).

44 Directive 2014/52/EU, OJ 2014 L124/1, of the European Parliament and of the Council of 16 April 2014 amending Directive 2011/92/EU on the assessment of the effects of certain public and private projects on the environment (Impact Assessment Directive).

45 Council Directive 92/43/EEC, OJ 1992 L206/7, of 21 May 1992 on the conservation of natural habitats and of wild fauna and flora (The Habitats Directive).

46 Directive 94/22/EC, OJ 1994 L164/3, of the European Parliament and of the Council of 30 May 1994 on the conditions for granting and using authorizations for the prospection, exploration and production of hydrocarbons.

47 Ibid, Article 6(2). See also Finn Arnesen, "The Adoption of the Production Licence Directive", in Peter-Christian Műller-Graff, Erling Selvig (eds) *The European Economic Area –Norway's Basic Status in the Legal Construction of Europe*, Berlin Verlag, Tano Aschehoug, 1997, pp. 81–96.

system of Member States (endorsed in the Directive) is one of the tools to give effect to the preventive measures and the Directive is closely linked to the Offshore Safety Directive and other relevant directives.

The timeline of the EU environmental measures in the offshore sector shows that the adoption of the sector-specific measures were significantly speeded up after the Macondo blow out (Deepwater Horizon) in 2009. Such a major accident at an offshore installation is obviously capable of damaging the environment to such a significant extent that, in the absence of sufficiently effective rules, the objective of achieving 'good environmental status' would be seriously compromised.[48]

After the Macondo, in 2012, the EU acceded to the Barcelona Offshore Protocol.[49] The next step was the adoption of the Offshore Safety Directive (OSD) in 2013.[50] Both instruments are concerned particularly (but not limited to) with *transboundary* pollution damage.

The OSD addresses only major offshore accidents and has a general geographic application. The Directive recognizes that major accidents are likely to have devastating and irreversible consequences on the marine and coastal environment.[51] 'Major environmental incidents' are included in the scope of the Directive only to the extent they result from such incidents and result, or are likely to result, in 'significant adverse' effects on the environment in accordance with the Environmental Liability Directive (ELD).

While the OSD aims at preventing major accidents in the oil and gas sector, or minimizing risks of such accidents, it does not include more subtle but systematic pollution arising from normal offshore operations. A detailed discussion of the issues relating to the EU rules on protection from such pollution is outwith the scope of this chapter. Regular discharges can be addressed through the emissions regulation.[52] It should be noted in this respect, however, that the two regional instruments mentioned earlier – Barcelona Offshore

48 It should be noted that, while environmental concerns are important, the Offshore Safety Directive is not designed to deal exclusively with the environmental effects of an accident; it includes all aspects including safety of life, health etc.

49 Above (n31).

50 Directive 2013/30/EU, OJ 2013 L178/66, of 12 June 2013 on safety of offshore oil and gas operations and amending Directive 2004/35/EC.

51 A 'major' accident is defined in Article 2(1) as a very serious incident in relation to the installations and connected infrastructure.

52 Directive 2010/75/EU, OJ 2010 L334/17, on industrial emissions (integrated pollution prevention and control). Only one provision specifically excludes offshore sector (see Article 28).

Protocol and OSPAR Convention – contain a framework for addressing pollution in a broader meaning than the OSD.[53]

Furthermore, the OSD only applies to oil and gas (hydrocarbon) activities, and thereby excludes activities aimed at exploring and extracting other seabed minerals.[54] The Directive applies only to activities within the limits of Member State's continental shelf.[55]

The preamble of the OSD expressly confirms the obligation for the EU environmental action to be supported by the high level of protection based on the principles referred to in Article 192 TFEU.[56] The Directive does not define these principles for the purposes of the offshore sector, but rather gives them practical effect through the provisions envisaging specific obligations for the offshore sector players and authorities. Member States are free to introduce more stringent requirements than those envisaged in the Directive.[57]

The common feature of the EU environmental regulation is the focus on incidents which may cause 'significant adverse' effects to the environment. These are determined in line with the criteria provided for in the Environmental Liability Directive (ELD).[58] However, the ELD criteria are open as they do not establish any standards or thresholds above which adverse effects on the environment become significant. Only damage with a proven effect on human health must be always classified as significant damage.[59]

The main rationale for limiting the scope of the OSD to major accidents with serious impact is based on the cost efficiency considerations inherent in the risk prevention and management strategies imposed by the Directive: preventive measures must be practicable and economically feasible, 'suitable' to prevent or limit consequences for human health and for the environment of major accidents in offshore oil and gas operations.[60] This approach takes into consideration the realities faced by those entities that have the main responsibility for the safety of the offshore operations, i.e. operators.

---

53 Offshore Protocol regulates pollution within the meaning of UNCLOS Article 208, including wastes and harmful and noxious substances.

54 Article 1(1) and 2(3) of the Directive. Cf. Seveso-III Directive which excludes offshore sector from its application scope: Article 2(2)(f) of Directive 2012/18/EU, OJ 2012 L197/1, of 4 July 2012 on the control of major-accident hazards involving dangerous substances, amending and subsequently repealing Council Directive 96/82/EC.

55 Article 2(3) of the Offshore Safety Directive.

56 Above Section 1.

57 OSD is a minimum harmonization directive (Article 1).

58 See Annex I and Article 2 of ELD.

59 See Annex I.

60 Article 3(1); Recital 26 and Article 3(3) ; Article 2(6).

The approach based on 'suitability' criterion reflects the idea of the preventive principle generally: measures to be taken to prevent certain, identifiable risks should be limited by the 'feasibility' threshold. A very high level of environmental protection envisaged may not be feasible to maintain, but it is better to have realistic, workable standards instead of declaratory high ones.

Whereas some risks brought about by the offshore activities are certain and thereby require specific (i.e. 'suitable') measures, it is much more difficult to determine what measures are appropriate to address uncertain, scientifically yet unidentified, risks. It is unclear how much effort one should invest to avoid these unknown risks, but the precautionary principle requires that at least some proportionate efforts are undertaken to this end. At the same time this also means that we need to tolerate some level of uncertainty in the environmental risks if we want to benefit from the natural resources of the seabed.

While admitting that 'risks relating to major offshore accidents are significant', OSD is aimed at risk minimization, not elimination.[61] OSD introduces the concept of 'acceptable risk' which means that the operators should reduce the risk of a major accident as low as reasonably practicable, to the point where the cost of risk reduction would be grossly disproportionate to the benefits of such reduction.[62]

OSD is important because it imposes quite specific duties on Member States to introduce risk management and prevention requirements in the national laws. The OSD applies the standard of 'best practice' to determine the acceptable risk levels and to the accident prevention systems.[63] The OSD accepts that the concept of a 'best' practice is dynamic and must evolve in the light of new knowledge, invention and technology developments.[64] An important quality of 'the best' practice for major accident prevention in offshore is the goal-setting approach based on the risk assessment and reliable management system. This will particularly benefit Member States with poorly developed offshore safety culture.[65]

In addition, a Major Hazard Report must be elaborated and kept up-to-date by the operators in order to meet the requirements of OSD relating to the preparing and carrying out offshore oil and gas operations.[66] OSD facilitates the

61 Recital 6. The OSD defines risk as 'the combination of the probability of an event and the consequences of that event.' Article 2(4).

62 Article 3(4). See also Sadeleer at p. 91 et seq.

63 Article 2(8).

64 Recital 30.

65 The North Sea States are often given as a good example to follow (see Impact Assessment).

66 Article 11(1).

application of the Environmental Impact Assessment Directive to the offshore sector-specific challenges.[67]

In addition to the operator's own control based on the risk management and emergency response systems (which is also subject to independent verification), OSD requires that these two tasks are performed by different public bodies to ensure that the competent authority controlling offshore operations is independent from the licensing authority.[68]

The main task of the licensing authorities is to make sure that the applicants for such licences are capable of meeting the requirements for the corresponding offshore activities.[69] The OSD establishes an important connection to Directive 94/22/EC (the Hydrocarbons Directive) clarifying the considerations to be taken into account by the licensing authority when assessing the technical and financial risks of the offshore operations. The licensing authority must consider the operator's ability to ensure 'continued safe and effective operations under all foreseeable conditions.'[70]

One of the principles set out in Article 191(2) TFEU is the polluter-pays principle. The principle addresses situations when environmental damage has already been caused. The costs resulting therefrom, including the costs of restoring the environment, must be borne by the polluter.[71]

The polluter-pays principle was incorporated into the EU law in the Environmental Liability Directive (ELD) which establishes the framework of the environmental liability based on the 'polluter-pays' principle, to prevent and remedy environmental damage. OSD refers consistently to the polluter-pays principle and to the ELD throughout the preamble and the main text, removing any doubt as to the applicability of environmental liability to damage caused by offshore activities (in the sector specified by OSD and ELD).[72]

The OSD has introduced several important adjustments to the ELD regime, making it stronger to deal with remedial issues in the offshore sector.

---

67 Recital 5 of the EIA Directive.

68 To strengthen control over offshore activities, the OSD also envisages public participation in the decision-making processes (EIA Directive) and public access to the supervision of offshore operations with potentially significant effects on the environment: see Recitals 15, 16 and Articles 5, 8(4), 9.d and 24.

69 Article 4 (OSD).

70 Article 4(2) specifying the aspects which the authority must take into consideration when making such assessment.

71 See generally, David Langlet and Said Mahmoudi, *EU Environmental Law and Policy*, OUP, 2016, p. 55.

72 'Environmental damage' includes damage to protected species and habitats and water damage (Article 2(1), but excludes damage caused by pollution of diffuse character (Article 4(5)).

First, the essential improvement in the liability regime for environmental damage caused by offshore accidents introduced in the OSD is the extension of the territorial scope of liability for water damage under the ELD. This is now brought in line with the definition of 'marine waters' in the Marine Strategy Framework Directive and stretches to the outmost reach of the area where a Member State has and/or exercises jurisdictional rights, in accordance with UNCLOS, namely, the continental shelf.[73] Considering that offshore operations occur mostly outside territorial sea limits, such an extension was very necessary.

Secondly, OSD ensures that the responsible entity is clearly identifiable and accessible, something which is identified as a problem in the ELD.[74] With respect to the accessibility, the OSD assigns the main responsibility to the licensee, irrespective of whether operations are carried out by the licensee or, or on behalf of, the licensee or the operator. However, the operator is also assigned a central role and must always be the entity with the primary responsibility for safety of operations, to ensure that the environmental liability regime is functioning properly.[75]

Importantly, Member States must ensure that the licensees are financially capable to remedy the damage caused by an offshore accident. OSD emphasizes in a number of provisions that the ability to provide financial compensation for damage is an essential requirement which Member States must impose on licensees and operators before granting the licence.[76] As to the licensee's financial ability to remedy the potential or real damage, Member States are required to ensure that the licensee is financially liable for the prevention and remediation of environmental damage, caused by offshore oil and gas operations.

Thus the OSD contributes to a more adequate and comprehensive mechanism for the application of the polluter-pays principle in the offshore sector. However, ELD and consequently OSD apply only to pure ecological damage and do not regulate the civil liability or affect rights of compensation for traditional damage granted under any relevant international agreement regulating

---

73 Article 7(1)(b)(ii) OSD and Article 3(1)(a) of Directive 2008/56/EC cited in fn. 41. Before the OSD was adopted, a significant 'limitation had followed from the restricted definition of 'water damage' of Directive 2000/60/EC, OJ 2000 L327/1, (Water Damage Directive) applicable only to territorial waters within 12 nautical miles. This definition excluded the larger areas of the continental shelf with most offshore activities.

74 Also C-188/07 *Mesquer* [2008] ECR I-4501 (holders or previous holders of waste).

75 Article 2(5) and Article 3(2) and Recital 13 OSD. OSD defines the operator as 'the entity appointed by the licensee or licensing authority to conduct offshore oil and gas operations, including planning and executing a well operation or managing and controlling the functions of a production installation'. (Note that ELD refers to the 'operator' generally).

76 Article 4(3).

civil liability.[77] So, by contrast to shipping, where there exist liability mechanisms allowing for financial compensation for damage caused by the pollution from ships, no equivalent mechanisms at the EU (or international) level exist for pollution damage resulting from offshore activities.

4 Conclusions and Challenges Ahead

There are several general conclusions that are clear as far as the application of EU law to the seabed is concerned. First, EU law applies to the seabed only where EU member States have sovereignty in accordance with international law. Secondly, EU policy papers in respect of seabed mining indicate that the EU clearly intends to play a role in shaping seabed mining as such activity is perceived as economically important for the EU market. Thirdly, even though several EU environmental legislative measures may be applied to seabed activities, uncertainty remains as to whether it is the EU, the EU Member States, or both that have responsibility for environment protection given that the EU Member States have individually assumed obligations under international law. Finally, although general principles of environmental protection are embedded in international law, the EU provides more precise context and has an effective enforcement mechanism to ensure adherence by the EU Member States.

As far as the application of EU law is concerned, it is submitted that in the context of the EEA, EEA law applies only to the territory of the parties to that agreement and therefore the position may be different from the one under EU law. Furthermore, it is uncertain whether the EU has competence to adapt measures requiring EU Member States to act, for example, by requiring them to carry out seabed activities or imposing specific conditions as to how EU Member States decide to exploit the seabed over which they have international sovereignty. Nevertheless, should coastal EU Member States adopt national legislative measures which may give rise to different national standards, then the challenge for the EU is to decide whether the risk to the internal market warrants legislation to be adopted under Article 114 TFEU on the basis of necessity to safeguard the integrity of the internal market.

---

77 ELD also excludes personal injury, damage to private property or to any economic loss; it does not affect any right regarding these types of damages and it does not give private parties a right to compensation as a consequence of environmental damage or of an imminent threat of such damage (Article 3(3)).

As to the regulation of seabed mining activities, the EU considers that international institutions have not delivered and that it is well paced to play a significant role in management of the seabed. Already, where EU Member States have jurisdiction to regulate seabed mining activities under the UNCLOS regime, general principles of EU law apply[78] and general EU rules adopted to regulate the activities of market operators, such as reporting obligations imposed on undertakings, will be applied equally to operators seeking mining licences. A problematic scenario may arise if the EU decides not to wait for the international community to act and adopts specific legislation to regulate seabed mining activities carried out within the jurisdiction of the EU Member States. Should such a situation arise, then there is a great risk that the EU may adopt its own standards for seabed mining operations ahead of the international regime.[79] The challenges will then be to establish an international regime, as well as to ensure that one regime prevails.

Similarly, as far as environmental protection of the seabed is concerned, there are clearly gaps and inconsistencies which the EU, irrespective of whether its competence is exclusive or shared, is willing to act and impose a minimum standard of safety in all activities in the offshore sector. The OSD has strengthened ELD in the oil and gas sector but it would be an advantage for some further action in respect of compensation and risk pooling arrangements as suggested in Recital 63 of the OSD.

There is no doubt that disposal and management of waste produced as a result of exploration and extraction of minerals of the seabed as well as dumping of de-commissioned installations, is a serious challenge. A comprehensive framework for the safe management of waste from extractive industries at EU level is laid down in the 2006 Mining Waste Directive.[80] However, waste resulting from the offshore prospecting, extraction and treatment of mineral resources is excluded from the scope of this Directive.[81] To the extent of this

78 E.g. the principle of no discrimination on the grounds of nationality, origin and destination.

79 A similar situation arose in 2003 concerning the thickness of the hull of oil tankers. The slowness of the international IMO regime to take a decision to amend MARPOL (International Convention on Prevention of Pollution from Ship) and phase-in double hull oil tankers, led the EU to take unilateral action and adopt Regulation 1726/2003, OJ 2003 L249/1, of the European Parliament and of the Council of 22 July 2003 amending Regulation 417/2002/EC on accelerated phasing-in of double hull oil tankers. At the end of 2003, IMO took the decision to amend MARPOL.

80 Directive 2006/21/EC, OJ 2006 L102/15, of the European Parliament and of the Council of 15 March 2006 on the management of waste from extractive industries and amending Directive 2004/35/EC.

81 Ibid, Article 2(2)(b).

exclusion, the general Waste Directive applies to waste produced on offshore installations and brought on shore,[82] leaving the operator with the responsibility to have the waste treated properly in accordance with the requirements of the Directive.

It seems fair to conclude the endeavours of the EU with regard to seabed mining in many respects resembles a voyage into unchartered waters, and that it is called for the vigilance required when sailing in such waters.

---

82 By virtue of Article 2(2)(d) of Directive 2008/98/EC of the European Parliament and of the Council of 19 November 2008 on waste (Waste Framework Directive).

CHAPTER 15

China's Domestic Law on the Exploration and Development of Resources in Deep Seabed Areas

Chelsea Zhaoxi Chen

1 Introduction

The "Law of the People's Republic of China on Exploration for and Exploitation of Resources in the Deep Seabed Area" (the law, or the Deep Seabed Law) was adopted on 26 February 2016 and went into force on 1 May 2016.[1] This represents China's first special law dealing with the exploration for and exploitation of deep seabed resources. "Resources" here however was not defined. It is speculated that currently the definition only refers to non-living resources such as minerals but is intentionally silent regarding living seabed resources. This limitation to non-living resources in China's Deep Seabed Law is inferred from the limitation of the 1982 United Nations Convention on the Law of the Sea (UNCLOS, or the Convention), and the definition given thereunder for the term "resources".[2]

In addition, the law applies to activities conducted by Chinese entities in the Area. Pursuant to Article 1(1) of the UNCLOS, "Area" means the seabed and ocean floor and subsoil thereof, beyond the limits of national jurisdiction. In other words, the applicable area of the Deep Seabed Law is not the traditional "within the territory of the People's Republic of China and other sea areas under the jurisdiction of China".[3] Instead, it aims to effectively govern the behaviour of Chinese citizens, legal persons or organisations from territories outside the

---

1 Zhonghua Renmin Gonghe Guo Shenhai Haidi Quyu Ziyuan Kantan Kaifa Fa (中华人民共和国深海海底区域资源勘探开发法) [Law of the People's Republic of China on the Exploration and Development of Resources in Deep Seabed Areas] (promulgated by Standing Committee of the National People's Congress, February 26, 2016, effective May 1, 2016) (Chinalawinfo) [hereinafter Deep Seabed Law], https://isa.org.jm/nld/china.

2 Article 133 of UNCLOS states, 'For the purposes of this Part: (a) "resources" means all solid, liquid or gaseous mineral resources *in situ* in the Area at or beneath the seabed, including polymetallic nodules; (b) resources, when recovered from the Area, are referred to as "minerals".'

3 Article 2 of Zhonghua Renmin Gonghe Guo Haiyang Huanjing Baohu Fa (中华人民共和国海洋环境保护法)[Marine Environmental Protection Law of the People's Republic of China] (promulgated by the National People's Congress, Aug. 23, 1982, amended Dec. 25,

DOI:10.1163/9789004391567_017

jurisdiction of China and other countries, i.e. a law governing extraterritorial issues.

Prior to the adoption of the Deep Seabed Law, China had already adopted laws, rules and regulations regarding the exploration for, and development of, oceanic mineral resources located within marine areas under its national jurisdiction:

- *The Mineral Resources Law of the People's Republic of China*, adopted at the 15th meeting of the Standing Committee of the Sixth National People's Congress on 19 March 1986, and revised on 29 August 1996;
- *Rules for Implementation of the Mineral Resources Law of the People's Republic of China*, promulgated by Decree No. 152 of the State Council on 26 March 1994, and effective as of the date of promulgation;
- *Marine Environmental Protection Law of the People's Republic of China*, adopted at the twenty-fourth session of the Standing Committee of the Fifth National People's Congress on 23 August 1982, effective as of 1 March 1983, and further revised on 25 December 1999;
- *Administrative Regulation on the Prevention and Treatment of the Pollution and Damage to the Marine Environment by Marine Engineering Construction Projects*, promulgated by the State Council on 25 March 2010 and entered into force on the same day.

Inter alia, the laws and regulations mentioned above govern only waters under Chinese jurisdictions. Therefore, the Deep Seabed Law is an innovative piece of legislation which completes China's legal system governing deep-sea mining activities.

This chapter intends to demonstrate how China has complied with the UNCLOS by the enactment of the Deep Seabed Law. It also aims to review how the law regulates deep-sea mining activities outside the territorial waters of any nation. In the following parts, Section 2 begins by introducing the favourable national circumstances leading towards its enactment. Section 3 then proceeds to answer two questions: why does China need this law and why in 2016. The main part of this chapter compares China's Deep Seabed Law with laws of other countries with respect to (a) application scope, (b) competent national authority, (c) State sponsorship scheme, (d) rights and obligations of contractors, (e) environmental protection mechanism and (f) information sharing and cooperation respectively (Section 4). Section 5 comments on the three core legislative intents of the law, namely to control deep-sea mining activities, to protect lawful rights and interests, and to prepare for greater capability.

1999), English translation available at: http://english.court.gov.cn/2016-04/15/content_24562836.htm.

Section 6 examines the significance of the Law, while Section 7 concludes by evaluating its implementation and impact on China's other areas of law.

2 National Circumstances Paving the Way for the Enactment of the Deep Seabed Law

Being consistent with Chinese legislative tradition, the Deep Seabed Law is a short but concise piece of legislation consisting of 7 chapters with 29 articles. It is the first law governing activities of exploration and exploitation of the seabed, ocean floor, and subsoil thereof beyond the limits of China's national jurisdiction. Preliminary research work for the drafting of the law has been conducted since 2011, and upon completion of the preparatory phase, the government proceeded with the legislative protocol straight away.[4] During the two rounds of voting at the National People's Congress (NPC), the NPC Standing Committee deliberated on the first draft of the law in its bimonthly session in October 2015, which contained 32 articles. The draft was then further revised and submitted to the 19th Session of the Standing Committee of the Twelfth National People's Congress for a second reading in February 2016.[5] This led to the law being adopted on 26 February 2016, and only 3 articles were deleted for its final version, which indicated that the law had been well received and

4 Legislative procedure of the National People's Congress (NPC) involves three stages: (1) introduction of Bills; (2) deliberations of Bills; (3) passage and promulgation of laws. Bills sponsored by the NPC Standing Committee, State Council, and/or the NPC special committees such as the NPC Environmental and Resources Special Committee which in this case sponsored the Deep Seabed Law, usually are introduced through the Presidium and are able get to the NPC legislation agenda automatically. A formal deliberation begins with an introduction to the bill at a NPC plenary session. Either the NPC Standing Committee or the sponsor explains the bill to the full session of the NPC.

5 Upon being introduced, the bill was to receive examination by the deputations, relevant special committees, the Legislative Work Committee respectively. Deputations' consideration of the bill submitted to the Standing Committee is the basic form of deliberation. The sponsor's representatives are expected to be present at the deputation meetings, to listen to their opinions and respond to inquiries about the bill. It is intended to let examiners and sponsors communicate directly. The Legislative Work Committee views the bill unitarily on the basis of the result of deputation and special committee examinations. A report on the deliberation, along with a revised draft, then goes back to the Presidium. The report is expected to include a description of major controversies, an assessment on the maturity and feasibility of the legislation, and an explanation to the amendments if there is any. The report and the revised draft, after viewed by the Presidium, will be printed and distributed for another round of deputations' consideration. The Legislative Work Committee would revise the draft again according to the deputations' suggestions. The output is then submitted by the Presidium to a plenary vote. It would be passed to become a law if the majority goes for it.

did not face any major controversies; thus, few changes had been made to the original proposal. Several favourable national circumstances have contributed to its smooth and speedy passing.

2.1 *Support from Top Leadership*

China is a coastal state with 2.997 million square kilometres of ocean area.[6] However, China is not yet viewed as a strong maritime power. The national strategy of marine development began on 8 November 2012 when Hu Jintao, the then general secretary of the Central Committee of the Communist Party of China (CPC) and Chinese president, delivered a work report to the 18th CPC National Congress.[7] The report vowed to develop China into a "maritime power": "*We should enhance our capacity for exploiting marine resources, develop the marine economy, protect the marine ecological environment, resolutely safeguard China's maritime rights and interests, and build China into a maritime power.*" In addition, the current Chinese president, Xi Jinping, has also stressed the importance of marine power development and of establishing China's marine economy as a new growth point in his speech addressing the 8th Group Study for the Political Bureau of the Central Committee of the CPC in 2013.[8] Thus, with firm support from the top leadership, the Twelfth National People's Congress efficiently moved to adopt the Deep Seabed Law.

2.2 *Four Exploration Contracts from the ISA*

A second factor that facilitated the passing of the Deep Seabed Law was that China is becoming increasingly active with international seabed mining activities. The China Ocean Mineral Resource R&D Association (COMRA) was established in 1990 and has since been instrumental in spearheading China's ambitions for deep seabed development. COMRA is a governmental organisation directly under the guidance of the Ministry of Natural Resources.[9] It

---

6 中国的海洋国土面积有多大[How big is China's Ocean Area],ChinaIRN.com, Oct 18, 2018, http://www.chinairn.com/news/20181018/141721227.shtml.

7 中国共产党十八大报告全文 [Full Report to the 18th CPC Congress], China Daily, August 01, 2013, http://www.mj.org.cn/zsjs/wsxy/201211/t20121126_145927.htm.

8 Sheng Yueyue (沈跃跃), *Renzhen Xuexi Quanmian Shishi Shenhai Haidi QUyu Ziyuan Kantn Kaifa Fa* (认真学习全面实施深海海底区域资源勘探开发法) [*Study and Implementation of the law on Exploration for and Exploitation of Resources in the Deep Seabed Area*], People's Congress of China Journal (中国人大) 8 (May 2016); 关于"海洋"十九大报告提到这些内容[Highlights of the19th CPC Congress Report in relation to marine industry with full text attached], Sohu News, https://www.sohu.com/a/198905707_726570.

9 COMRA used to be under the guidance of the State Oceanic Administration which is now a part of the Ministry of Natural Resources. China's Ministry of Natural Resources was newly formed in March 2018.

undertakes both research and commercial activities related to exploration and exploitation in the Area, in accordance with the applicable provisions of UNCLOS and the 1994 Agreement relating to the Implementation of Part XI of UNCLOS (the Implementation Agreement). COMRA was registered as one of the seven pioneer investors for the preparatory committee for the International Seabed Authority (ISA, the Authority) and the International Tribunal for the Law of the Sea (ITLOS) in 1991.

Via COMRA, China signed its first exploration contract for polymetallic nodules with the ISA in 2001, and has since acquired the exclusive exploration rights as well as preferential rights for exploitation with respect to those minerals in the contracted seabed area. In 2011 and 2014, COMRA signed additional exploration contracts for polymetallic sulphides and cobalt-rich ferromanganese crusts with the Authority, respectively.

Meanwhile, Chinese scientists have predicted that China is facing a severe problem related to a major mineral shortage. By 2020, 39 out of 45 major types of minerals produced within the nation could be inadequate for meeting China's domestic demand.[10] In order to supply the country's rising demand for copper, cobalt, gold and the rare-earth elements that are vital in the manufacturing of smartphones and other high-tech products, China Minmetals Corporation, a state-owned company and now the country's largest metals trader,[11] has also delved into deep-sea mining research and development. In 2015, the ISA authorised an exploration contract, a 72,745 square kilometre (28,087 sq. mile) permit in the Pacific Ocean, to China Minmetals. Then on 12 May 2017, a contract for exploration for polymetallic nodules between the ISA and China Minmetals Corporation was signed on 17 May 2017. Together with the COMRA contracts, China is now a Party State which sponsors the largest number of seabed exploration contracts issued by the Authority covering all three types of deep-sea minerals (i.e., polymetallic nodules, polymetallic sulphides and cobalt-rich ferromanganese crusts).

The exploration and exploitation of resources in the Area are considered a difficult, risky, pioneering, and comprehensive project. It is also an international undertaking of tremendous political, economic, and strategic significance.

---

10 Zhang Xianglan & Ye Quan (张湘兰&叶泉), *Zhongguo Guoji Haidi Quyu Kaifa Lifa Tanxi* (中国国际海底区域开发立法探析) [*An Analysis on China's Legislation on Internationl Seabed Area*], 8 Law Science Magazine (法学杂志) 73 (2012).

11 China Minmetals Corporation is China's biggest metal ore mining company and metallurgical engineering service provider. China Minmetals is born out of the strategic merger of the former China Minmetals Limited Company (founded in 1950) and the Metallurgical Construction Corporation of China (MCC Group, founded in 1982) in 2015. Both were state-owned and Fortune Global 500 companies.

With the perspective that more national entities would assume a part in the project, and more ISA contracts would be signed in the future, China's central government was in need of a competent legal tool to strengthen its supervision and management of activities conducted in the Area, as well as to ensure compliance of entities operating in the Area. The promulgation of the Deep Seabed Law fulfils China's sponsoring State obligations and protects the legal interests of the sponsored contractors.

2.3 *Technological Advancement*

Thirdly, China has seen remarkable achievements in deep-sea exploration technology. For example, the successful launch of the "Three Dragons" equipment system symbolised China's ambition to make full use of the deep seabed areas.

Representing a technological milestone and an important part of the national marine strategy, the "Three Dragons System" consists of:

– Human Occupied Vehicles (HOVs) known as "Jiaolong";
– Autonomous/Unmanned Underwater Vehicles (AUVs) called the "Qianlong" series;[12] and
– Remotely Operated Vehicles (ROVs) designated the "Sea Dragon" series.[13]

Benefiting from a variety of technological advancements, China has undertaken over 50 ocean voyages to conduct comprehensive surveys and scientific research on the deep sea environment and has acquired valuable first-hand knowledge in relation to mining activities in the Area.[14] The latest 52nd ocean voyage set sail from Qingdao on 10 December 2018 and finished on 25 July 2019. Equipped with the most advanced "Jiaolong", "Qianlong No. 3" and "Sea Dragon No. 3", this voyage was an important task and a major part of China's marine development project under the 13th Five-Year Plan for economic and social development (2016–2020). The voyage was designed in two parts consisting of a total of 228 days. Part A was 150 days, covering the Indian Ocean and Atlantic Ocean, conducting surveys of the marine environment, prospecting biological genetic resources, and collecting environmental baseline data and samples in the surveyed area. Part B was 78 days, mainly conducting a comprehensive investigation of polymetallic sulphides in the southwest Indian Ocean in order

12 According to COMRA website information, China now has 'Qianlong No. 1', 'Qianlong No. 2' and 'Qianlong No. 3' in operation. http://www.comra.org/2019-02/11/content_40660825.htm.

13 *Ibid.* 'Sea Dragon No. 1', 'Sea Dragon No. 2' and 'Sea Dragon No. 3'.

14 *See* "大洋一号"起航执行中国大洋52航次科考任务 ["Ocean No. 1" sets sail to carry out China's 52nd oceanic voyage for scientific research mission], People's Daily, December 11, 2018, http://www.comra.org/2018-12/11/content_40608711.htm.

to collect samples and information to further comprehend the resources and environmental status.[15]

Above all, technological developments have provided a solid foundation for expanded research and served as important precondition for China's legislation concerning the exploration and development of resources located in the international seabed area.

3 The Need for a Deep Seabed Law

3.1 *Why Does China Need the Deep Seabed Law?*

The most important reason for China to pass the Deep Seabed Law was to proactively fulfil its obligations under international laws. China ratified UNCLOS in 1996 that consisted of three layers of obligations of which China is required to comply under the Convention: as a State Party, as a sponsoring State, and as a responsible member of the international community.

First, the Deep Seabed Law fulfils China's responsibilities as a State Party. This originates from Article 139, paragraph 1 of UNCLOS, which establishes the obligations of States Parties. Pursuant to that Article, States Parties shall have the responsibility to ensure that activities carried out by their sponsored contractors in the Area shall be in conformity with Part XI of UNCLOS.[16] Additionally, Article 153, paragraph 4 of Part XI provides that the obligation of the States Parties in accordance with Article 139 entails "taking all measures necessary" to ensure compliance by the sponsored contractor, and Annex III, Article 4, paragraph 4 of UNCLOS further states that such sponsoring States have the "responsibility to ensure" that "laws and regulations" should be adopted "within their legal systems".[17] The purpose of these provisions is to achieve a result where the obligations set out in the Convention, a treaty under international law which binds only States Parties thereto, are fully complied with by entities that are subjects of domestic legal systems.

The process of how international agreements become a part of domestic laws of a sovereign state varies from State to State. Some States follow a monist

15 利器加持：中国大洋第52航次启航[Equipped with powerful tools, the 52rd ocean voyage embarks], Science and Technology Daily, December 11, 2018, http://www.comra.org/2018-12/11/content_40608572.htm.

16 Part XI entitled 'the Area' containing Articles 133–191 of the UNCLOS, governs all activities taking place in the Area, available at: http://www.un.org/Depts/los/convention_agreements/texts/unclos/closindxAgree.htm.

17 Annex III of UNCLOS, available at: http://www.un.org/Depts/los/convention_agreements/texts/unclos/closindxAgree.htm.

system where treaties can become domestic law without incorporation, such as the Netherlands and France. Moreover, domestic courts in these countries could directly apply self-executing treaties if the provisions are considered sufficiently self-explanatory and these countries generally have constitutional provisions to provide legitimacy. In contrast, dualist states such as the United Kingdom require all treaties to go through a process of incorporation of international law before treaties can have any domestic legal effects.

China's constitution is silent concerning the relationship between international treaties and domestic laws. However, Judicial practice has demonstrated that China follows a treaty ratification method somewhere between these two extremes. Automatic incorporating takes place only when the domestic law contains specific provisions. For example, Article 142 of China's General Principles of the Civil Law has indicated that *"If any international treaty concluded or acceded to by the People's Republic of China contains provisions differing from those in the civil laws of the People's Republic of China, the provisions of the international treaty shall apply, unless the provisions are ones on which the People's Republic of China has announced reservations."*[18] Therefore, with respect to civil laws, when international treaties contradict domestic laws, international laws prevail. In other cases, China incorporates international treaties by passing a law that gives effect to that treaty at the domestic level, such as in the case of incorporating UNCLOS. The enactment of the Deep Seabed Law is China's response to its "responsibility to ensure" in order to comply with its requirements as a State Party.

Second, the enactment of the Deep Seabed Law aims to relieve China from its liability as a sponsoring State, namely, the consequences of a breach of the sponsoring State's primary obligations. In practical terms, as pointed out by Ximena Hinrichs Oyarce, the issue to be considered is whether, in the event of a single incident causing damage to the Area or its resources, the liability either of the sponsoring State or of the sponsored contractor would arise. (See Ximena Hinrichs Oyarce, Sponsoring States in the Area: Obligations, Liability and the Role Development States, Marine Policy 95 (2018) 317–323) The special advisory opinion of the Seabed Disputes Chamber of ITLOS in 2011 (the Advisory Opinion) has given clarification to that point. The Advisory Opinion indicates that the failure of a sponsoring State to carry out its obligations gives rise to liability only when such failure results in damage by the contractor. The sponsoring State is absolved from liability if it has taken the necessary and appropriate measures.

18 General Principles of the Civil Law, available at: http://www.npc.gov.cn/englishnpc/Law/2007-12/12/content_1383941.htm.

However, this exemption does not apply automatically, as the "responsibility to ensure" is an obligation of "due diligence". For it to apply, the sponsoring State must have taken "all necessary and appropriate measures to secure effective compliance" by the sponsored contractor with its obligations,[19] while the exception does not apply to the failure of the sponsoring State to carry out its direct obligations.[20]

China is a sponsoring State for four ISA contracts. The promulgation of the Deep Seabed Law and the implementation of administrative control with regard to the exploration and exploitation of deep seabed resources therefore serves two functions. It is a measure not only through which Chinese government could ensure compliance by its sponsored contractors (currently COMRA and China Minmetals) with their obligations under the relevant contracts and UNCLOS, but also an indispensable conditions that China must satisfy if it intends to be exempted from liability for damage.

Third, the Deep Seabed Law reflects China's pledge to protect the marine environment as a rule-abiding and responsible member of the international community. Both UNCLOS Part XII and the Authority's three regulations governing mineral-related exploration activities have provided special emphasis on protecting the marine environment from any harmful effects arising from mining and exploration activities.[21] The Area and its resources are the common heritage of mankind; thus, all rights associated with the resources of the

19 Responsibilities and Obligations of States with respect to activities in the Area, Advisory Opinion, 1 February 2011, ITLOS Reports, 2011 (hereinafter Responsibilities and obligations of States), pp. 60–61, para. 186.

20 *Id.* Responsibilities and obligations of States, p. 65, para. 207. The Advisory Opinion also indicates that the most important direct obligations of sponsoring States are: 1) the obligation to assist the ISA in the task of controlling activities in the Area pursuant to article 153(4) of UNCLOS; 2) the precautionary approach based on Principle 15 of the Rio Declaration, requiring actions where scientific evidence is insufficient but "where there are plausible indications of potential risk"; 3) best environmental practices (i.e., more than just best available technology); 4) the obligation to take measures to ensure the provision of guarantees in the event of an emergency order by the ISA for protection of the marine environment; 4) requirements to provide recourse for compensation; and 5) the obligation to conduct an environmental impact assessment. These "direct obligations" exist independently of the obligation "to ensure". The "obligation of ensuring compliance" (Article 139 of UNCLOS) is a duty of "due diligence". Compliance with these direct obligations can be considered a relevant factor in determining whether a State has acted with due diligence. Responsibilities and obligations of States, p. 44, para. 122.

21 To date, the ISA has issued Regulations on Prospecting and Exploration for Polymetallic Nodules in the Area (adopted 13 July 2000) which was later updated and adopted 25 July 2013; the Regulations on Prospecting and Exploration for Polymetallic Sulphides in the Area (adopted 7 May 2010) and the Regulations on Prospecting and Exploration for Cobalt-Rich Crusts (adopted 27 July 2012).

Area are vested in mankind as a whole. Therefore, protecting the marine environment of the Area is essentially for the benefit of all mankind and future generations. Adopting the Deep Seabed Law reflects the strong commitment and effort that China is making in order to maintain the international seabed order, to promote the sustainable and healthy development of the marine industry, and to advocate for the peaceful exploitation of deep-sea resources.

3.2 *Why Was It Promulgated in 2016?*

While China has become an active State Party to UNCLOS, it was not among the first few countries to pass national legislation to incorporate UNCLOS requirements. Prior to the enactment of the Deep Seabed Law in March 2016, 16 countries throughout the world had already adopted legislation (Act or Law) governing their engagement in seabed mineral activities undertaken in the Area. Consistent with the timing of their promulgation and against the benchmark event of UNCLOS entering into effect, the legislation may be divided into two specific groups (see table below).

One could say that the promulgation of the Deep Seabed Law represented a way for China to follow the regulatory trend. However, a crucial reason for the enactment of the Deep Seabed Law in 2016 was based primarily on the fact that one of China's exploration contracts was due to expire.

Among the twenty-nine 15-year contracts the ISA has entered into for exploration for polymetallic nodules, polymetallic sulphides and cobalt-rich ferromanganese crusts in the deep seabed, China has acquired the following four contracts:

- 2001, COMRA, polymetallic nodules, 75,000 km^2 of the North-east Pacific Ocean;
- 2011, COMRA, polymetallic sulphides, 10000 km^2 of the South-west Indian Ocean;
- 2014, COMRA, cobalt-rich ferromanganese crusts, 3000 km^2 of the North-east Pacific Ocean;
- 2017, China Minmetals, polymetallic nodules, 72,745 km^2 of the East Pacific Ocean

As provided for in Regulation 26 of the 2000 Regulations on Prospecting and Exploration for Polymetallic Nodules in the Area (amended and adopted 25 July 2013), the duration of a plan of work for an exploration contract is for 15 years. Upon expiration of that period, the contractor shall apply for a plan of work for exploitation, unless the contractor has already done so, or has obtained an extension for the plan of work for exploration, or decides to renounce its rights in the area covered by the plan of work for exploration.

TABLE 15.1

Specific national legislation enacted before UNCLOS	Specific national legislation enacted after UNCLOS
1. **France** (Law on the Exploration and Exploitation of Mineral Resources of the Deep Seabed 1981; In addition, France incorporated an article in an ordinance relating more generally to all maritime spaces under its sovereignty or jurisdiction.[22])	1. **Australia** (The Commonwealth Offshore Minerals Act 1994), 2. **Belgium** (Act on prospecting and exploration for, and exploitation of, resources of the seabed and ocean floor and subsoil thereof, beyond the limits of national jurisdiction, adopted on 17 August 2013), 3. **Cook Islands** (Seabed Minerals Act 2009, Amended in 2015; Seabed Minerals (Prospecting an Exploration) Regulations 2015), 4. **Czech Republic** (Act No. 158/2000 of 18 May 2000 on Prospecting, Exploration for and Exploitation of Mineral Resources from the Seabed beyond Limits of National Jurisdiction 2000),
2. **Japan** (Act on Interim Measures for Deep Seabed Mining 1982)[23]	5. **Fiji** (International Seabed Mineral Management Decree (Decree No. 21, 12 July 2013), 6. **Germany** (Seabed Mining Act of 6 June 1995 (the Act), Amended by article 74 of the Act of 8 December 2010),
3. **the United States** (Deep Seabed Hard Mineral Resources Act 1980)	7. **Nauru** (International Seabed Minerals Act 2015), 8. **New Zealand** (United Nations Convention on the Law of the Sea Act 1996), 9. **The Russian Federation** (Decree of the President of 22 November 1994 No. 2099 "About activities of the Russian physical and legal entities for exploration and development of mineral resources of the seabed outside the continental shelf"; 10. **Singapore** (Deep Seabed Mining Act 2015), 11. **Tonga** (Seabed Minerals Act 2014), 12. **Tuvalu** (Seabed Minerals Act 2014), 13. **the United Kingdom** (Deep Sea Mining (Temporary Provisions) Act 1981, amended by the Deep Sea Mining Act 2014).

22 Ordinance No. 2016–1687 of 8 December 2016 relating to the maritime areas under the sovereign-ty or jurisdiction of the Republic of France.

23 The Act provides for the regulations of mining activities by Japanese person in the Area. The Act is implemented by the Ordinance for Enforcement, which was enacted also in 1982 and last amended in 2013.

China's first exploration contract signed in 2001 was due to expire in 2016. Thus, if the commercial development conditions are ripe, the contract may need to be transferred to the development stage. However, the current domestic laws, such as the "Mineral Resources Law", as mentioned above, were not sufficient for this purpose. Not only did the application scope of domestic laws not cover international seabed areas beyond the limits of national jurisdiction, but the legal language was also incompetent to govern increasingly active and complex activities conducted in the Area, which now involve huge investments, a long contractual period and high risks. Therefore, China's Deep Seabed Law was a necessary measure to follow up on current activities conducted in the Area.

4 The Deep Seabed Law Compared with Laws of Other Countries

The Deep Seabed Law has a total of 29 articles and divided into 7 chapters.[24] Chapter I on General Provisions is the most extensive part of the law and comprised of 6 articles, which state the purposes and objectives, the scope of the application, as well as the general principles under which the mining activities should be conducted.

China's Deep Seabed Law was developed to serve five specific purposes: (a) regulating the exploration and development of resources in the Area; (b) promoting deep sea scientific and technological research and resource investigation; (c) protecting the marine environment; (d) promoting the sustainable utilization of resources in deep seabed areas; and (e) protecting the common interests of mankind. The law also states that the exploration and development of resources in deep seabed areas shall be conducted under the principles of peaceful utilisation, cooperation and sharing, environmental protection, and the protection of the common interests of mankind. Other countries such as the Czech Republic, Nauru, Tonga and Tuvalu have also recognised the principle of the "common heritage of mankind" in their respective national laws.

The law was adopted at the 19th Session of the Standing Committee of the Twelfth National People's Congress of the People's Republic of China on 26 February 2016 and came into force on 1 May 2016. Whereas Chapter VII addresses the terms and interpretation of the law, Chapters II through VI establish five primary mechanisms regulating different aspects of the mining activities in the Area:

24 Deep Seabed Law, *supra* note 1.

1) a licensing procedure with a dual-track approval mechanism for exploration and development activities;
2) a marine environmental protection mechanism;
3) a sharing and cooperation mechanism for scientific and technological research and resource investigation;
4) a supervision and administration mechanism; and
5) a legal liability mechanism.

4.1 *Scope of Application*

The scope of the application contains three general aspects: eligible subjects; applicable activities; and applicable area.

4.1.1 Eligible Subjects

Pursuant to Article 2, the eligible subjects of the law apply to Chinese citizens, legal persons or organisations of the People's Republic of China. Article 153, paragraph 2 (b) of UNCLOS requires all persons interested in carrying out activities in the Area – whether natural or juridical – to possess the nationality of States Parties or be under the effective control of them or their nationals. Therefore, China and other countries such as Belgium, the Czech Republic, Fiji, Japan, Nauru, New Zealand, Singapore, Tonga and Tuvalu have all adopted this criterion of nationality or registration/residence within the domestic jurisdiction as a necessary requirement to be entitled to apply for State sponsorship.

Second, thus far, COMRA and China Minmetals are the only two Contractors sponsored by the Chinese government. However, law here clearly suggests that the identity of a Contractor is not restricted to state-owned enterprises, and any Chinese individual or organisation that passes the qualification screening would be eligible. Essentially, the government is opening the door for private enterprises and individuals to participate in deep seabed exploration in the future, stating that the applicant can qualify if "*the application is in the national interest and the applicant has funds, technologies, equipment and other capabilities and conditions.*" Thus, the qualification requirements of an applicant is rather low and wide-ranging under China's Deep Seabed Law. Some other countries such as Japan and Fiji have stipulated more stringent rules governing the identity and qualification of the applicant. For example, Japan has a requirement regarding criminal records and the applicant is not eligible to apply for State sponsorship within two years after serving a sentence for criminal offences.

4.1.2 Applicable Activities

In relation to the range of activities it governs, Article 2 states that the law shall apply to activities in relation to the exploration and development of resources, environmental protection, scientific and technological research, and resource investigation. China defines "resources investigation" as "the searching for resources in deep seabed areas, including estimates of the components, quantity, distribution and economic value of resources", which is arguably equivalent to prospecting activities. However, there is no explanation as to what constitutes "resources" in the Deep Seabed Law, although it is speculated that "resources" should be interpreted in a generic way but only refers to non-living resources. The legal status of living resources, such as marine or genetic resources, is a complex issue that prompts an international legal response. Therefore, the Chinese legislators may be of the opinion that the law should be intentionally silent on the definition of "resources" and that the adoption of rules dedicated to living resources should first be discussed at the international level.

When compared with the other 16 countries which have passed domestic laws relating to deep seabed mining ahead of China, China's Deep Seabed Law offers the broadest range of applicable activities, including environmental protection.[25] The laws of Germany, the Czech Republic, and Belgium regulate prospecting, exploration, and development activities, while the laws of the United Kingdom and Japan primarily regulate exploration and development activities. Other countries such as the United States and Fiji focus more on the exploitation/development phase than the exploration/prospecting phase, without establishing any requirements for prospecting in their laws.

4.1.3 *Applicable Area*

Whereas the scope of the subject in other Chinese national laws concerning marine activities is generally connected to "territories", or governs activities in accordance with where they take place, the Deep Seabed Law, on the other hand, revolves around "one's activities", i.e., what the eligible subject actually does. Thus, it regulates the behaviour of the entity undertaking exploration and development activities within the Area. Moreover, it is the first extra-territorial law passed in China's legal history that governs activities outside its territorial waters.

With respect to the applicable area, the term "deep seabed area" is used in the law, instead of the term "the Area" as used in UNCLOS. For the purpose of the Deep Seabed Law, "deep seabed areas" means seabed, ocean floor, and the subsoil outside the jurisdiction of the People's Republic of China and other

25 Arguably the applicable activities do not include prospecting activities, although the definition given to "resource investigation" almost covers all prospecting activities.

countries, which is exactly the same definition of "the Area" as defined by UNCLOS.[26]

4.2 *Competent National Authority*

Article 5 of the Deep Seabed Law establishes China's competent national authority for mining activities in the Area by stating, "*the oceanic administration of the State Council is responsible for the supervision and administration of the exploration, development, and investigation of resources in deep seabed areas. Other relevant departments of the State Council shall be responsible for the relevant administration, according to the functions prescribed by the State Council*".

Although not explicitly provided for, the "oceanic administration of the State Council" at large refers to China's State Oceanic Administration (SOA). However, it should be noted that the functions and responsibilities of the SOA have now been divided and separated into two different, newly established ministries. Pursuant to the "Institutional Reform Plan of the State Council" which was approved at the first session of the 13th National People's Congress in March 2018, oceanic administration functions of the SOA now falls under the regime of the new Ministry of Natural Resources. Meanwhile, the marine environmental protection responsibilities of the SOA are now integrated into the regime of the new Ministry of Ecology and Environment. Although for convenience reason the Ministry of Natural Resources still retains the brand of State Oceanic Administration for the public, the SOA is now officially an office/a subordinate department within the Ministry of Natural Resources. The SOA/Ministry of Natural Resources functions as the inspection and regulatory body of the Deep Seabed Law and is responsible for the supervision and administration of activities conducted by Chinese nationals in relation to the exploration, development, and investigation of resources in deep seabed areas. the Ministry of Ecology and Environment is involved with respect to marine environmental protection issues. However, in practice, COMRA has a direct operational role in the implementation of the Deep Seabed Law as well as general oceanic administration affairs. For example, at the media conference for the promulgation of first implementing regulation of the Deep Seabed Law, it was the Secretary-General of COMRA, Mr. Liu Feng, rather than any government officials from the SOA to address the public in relation to the regulation.

Ideally, the institutional structure of governance should include the following: a legislative body, an executive branch, a regulatory and monitoring agency, as well as an oversight or appeal committee for agency decision-making. The

26 Art. 2, Deep Seabed Law, *supra* note 1.

National People's Congress is the legislative body. The SOA plays the role of regulatory and monitoring agency but without a supervision entity for its decision-making. COMRA is both the regulatory and monitoring agency and a sponsored contractor. Responsibilities of the SOA and the COMRA are not defined, while COMRA is both an athlete and a referee. Arguably, procedural fairness cannot be guaranteed and this institutional arrangement should be improved in the future.

4.3 *Procedural Mechanism for Licensing*

Except for France and the Russian Federation, 14 out of the 16 countries which have enacted national laws to incorporate UNCLOS have established a procedural mechanism to control and grant a license or State sponsorship for mining activities in the Area. All of them grant a license/certificate of sponsorship based on the application.

Their respective regulatory approach can be roughly grouped into four types:[27]

a) A certificate of sponsorship without reference to a license/approval/permit (the Czech Republic, Fiji and Nauru)

The national law provides that a certificate of sponsorship is to be entered into by the sponsoring State and the sponsored applicant. Obtaining State sponsorship is a pre-condition for a contract application to the ISA. The Czech Republic requires an applicant to get a certificate of sponsorship before negotiations with the Authority; however, there is no reference to a licensing mechanism in the law. This approach is followed by countries such as Fiji, Nauru, and the Czech Republic.

b) A license (approval or permit) for seabed mining activities within national jurisdiction and a certificate of sponsorship for activities in the Area (Tonga and Tuvalu)

Both the States' sponsorship scheme and the licensing procedure are provided in detail via the domestic laws, such as can be seen in Tonga and Tuvalu.

---

27 Wang Lan (王岚), *Guoji Haidi Quyu Kaifa Zhongde Guojia Danbao Zhiadu Yanjiu* (国际海底区域开发中的国家担保制度研究 – 中国国家担保制度立法回应) [Research on the National Guarantee System for International Seabed Area Development], 223 Xue Shu Jie (学术界) [Academics] 209 (2016); and ISA, *Comparative Study of the Existing National Legislation on Deep Seabed Mining* (2017), [hereinafter ISA comparative study].

c) A certificate of sponsorship is to be issued after a license has been granted (Singapore)

Singapore's State sponsorship scheme is only lightly touched upon in the Deep Seabed Mining Act 2015, with great emphasis attached to the licensing procedure. Thus, Singapore fulfils its sponsoring State obligations mainly through the licensing mechanism.

d) A license (approval or permit) without explicit reference to the certificate of sponsorship (Germany, Japan and the United Kingdom), but in practice, a certificate of sponsorship is issued for sponsored contractors.

No direct State sponsorship scheme is addressed in the domestic law; however, the sponsoring State commitment is integrated into the licensing system under the specific Deep Seabed Law. Such an approach is found in Germany and the UK.

China adopts the last approach. A licensing scheme is detailed in the Deep Seabed Law without explicit reference to the certificate of sponsorship. In accordance with Articles 7 and 8, before any Chinese citizen, legal person or organisation can apply to the ISA for an exploration and development contract, the entity shall first file an application with the SOA. The SOA will then examine the application materials submitted and conduct a qualification review. If successful, the SOA will grant a permit to the applicant within 60 working days. After obtaining the permit, the entity is then eligible to enter into an exploration and development contract with the ISA and become a Contractor. A duplicate copy of the ISA contract must be submitted to the SOA for record purposes no more than 30 days from the date of the signature.

In essence, China's Deep Seabed Law has established a dual-track approval mechanism for exploration and development activities in the Area, namely, a valid license (approval or permit) issued by the SOA under domestic law, and a valid contract signed with the Authority under international law, prior to the applicant commencing activities in the Area. A similar practice can be seen in other countries such as the Czech Republic, Fiji, Germany, Japan, Nauru, New Zealand, Singapore, Tonga, Tuvalu, and the United Kingdom, which similarly provides a general prohibition on deep seabed mining activities in the Area unless the applicant holds a valid national license/State sponsorship. China further requires a valid contract with the Authority for activities in the Area.

This dual-track (national and international) approval procedure means that sponsored contractors are bound by Chinese domestic laws, as well as contractual terms with the ISA. The former domestic procedure is the prerequisite for the latter. The SOA validates and approves the qualification of the applicant and provides its consent for the applicant to enter into contracts with the ISA.

Details of the contractual terms still depend on the negotiations between the contractor and the ISA. If a contract is not successfully concluded with the ISA within 3 years after the grant of the license, the license automatically expires.[28] In addition, the effective period of the license spans from the date of grant to the expiry date of the ISA contract.[29] If the ISA contract requires an extension, before the contractor files an extension application to the ISA, the contractor should first apply for an extension of the license with the SOA. The SOA has 60 working days to make a decision whether or not to allow the extension of the license and each extension period is no more than 5 years.[30] Essentially, the SOA could unilaterally withdraw State sponsorship and terminate the ISA contract if it refuses to extend the license. This type of governance system is new to Chinese authorities, which requires the SOA to fully consider the relevant international rules, while understanding the limits of their functions and powers. In that respect, the interaction between international laws and the provisions of the Deep Seabed Law affecting the rights and duties of the contractors may stress some areas of conflict and, thus, represent a challenge for Chinese judicial authorities in the future. The implementation impact of this practice is yet to be seen.

4.3.1 *Application Documentation*

With respect to the application documentation, Article 7 of the Deep Seabed Law provides that an application with the SOA must be in writing, and should include the following:

a) Basic information about the applicant, such as nationality, residency, and company registration information;
b) An explanation regarding the location and area of the region that the applicant intends to explore and develop, and the categories of minerals;
c) Certificates of financial status and investment ability, and an explanation concerning technical capability;
d) An exploration and development work plan, including the materials relating to the possible impact of exploration and development activities on the marine environment, and an emergency response plan for serious damage to the marine environment, among others; and
e) Other materials required by the oceanic administration of the State Council.

28 2017 Implementing Regulation on Licensing, *Infra* note 30, Art.11.
29 *Id.*
30 2017 Implementing Regulation on Licensing, *Infra* note 30, Art.15.

Whereas "*materials on the possible impact of exploration and development activities on the marine environment*" is vaguely mentioned in the Deep Seabed Law, the implementing regulation on licensing issued by the SOA in 2017 specifically requires an environmental impact assessment to be produced and an establishment of environmental baseline data.[31]

4.3.2 *Criteria for Granting a License*

In China, a number of eligibility criteria are required prior to the SOA granting the license: the application must be in the national or public interest; and the applicant must have sufficient funds and possess competent technical and technological capabilities. Although not manifested in the Deep Seabed Law, the implementing regulation on licensing further requires an assessment of the credibility and integrity of the applicant, and that the applicant is equally satisfying the eligibility criteria established by the ISA regarding the exploration and development of marine resources.[32]

4.3.3 *Scope of the License*

China's Deep Seabed Law foresees the scope of a national license covering two stages of deep-sea mining activities, namely exploration and development. Article 27 of the Deep Seabed Law defines "Exploration" as "*the searching and analysis of resources in deep seabed areas, use and testing of resource collection systems and equipment, processing facilities and transport systems, as well as the research on environment, technologies, economy, commerce and other relevant factors that shall be considered at the time of development.*" Moreover, "Development" is defined as "*the recovery and selection of resources in deep seabed areas for business purposes, including the construction and operation of collection, processing and transport systems serving the production and sale of resources.*"

Other countries such as Japan, New Zealand, Singapore, Tonga, Tuvalu and the United Kingdom may grant two types of licenses/certificates regarding exploration and exploitation, respectively.[33] For example, Singapore's license must specify whether they are for exploration or exploitation, the type of

31 Shenhai Haidi Quyu Ziyuan Kantan Kaifa Xuke Guanli Banfa (深海海底区域资源勘探开发许可管理办法) [Measures for the Management of Permits on Exploration for and Exploitation of Resources in the Deep Seabed Area] (promulgated by the State Oceanic Administration, Apr. 27, 2017, effective Apr. 27, 2017) Art.6(6) [hereinafter 2017 Implementing regulation on licensing], http://www.gov.cn/gongbao/content/2017/content_5227831.htm.

32 *Id.* Art.8.

33 Permits for exploration activities and mining activities in the case of Japan.

resource, and the part of the Area and the license must relate to only one type of resource.[34] In the Cook Islands, four types of certificates are issued: a prospecting permit, an exploration license, a mining license, and a retention lease.[35]

4.3.4 *Transfer, Variation, Revocation, Termination of License/Certificate of Sponsorship/the ISA Contracts*

The Deep Seabed Law allows a contractor to transfer, modify, and terminate the ISA contract as long as it informs the SOA for approval. The contractor shall, within 30 days of the transfer, modification or termination of the exploration and development contract, report the issue to the SOA for recordation.[36] A major modification to the license is allowed, but is governed by the implementing regulation on licensing instead. Under the circumstances of a major modification to the exploration and development work plan, a major modification or revision of the ISA contract, or a transfer (all or partly) of the obligations under the ISA contract, a modification or reissuance of the license may be approved. However, whereas ISA contracts can be transferred under the Deep Seabed Law, the implementing regulation is not clear with respect to whether a license could be transferred equally. As a comparison, in its 2010 Seabed Exploration Act, Germany does not allow a contractor to transfer its rights and obligation under an approval (license).

Moreover, many States have provided in their national law for the competent national authority to vary, suspend or revoke the license/certificate of sponsorship in certain cases. For example, the corresponding contract with the Authority has been suspended for any reason or it is in the national interest to suspend or revoke the license (Singapore), and to avoid a conflict with any obligation of the sponsoring State arising out of any international agreement or instrument in force (Nauru, Tonga and the United Kingdom).

Although not provided in the Deep Seabed Law, the 2017 implementing regulation on licensing has established situations under which the SOA may revoke or terminate the license if the licensed:

a) Provided false or misleading information to obtain the license (Article 23(1));

---

34 ISA comparative study, *supra* note 27, at para.40.

35 Zhang Zitai & Sheng Hao (张梓太&沈灏), 深海海底区域资源勘探开发立法研究 – 域外经验与中国策略 [Research on Laws Governing Exploration and Development of Deep Seabed Resources – Foreign Experience and China' Strategy] conference paper in National Symposium on Environmental and Resource Law, edited by Chinese Society of Environmental and Resource Law, 700, Guangzhou, 2014.

36 Deep Seabed Law, *Supra* note 22, Art.10.

b) Failed to perform the ISA's contractual obligations or perform obligations in any manner other than complying with the terms of the contract (Article 23(2));
c) Has transferred or significantly modified the rights and obligations under the ISA contract without the SOA's approval (Article 23(3));
d) Does not apply for an extension upon expiry of the license (Article 24(1));
e) No longer meets the qualification criteria for the exploration and development of deep seabed marine resources (Article 24(2));
f) Upon the bankruptcy, insolvency or receivership of, or ceasing to exist as a legal entity of, the sponsored party (Articles 24(3) and 24(4)); and
g) Under other situations where the license is lawfully terminated (Articles 24(5) and 24(6)).

4.4 *Rights, Obligations and Liabilities of a Sponsored Contractor*

The rights, obligations and responsibilities of the licensee/sponsored contractor are directly set out in Articles 9–11 of China's Deep Seabed Law. Some other countries such as Singapore and the United Kingdom provide for rights and obligations of the licensee under the terms and conditions of the license, which may have the inconvenience of having a lower legislative ranking and which may have less constraint on the contractors.

With respect to rights, Article 9 of the Deep Seabed Law affirms that a contractor has the exclusive rights to explore and develop specific resources within the region specified in the exploration and development contract. "Resources" here was not defined in the legislation. However, as the Deep Seabed Law directly incorporates the UNCLOS, it can be inferred from Article 133 of the UNCLOS that "resources" means all solid, liquid or gaseous mineral resources *in situ* in the Area at or beneath the seabed.[37]

With respect to obligations of the sponsored contractor, they amount to a comprehensive list in the Deep Seabed Law and may be classified into the following categories:
a) to perform obligations under the exploration and development contract;
b) to guarantee the personal safety of employed operation persons subject to Chinese laws and the administrative regulations on work safety and labour protection;
c) to ensure the safety of the vessels, installation, equipment and operations in the Area;
d) to protect the marine environment, among others, including taking necessary measures to protect and preserve rare or fragile ecological systems

37 Art. 133(a), Part XI, UNCLOS.

and the living environment of weakened, threatened or endangered species and other marine organisms;

e) to safeguard the objects of archaeological and historical nature in the Area;

f) to periodically update the Authority and the SOA on the performance of exploration and development contracts; and

g) to assist and cooperate in the supervision and inspection conducted by the SOA.

On the other hand, countries such as Fiji, Tonga and Tuvalu have provided in their national laws a number of responsibilities of the sponsoring State for the benefits of the sponsored contractor. These responsibilities, among others, include communicating and providing any assistance to the Authority to give effect to the State's sponsorship and not imposing unnecessary and disproportionate regulatory burdens on sponsored parties. China however limits its role and responsibilities of sponsoring States only to review and approve the transfer of license/ISA contracts and keep corresponding records. Even though Article 3 of the Deep Seabed Law unequivocally states that the State shall protect the lawful rights and interests of Chinese citizens, legal persons or other organisations that conduct the exploration, development and investigation of resources in the Area, the effectiveness of that is doubtful.

4.4.1 *Supervision and Inspection*

Like the Czech Republic, Fiji, Germany, Japan, Nauru, Singapore, Tonga, Tuvalu and the United Kingdom, China has established a monitoring, supervision and inspection mechanism by the competent national authority in Chapter V of the Deep Seabed Law.

The measures include sending an inspector and obtaining access to the contractor's vessels, facilities and equipment; examining relevant documentation, including logbooks, records and data; requiring the contractor to report regularly on the implementation of the exploration and development contract; and ensuring the contractor provides assistance and cooperate with the inspection team.

4.4.2 *Legal Liabilities*

Except for France and Russia, all countries which have domestic laws in place for mining activities in the Area have explicitly stipulate sanctions in case of a breach of contractors' obligations. Chapter VI of China's Deep Seabed Law defines a sanction regime in the form of contractors' legal liabilities in case of non-compliance. In most cases, a non-compliance situation will result in fines, orders to take corrective actions or the revocation of the license by the SOA. The contractor may be exposed to criminal liabilities only for acts which cause pollution or damage to the marine environment or to the cultural relics and paving things within the operation area. If any crime is constituted, it shall be subject to criminal liability in accordance with the law.[38] This again has demonstrated China's firm commitment to marine environmental protection.

Other countries have also established financial penalty mechanisms in their respective national laws. The amounts of the fines depend on the type of offence, but the severity of the sanctions in relation to offences varies from one country to another. The Czech Republic, for instance, sanctions the acts of undertaking seabed mining activities without State sponsorship or without a contract for over 4 million USD, while Tonga's fines for an equivalent offence can be up to 1 million USD. In comparison, China imposes a fine of not less than 500,000 yuan but not more than 1 million yuan (around 148,000 USD) for even its most serious type of offence. China can also confiscate, in addition to the fines, the gains and products derived from the illegal mining activities. It is speculated that China's sponsored contractors often have close ties to the central government or directly under the control of the government; hence, a low penalty scheme is sufficient to serve as deterrence for non-compliance. With the increasing cohort of private entities sponsored by States Parties, China's penalty mechanism might be revised in the future.

4.5 *Environmental Protection Mechanism*

Unlike the sponsorship requirement which only the State Parties willing to cooperate with the ISA will implement into domestic laws, the protection of the marine environment with respect to deep seabed mining activities has received global consensus. Many countries, including the United States which currently is an observer to the ISA and a non-party to the UNCLOS, have adopted a compensation and penalty mechanism for pollution and damage to marine environments. Some States set the element of marine environmental protection as one of criteria for granting a license/certificate of sponsorship (Nauru, Tonga, Tuvalu and the United Kingdom), while others set out in national laws the general or specific obligations of the licensee/contractor to protect the marine environment (Belgium, China, Czech Republic, Germany, Japan and the United Kingdom).[39]

China has prioritised marine environment protection and made it a full part of the country's marine strategy. This priority focus is reflected in the Deep Seabed Law, where the word "environment" is mentioned 20 times and the law dedicates a whole chapter (Chapter III) to environmental protection. As to

38 However, the law refers to the applicability of criminal law without specifying minimum or maximum imprisonment time.

39 ISA comparative study, *supra* note 26, at para.57.

the content of the requirements, Chapter III provides that a contractor shall apply the available advanced technologies and take necessary measures as far as reasonably practical: (i) to investigate and conduct research on the oceanic conditions; (ii) to establish environmental baselines and assess the possible impacts of mining activities; and (iii) to make and implement an environmental monitoring plan in order to prevent, reduce and control pollution and other damage to the marine environment caused by activities in the exploration and development area.

The Deep Seabed Law also addresses emergency situations by providing an emergency response plan. If an accident causing serious damage to the marine environment occurs or may occur, the contractor shall immediately sound an alarm, report to the SOA, and initiate the emergency response plan in accordance with Article 11. Similarly, the Acts of Fiji, Nauru, Tonga and Tuvalu also contain contingency plans of "incident" reporting, response and inquiry processes. Such environmental contingency plans are triggered by situations of pollution or serious harm to the environment.

In addition, Chapter I on General Provisions and Chapter II on Exploration and Development contain supplementary provisions related to marine environmental protection:

a) Article 1 stipulates that one of the objectives of the law is to protect the marine environment;
b) Article 3 requires that the exploration and development of resources in deep seabed areas shall be conducted in accordance with the principles of, *inter alia*, environmental protection and protection of the common interests of mankind;
c) Article 4 requires the State to adopt economic and technical policies and measures to enhance the capability of the protection of the marine environment;
d) Article 6 states that the government should encourage and support international cooperation in relation to environmental protection, scientific and technological research, education and training, among others;
e) Article 7 requires that the exploration and development plan submitted to the SOA for permit application should include background materials on the possible impact of exploration and development activities on the marine environment, including an emergency response plan for serious damage to the marine environment; and
f) Article 9 provides that the contractor shall perform the exploration and development contract under the obligation, *inter alia*, to protect the marine environment.

Compared to the other 15 countries which have passed domestic laws to regulate the deep seabed mining activities, China has demonstrated a high degree of commitment, and probably it is the only country prioritising the issue of marine environmental protection in its domestic law.

Besides national requirements, the three ISA Regulations have set out six environmental protection measures.[40] Some of them provide the basis for the national provisions mentioned above:

a) The contractor should "*take necessary measures to prevent, reduce and control pollution and other hazards to the marine environment ... as far as reasonably possible using the best technology available to it ...*", and "*apply a precautionary approach, as reflected in Principle 15 of the Rio Declaration, and best environmental practices*";[41]
b) The contractor should establish and implement programs for monitoring and evaluating the impacts of deep seabed mining on the marine environment;[42]
c) The contractor should gather environmental baseline data and establish environmental baselines;[43]
d) Prior to the commencement of its programme of activities, the contractor should submit a contingency plan to the Secretary-General to respond effectively to incidents that are likely to cause serious harm or constitute serious threats to the marine environment when performing contractor's activities;[44]
e) The contractor shall promptly report to the Secretary-General any incident arising from its activities that has caused, is causing or poses a threat of serious harm to the marine environment;[45] and
f) The contractor should comply with emergency orders issued by the Council of the ISA, as well as the immediate measures issued by the Secretary-General, to prevent, contain, minimise or repair serious harm or the threat of serious harm to the marine environment. This may

40 Namely, the Regulations on Prospecting and Exploration for Polymetallic Nodules in the Area (adopted 13 July 2000, updated and adopted 25 July 2013) [hereinafter Polymetallic Nodules Regulations]; Regulations on Prospecting and Exploration for Polymetallic Sulphides in the Area (adopted 7 May 2010) [hereinafter Polymetallic Sulphides Regulations]; and the Regulations on Prospecting and Exploration for Cobalt-Rich Crusts (adopted 27 July 2012) [hereinafter Cobalt-rich Crusts Regulatons], https://www.isa.org.jm/mining-code.

41 For example, Polymetallic Sulphides Regulations, https://ran-s3.s3.amazonaws.com/isa.org.jm/s3fs-public/files/documents/isba-16a-12rev1_0.pdf, similar provisions are also provided in other two regulations.

42 *Id.* Part 5, Regulation 33, para.6.

43 *Id.* Annex4, section 5.3.

44 *Id.* Annex4, section 6.1.

45 *Id.* Annex 4, section 6.2.

include orders to the contractor to immediately suspend or adjust any activities in the exploration area.[46]

Some States emphasise the application of the precautionary approach and best environmental practice (Fiji, Nauru, Tonga and Tuvalu), while others implicitly include these principles in the national laws by endorsing the Authority's rules, regulations and procedures.[47] China has fully incorporated these measures into its Deep Seabed Law and sets out standards, which are no less stringent than the requirements of ISA Regulations. In addition, the preservation of the ecosystem, the protection of endangered marine species, the preservation of marine biodiversity, and the sustainable use of marine resources required under Article 14 of the law also embody the principles of the resolution adopted by the United Nations General Assembly at RIO+20 on 27 July 2012.[48]

Moreover, other Chinese laws and regulations have also provided a solid legal foundation for marine environmental protection. Domestic laws such as the "Marine Environmental Protection Law of the People's Republic of China" and the "Administrative Regulation on the Prevention and Treatment of the Pollution and Damage to the Marine Environment by Marine Engineering Construction Projects" address environmental pollution and ecological damage caused by the exploitation and use of marine resources. Nevertheless, these laws only govern activities taking place in areas within China's jurisdiction. The promulgation of the Deep Seabed Law further complements and strengthens the legal system for marine environmental protection in China.

However, although it is applauded that China's Deep Seabed Law emphasises the ultimate importance of marine environment protection, relevant provisions in the law are quite vague and general. For instance, specific rules regulating environment impact assessment are still found wanting. Supplementary laws and administrative regulations shall be rolled out in the future to provide details and clarification. According to news report, the SOA previously had plan for four implementing regulations to be issued in 2017. The first three have already been promulgated in April 2017, December 2017, with the last "Measures for the Environmental Investigation and Assessment of Resources in Deep Seabed Areas" still in the pipeline.

46 *Id.* Annex 4, section 6.3.

47 ISA comparative study, *supra* note 27, para.58.

48 *The Future We Want*, UNGA Res. 66/288, para. 158, U.N. Doc. A/RES/66/288 (Sept. 11, 2012), http://www.un.org/ga/search/view_doc.asp?symbol=A/RES/66/288&Lang=E.

4.6 *Information Sharing and Cooperation*

Few countries have stressed the importance of the sharing and cooperation mechanism contained in their respective national laws. As a general rule, Belgium, Fiji, Nauru, Tonga, Tuvalu and the United Kingdom regard data and information received as confidential or to be held with appropriate confidentiality, with only a few prescribed exceptions specified under their laws.[49] China, on the other hand, requires the SOA to register and retain the information and samples submitted by the persons or entities that conduct deep sea mining activities, and make such information and samples available for public use pursuant to relative rules. In fact, China takes this step further by stipulating in the Deep Seabed Law that it supports the construction and operation of a "deep sea public platform" for the purpose of sharing the benefits arising from its research as related to deep sea technology and the exploration for deep sea resources. Again, this tends to further demonstrate China's commitment to the peaceful and sustainable utilization of deep-sea resources for the benefit of all mankind.

5 Comments on the Core Legislative Intent of the Law

At its core, China's Deep Seabed Law is about controlling, securing and preparing, as will be elaborated below.

First, the main theme of the law is to regulate and pursue an effective control mechanism for the activities in the Area. The very first sentence of Article 1 has stipulated that this law, among others, is developed for the purpose of regulating the exploration and development of resources in deep seabed areas. This stipulation fully reflects the law's basic value judgement as a tool to command and control. Chapter II then dedicates a whole chapter on exploration and development regulation, and it includes a licensing mechanism, eligibility criteria for applicants, comprehensive obligations for sponsored contractors and stringent restriction on modification and transfer of the contracts. Chapter III also dedicates a whole chapter on regulating exploration and development activities; however, it does this via a different angle of imposing marine environment protection obligations. To reinforce the control mechanism, Chapter V grants the SOA supervision and inspection rights with respect of mining activities conducted in the Area, and it is an obligation for contractors to assist and cooperate with the inspection team. Moreover,

49 ISA comparative study, *supra* note 27, at para.64.

the non-compliance of a contractor's obligations under the law attracts a financial penalty or even criminal imprisonment as detailed in Chapter VI. In summary, the law is devised more for the benefit of the government to carry out the State's "responsibility to ensure". Therefore, it is understandable that the reciprocal obligations of sponsoring States are not emphasised.

Second, the law is to secure and protect the lawful rights and interests of citizens, legal persons or other organizations of the People's Republic of China that conduct mining activities in the Area, as enshrined in Article 3 of the law. Since its launch in 1994, the ISA has issued twenty-nine (29) 15-year contracts for the exploration of deep sea minerals, with China holding four of them. In the future, China is expected to continue more intensive engagement and cooperation with the ISA. The provisions of the Deep Seabed Law are largely rooted in the provisions of the UNCLOS and ISA Regulations, which demonstrates China's support for the international legal regime of the Area, thus consolidating ISA's function further. China's promulgation of the Deep Seabed Law contributes to its own discourse power and discourse rights within the Council of ISA; hence this measure ensures effective protection of Chinese citizens' rights in the Area. However, it is arguable whether the law *per se* could effectively protect rights of sponsored contractors with only one Article (Article 9) stipulating vaguely and generally that "*a contractor shall enjoy corresponding exclusive rights to explore and develop specific resources within the region specified in the exploration and development contract*".[50] Moreover, the responsibilities of China as a sponsoring State were only restricted to review and approve the transfer of license/ISA contracts and keep corresponding records. The law also lacks provisions regarding the rights of relief and the rights of appeal if a contractor's rights are infringed by the sponsoring State under the law. Given the legal uncertainty, further implementing regulations or judicial interpretations relevant to securing contractors' rights should be supplemented to the Deep Seabed Law.

Third, to prepare China for greater capability in relation to deep seabed mining activities in the Area, with reference to Articles 15–17, a number of incentive measures have been put in place to advance China's deep-sea scientific and technological research and to facilitate China's capacity-building for mining activities in the Area. Encompassing these measures, Chapter IV of the Deep Seabed Law dedicates a whole chapter on "Scientific and Technological Research and Resource Investigation".[51]

---

50 Deep Seabed Law, *supra* note 1, Art.9.

51 'Resource investigation' is defined under the Deep Seabed Law as the search for resources in deep seabed areas, including the estimates of the components, quantity, distribution

a) The State supports the training of professional talents and encourages research cooperation between relevant industries;
b) The State lists deep-sea science and technology among the priority fields of scientific and technological development, meaning that more government spending would be devoted to this Area;
c) The State encourages private enterprises to carry out deep-sea scientific research and marine technology equipment research, essentially opening doors for a diversity of participants;
d) The State supports the construction and operation of a "deep-sea public platform", which is a mechanism to promote information exchange, cooperation and result sharing;
e) The State encourages entities and individuals to conduct deep-sea science popularization activities such as opening vessels for exhibitions, holding seminars and providing relevant consultation services; and
f) The State shares with the public the physical samples and relevant materials obtained by contractors from investigation, exploration and development activities in accordance with relevant provisions.

China is the only country that promotes resources investigation and marine scientific research as one of its legislative objectives under the law. Chinese legislators are of the opinion that capacity-building is the foundation of the preceding two objectives mentioned previously. A State does not even possess the competent ability to conduct deep sea mining activities let alone to effectively regulate its sponsored contractors and to protect their legal rights and interests. Competent ability relates to the technology, the equipment and the qualified personnel. By stipulating such content of national policy/strategy in its national law on deep seabed mining, China is preparing itself to move onto the next stage of commercialization of deep seabed resources once the conditions are matured.

6 Significance of the Law

With the enactment of China's Deep Seabed Law in 2016, it is worth assessing its significance and the influence it may have on the international regulations governing deep seabed resources. Three main points can be made.

---

and economic value of resources. Arguably, it is similar to the definition of prospecting under the Mining Code.

6.1 *A Regulatory Blueprint for Other National Legislation*

First, the law could be viewed as an effective precedent for countries who are yet to pass domestic legislation governing international seabed activities.

The Deep Seabed Law, in many ways, is a direct incorporation of the contents and requirements of the relevant international rules. Many provisions under the Deep Seabed Law are resonant with the legal wording formulated under the UNCLOS and other supplementary legal instruments (together referred as the "Mining Code").[52] For example, the definition of "deep seabed areas" established under the Deep Seabed Law has similar wording as the "Area" as defined under UNCLOS Article 1.

On the other hand, the law contains some additional requirements that are not covered by the Mining Code, such as the protection of marine biodiversity and relevant financial arrangements with respect to the development of resources in deep seabed areas. The Deep Seabed Law also emphasises the importance of international cooperation as well as the peaceful exploitation and certain human rights protection issues, which mirrors the principles established in the Rio+20 UN Resolutions and the Mining Code.

Moreover, words and terms are carefully crafted in China's Deep Seabed Law and many represent legal innovations from Chinese legislators. For example, unlike other definitions, the Deep Seabed Law does not directly transplant the term "common heritage of mankind" as prescribed in Article 136 of the UNCLOS, but rather has adopted the term "common interests of mankind" derived from the term or concept of "benefit of mankind as a whole" as prescribed in Articles 140(1), 143(1), 149 and 150(i) of the Convention. Article 136, Part XI of the UNCLOS stipulates "*the Area and its resources are the common heritage of mankind*", whereas Article 140(1) and 150(i) of the UNCLOS state that "*Activities in the Area shall ... be carried out for the benefit of mankind as a whole*". It, therefore, signifies that the term "common heritage of mankind" is closely linked with "the Area and its resources", while the concept of "benefit of mankind as a whole" governs "activities" in the Area. As Article 2 of the Deep Seabed Law provides that the law shall apply to "activities" conducted by Chinese entities in the Area, it is more appropriate to use the term of "common interests" rather than "common heritage", while not frustrating

52 The "Mining Code" refers to the whole of the comprehensive set of rules, regulations and procedures issued by the International Seabed Authority to regulate prospecting, exploration and exploitation of marine minerals in the international seabed Area (defined as the seabed and subsoil beyond the limits of national jurisdiction). At its core, it includes the UNCLOS, the Implementation Agreement and the three ISA Regulations, among others. See Chapter 7 of this book, J. Dingwall, 'Commercial Mining Activities in the Deep Seabed Beyond National Jurisdiction: The International Legal Framework'.

the well-established international legal status of "the Area and its resources". However, although "common heritage of mankind" is not specifically articulated in the Deep Seabed Law, the principle of "protection of the common interests of mankind" does not deviate from the common heritage principle under the UNCLOS and is precisely in-line with the Convention. At the same time, it allows more flexibility in applying the Chinese law to more Area activities in the future, such as the sharing and cooperation mechanism of the undefined "deep seabed public platform".[53]

Another example is the definitions of "exploration" and "exploitation" ("development" as in the Deep Seabed Law). China's Deep Seabed Law largely directly incorporates the two definitions given under the ISA Regulations. The only nuanced difference between the two is that the latter has specifically designated the resources as the three kinds of minerals in the Area,[54] whereas China's Deep Seabed Law uses a more generic meaning of "resources".[55] The intentionally undefined term "resources" has thus left room for further interpretation and clarification, which also provides the law with some flexibility for application.

In summary, this law has enshrined the fundamental spirit of international laws in relation to the exploration and exploitation activities in the Area, and at the same time preserved a flexibility to cater to national circumstances. In view of the growing interest in the development of marine minerals in the

53 G. Zhang & P. Zheng, *A new step forward: Review of China's 2016 legislation on international seabed area exploration and exploitation*, 73 Marine Policy 251 (2016).

54 *Supra* note 40. For the purposes of these Regulations: (a) "exploitation" means the recovery for commercial purposes of polymetallic nodules, polymetallic sulphides and cobalt-rich ferromanganese crusts in the Area and the extraction of minerals therefrom, including the construction and operation of mining, processing and transportation systems, for the production and marketing of metals; (b) "exploration" means searching for deposits of polymetallic nodules, polymetallic sulphides and cobalt-rich ferromanganese crusts in the Area with exclusive rights, the analysis of such deposits, the testing of collecting systems and equipment, processing facilities and transportation systems, and the carrying out of studies of the environmental, technical, economic, commercial and other appropriate factors that must be taken into account in exploitation.

55 Deep Seabed Law, *supra* note 1, Article 27(1) and (2): "Development" means the recovery and selection of resources in deep seabed areas for business purposes, including the construction and operation of collection, processing and transport systems serving the production and sale of resources. "Exploration" means the searching and analysis of resources in deep seabed areas, use and testing of resource collection systems and equipment, processing facilities and transport systems, as well as the research on environment, technologies, economy, commerce and other relevant factors that shall be considered at the time of development. It should be noted that the English translation of China's Deep Seabed Law may not be accurate and therefore may not be the exactly same wording as the ISA Regulations definitions.

deep seabed, both within the limits of a national jurisdiction and in the Area, China's Deep Seabed Law could serve as an appropriate reference for the follow-up legislative work of other States Parties.

6.2 *A Contribution to the Development of International Rules*

Second, the law facilitates the legislative work at an international level. The international community is increasingly discussing matters related to the Area, and with a particular focus related to the establishment of supplementary legal rules to UNCLOS. As the 5th pioneer investor of the ISA, China is a major participant in all ISA meetings and is an important opinion leader for the ISA Assembly resolutions. In the past, Chinese delegates have proposed numerous insightful and productive opinions which have helped to shape the overall legal regime of the international law of the sea. It was a major shortcoming that China previously lacked a domestic legislation in relation to deep seabed issues. Therefore, the introduction of the Deep Seabed Law has had immeasurable and far-reaching significance to strengthen China's discourse in vying for the influence on other countries.

In addition, with its first set of 15-year exploration contracts ending in 2016–2017, there is an urgent need for the ISA to make progress on the development of an appropriate fiscal and legal regime that would allow those contractors, including Chinese contractors, to proceed toward commercial development, while at the same time safeguarding the interests of the members of the Authority as a whole. China's domestic Deep Seabed Law includes not only the content governing exploration activities, but also establishes the rules relating to developmental activities. These provisions could essentially provide further references for the ISA, and contribute to the relevant rule-making relating to the commercial recovery of mineral resources from the Area.

6.3 *Regulating a Sphere of Intense Competition between States*

Third, the law evidenced China as a rule-abiding international community member. Most notably, China is now in a position to mediate various conflicting interests between industrialized states and developing countries for the exploration and development of deep seabed resources. China is not the first country with high ambitions in the Area and the deep seabed in general. Since the 1960s, industrialized states such as the United States, the United Kingdom, Germany, Japan and Russia have been conducting regular deep-sea surveys of marine resources. By the early 1980s, the Pacific Ocean deep seabed area, which had the most commercial potential consisting of nearly 2.8 million square kilometres, had almost been divided among these countries who regard their active presence as a legitimate exercise of high seas freedom.

In order to challenge excessive maritime claims, the UNCLOS and the 1994 Implementation Agreement have since been ratified by 168 states.

As a developing country, China was granted the largest number of exploration contracts by the ISA. On the other hand, China's marine technology is rapidly improving and fast approaching the technological level of developed countries. Nevertheless, with China's rising international influence and its increasing activities in the Area in recent years, it has attracted some criticism voiced as the "deep-sea threat from China". For example, when COMRA signed the second exploration contract with the ISA in 2011, other developing countries such as India reacted with fierce criticism. Thus, there might be some elements of competition between States with respect to these reactions. The Deep Seabed Law, especially Article 16, has, therefore, addressed the conflict by stating that China supports the construction and operation of a "deep sea public platform" for the purpose of sharing the benefits and knowledge generated from its research focusing on deep sea technology and the exploration of marine resources.

Meanwhile, there is always a fragile balance of interests to accommodate national ambitions to access deep seabed resources and common interests surrounding the exploitation and protection of those marine areas. By adopting the Deep Seabed Law, China pledges its commitment to the peaceful and sustainable utilisation of deep-sea resources with the associated goal of motivating more countries to ratify the UNCLOS agreement and cooperate with the ISA. Moreover, it reinforces China's stance that all activities in the Area shall be carried out for the benefit of mankind as a whole, while taking into particular consideration the interests and needs of developing States.

7 Conclusion: Looking Ahead towards Implementation

Although a relatively comprehensive and intricately crafted law, the Deep Seabed Law is not without its limitations.

The first limitation relates to the lack of an adequate definition for "prospecting".[56] In other words, China's Deep Seabed Law does not distinguish the "prospecting" stage from the "exploration" stage, although prospecting is a

56 Arguable, the term "Resource investigation" used in the Deep Seabed Law equals to "Prospecting" but is not given the rights as tended to "Prospecting" under the ISA Regulations. "Prospecting" under the ISA Regulations refers to "*the search for deposits of polymetallic nodules, polymetallic sulphides, cobalt-rich ferromanganese crusts in the Area, including estimation of the composition, sizes and distributions of deposits and their economic values, without any exclusive rights*". "Resource investigation" is defined in Article 27(3) of China's Deep Seabed Law as "*the searching of resources in deep seabed*

separate preparation stage of exploration and is entirely different from exploration in nature. Meanwhile, the legal importance of distinguishing prospecting from exploration is that the rights and obligations of "prospecting" under international laws are quite different from that of "exploration". Pursuant to the ISA Regulations,[57] there are no regional and time restrictions on prospecting, and the prospector is not required to sign a contract with the ISA.[58] Although no exclusive property rights are granted to the prospector, a prospector has the right to recover "a reasonable quantity of minerals", as long as the quantity is necessary for testing and is not collected for commercial use. China is actively engaged in prospecting activities in the Area; thus, the absence of rules prescribed to prospecting may lead to confusion and create legal uncertainty for Chinese contractors. Therefore, the formulation of relevant implementing regulations or judicial interpretations is essential to clarify this issue.

Second, except for the relevant liability and penalty rules as prescribed in Article 26, the law does not establish financial arrangements for compensation or remediation as a result of the contractor's failure to perform certain obligations. All national laws studied here, including the Deep Seabed Law, have required a sponsored contractor to provide evidence or written undertaking as to the financial capacity at the application stage. Some countries expressly include their capacity for adequate compensation for any environmental damage related to this requirement. Several countries (Singapore, Fiji, Nauru, Tonga and Tuvalu) follow this through with required financial guarantees or security after licence/sponsorship has been granted, so as to address potential damage or take measures toward rectifying the contractor's potential failure to fulfil its obligations.[59] Fiji, Nauru, Tonga and Tuvalu may require a sponsored party to deposit a security as a guarantee of performance concerning its obligations or rectify any damage or loss caused as a result of failure to perform certain obligations.[60] Tonga and Tuvalu further specify that the security may be used for clean-up or compensation costs in respect of any damage caused by pollution or other incidents occurring as a result of seabed mineral activities.[61] The absence of the funds and security provisions in China's Deep Seabed Law may, therefore, render the controlling function of the law less effective or enforceable. Similar inadequacies also exist in other provisions of the law such as the

areas, including estimates of the components, quantity, distribution and economic value of resources".

57 ISA Regulations, *supra* note 40.

58 Polymetallic Sulphides Regulations, *supra* note 41, at Regulation 2.

59 ISA comparative study, *supra* 27, at para.71.

60 *Id.*

61 *Id.*

elaboration on "deep sea public platform", and the uncertainty as to whether foreign investment is allowed to engage in China's deep seabed activities in the Area. All of these should be supplemented by more specific and detailed implementing regulations, administrative orders or judicial interpretations in the future.

The Deep Seabed Law was effective from 1 May 2016 and has only been in force for a few years. Thus far, it has been implemented by three administrative regulations pertaining to licensing, sample management, and data and information management, respectively. The first implementing regulation, "*Measures for the Management of Permits on the Exploration and Development of Resources in Deep Seabed Areas*", was promulgated on 27 April 2017.[62] The "*Interim Measures for the Management of Marine Samples from Exploration and Development of Resources in Deep Seabed Areas*",[63] as well as the "*Interim Measures for the Management of Data on the Exploration and Development of Resources in Deep Seabed Areas*" were issued by the SOA on 29 December 2017,[64] as two further supplementary regulations to the Deep Seabed Law. Moreover, a national policy paper, the "13th Five-Year Plan for Exploration and Development of Deep Seabed Resources" was jointly issued by the State Oceanic Administration, the Ministry of Foreign Affairs, the Development and Reform Commission, the Ministry of Science and Technology, the Ministry of Finance, and the Ministry of Land and Resources on 16 April 2017. The paper stipulates that, during the 13th Five-Year Plan period (2016–2020), deep-sea ocean development work shall be guided by the goal of "building a maritime power"; it shall actively promote major engineering projects; and it shall further enhance China's capabilities in relation to resources investigation, exploration, development, and environmental protection. As one of the four major engineering projects in the marine field, "Jiaolong Exploration of the Sea" (蛟龙探海) is of significant importance toward advancing China's scientific and technological ability, and safeguarding Chinese entities' rights and interests in the deep-sea areas.

---

62 海洋局关于印发《深海海底区域资源勘探开发许可管理办法》的通知 [Announcement from the SOA on the issuance of the 2017 Implementing Regulation on Licensing], http://www.gov.cn/gongbao/content/2017/content_5227831.htm.

63 海洋局关于印发《深海海底区域资源勘探开发样品管理暂行办法》的通知 [Announcement from the SOA on the issuance of the 2017 Implementing Regulation on Management of Marine Samples], http://www.gov.cn/gongbao/content/2018/content_5299621.htm.

64 海洋局关于印发《深海海底区域资源勘探开发资料管理暂行办法》的通知 [Announcement from the SOA on the issuance of the 2017 Implementing Regulation on Management of Data], http://www.comra.org/2018-01/04/content_40131061.htm.

The forth implementing regulation, "*Measures for the Environmental Investigation and Assessment of Resources in Deep Seabed Areas*", is still in the pipeline and plans to be rolled out in 2020. Thus, it is yet to be determined what impact the Deep Seabed Law and its implementing regulations might have on China's domestic legal system, especially in relation to issues of environmental protection.

Moreover, China is a State committed to contributing to the international legal order by adhering to the rules of international law. The dual-track licensing approval mechanism means a contractor benefiting from China's State sponsorship is subject to both international and national rules. Using a domestic law to govern activities of Chinese citizens, juridical persons or organisations in an area outside the territorial waters of any nation describes an approach that is new to the Chinese administration. It remains to be seen how this arrangement will actually impact the behaviour of the industry, given the SOA and Chinese judicial authorities need to fully consider the relevant international rules, cooperate with the ISA, and understand the limits of their functions and powers.

Nevertheless, the Deep Seabed Law, in many aspects, represents a legal innovation by Chinese legislators and may serve as a guiding reference for other States wishing to engage in mining activities in the Area. China's Deep Seabed Law is an indispensable part of the overall comprehensive legal regime concerning the world's oceans and seas. It enshrines the notion that all problems related to ocean space are closely interrelated and need to be addressed in a holistic manner.

CHAPTER 16

Implementation of Article 82 of the United Nations Convention on the Law of the Sea: the Challenge for Canada

Aldo Chircop

1 Introduction

The United Nations Convention on the Law of the Sea, 1982 (LOSC)[1] contains an obscure, unprecedented and unique provision in international law concerning equitable sharing of ocean resources, as follows:

> *Article 82*
> *Payments and contributions with respect to the exploitation of the continental shelf beyond 200 nautical miles*
> 1. The coastal State shall make payments or contributions in kind in respect of the exploitation of the non-living resources of the continental shelf beyond 200 nautical miles from the baselines from which the breadth of the territorial sea is measured.
> 2. The payments and contributions shall be made annually with respect to all production at a site after the first five years of production at that site. For the sixth year, the rate of payment or contribution shall be 1 per cent of the value or volume of production at the site. The rate shall increase by 1 per cent for each subsequent year until the twelfth year and shall remain at 7 per cent thereafter. Production does not include resources used in connection with exploitation.
> 3. A developing State which is a net importer of a mineral resource produced from its continental shelf is exempt from making such payments or contributions in respect of that mineral resource.
> 4. The payments or contributions shall be made through the Authority, which shall distribute them to States Parties to this Convention, on the basis of equitable sharing criteria, taking into account the interests and

1 United Nations Convention on the Law of the Sea, Montego Bay, 10 December 1982 (adopted 10 December 1982, entered into force 16 November 1994), 1833 UNTS 3 [hereinafter LOSC].

DOI:10.1163/9789004391567_018

> needs of developing States, particularly the least developed and the land-locked among them.[2]

After its adoption at the Third United Nations Conference on the Law of the Sea in 1973–82 (UNCLOS III) the provision hardly received further attention. This was possibly because the prospects of technological and commercial feasibility of deep sea drilling were remote at the time and the inconvenience it entailed for domestic revenue generation regimes. It is unprecedented because it was the first provision in a multilateral treaty to introduce an international royalty on natural resource production within national jurisdiction. The discourse on equitable sharing of ocean resources was a dominant theme at UNCLOS III and in subsequent instruments of international environmental law, and currently in the discourse on the conservation of areas beyond national jurisdiction.[3] Article 82 stands out in the discourse on international equity as it sets out a novel framework for equitable sharing of production from non-living resources within national jurisdiction.

The literature on the diplomatic history of Article 82 explains how this provision secured a key compromise in the UNCLOS III package deal and needs no revisit.[4] Articles 82 and 76 constituted a *quid pro quo*.[5] There would have been no agreement on key aspects of Article 76 concerning the extended continental shelf, i.e., that part of the shelf defined in the LOSC as extending beyond the limits of the exclusive economic zone (EEZ) and reaching out seaward to the outer limits of the continental margin, without a counterpart

2 Ibid, Art 82.

3 Development of an International Legally Binding Instrument under the United Nations Convention on the Law of the Sea on the Conservation and Sustainable Use of Marine Biological Diversity of Areas beyond National Jurisdiction, UN General Assembly Resolution 292 (LXIX), 19 June 2015, available at < https://documents-dds-ny.un.org/doc/UNDOC/GEN/N15/187/55/PDF/N1518755.pdf?OpenElement>.

4 SN Nandan & S Rosenne, vol. eds., *United Nations Convention on the Law of the Sea 1982: A Commentary*, Vol. 2 (Dordrecht: Nijhoff, 1993), commentary on Part VI, Article 82, at 930–947 [hereinafter Nandan & Rosenne]; RJ Dupuy & D Vignes, *A Handbook on the New Law of the Sea* (The Hague: Nijhoff, 1991), 375–381; A Chircop & B Marchand, 'International Royalty and Continental Shelf Limits: Emerging Issues for the Canadian Offshore,' (2003) 26 *Dalhousie Law Journal* 273 at 283–293; A Chircop, 'Operationalizing Article 82 of the United Nations Convention on the Law of the Sea: A New Role for the International Seabed Authority?' (2004) 18 *Ocean Yearbook* (2004) 395 at 395–412; *Issues Associated with the Implementation of Article 82 of the United Nations Convention on the Law of the Sea*, ISA Technical Study no. 4 (Kingston, Jamaica: International Seabed Authority, 2009) [hereinafter Chatham House Report].

5 Ibid, Nandan & Rosenne, 932.

agreement on the equitable sharing of non-living resources of the extended shelf. The compromise was difficult because States enjoying broad margins (known as margineers) were confronted by land-locked and geographically disadvantaged States who felt that the margineers' negotiating position was being advanced at the expense of the International Seabed Area (the Area). An earlier United Nations General Assembly resolution had declared the Area as common heritage of mankind, a unique legal status for the global commons and not susceptible to appropriation by any State.[6] In the event that a natural resource deposit straddles the extended continental shelf and the Area, further issues arise in the implementation of Article 82.[7]

Although not a beneficiary of Article 82, the International Seabed Authority (the Authority), a new intergovernmental organization established by the LOSC, is a stakeholder of the provision because it is tasked with receiving payments and contributions in kind, developing equitable criteria for distribution to beneficiaries, and the administration of the funds. The Authority cannot perform its distributive tasks unless extended continental shelf States perform their respective responsibilities. Moreover, as the organization with the exclusive mandate for the administration of deep sea mining activities in the Area, the Authority could conceivably be engaged in dialogue with the coastal State over resource production and management of non-living resources that straddle the outer limits of the continental shelf and the Area.

Although the deep water market for offshore hydrocarbons is depressed by prolonged low oil prices, investments are still substantial and driven predominantly by expenditures in Africa and the Americas.[8] There are relatively few activities in the deep waters of extended continental shelves and these include Angola, Brazil, eastern Canada, northern Norway and the United States' Gulf of Mexico. Most activities are prospecting or exploratory in nature. The United States is not a State Party to the LOSC and to date has not issued discovery licenses on the extended continental shelf. However, it was the first State to implement Article 82 in the calls to tender in the Central Gulf of Mexico, by

6 Declaration on Principles Governing the Sea-Bed and the Ocean Floor, and the Subsoil Thereof, beyond the Limits of National Jurisdiction, UN General Assembly Resolution 2749 (XXV), 17 December 1970, available at < http://www.un-documents.net/a25r2749.htm>. The principle was subsequently codified in LOSC, n 1, Art 137.

7 A Chircop, 'Managing Adjacency: Some Legal Aspects of the Relationship between the Extended Continental Shelf and the International Seabed Area,' (2011) 42 *Ocean Development and International Law* 307 at 312–313; Chatham House Report, n 4, 59–63.

8 For the current state of the deep water industry and forecasts, see: Douglas-Westwood, *World Deepwater Market Forecast 2016–2020* (Kent UK 2016).

putting bidders on notice that a contingent Law of the Sea Royalty would apply in the event that the United States became a Party to the LOSC.[9] The position taken by the United States is that bidders require notice of the existence and contingent application of the Article 82 royalty at the stage of tendering.[10]

Of more substantial note at this time is Canada's practice. Canada stands out as a pioneer State because it is the first LOSC Party to issue discovery licences, which it has done with respect to developments in the deep waters off the coast of Newfoundland in the Northwest Atlantic Ocean. If economic and technical feasibility are confirmed, the licence holders will be expected to apply for production licences. The prospective issuance of the first production license will most likely occur before Canada will have defined the outer limits of its continental shelf in the Northwest Atlantic. Canada ratified the LOSC on 6 November 2003 and on 6 December 2013 made a partial submission to the Commission on the Limits of the Continental Shelf (CLCS) in accordance with Article 76 to define the outer limits of the extended continental shelf in the Northwest Atlantic.[11] Under the LOSC, the Commission reviews such submissions and issues recommendations concerning the definition of the outer limit of the shelf. If Canada receives a positive recommendation, the submission will add 1.2 million square kilometres of seabed areas to national jurisdiction. Canada issued the first significant discovery licence to Norwegian company Equinor (formerly Statoil until 2018) for the Mizzen field on the Flemish Cap,

9 For example Stipulation No. 3: Law of the Sea Convention Royalty Payment in Lease Stipulations, Western Planning Area, Oil and Gas Lease Sale 233, Proposed Notice of Sale (NOS), BOEM, August 2013, available at <http://www.boem.gov/Oil-and-Gas-Energy-Program/Leasing/Regional-Leasing/Gulf-of-Mexico-Region/Lease-Sales/233/pstips233.aspx>.

10 'If the United States becomes a party to the 1982 Law of the Sea Convention (Convention) prior to or during the life of a lease issued by the United States on a block or portion of a block located beyond the U.S. Exclusive Economic Zone (EEZ) and subject to such conditions that the Senate may impose through its constitutional role of advice and consent, then the following royalty payment lease provisions will apply to the lease so issued, consistent with Article 82 of the Convention …' Ibid. The United States' position on the LOSC is that the bulk of it represents customary international law. However, the characterization of Article 82 as having a contingent application, i.e., if and when the US becomes a party, suggests that this particular provision might not be considered a customary rule.

11 Government of Canada, Partial Submission of Canada to the Commission on the Limits of the Continental Shelf regarding its Continental Shelf in the Atlantic Ocean: Executive Summary (Government of Canada, 2013), available at <http://www.un.org/depts/los/clcs_new/submissions_files/can70_13/es_can_en.pdf>. The Convention entered into force for Canada on 6 December 2013, 30 days after the deposit of its instrument of ratification in accordance with LOSC, n 1, Art 308(2). With this partial submission Canada managed to meet the ten-year deadline for extended continental shelf submissions in accordance with Annex II, Art (4) on the last day.

located well beyond 200 nautical miles off Newfoundland.[12] In the meantime, offshore operators have been notified that the boundaries and outer limits of licences may be reviewed after Canada establishes the outer limits of the continental shelf following recommendations from the CLCS.[13]

This chapter discusses the issues of implementation of Article 82 for coastal States with a special focus on Canada against the backdrop of initiatives of the Authority aimed at clarifying its role and being of assistance to LOSC State Parties. The chapter starts by discussing the Authority's initial steps to date in facilitating the implementation of Article 82 (Section 2), followed by brief analysis of textual ambiguities and uncertainties and possible interpretations in the first two paragraphs of the provision, as these constitute the core legal obligation to be implemented by Canada (Section 3). The chapter does not discuss exemptions for developing States in the third paragraph and the Authority's tasks in the fourth paragraph of Article 82. The chapter also considers issues that might arise in a hypothetical situation of production from a non-living resource straddling the shelf and the Area (Section 4). Thereafter the discussion moves to the domestic issues confronting the legal, political and economic mechanics of implementation of the provision in Canada (Section

12 'Statoil confirms up to 200M barrels at Mizzen prospect', CBC News, 20 June 2012, available at <http://www.cbc.ca/news/canada/newfoundland-labrador/statoil-confirms-up-to-200m–barrels-at-mizzen-prospect-1.1251285>. In addition to the Mizzen discovery, first made in 2009, Equinor has four other discoveries in the Flemish Pass off Newfoundland: Harpoon, A light hydrocarbon discovery in 2013; Bay du Nord, a 2013 discovery estimated to hold around 300 million barrels of light high-quality oil; Bay de Verde, a light high-quality oil discovery in 2016; and Baccalieu, a light high-quality oil discovery in 2016. See Equinor (Canada), available at <https://www.equinor.com/en/where-we-are/canada.html>.

13 The following caveat was reproduced by the licensing authority for areas off Newfoundland and Labrador: 'Any sector, parcel or licence depicted on the following map, (found on pages 24 and 25), beyond 200 nautical miles off the coast of Newfoundland and Labrador is not represented by the Board to reflect the full extent of Canada's continental shelf beyond 200 nautical miles. Canada has filed a submission regarding the limits of the Outer Continental Shelf in the Atlantic Ocean with the Commission on the Limits of the Continental Shelf, the review of which is pending. Any call for bids based on a sector or parcel identified in this map and any licences issued in those areas will be subject to approval as a Fundamental Decision under applicable legislation. The boundaries of sectors, parcels or licences in areas beyond 200 nautical miles may be revised to reflect the limits of the Outer Continental Shelf established by Canada. All interest holders of production licences containing areas beyond 200 nautical miles may be required, through legislation, regulation, licence terms and conditions, or otherwise, to make payments or contributions in order for Canada to satisfy obligations under Article 82 of the United Nations Convention on the Law of the Sea.' C-NLOPB 2014–2015 Annual Report, 23, available at <http://www.cnlopb.ca/pdfs/ar2015e.pdf?lbisphpreq=1>.

5). The chapter concludes with observations on the precedent-setting challenges facing Canada, as the pioneer implementing State, and the Authority in operationalizing Article 82 (Section 6).

2 The Road to Interpretation

The short and simple text of Article 82 belies the complexity of implementation. The Authority was the first to take initial steps encouraging States to consider the implementation needs of Article 82. The Authority was well-placed to take this initiative. The Council and Assembly of the Authority are empowered by the LOSC to adopt

> … rules, regulations and procedures on the equitable sharing of … the payments and contributions made pursuant to article 82, taking into particular consideration the interests and needs of the developing States and peoples who have not attained full independence or other self-governing status …[14]

The regulatory power of the Council and Assembly is limited, and while the Authority is in a position to take steps to commence implementation of its own responsibilities in Article 82, it does not have express power to regulate the discharge of the obligation by coastal States. However, the respective obligations of the coastal State to make payments or contributions in kind and the Authority's responsibility to make arrangements to receive them are intertwined, as will be discussed below.

Using this power, in 2008 the Authority commissioned an expert study which was tabled for discussion in 2009 at a preliminary meeting of experts in Chatham House. The meeting was convened to explore and scope out legal issues of implementation and the role and tasks of the Authority.[15] The meeting

14 The Council is empowered to make recommendations to the Authority to adopt rules, regulations and procedures, LOSC, n 1, Art 162(2)(o)(i). In turn, the Assembly is empowered to consider and approve them and if it does not it remands them to the Council for reconsideration in the light of the views expressed by the Assembly. Ibid, Art 160(2)(f)(i).

15 Draft Paper for Discussion at the Chatham House/International Seabed Authority Seminar on Issues Associated with the Implementation of Article 82 of the Law of the Sea Convention, Chatham House, London, 11–13 February 2009, Fourth Draft prepared by Aldo Chircop.

resulted in a ground-breaking report.[16] The discourse was guided by principles and rules of the Vienna Convention on the Law of Treaties, 1969,[17] the context and purposes of the LOSC, and the understanding on the package deal reached at UNCLOS III, in particular the relationship between Articles 76 and 82, the relationship between the extended continental shelf and the Area,[18] and the powers of the Authority stemming from Articles 82, 160 and 162. The discourse was also informed by other provisions of the LOSC, including the duty of States Parties to 'fulfil in good faith the obligations assumed under this Convention' and to 'exercise the rights, jurisdiction and freedoms recognized in this Convention in a manner which would not constitute an abuse of right.'[19] The meeting report identified several issues and unanswered questions that coastal States and the Authority would need to address, including clarification of terminology, the scope of the Authority's tasks and the relationship between those coastal States and the Authority in the administration of Article 82. The report led the Assembly of the Authority to place the study of Article 82 on the formal programme of work.[20] This led to the convening of a larger follow-up workshop of experts in Beijing in 2012, jointly convened by the Authority and the China Institute for Marine Affairs of the State Oceanic Administration of China, and resulted in a second report.[21] The workshop had a practical orientation and purposed to produce draft proposals for consideration by coastal States and the Authority.[22] Its conclusions and recommendations setting out

16 Chatham House Report, n 4. This was accompanied by a second study *Non-Living Resources of the Continental Shelf beyond 200 Nautical Miles: Speculations on the Implementation of Article 82 of the United Nations Convention on the Law of the Sea*, ISA Technical Study no. 5 (International Seabed Authority, 2010).

17 Convention on the Law of Treaties (adopted 23 May 1969; in force 27 January 1980), UN Doc. A/ Conf.39/27; 1155 UNTS 331 [hereinafter Vienna Convention].

18 Chatham House report, n 4, 12–13, 21–24.

19 LOSC, n 1, Art 300.

20 Report of the Secretary-General in *International Seabed Authority*, Selected Decisions and Documents of the Sixteenth Session (26 April-7 May 2010), (Kingston, 2010), para 75, available at <http://www.isa.org.jm/en/documents/publications>.

21 *Implementation of Article 82 of the United Nations Convention on the Law of the Sea*, Report of an International Workshop convened by the International Seabed Authority in collaboration with the China Institute for Marine Affairs in Beijing, the People's Republic of China, 26–30 November 2012, ISA Technical Study no.12, at 24, available at <http://www.isa.org.jm/files/documents/EN/Pubs/TS12-web.pdf> [hereinafter Beijing Workshop Report].

22 Report of the Secretary-General of the International Seabed Authority under article 166, paragraph 4, of the United Nations Convention on the Law of the Sea, ISA Doc ISBA/19/A/2, 22 May 2013, 14–15.

the roadmap for the next steps in implementation were submitted for consideration by the Assembly.[23]

The Beijing workshop produced several clarifications and recommendations pertinent to the obligation of coastal States. First, coastal States that have issued or plan to issue offshore licenses for non-living resources on the extended continental shelf were encouraged to consider and anticipate the implementation needs of Article 82.[24] The report suggested that coastal States had an obligation to notify the Authority 'of the anticipated date of commercial production' which, while not an express requirement in Article 82, is a reasonable implied requirement.[25] It should be recalled that the coastal State has a duty to deposit charts or lists of geographical coordinates of the outer limits of the extended continental shelf with the Secretary-General of the Authority.[26] Also, the Authority needs to be in a position to receive payments and contributions in kind from the coastal State.

Second, the report noted that coastal States have the exclusive right to determine whether the obligation is to be performed through payments, i.e., monetary transfers, or contributions in kind, presumably through a portion of the resource corresponding to the required percentage. However, given the complexities of managing receipts of contributions in kind first, a problem first identified at the Chatham House meeting, the experts recommended that States be encouraged 'to make payments in the interest of simplicity and efficiency of implementation.'[27] It was noted that such a recommendation to States might have to be made by State Parties to the Convention.

Third, the uncertainly of the terminology of Article 82 led to the recommendation for a more detailed study of key terms as 'they are used in contemporary regulatory and industry practices across different jurisdictions.'[28] This is important because Article 82 was conceived as a royalty, a type of revenue generation

23 Outcomes of the International Workshop on further Consideration of the Implementation of Article 82 of the United Nations Convention on the Law of the Sea, ISA Doc ISBA/19/A/4, 6 May 2013.

24 Beijing Workshop Report, n 21, at 24: 'The ISA should encourage OCS States, in particular those that are issuing or plan to issue offshore licences for the exploitation of the non-living resources of the outer continental shelf, to consider and anticipate the implementation needs of Article 82 within their respective jurisdictions.'

25 Chatham House Report, n 4, at 51; Beijing Workshop Report, n 21, at 22–23.

26 LOSC, n 1, Art 84(2).

27 Beijing Workshop Report, n 21, at 24: 'OCS States, while enjoying the exclusive choice to make payments or contributions in kind, should be encouraged to opt only for payments in the interests of simplicity and efficiency of implementation. It is conceivable that a SPLOS resolution may be needed to move this recommendation forward.'

28 Ibid, 24.

instrument applied to production from natural resources, long discarded in some jurisdictions, but retained by others like Canada, as will be seen below.

Fourth, given that coastal States making payments or contributions will be engaged in a long-term relationship with the Authority, the experts recommended a voluntary memorandum of understanding to help structure the relationship between the two, or a guidance document to provide practical advice in the implementation of Article 82.[29] Experts preferred an informal and non-binding instrument to a formal instrument, such as an agreement between a State and an international organization.

Fifth, it was underscored that the term 'through the Authority' clearly did not mean 'to the Authority', since the function performed by the Authority is that of a receiver rather than a collector of the royalty. Clearly, the Authority needs to have a mechanism for receiving royalties and for distributing them to beneficiaries designated by the Convention. There would be issues of costs of administration of the funds that would need to be considered. Further studies and technical reports would need to be produced by the Authority. Consideration of the report at the 19th Session in 2013 anticipated continued discussions within the Authority 'to achieve a system for the pragmatic and functional implementation of article 82.'[30]

The development of regulations for deep seabed mining in the Area occupied much of the Authority's agenda in recent years and Article 82 was largely placed on a backburner. Some Member States felt that the Authority should give priority to its work regarding activities in the Area. However, the annual reports of the Secretary-General of the Authority consistently reminded Member States of the organization's responsibilities in Article 82. At the 21st Session of the Assembly in 2015, the Beijing meeting of experts was recalled, in particular the recommendation that a study of terminological issues be undertaken and that the Secretariat hoped to progress that work in 2015, but with 'respect to the implementation of article 82, paragraph 4, of the Convention',

29 Ibid, 24: 'The ISA should explore further the concept of a Memorandum of Understanding between an OCS State and the ISA, or a guidance document, and take steps to prepare a draft for discussion, bearing in mind that such instruments will be essentially voluntary and aim to provide practical guidelines and advice to assist OCS States in the implementation of Article 82. The content should reflect terminological matters, functions and tasks, and other appropriate implementation matters discussed in this report. It could be undertaken in three sections, namely: (a) practical and administrative arrangements; (b) provisions regarding contributions in kind; and (c) considerations for OCS States to take into account.'

30 Press Release, Nineteenth Session Kingston, Jamaica 15–26 July 2013, SB/19/18, 26 July 2013, 10.

i.e., the Authority's task in receiving and distributing benefits.[31] This suggests that the Authority was under pressure to limit its work specifically to its own responsibilities in the provision and less so on those of coastal States. However, in 2015 the Authority proceeded with the recommendation of the Beijing Workshop to conduct an expert study of the terminology of Article 82 as used in the oil and gas industry on a comparative basis in order to help broaden coastal States' understanding of the issues. The study was completed in 2016.[32] More recently, the Strategic Plan of the International Seabed Authority for the Period 2019–2023 adopted at the 24th Session of the Assembly includes the development of equitable sharing criteria for Article 82.[33] The ISA's Finance Committee observed that the development of the criteria should be undertaken in parallel with the development of equitable sharing of financial and other economic benefits from activities in the Area to avoid duplication of work.[34]

3 Key Terminological Issues of Concern to Coastal States

The ambiguities and gaps in Article 82 posing major challenges to implementation have been studied in the literature and addressed in the Authority's technical reports and therefore need not be recounted in detail. Some issues are purely terminological, whereas others relate to the *modus operandi* of implementation and are of potential concern for Canada.

3.1 *Paragraph (1)*

The paragraph (1) core rule concerning the duty of the coastal State to make payments or contributions in kind on the production from non-living resources is clear. That the choice of discharge procedure is the prerogative of the coastal

31 Report of the Secretary-General of the International Seabed Authority under article 166, paragraph 4, of the United Nations Convention on the Law of the Sea, ISA Doc ISBA/21/A/2, 3 June 2015, 3.

32 *A Study of Key Terms in Article 82 of the United Nations Convention on the Law of the Sea*, ISA Technical Study No. 15, authored by W Spicer/E McIsaac, (Kingston, Jamaica: International Seabed Authority, 2016). The study focused on the use of key terms in the following jurisdictions, according to whether their revenue generation systems are based on royalty or taxation regimes: (a) royalty regimes: Australia; Brazil; Canada (Alberta; Newfoundland and Labrador); Nigeria; and United States; taxation regimes: Norway; Russian Federation; and United Kingdom [hereinafter Key Terms Report].

33 Decision of the Assembly of the International Seabed Authority relating to the strategic plan of the Authority for the period 2019–2023, ISBA/24/A/10 (27 July 2018), para 21.

34 Report of the Finance Committee, ISBA/24/A/6 – ISBA/24/C/19 (13 July 2018), para 27.

State is also clear. The term 'contribution in kind' could give rise to alternative meanings. The question that arises is whether this term is meant to refer to a share of the resources or potentially, and more broadly to other contributions of equivalent value, such as technical assistance, technology transfer, and so on.[35] A recent paper revived an earlier idea by this author that a contribution in kind could be interpreted to include national official development assistance.[36] The deliberations at UNCLOS III suggest that negotiators may have had in mind a share of the resource, because at the time developing States wanted to ensure they had access to oil production. The Beijing Workshop appeared to lean towards an interpretation in favour of a share of the resource.[37]

Assuming an actual share of the natural resource is the correct legal interpretation of contribution in kind, a related legal, administrative and economic issue would arise, namely where and when does the Authority take delivery of the produced resource and who is responsible for the delivery and costs (e.g., transportation, brokerage, insurance, etc.)? The Authority is not equipped with the tools and capacities necessary to handle these administrative tasks. It is not mandated to raise the funds needed for this purpose and nor is it authorised to levy an overhead charge or fee.[38] The obligation of coastal States does not expressly extend to anything more than making the contribution in kind. In any case, if coastal States act on the advice of the Chatham House and Beijing Workshop reports that payments are easier, more efficient and less costly to administer, then the vagueness and complexities of in kind contributions would be avoided altogether.

3.2 *Paragraph* (2)

The major difficulties for coastal State implementation are mostly with regard to paragraph 2. The general rule in paragraph 2 is that payments

35 Chatham House Report, n 4, 51–52.

36 W Spicer, 'Canada, the Law of the Sea Treaty and International Payments: Where Will the Money Come From?' SPP Research Papers vol 8(31) (University of Calgary, 2015), 20.

37 Beijing Workshop Report, n 21, at 20: 'When negotiated at the Third United Nations Conference on the Law of the Sea, 1973–1982 (UNCLOS III), the intention behind insertion of contributions in kind was to secure resource access to State Party beneficiaries.' Also ibid at 22: 'The difficulties to be encountered when an OCS State opts to make 'contributions in kind' have already been addressed in this report. In addition to those difficulties, an understanding of what is acceptable as a contribution in kind within the letter and spirit of the provision is necessary. Discussions in the Working Group proceeded on the assumption that this phrase refers to a share of the resource, but other possible interpretations were not discussed.'

38 Chatham House Report, n 4, 37–38; Beijing Workshop Report, n 21, at 23 in relation to contributions in kind, and suggesting a further study.

and contributions are to be made annually, subject to a grace period, i.e., a royalty-free period during the first five years of production, to enable recovery of expenses incurred in bringing the resource to production. On the sixth year of production, a royalty of one percent applies and will increase by one percent annually until seven percent is reached on the thirteenth year, and which will remain the royalty's ceiling thereafter. The principles behind the royalty are clear.

Clarity diminishes on a closer look at the details. For example, can the grace period be interrupted if production is suspended and reactivated on resumption of production?[39] Presumably yes, as a matter of fairness, but this is a matter of interpretation of text which is not directly on point. What is the meaning of 'annually'? Financial years vary by jurisdiction.[40] It could refer to the financial year in the jurisdiction concerned or to the anniversary date of first production, and if production is suspended the calculation of the year may have to vary accordingly.

A difficult term is 'all production'. During negotiations this term appears to have meant gross production, less any portion of the resource used to enhance production according to Article 82(2) (e.g., use of the resource for production of energy for operations, re-injection of gas into a well to enhance production, flaring for safety purposes, etc.).[41] The meaning of all production as gross production was intentional because this approach simplified the royalty and was a common practice in those jurisdictions deriving royalty-based revenues in the oil and gas industry. The alternative interpretation of calculating the royalty on the basis of net production would have engaged a complex discussion of accounting practices that varied significantly among jurisdictions.[42] In the contemporary context of deep water drilling and production, where costs are significantly higher than offshore production elsewhere, an additional levy to the applicable royalty or tax regime under national law could impact on the feasibility of the entire operation. However, this is a domestic concern and it does not excuse or lessen the international legal obligation.[43]

39 Chatham House Report, n 4, 56–57.

40 Ibid, 32.

41 Chircop & Marchand, n 4, 297–298; International Law Association, Committee on the Outer Continental Shelf, 'Report on Article 82 of the UN Convention on the Law of the Sea (UNCLOS),' Rio De Janeiro Conference, 2008, para 2.8, available at <http://www.ila-hq.org/en/committees/index.cfm/cid/33>.

42 MH Nordquist and C-H Park eds, *Reports of the United States Delegation to the Third United Nations Conference on the Law of the Sea* (Honolulu: Law of the Sea Institute, 1983), 98–99.

43 Vienna Convention, n 17, Art 27: 'A party may not invoke the provisions of its internal law as justification for its failure to perform a treaty.'

The next set of issues in the second paragraph concern the meaning of 'value or volume of production at the site.' Is the value expected to be determined at the well-head or at the market location at the time the product is sold?[44] There are significant differences between oil and gas because they are transported and priced differently. While oil may be priced at the well-head, gas tends to be priced at the point of distribution, which may be at the terminal point of a pipeline. The reference to 'site' adds another layer of ambiguity as it could be interpreted to mean 'resource field, geological structure, well site, license area and a whole development area subject to multiple licenses'.[45] Some deposits may be layered over each other in the same field.

Negotiated through four successive versions of early drafts of the future LOSC,[46] Article 82 was drafted on the basis of information, assumptions and an understanding of the petroleum industry and government revenue generation practices of the 1970s. A key assumption was that the first five years constituted the period for the recovery of developmental costs.[47] Today, deep water drilling (on extended continental shelves as compared to shallow inshore waters) and cost models are riskier and fundamentally different, such that the period of cost recovery may be significantly longer than the Article 82 grace period.[48] Article 82 follows a royalty model.[49] The concept of royalty meant essentially a share of the produced resource to reflect a rent payable to the owner of the resource. Today government revenue generation in this industry also includes a range of fiscal tools such as corporate income tax, profit taxes, resource rent taxes, value-added taxes, excise taxes, etc.[50] Revenue generation practices have

---

44 Chatham House Report, n 4; 33–34; Beijing Workshop Report, n 21, 20–21.

45 Beijing Workshop Report, n 21, 21–22. The meaning of site was also discussed at the Chatham House Workshop. Chatham House Report, n 4, 61–63.

46 The earlier iterations of Article 82 were contained in the following drafts: ISNT, 1975, Art 69; RSNT, 1976, Art 70; ICNT, 1977, Art 82. The final text remained largely unchanged in the following drafts: ICNT Rev 1, 1977; ICNT Rev 2, 1980; Draft Convention (Informal Text), 1980; and the Draft Convention, 1981, which was the last version of the text before the LOSC was adopted in 1982. Chatham House Report, n 4, 19–20.

47 Nordquist & Park, n 42.

48 'Deep water drilling and production involves long-term, multi-billion dollar projects that take several years to complete and are less impacted by short-term fluctuations in oil prices. Offshore operators often have major project budgets for years and most projects are completed with the anticipation of higher oil prices in the future.' US Energy Information Administration, Trends in U.S. Oil and Natural Gas Upstream Costs, March 2016, at 23 <https://www.eia.gov/analysis/studies/drilling/pdf/upstream.pdf>.

49 Chircop & Marchand, n 4.

50 EM Sunley, T Baunsgaard and D Simard, 'Revenue from the Oil and Gas Sector: Issues and Country Experience,' (World Bank, 2002), 2, available at <http://siteresources.worldbank.org/INTTPA/Resources/SunleyPaper.pdf>; J Mintz and D Chen, Capturing Economic

made fundamental shifts, with some jurisdictions moving away from royalty-based levies in favour of taxation-based regimes, such as major petroleum producing States Norway and the United Kingdom.[51] Revenue is thus generated in more complex ways than, or instead of, royalties. Others, like Canada and the United States, have retained royalties side-by-side other fiscal tools.[52] Naturally, Article 82 is a levy on the production of the resource and not on the entirety of the revenue generated by the operation. This is important to note as the producer may significantly enhance the original product to produce by-products. Even so, for those jurisdictions where royalties are no longer levied, there is concern regarding how the payments to be made through the Authority are to be generated. Thankfully, the latter is not a Canadian problem, but Canada has other, perhaps more complex domestic issues to navigate.

4 Production from Resources Straddling the Outer Limits

The LOSC does not require a coastal State to refrain from activities on the extended continental shelf until the process of definition of the outer limits pursuant to Article 76 has been completed. The precise limits of the international seabed area will remain uncertain until outer limits of extended continental shelves are fully defined in accordance with the requirements of Article 76. The LOSC requires coastal States to give due publicity to charts or lists of geographical coordinates of the outer limit lines of the continental shelf and to deposit a copy of such charts or lists with the Authority's Secretary-General.[53] At the time of writing, there appears to be a low rate of compliance by coastal States that have received recommendations from the Commission regarding their submissions on the outer limits. By 2015 only Australia, Ireland, Mexico, Niue and Philippines had deposited charts or lists with the Secretary-General.[54]

Rents from Resources through Royalties and Taxes, SPP Research Papers vol 5 (University of Calgary, 2012), 5.

51 Key Terms Report, n 32.

52 Ibid.

53 LOSC, n 1, Art 84(2). This duty is in addition to the Art 76(9) duty of a State to deposit charts, lists and other relevant information regarding the outer limits of the continental shelf with the UN Secretary-General.

54 Report of the Secretary-General of the International Seabed Authority under article 166, paragraph 4, of the United Nations Convention on the Law of the Sea, ISA Doc ISBA/21/A/2, 3 June 2015, 2–3.

The Secretary-General renewed calls for States to submit charts or lists of coordinates in his most recent annual report.[55]

The definition of the outer limits in a final and binding manner, on the basis of recommendations from the Commission, has the benefit of legitimizing the boundary between the extended continental shelf and the Area. This is important for the offshore oil and gas industry because it provides greater certainty regarding the legality of their licenses. The occurrence of single geological structures straddling international maritime boundaries is a realistic possibility, as demonstrated by the several instances of offshore joint development and unitization activities around the world.[56] Where a deposit is known to straddle a licence area or maritime boundary, a longstanding good oilfield practice is for the development of the resource to be unitized, that is developed in an integrated manner.[57] Licence holders would then come to agreement on their respective shares of production and a model agreement will be developed for this purpose.[58]

If such a situation were to arise, for example where a commercial hydrocarbon find straddles areas within and beyond national jurisdiction, there could be three issues to consider. First, there would need to be consideration of whether the resource should be developed on the basis of a joint development zone or, alternatively, by being unitized through the direct involvement of the licence holders. In either case, there would need to be approval by the Authority insofar as licence holders in the Area are concerned. The Authority would likely need to develop regulations for this purpose, as it has for other mineral resources. The Authority has an international governance structure for licencing activities in the Area governed by Part XI of LOSC, and is overseen by Member States, who are also State Parties and include the coastal State concerned. In turn, the coastal State could have sub-national units it would need to involve in negotiations. Canada, a federal State, is a good example of

55 Report of the Secretary-General of the International Seabed Authority under article 166, paragraph 4, of the United Nations Convention on the Law of the Sea, ISBA/24/A/2 (29 May 2018), para 8.

56 J Lang Weaver and DF Asmus, 'Unitizing Oil and Gas Fields around the World: A Comparative Analysis of National Laws and Private Contracts,' (2006) 28 *Houston Journal of International Law* 4. See also DM Ong, 'Joint Development of Common Offshore Oil and Gas Deposits: "Mere" State Practice or Customary International Law?', 93 *American Journal of International Law* 771.

57 JC Woodliffe, 'International Unitization of an Offshore Gas Field,' (1977) 26 *International and Comparative Law Quarterly* 338.

58 Lang Weaver & Asmus, n 52, 22 et seq. For the model agreement, see Association of International Petroleum Negotiators, available at <www.aipn.org/modelagreements/>.

this because the regulation of the offshore is conducted by federal-provincial boards.[59] A consequence of a joint development zone or unitization scenario is the need for regulators on both sides of the boundary to revisit the licences in the areas adjacent to the boundary and to explore whether production policies and regulatory approaches require adjustment. Moreover, environment protection standards may differ and would need to be reconciled for the operation in question.

Second, the payments or contributions in kind to be made by the coastal State would logically apply to only a portion of the produced resource. If the resource is unitized, the production units will be located in a manner to optimize production and it is conceivable that most of these might be placed on one side of the boundary, rather than be evenly distributed on the shelf and in the Area. The consequence is that the meaning of 'site' might not be clear where most, if not all of the production units are placed in the Area, for the purpose of determining the Article 82 eligible production.

Third, Article 142 of LOSC safeguards the rights of the coastal State by providing that '[A]ctivities in the Area, with respect to resource deposits in the Area which lie across the limits of national jurisdiction, shall be conducted with due regard to the rights and legitimate interests of any coastal State across whose jurisdiction such deposits lie.'[60] The duty in the Convention is not reciprocal (and the Authority does not enjoy a counterpart right of protection), but this author is of the view that there might be a duty concerning transboundary resources at customary international law which could apply to the unilateral actions of the coastal State, such as the duty to inform, consult and invite to cooperate.[61] Accordingly, although the Authority would be expected to consult the coastal State with regard to activities in the Area, an analogous duty could be expected for the coastal State on the basis of general international law. From the coastal State's perspective, the type of activity in the Area adjacent to the boundary that could give rise to concern is where operations pose a threat to the marine environment and resources of the EEZ or possibly even to the sedentary species of the continental shelf. From the Authority's perspective, perhaps the major concern is where the coastal State issues licences for unilateral production from a transboundary resource.

59 Spicer, n 36.

60 LOSC, n 1, Art 142(1).

61 Suriname/Guyana Arbitration, Permanent Court of Arbitration (Award), 17 September 2007, available at <www.pca-cpa.org/ showpage.asp?pag_id=1147>.

5 Canada: First Past the Post?

Although Canada ratified the LOSC in 2003, it was unprepared for the implementation of Article 82. This is ironic because during UNCLOS III Canada played an active role in the negotiation of the provision, both as a margineer State seeking national gain while at the same time exploring options for a potential compromise. In 1975, Secretary of State for External Affairs Allan J MacEachen stated Canada's commitment to revenue-sharing.[62] This position was pursued consistently by the Canadian Delegation led by Ambassador Alan Beesley, which joined ranks with the United States' and other delegations in exploring options for a compromise through the establishment of a revenue-sharing scheme primarily for the benefit of developing countries, in exchange for agreement on Article 76.[63] Negotiations were difficult and Canada, like other delegations, had to compromise. In 1979 Ambassador Beesley underscored that revenue-sharing must be respectful of the coastal State's sovereign rights over the resources of the continental shelf and must not be set at a level which is unreasonable, considering the high costs of deep water activities.[64] In 1980, when compromise was not yet fully reached in the late hours of UNCLOS III, Ambassador Beesley reiterated that position and was critical of the raising of the royalty ceiling from five percent to seven percent. He therefore reserved his delegation's position '... not out of lack of generosity but because the suggested rate could make it uneconomic for Canada to explore and exploit its continental margin in deep, cold water areas unless some safeguard provision was developed to ensure that any revenue-sharing we could agree upon would be practicable.' The Canadian Delegation's position was that Canada '... was prepared to explore the possibility of financial contributions related to the net revenues derived from the resources of the continental shelf beyond 200 miles from shore.'[65] However, '... the question of revenue-sharing will require further discussion with a view to ensuring that the formula and the rate of contribution

62 Secretary of State MacEachen stated that '... we are prepared to explore the possibility of financial contributions related to the net revenues derived from the resources of the continental shelf between 200 miles from the shore and the seaward edge of the continental margin. We are prepared to explore that possibility and we are prepared to support that principle in order to promote an accommodation.' Statement by the Delegation of Canada, 2 April 1980, Doc A/CONF.62/WS/4, 10 April 1980, in Third United Nations Conference on the Law of the Sea: Official Records, 9th Session, 3 March-4 April 1980 (United Nations, 1981), 101 at 102.

63 Ibid.

64 Ibid, 102.

65 Ibid, Plenary, 125th meeting, 2 April 1980, at 8.

will be both equitable and viable from the standpoint of both potential contributors and beneficiaries …'[66] On the eve of the vote on the final package of the LOSC in 1982, Secretary of State MacEachen stated that '[W]e must recognize, however, that there will be funds to dispense only if these resources prove to be commercially exploitable.'[67] This was a difficult compromise for the margineers, because other delegations were of the view that the royalty rate was not proportionate to the margineers' continental margin gains.

The difficulties Canada faced in negotiating Article 82 were only the beginning of a series of implementation challenges at the domestic level. According to the Constitution of Canada, a province's boundaries may not be changed without its consent[68] and that all lands, resources and royalties in provincial territory at the time of Union belonged to the province.[69] With the growing interest in offshore development, the question arose as to whether the boundaries of coastal provinces included adjacent ocean space. Complicating this question was that the various provinces joined confederation at different dates, raising an additional issue with regard to their entitlements under international law at the time to determine what they imported into and when they joined the Union.[70] A case by cases analysis was essential. In 1967 the Supreme Court of Canada determined that the territorial sea, its seabed and subsoil off British Columbia (Vancouver Island) belonged to Canada, not the province.[71] In 1984 the waters in Juan de Fuca Strait and Strait of Georgia between Vancouver Island and the British Columbia mainland were determined to be

66 Ibid, 9th Session, at 102.

67 Third United Nations Conference on the Law of the Sea: Official Records, Resumed 11th Session, 185th meeting, 6 December 1982 (United Nations, 1982).

68 Constitution Act, 1867, 30 & 31 Victoria, c 3 (UK), s 43: 'An amendment to the Constitution of Canada in relation to any provision that applies to one or more, but not all, provinces, including (a) any alteration to boundaries between provinces, … may be made by proclamation issued by the Governor General under the Great Seal of Canada only where so authorized by resolutions of the Senate and House of Commons and of the legislative assembly of each province to which the amendment applies.'

69 Ibid, s 109: 'All Lands, Mines, Minerals and Royalties belonging to the several Provinces of Canada, Nova Scotia, and New Brunswick at the Union, and all Sums then due or payable for such Lands, Mines, Minerals, or Royalties, shall belong to the several Provinces of Ontario, Quebec, Nova Scotia, and New Brunswick in which the same are situate or arise, subject to any Trusts existing in respect thereof, and to any Interest other than that of the Province in the same.'

70 For example while Nova Scotia was a founding province of Confederation on 1 July 1867, British Columbia joined confederation on 20 July 1871 and Newfoundland on 31 March 1949. Newfoundland became Newfoundland and Labrador after a constitutional amendment in 2001.

71 Reference Re Ownership of Off Shore Mineral Rights [1967] SCR 792.

encompassed within the limits of British Columbia when it joined the Union in 1871 and were consequently provincial.[72]

Newfoundland (Newfoundland and Labrador since 2001) was the last province to join Canada, in 1949. This is significant because the province joined the Union shortly *after* the Truman Proclamation on the Continental Shelf of 1945. It will be recalled that the Proclamation triggered the development of pertinent general international law to the effect that 'the continental shelf may be regarded as an extension of the land-mass of the coastal nation and thus naturally appurtenant to it' and with the consequence that 'the natural resources of the subsoil and sea bed of the continental shelf beneath the high seas but contiguous to the coasts of the United States as appertaining to the United States, subject to its jurisdiction and control.'[73] In a few years this concept attained the status of customary international law so that the rights of the coastal State over the continental shelf belonged to that State *ipso facto, ipso iure* and *ab initio*, that they were exclusive to it and that its title did not depend on express claim, occupation or use.[74] Newfoundland was of the view that it joined Canada while possessing the territorial and other rights at international law appertaining to a State, and that these included the territorial sea and continental shelf. In 1983 a Newfoundland Court of Appeal decision found that the province had a territorial sea of three nautical miles when it joined the Union.[75] Continental shelf rights remained uncertain. A year later, the Supreme Court of Canada decided that Newfoundland had not acquired continental shelf rights prior to joining the Union in 1949.[76] The reasons were that continental shelf rights granted by international law to a coastal State are not in pith and substance proprietary, but are rather 'an extraterritorial manifestation of external sovereignty.'[77] In 1949 Newfoundland did not enjoy external sovereignty as this power and right were enjoyed by the Crown in right of the United Kingdom. The Supreme Court went on to hold that if Newfoundland

---

72 Reference Re Ownership of the Bed of the Strait of Georgia and Related Areas, [1984] 1 SCR 388.

73 Policy of the United States With Respect to the Natural Resources of the Subsoil and Sea Bed of the Continental Shelf, 150 – Proclamation 2667–28 September 1945.

74 North Sea Continental Shelf, Judgment, ICJ Reports 1969, 3 at para 19. For a commentary on how the continental shelf doctrine attained the status of customary international law, see ZJ Slouka, *International Custom and the Continental Shelf: A Study in the Dynamics of the Customary Rules of International Law* (Martinus Nijhoff, 1968).

75 Reference Re Mineral and other Natural Resources of the Continental Shelf (1983), 145 DLR (3d) 9 (Nfld CA).

76 Reference Re the Seabed and Subsoil of the Continental Shelf Offshore Newfoundland, [1984] 1 SCR 86.

77 Ibid.

had acquired continental shelf rights prior to joining Canada, the Terms of Union perpetuating natural resource rights in the Province at the date of Union did not extend to continental shelf rights. Moreover, the Court went on to hold that international law did not confer continental shelf rights in 1949 and that such rights 'were not indisputably recognized before the Geneva Convention of 1958,' and while coastal States enjoyed these rights retroactively, they 'do not have a retroactive effect capable of assisting Newfoundland.'[78]

The Supreme Court's decision did not sit well with the Province. While the legal basis of the dispute appeared to have been resolved by the Supreme Court, the issue of rights to the shelf's resources continued as a major political problem. During judicial proceedings, other provinces, including neighbouring Nova Scotia, adjacent to which is a large extended continental shelf, intervened to advocate in favour of provincial rights. The differences between Newfoundland and Labrador and the federal government were finally resolved in a political settlement that was mirrored in federal and provincial legislation in 1987 and 1990.[79] The federal government agreed that the regulation of offshore activities would occur through federal-provincial boards. Canada agreed that the royalties from offshore production would be levied by the federal government and equal amounts would be paid to the provinces, as if production occurred in the provinces.[80] The consequence of this legal fiction is that the Province of Newfoundland and Labrador legislated the levying of royalties and which it retains.[81]

Somehow Article 82 did not feature in the history of the federal-provincial disputes, with the consequence that the political and legal settlements omitted reference to the potential role of the LOSC provision. Several rounds of calls to bid were issued as a matter of course with no mention of the royalty applying to licence areas outside 200 nautical miles.[82] The 2012 Mizzen discovery on

78 Ibid.

79 Canada-Newfoundland and Labrador Atlantic Accord Implementation Act, SC 1987, c 3; Canada-Newfoundland and Labrador Atlantic Accord Implementation Act, RSNL 1990, c C-2. There was a similar settlement with Nova Scotia: Canada-Nova Scotia Offshore Petroleum Resources Accord Implementation Act, SC 1988, c 28; *Canada*-Nova Scotia Offshore Petroleum Resources Accord Implementation (Nova Scotia) Act, 1987 c 3.

80 The responsible federal government department is Natural Resources Canada. See Natural Resources Canada, *2014–2015 Departmental Performance Report* (Canada, 2015), 57, available at <http://www.nrcan.gc.ca/sites/www.nrcan.gc.ca/files/performancereports/files/dpr-2014-15/NRCan-DPR2014-15_e.pdf>.

81 Petroleum and Natural Gas Act, RSNL 1990, c P-10.

82 Early licences on the extended continental shelf are discussed by Chircop and Marchand, n 4, at 281.

the extended continental shelf off Newfoundland and Labrador was on the basis of a licence issued by the Canada-Newfoundland & Labrador Offshore Petroleum Board and without including any reference to the Article 82 royalty.

The significance of Article 82 in Canada was flagged early in the scholarly literature.[83] It was observed that Canada could face difficulties in the implementation of the provision, not only due to the uncertainty of the terminology in the provision, but also, and most especially, because of the political understanding reached between the federal and provincial governments with respect to the offshore royalty regime. The oil and gas industry enters into agreements for prospecting, exploration, development and production, in response to original calls to tender, and production agreements would be expected to subsist for decades.

The first federal official notice of the potential application of Article 82 with regard to licenses on the extended continental shelf in the Northwest Atlantic was made only in 2013,[84] likely as a result of the Beijing Workshop of experts and the recommendation to the Authority to call upon State Parties to consider taking steps for implementation.[85] While differences between the federal and Newfoundland and Labrador governments remained behind closed doors for much of the time until then, they are now out in the open, in dispute and with no solution announced, or in sight.[86] The federal position is that the Article 82 royalty should be levied from royalties on production, which at this time are enjoyed exclusively by the province.[87] Newfoundland and Labrador considers this as Canada's obligation.[88] Natural Resources Canada, the federal government organization primarily responsible with regard to royalty management, has made the issue an object of focused study. Its plans and priorities

83 A Chircop and B Marchand, 'Oceans Act: Uncharted Seas for Offshore Development in Atlantic Canada?' (2001) 24 *Dalhousie Law Journal* 23 at 30–32.

84 'The Board [Canada-Newfoundland & Labrador Offshore Petroleum Board] informs prospective bidders for these parcels, which are entirely or partially beyond Canada's 200 nautical mile zone, that it has been advised by the Government of Canada that, in order to meet obligations arising pursuant to article 82 of the *United Nations Convention on the Law of the Sea*, additional terms and conditions may be applied through legislation, regulations, amendments to licences or otherwise.' Call for Bids No. NL13–01 (Area 'C' – Flemish Pass), *The Newfoundland and Labrador Gazette*, vol 88(23), 7 June 2013, at 213.

85 Beijing Workshop Report, n 21, 24.

86 'Ottawa, N.L. unsure who will pay UN tax on new oil finds', CBC News, 28 October 2013, available at <http://www.cbc.ca/news/canada/newfoundland-labrador/ottawa-n-l-unsure-who-will-pay-un-tax-on-new-oil-finds-1.2251838>.

87 Ibid.

88 Ibid.

for 2015–2016 included 'developing a domestic mechanism to implement Article 82 of the United Nations Convention on the Law of the Sea.'[89] However, the current 2018–2019 departmental plan makes no mention of continuing work on Article 82.[90] Deliberations on a solution appear to be on hold, possibly because production on the extended continental shelf is not imminent.

The performance of Article 82 as a treaty obligation is clearly Canada's responsibility as a State in international law, and this is confirmed in the country's constitutional law.[91] As observed earlier, domestic law issues that impede implementation may not be invoked to avoid the international obligation.[92] Thus the onus to develop a solution lies with the federal government. The federal government is bound by a political accord and related legislation concluded three decades ago. The domestic solution would have to be more than a simple amendment of pertinent federal and provincial legislation. Canada may well have to explore a political solution first, followed by legislative steps, perhaps not dissimilar to the political settlement concluded in the wake of the Supreme Court of Canada Reference that rendered a decision unpalatable to Newfoundland. In that settlement the federal government was generous and relinquished royalty receipts to the province, thus enabling the offshore industry in the Northwest Atlantic to be launched. This renewed scenario means the federal government would bear the principal responsibility for Article 82. This route could provoke negative reactions from other provinces, in particular oil-producing provinces, because it could be perceived as a subsidy for provincial production of offshore oil and gas that is not similarly enjoyed by land-based producers.

If the federal government insists on clawing back domestic royalties currently levied by Newfoundland and Labrador or by introducing a surcharge royalty or levy on current operators, it is possible additional issues could arise. In addition to increasing uncertainty in the Atlantic offshore industry at a time when revenues are low because of low oil prices, there could be the realistic prospect of prolonged federal-provincial litigation. In respect to the latter,

89 Natural Resources Canada, Reports on Plans and Priorities, 2015–2016, Strategic Outcome 1, Sub-Program 1.1.3: Energy Market Regulation and Information, available at <http://www.nrcan.gc.ca/plans-performance-reports/rpp/2015-16/17006>.

90 Natural Resources Canada, 2018–19 Departmental Plan, available at <https://www.nrcan.gc.ca/sites/www.nrcan.gc.ca/files/plansperformancereports/dpr/2018-19-DP-NRCan-EN.pdf>.

91 Constitution Act, n 68, s 132.

92 Vienna Convention, n 17, Art 27.

operators would be faced with an additional expense that could weigh on the commercial viability of expensive operations in the deep waters of the extended continental shelf, essentially recalling Secretary of State McEachen's warning about the challenges of Article 82.

It is possible that in the larger scheme, the Article 82 levy, applied over and above existing royalties, might not be perceived as a major issue by the first industry movers on the Atlantic extended continental self. In the alternative, i.e., if it were to become an issue, a practical way forward could be to distinguish between pre-2013 and post-2013 calls to bid licensees. Tenderers to the latter call received notice of Article 82 and can be considered as forewarned,[93] as have subsequent responders to subsequent calls for bids, most recently in 2016.[94] They would be in a position to anticipate future production cost scenarios against the backdrop of domestic and Article 82 royalties. Pre-2013 calls to bid licensees could be addressed differently because they were not put on notice at the time they placed their bids. The federal government would either absorb the cost or enter into discussions with Newfoundland and Labrador to explore a cost-sharing arrangement. However, the latter is not likely politically and economically realistic because the economy of Newfoundland and Labrador has consistently had the highest unemployment levels in the country and its heavy reliance on single resource industries, initially fisheries and more recently oil, makes it especially vulnerable to external market and environmental factors.

6 Conclusion

Article 82 poses implementation challenges for both coastal States and the International Seabed Authority. This chapter has discussed the challenges faced by coastal States, with a particular focus on Canada. Although offshore exploration licences have been issued by a few States, Canada appears to be the first LOSC Party to issue significant discovery licences for areas on the extended continental shelf. While Canada has taken domestic steps to study how domestic implementation will affect the internal royalty arrangements, it will need to address the political and economic impacts in a fair and practical

93 Call for Bids No. NL13–01, n 84.

94 The most recent call repeated the notice in the 2013 call almost verbatim. See 'C-NLOPB Announces 2016 Calls for Bids,' 2 April 2016, available at <http://www.cnlopb.ca/news/nr20150331.php>.

manner, and to do so in the near future to ensure that there will be a smooth transition from discovery to production licences.

Being first entails a special responsibility for Canada. It places Canada in a leadership role and implies a responsibility to establish a good implementation precedent. The precedent has two dimensions. First, the construction of key terms in the first two paragraphs of Article 82 is a matter 'concerning the interpretation or application' of LOSC, and while the coastal State may develop a particular interpretation, it has to be reasonable and acceptable to other LOSC State Parties, in other words it reflects the *quid pro quo* of Articles 76 and 82. Second, the first State to implement Article 82 will need also to develop a good working relationship with the Authority, because that State and the Authority will be engaged in a long-term relationship.

PART 6

Building, Operating and Removing Installations on and under the Seabed

CHAPTER 17

The Use of Sub-Seabed Transboundary Geological Formations for the Disposal of Carbon Dioxide

Nigel Bankes

1 Introduction

This chapter examines the legal issues associated with the use of sub-seabed transboundary geological structures including saline aquifers for storage or disposal purposes. It focuses on carbon capture and storage (CCS), specifically the disposal element in the CCS chain.

The chapter proceeds as follows. Section 2 provides a brief discussion of CCS focusing on the disposal or storage part of the chain, as well as the injection of carbon dioxide (CO_2) as part of enhanced oil recovery (EOR) operations (CO_2\EOR). Section 3 deals with CCS and CO_2\EOR operations in transboundary formations, reviewing applicable international law requirements, such as the general duty to cooperate as well as the more explicit rules developed under the auspices of the London Dumping Convention (LDC)[1] and its Protocol (PLDC).[2] It also examines the implications of Articles 74(3) and 83(3) of UN Law of the Sea Convention (UNCLOS)[3] for the situations in which one party may be proposing to engage in CCS operations in an area of overlapping maritime entitlement claims in the absence of an agreed delimitation. Section 4 examines existing boundary delimitation agreements and joint development agreements in order to assess the extent to which they may apply to CCS and CO_2\EOR operations. Section 5 recognizes that straddling hydrocarbon deposits are frequently developed through unitization agreements between the

1 Convention on the Prevention of Marine Pollution by Dumping of Wastes and Other Matter, London, 13 November 1972.

2 Protocol to the Convention on the Prevention of Marine Pollution by Dumping of Wastes and Other Matter, London, 7 November 1996.

3 United Nations Convention on the Law of the Sea 1982, 1833 UNTS 3 and see generally British Institute of International and Comparative Law, *Report on the Obligations of States under Articles 74(3) and 83(3) of UNCLOS in respect of Undelimited Maritime Areas*, (2016) (hereafter BIICL report) available at <http://www.biicl.org/documents/1192_report_on_the_obligations_of_states_under_articles_743_and_833_of_unclos_in_respect_of_undelimited_maritime_areas.pdf?showdocument=1>.

DOI:10.1163/9789004391567_019

parties who hold the hydrocarbon exploitation rights from each state on either side of the maritime boundary. This section considers how this model might be applied to the situation where a shared formation is being used on one or both sides of the maritime boundary for CCS operations. Section 6 provides conclusions.

With its focus on the use of shared geological structures for disposal purposes, particularly for the disposal of CO_2 the chapter does not review the general laws dealing with seabed disposal except to the extent that they specifically touch on the use of shared geological structures of the seabed. These general laws include provisions of UNCLOS, the LDC, the PLDC and the Convention for the Protection of the Marine Environment of the North-East Atlantic (OSPAR)[4] and the Basel Convention.[5] Both the PLDC and OSPAR have been amended to take account of CO_2 disposal and both have developed regulatory guidance for CCS activities. There are a number of good accounts of how these instruments may apply (or not) to CCS activities in marine areas.[6] Individual states have also adopted rules for CCS projects and, in the case of European Union (EU) Member States, these rules must comply with the terms of Directive 2009/21/EC on the Geological Storage of Carbon Dioxide.[7] Article 24 of the Directive specifically addresses the issue of transboundary storage and transboundary transport[8] and provides that 'in such case the competent authorities of the

4 Convention for the Protection of the Marine Environment of the North-East Atlantic, Paris, 22 September 1992, in force 25 March 1998.

5 The Basel Convention on the Control of Transboundary Movements of Hazardous Waste and their Disposal, 22 March 1989, in force 1992.

6 Chiara Armeni, 'Legal Developments for Carbon Capture and Storage under International Regional Marine Legislation' in Ian Havercroft, Richard Macrory & Richard B Stewart (eds) *Carbon Capture and Storage: Emerging Legal and Regulatory Issues*, (Hart, Oxford, 2011) 145–159; Tim Dixon, Sean McCoy and Ian Havercroft, 'Legal and Regulatory Developments on CCS' (2015), 40 International Journal of Greenhouse Gas Control 431; UNFCCC Secretariat, Transboundary carbon capture and storage project activities, Technical Paper, 1 November 2012 UNFCCC/TP2012/9, (hereafter UNFCCC, Transboundary CCS Technical Paper) available at <http://unfccc.int/resource/docs/2012/tp/09.pdf>; Carbon Sequestration Leadership Forum, Technical Group, 'Final Report from Task Force on Technical Barriers and R & D Opportunities for Offshore, Sub-Seabed Geologic Storage of CO_2', CSLF-T-2015–06, 20 October 2015 (hereafter CSLF Offshore Geologic Storage) available at <http://www.cslforum.org/publications/documents/riyadh2015/tg_OffshoreSubSeabedStorageTaskForceFinalReport-Riyadh1115.pdf>

7 Directive 2009/31/EC of the European Parliament and of the Council of 23 April 2009 on the geological storage pf carbon dioxide, as amended.

8 Article 2, ibid, prohibits the storage of CO_2 in a storage site with a storage complex extending beyond the territories, exclusive economic zones and continental shelves of Member States.

Member States concerned shall jointly meet the requirements of this Directive and of other relevant Community legislation.'[9]

Neither does this chapter deal with the much-discussed question of the eligibility of CCS projects for crediting under the terms of the Clean Development Mechanism (CDM) of the Kyoto Protocol[10] of the UN Framework Convention on Climate Change (UNFCCC)[11] other than to note that when the Meeting of the Parties ultimately decided to make CCS projects CDM eligible it did so subject to exceptions for project activities which involve the transport of CO_2 from one country to another and which involve geological storage sites located in more than one country.[12]

2 Carbon Capture and Storage and CO_2\EOR

This Section discusses first CCS generally (2.1) and then the injection of CO_2 for EOR purposes (2.2). It concludes with some comments on the transboundary implications of CCS and CO_2\EOR operations (2.3).

9 The recent review of the Directive, Report on Review of Directive 2009/31/EC, COM (2015) 576 final, Annex 2, 18 November 2015 reported (at 5) there was no practical experience to date with transboundary issues but (at 7) that parties should consider cooperation to reduce costs and that the Connecting Europe Facility might "play a role in supporting cross-border transport networks and regional cooperation in this area." For discussion of the coordination issues that might arise see David Langlet, 'Transboundary Dimensions of CCS: EU Law Problems and Prospects' (2014) CCLR 196 at 206.

10 Kyoto Protocol to the UN Framework Convention on Climate Change, Kyoto, 11 December 1997.

11 UN Framework Convention on Climate Change, New York, 9 May 1992.

12 Decision 10/CMP.7, paragraph 6, http://unfccc.int/resource/docs/2011/cmp7/eng/10a02.pdf . Further consideration of transboundary projects were deferred until SBSTA 45 and CMP 12 in Marrakech, November 2016 at which point the CMP accepted SBSTA's recommendation to the effect that since no party had applied for registration of a CCS project as a CDM activity that it was appropriate "to conclude the consideration of the eligibility" of transboundary projects: FCCC/SBSTA/2016/L.19/Add.1. See generally Tim Dixon, Gregory Leaman, Paul Zakhour and Luke Warren, "CCS Projects as Kyoto Protocol CDM activities" (2013), 37 Energy Procedia 7596. A more specific article is Viviane Romeiro and Virginia Parente, "Carbon Capture and Storage and the UNFCCC: Recommendations to Address Trans-Boundary Issues" (2012), 3 Low Carbon Economy 130. Despite its title this paper is principally concerned with transboundary transportation of CO_2 for injection purposes and related accounting issues.

2.1 *Carbon Capture and Storage – Generally*

CCS is a recognized technology for mitigating greenhouse gas (GHG) emissions, and, in particular CO_2.[13] CCS involves the capture of CO_2 at large final emitters (LFEs),[14] the compression and transportation of the CO_2 and its injection under pressure into suitable geological formations.[15] The main targets are depleted oil and gas reservoirs and saline aquifers; the latter are much more significant in terms of volume. Two important examples of this are the Sleipner and Snøhvit projects on the Norwegian continental shelf. Sleipner commenced operations in 1996 and Snøhvit in 2008.[16] Both inject CO_2 into saline formations.

Injection sites may be onshore or offshore (as above).[17] There are positive and negative aspects associated with selecting an offshore disposal site. Positive aspects include the following: sediments of continental shelves frequently contain large volumes of high quality storage; in many cases prior oil and gas exploration on shelves provides a good geological understanding of the offshore; there is typically a single owner (the state) and no severance of surface and mineral titles; there is minimal conflict with freshwater aquifers (a major potential concern with onshore sites); the absence of resident populations and communities; there is frequently an existing pipeline and production\injection infrastructure; likely fewer wells offshore than onshore to serve as possible pathways to surface; it may be easier and cheaper to apply monitoring techniques and seismic imaging offshore.[18] In addition, it may simply be the case that a state has no suitable onshore sites. The principal disadvantages of offshore sites are the elevated costs and risks of offshore

13 A particularly useful policy oriented study is IEAGHG, *Interaction of* CO_2 *Storage with Subsurface Resources*, Report 2013–08, April 2013, (hereafter *IEAGHG Interaction Report*). The report is available here http://ieaghg.org/docs/General_Docs/Reports/2013-08.pdf.

14 The LFEs at which capture may occur may include power plants (coal or natural gas) and a range of industrial facilities including cement operations and a variety of hydrocarbon processing facilities.

15 In some cases natural gas reserves (methane, CH_4) may be contaminated with a significant percentage of CO_2 requiring processing in the field to produce pipeline quality gas. The separated CO_2 may then be reinjected.

16 The Sleipner project is located 240 km west of Stavanger. Processing and injection occur at the production site. The Snøhvit project involves production from three separate fields in the Barents Sea north of Norway northern coast. The gas is piped to shore to a processing and LNG facility. The CO_2 is separated and piped back to the field for injection. See: CSLF Offshore Geologic Storage (n 6) at 8–9.

17 In one case, the Tomakomai Demonstration Project (Japan), the drill site is on land but the injection wells are highly deviated and the targeted formations are between 3 and 4 km offshore. See CSLF Offshore Geologic Storage (n 6) at 11.

18 This section relies on CSLF Offshore Geologic Storage (n 6) at 2 & 5.

operations as well as concerns for the marine environment.[19] A recent report (2013) by the International Energy Agency concluded that CO_2 storage is not likely to take place in the near future at onshore locations in the five countries surrounding the North Sea and that the North Sea is a much more promising option both for those countries 'and indeed for many other nearby European countries.'[20]

Injection sites should be carefully characterized and located so as to ensure adequate and efficient storage capacity[21] as well as geological containment with appropriate trapping mechanisms.[22] Possible pathways to the surface (or sources of potable water) should be assessed, including natural fault lines and existing wells. While disposal into a depleted oil and gas reservoir involves injection into an underpressured formation (because of past production) injection into a saline formation will occur at natural formation pressure. Thus, the CO_2 will need to be injected 'at pressure higher than the natural formation pressures to be able to displace resident brines.'[23] Upon injection the

---

19 Ibid at 6.

20 *IEAGHG Interaction Report*, (n 13) at 37–38. There have been a number of regional scale European studies of offshore storage options. These studies include One North Sea, A study into North Sea CO2 cross-border transport and storage. Final Main report for the Norwegian Ministry of Petroleum and Energy and the UK Foreign and Commonwealth Office on behalf of the North Sea Basin Task Force (authors: Elements Energy Limited) 2010 available at <http://www.ccsassociation.org/docs/2010/OneNorthSea.pdf> Chapter 6 deals with legal and regulatory issues; Carbon Capture and Storage in the Skagerrak/Kattegat Region, Final Report, February 2102, available at <http://interreg-oks.eu/webdav/files/gamla-projektbanken/se/Material/Files/Kattegat/Skagerrak/Dokumenter+projektbank/CCS%20final%20report.pdf> the report contains a chapter on 'legal issues concerning CCS' authored by David Langlet; CCS in the Baltic Region – Bastor 2, Final Summary Report, Elforsk Report 14:50, 30 September 2014 available at <http://www.elforsk.se/Rapporter/?rid=14_50_> This report also contains a chapter on 'legal aspects' also authored by Langlet. A more detailed report Nils Rydberg and David Langlet, Work Package 4 – Legal and Fiscal Aspects (nd) is available here https://www.globalccsinstitute.com/publications/ccs-baltic-sea-region-bastor-2-work-package-4-legal-and-fiscal-aspects Most of the legal analysis in these reports is concerned with the EU Directive and the PLDC. There is more limited consideration of the problems posed by transboundary storage complexes.

21 Stefan Bachu, 'Review of CO_2 storage efficiency in deep saline aquifers' (2015), 40 International Journal of Greenhouse Gas Control 188. 'Storage efficiency' refers to the fact that not all available pore space will be accessed by CO_2 in the same way as not all estimated hydrocarbon reserves will be producible.

22 The literature refers to six trapping mechanisms which may operate cumulatively and over different time scales: (1) structural and stratigraphic trapping, (2) hydrodynamic traps (CO_2 entrained in groundwater, (3) residual gas trapping, (4) solubility trapping, (5) mineral trapping, (6) adsorption trapping. *IEAGHG Interaction Report* (n 13) at 7.

23 Jens T Birkholzer, Curtis M. Oldenberg and Quanlin Zhou, 'CO_2 migration and pressure evolution in deep saline aquifers' (2015), 40 International Journal of Greenhouse Gas Control 203, 209.

CO_2 will migrate within the target formation as a plume of CO_2. The CO_2 will be buoyant relative to the brine in the reservoir and will thus rise and spread laterally when trapped under cap-rock formations.[24] Injection will also elevate pressure conditions within the reservoir. The pressure effect of injection will initially be localized at the injection site but the resulting pressure front will extend beyond the actual CO_2 plume (in some cases as far as 100 km[25]) and slowly dissipate throughout the target formation. Once injection comes to an end, the CO_2 plume may continue to migrate within the target formation moving upwards and laterally (because of the buoyancy of the plume).[26] Migration will continue at a slow rate for centuries.

It is important to keep in mind the industrial scale of operations that will be required if CCS is to make a significant contribution to the mitigation of GHG emissions. Birkholzer et al provide a useful example noting that a 1000 MW power plant (enough electricity for one million people) will provide enough CO_2 over the life of the plant to create a subsurface CO_2 plume with linear dimensions of 10 km with pressure plume extending over 100 km.[27] As a result: 'In a future world with CCS being a fully employed technology, sedimentary basins with interconnected reservoirs might host multiple storage sites in which case pressure interference could be expected between individual projects.'[28]

The precise implications of the pressure changes in a formation as a result of injection will depend on the type of formation or 'storage system'[29] Open systems are likely the most desirable for large scale storage since they provide opportunities for pressure dissipation. Elevated pressures in a closed system or at the point of injection may impair cap rock integrity. It may be necessary

24 id, 206.

25 *IEAGHG Interaction Report* (n 13) at 67.

26 Bachu (n 21) at 198.

27 Birkholzer et al (n 23) at 204.

28 Bachu (n 21) at 197 and 198.

29 The literature distinguishes different types of storage system: closed systems (characterized by a formation surrounded laterally by impervious boundaries and vertically by sealing formations) and open systems where the large scale storage formation is overlain by a series of sealing formations but where the lateral boundaries are so far away that they will be unaffected by pressure changes. See Quanlin Zhou and Jens Birkholzer, 'On scale and magnitude of pressure build-up induced by large-scale geologic storage of CO_2' (2011), 11(1) Greenhous Gases Science and Technology 11, 12. The authors actually offer three categories – the middle category is a semi-closed system characterized by semi-pervious sealing units. See also Bachu (n 21) at 189 & 191.

for projects to adopt pressure management techniques to respond to this problem.[30]

The principal environmental concerns associated with CO_2 storage in saline aquifers, in addition to leakage to surface, include the migration of displaced brines causing potable groundwater contamination, not simply as a result of elevated salinity but also because brines may have elevated concentrations of heavy metals and natural radioactive materials.[31] This is far less likely to be a concern for ocean storage sites because there will generally be no potable groundwater sources that are at risk of contamination.

In addition it may also be necessary to assess resource interactions i.e. the competing uses of available geological storage space and the implications of one resource use for others. These implications may be positive or negative.[32]

2.2 *CO_2 Storage and Enhanced Oil Recovery*

Enhanced oil recovery using carbon dioxide (CO_2\EOR)[33] is a tertiary recovery technique for enhancing recovery of petroleum from some reservoir types.[34] Primary production (using the original pressure conditions in the reservoir) may result in the recovery of between 10 and 20% of the original oil in place in the reservoir (OOIP).[35] Waterflooding operations and other secondary recovery techniques may double the recovery rate. CO_2\EOR operations may allow the recovery of an incremental 5–15% of the OOIP and in some cases as much as an additional 25–40%. CO_2\EOR operations involve the injection of liquid

---

30 Bachu (n 21) at 199 and *IEAGHG Interaction Report* (n 13) at 47–52 discussing the Gorgon Project, Western Australia, which contemplates using, in addition to nine CO_2 injection wells, a pressure management scheme using four production wells for pressure relief and two injection wells for the produced water.

31 Birkholzer et al (n 23) 211.

32 The *IEAGHG Interaction Report* (n 13) esp at 17–69 assesses possible subsurface resource interactions between CCS projects and a variety of other uses.

33 This section draws on a number of sources including the Report Prepared for the Carbon Sequestration Leader Forum (CSLF) Technical Group by the CSLF Task Force on the Technical Challenges in the Transition from CO_2 -EOR to CCS, September 2013, (CSLF, CO_2 \EOR Technical Report) http://www.cslforum.org/publications/documents/CO2-EORtoCCS_FinalReport.pdf; Philip M. Marston, Patricia A. Moore and J. Greg Schnacke, "Carbon Dioxide Infrastructure: Pipeline Transport Issues and Regulatory Concerns – Past, Present and Future" (2015), 52 Rocky Mountain Mineral Law Foundation Journal 275–313 esp at 278–286, Nigel Bankes and Elizabeth Brennan, "Enhanced oil recovery and the geological sequestration of carbon dioxide", 2013, a paper prepared for Natural Resources Canada, http://www.law.ucalgary.ca/files/law/final_june7_enhanced-oil-recovery-and-the-geological-sequestration-of-carbon-dioxide.pdf.

34 Not all fields are suitable for CO_2 \EOR operations or even for CO_2 storage in a depleted reservoir. For an accessible account of relevant considerations see *IEAGHG Interaction Report* (n 13) at 12.

35 Marston et al, (n 33) 279.

or supercritical CO_2 into the producing formation where the CO_2 interacts with the oil remaining in the pores of the formation. Marston et al describe the process as follows:[36]

> … the dense-phase CO_2 becomes *miscible* with the oil (i.e. the two substances will flow together as if they were a single fluid). In the process, the oil droplets swell and the surface tension that held the oil droplets attached to the pore surfaces is reduced. This allows the oil to become mobile.

The combined fluids (oil, CO_2 and brine) flow towards the low pressure point of the producing well. As the substances are produced any entrained CO_2 returns to the gaseous phase. The fluids and gases are separated and collected at the surface. Captured CO_2 is dehydrated, pressurized and re-injected. The operator has an incentive to engage in re-capture because the costs of re-capture are lower than the costs of acquiring supplementary sources of CO_2 with which to continue the EOR operation. Not all of the CO_2 injected will be produced with the oil, some will remain permanently contained within the reservoir. Industry experience with EOR projects suggests that up to 90–95% of purchased CO_2 will eventually become 'incidentally trapped' (i.e. stored) within the reservoir over the life of the project.[37]

While CO_2\EOR operations may be relatively small scale, they offer some significant advantages when compared with CCS. The most important of which is that the incremental oil recovery provides a source of revenue to offset the expenses involved in capturing the CO_2.[38]

There may also be other non-carbon advantages for CO_2\EOR operations. For example, the use of CO_2\EOR in the North Sea should extend the producing lives of North Sea reservoirs and the basin as a whole thereby reducing the need for imports and maintaining employment.[39]

---

36 Ibid 281. Emphasis in original, references omitted.

37 Ibid 283 referring to various sources including CSLF, CO_2 \EOR Technical Report (n 33) and Global CCS Institute, "The Global Status of CCS: 2012" at 147.

38 The CSLF Technical Group report summarizes the advantages as follows:".... 1) it enables CCS technology improvement and cost reduction; 2) it improves the business case for CCS demonstration and early movers; 3) it supports the development of CO_2 transportation networks; 4) it may provide significant CO_2 storage capacity in the short-to-medium-term, particularly if residual oil zones (ROZ) are produced; 5) it enables knowledge transfer, bridging the experience gap and building and sustaining a skilled CCS workforce; and 6) it helps gaining public and policy-makers acceptance". See CSLF, CO_2 \EOR Technical Report (n 13), 1 & 58.

39 SCCS, CO_2 *Storage and Enhanced Oil Recovery in the North Sea: Securing a Low Carbon Future for the UK*, (2015) at 2. Available at <http://www.sccs.org.uk/images/expertise/reports/co2-eor-jip/SCCS-CO2-EOR-JIP-Report-SUMMARY.pdf>

Despite those prospects, there are very few offshore CO_2\EOR projects due to the limited availability of CO_2 and the cost of converting existing infrastructure.[40]

2.3 *Transboundary Implications of Saline Disposal and CO_2\EOR Projects*

It is apparent that the geographical implications of a saline disposal project will be more widespread and cover much larger areas than those associated with CO_2\EOR projects. This is because conventional oil and gas reservoirs are always closed systems[41] and the individual reservoirs will be much smaller than the regional scale, open geological formations that will likely be the target of saline disposal projects. A saline disposal project may raise transboundary concerns in one of two ways. First and most directly, the CO_2 plume itself may migrate across the boundary. Second, and more indirectly, the pressure front associated with injection into a formation on one side of the boundary may extend across the boundary even through the plume itself may never do so. Increased formation pressure may make these adjacent areas less useful for other CCS operations since pressure conditions will affect the injectivity of the formation. Brine migration induced as a result of the pressurization may be of concern for groundwater resources if present. CCS projects may have implications for other adjacent resource activities as noted above.[42]

CO_2\EOR projects on the other hand are unlikely to pose *new* transboundary problems for two reasons. First, since EOR operations occur in contained reservoirs that are well characterized there are less likely to be regional scale pressure implications.[43] Thus, unless the reservoir itself crosses the boundary there are unlikely to be transboundary effects. Second, if the reservoir does cross the boundary it will already be subject to some form of agreement whether a project specific treaty and unitization agreement or some form of framework or unitization agreement.[44] At most the question becomes one of asking how those arrangements might accommodate CO_2\EOR operations.

---

40 CSLF Sub-Seabed Geologic Storage (n 6) at 14. There is only one significant offshore CO_2 \EOR project and that is Lula Project, Brazil, id at 11. There is also one offshore CO_2 enhanced gas recovery (EGER) project. This is the experimental K12-B project in the Dutch sector of the North Sea about 150 km northwest of Amsterdam. For details see IEAGHG, *Review of Offshore Monitoring for CCS Projects*, 2015/02, July 2015 at 22–23, available at <http://www.ieaghg.org/docs/General_Docs/Reports/2015-02.pdf>.

41 Non-conventional hydrocarbon reserves (e.g. shale oil and gas) however represent accumulations of hydrocarbons that have not migrated into structural traps.

42 See also UNFCCC, Transboundary CCS Technical paper (n 6) at para 110.

43 There may be some exceptions if different pools are connected to a regional aquifer system.

44 This question is discussed further below in Section 4.

3 The Law Governing CCS Operations and CO_2\EOR Operations in Transboundary Formations

The present section considers the applicable law specifically with respect to CCS operations and CO_2\EOR operations *in transboundary formations*. While some may express doubts as "to the *correct ground* for exercising coastal state jurisdiction" with respect to CCS operations[45] there surely cannot be any doubt that a coastal state has exclusive jurisdiction to authorize CCS and CO_2\EOR operations within its territorial sea,[46] exclusive economic zone[47] and the continental shelf[48] including any extended continental shelf beyond 200 nm.[49]

The section summarizing the general rules pertaining to the duty to cooperate under UNCLOS as well as the more specific rules pertaining to shared resources including the work of the International Law Commission (ILC) in relation to shared groundwater resources (3.1). Section 3.2 examines the treatment of sub-seabed disposal of CO_2 in shared formations under the LDC and its Protocol. The section ends by reviewing the implications of Articles 74(3) and 83(3) of UNCLOS for the situations in which one party may be proposing to engage in CCS operations in an area of overlapping maritime entitlement claims in the absence of an agreed delimitation (3.3).

3.1 *The General Duty to Co-operate*

International law recognizes a general obligation to cooperate in relation to the use of shared resources.[50] Shared geological structures which can be used for storage or disposal purposes are examples of shared resources.

---

45 David Langlet, 'Transboundary Dimensions of CCS: EU Law Problems and Prospects' (2014) CCLR 196 at 201 (emphasis added). As Langlet concedes all seabed CCS operations require drilling.

46 UNCLOS Article 2 (n 3) the sovereignty of a coastal State extends to the territorial sea.

47 UNCLOS Article 56, sovereign rights for *inter alia* exploring, exploiting and managing the natural resources of the seabed and its subsoil, subject to "due regard" obligations with respect to the rights and duties of other States. UNCLOS Article 60 accords the coastal state the exclusive right to construct and exercise jurisdiction over installations and structures for the purposes of Article 56.

48 UNCLOS Article 77 extends to the coastal state the exclusive sovereign rights to explore and exploit the natural resources of the continental shelf which comprise the sea-bed and subsoil. UNCLOS Article 81 confirms that the coastal state has the exclusive right to authorize and regulate drilling on the shelf and Article 80 makes it clear that Article 60 applies *mutatis mutandis* to installations and structures on the shelf.

49 UNCLOS Article 76.

50 David M. Ong, 'Joint Development of Common Offshore Oil and Gas Deposits: "Mere" Sate Practice of Customary International Law' (1999), 93 AJIL 771 esp at 780–781. Ong refers to a number of sources including Article 3 of the Charter of Economic Rights and Duties of States, GARes 3281 (XXIX), UN GAOR, 29th Sess., Supp. No. 30, UN Doc. A/9030 (1974)

The ILC has endeavored to codify the law pertaining to a number of shared resources, specifically the non-navigational uses of international watercourses[51] and shared aquifers. The Commission's Draft articles on the Law of Transboundary Aquifers, with commentaries were adopted on second reading in 2008 (hereafter Draft Articles).[52] While these articles are principally concerned with the use of aquifers for potable drinking water a number of references in the articles and the commentary suggest that the ILC contemplated that the articles should also speak to the use of aquifers for disposal purposes including for geological sequestration of CO_2.[53] However, the commentary to the Draft Articles also makes it clear that the Commission concluded that the Articles should *not* apply to aquifers underlying the continental shelf on the grounds that 'Submarine geological formations under the continental shelf do not hold freshwater and accordingly such formations and water therein fall outside the scope of the draft articles.'[54] While the Commission seems inconsistent with respect to the applicability of the articles to the use of

---

and the Draft Principles of Conduct in the Field of the Environment for the Guidance of States in the Conservation and Harmonious Utilization of Natural Resources Shared by Two or More States, adopted by the United Nations Environment Programme in 1978, available at <http://www.unep.org/training/programmes/Instructor%20Version/Part_2/Activities/Interest_Groups/Decision-Making/Supplemental/Enviro_Law_Guidelines_Principles_rev2.pdf>. See also Tara Davenport, 'The exploration and exploitation of hydrocarbon resources in areas of overlapping claims' in in S. Jayakumar, Tommy Koh and Robert Beckman (eds), *The South China Sea Disputes and the Law of the Sea*, Edward Elgar, Cheltenham, 2014, 93–113 at 107.

51 The ILC's work on this topic resulted in the adoption of the UN Convention on the Non-Navigational Uses of International Watercourses, 21 May 1997, <http://legal.un.org/ilc/texts/instruments/english/conventions/8_3_1997.pdf>

52 International Law Commission, Draft articles on the Law of Transboundary Aquifers, with commentaries, adopted at the ILC 60th Session, (2008) online: <http://untreaty.un.org/ilc/texts/instruments/english/commentaries/8_5_2008.pdf> (ILC Draft Articles).

53 See e.g. the definition of 'utilization of transboundary aquifers or aquifer systems' (Article 2(e) ibid) which refers to 'extraction of water, heat and minerals, and storage and disposal of any substance'; furthermore the commentary to that definition notes that 'Extraction of freshwater is of course the main utilization of aquifers'. Other kinds of utilization, however exceptional and peripheral, should not be ignored. 'Utilization' is defined in a non-exhaustive manner to include not only extraction of water but also extraction of heat for thermo-energy, extraction of minerals that may be found in aquifers, as well as storage or disposal of waste, such a new experimental technique to utilize an aquifer for carbon dioxide sequestration." There is a further reference to geological sequestration in the commentary to article 4 dealing with the principle of equitable and reasonable utilization: 'an aquifer may also be used for disposal, in particular through a new experimental technique to utilize aquifers for carbon dioxide sequestration. This use is peripheral to the present draft articles.' It is an open question whether the balance of the text of the articles was really drafted with CCS and other disposal operations in mind.

54 Paragraph (2) of the commentary to Article 2.

aquifers for sequestration or disposal purposes the principles underlying these articles should still inform how we might think about the duty to cooperate in the context of shared geological formations in marine areas.[55]

The duties associated with the general obligation to cooperate[56] include both procedural and substantive duties. Procedural duties include the duty to exchange information about the resource,[57] the duty to notify as to proposed uses of the shared resource which may affect other parties,[58] the duty to conduct an environmental assessment to the extent that transboundary impacts are anticipated,[59] the duty to engage in consultations as to any concerns that are identified and if necessary to negotiate (or engage in other peaceful means of dispute resolution) to resolve any outstanding issues.[60] There may also be a duty to monitor the use and status of the shared resource and to share the resulting information.[61] The substantive duties include a duty of equitable and reasonable utilization of the shared resource,[62] the duty to protect the environment,[63] and the duty not to cause significant harm to the shared resource or the interests of other states in that resource.[64] What amounts to harm, and indeed to significant harm, must be contextualized and to some extent may depend upon the uses that others may be making of that resource. Thus, disposal of CO_2 into an aquifer may be a reasonable use of that aquifer if it is saline; it is likely not a reasonable use if the aquifer contains potable groundwater and certainly not if the aquifer is being used as a potable water source by another aquifer state.

The duty to cooperate with respect to shared resources may also result in the aquifer states agreeing to create appropriate institutions. Such institutions have proven to be particularly important in relation to international watercourses. The draft articles on aquifers refer *inter alia*, to joint mechanisms of

55 See also UNFCCC, Transboundary CCS Technical Paper (n 6) at para 63 noting that parties 'may wish to consider the cooperative and regulatory approaches adopted in relation to the management of water resources and oil and gas reserves' and see also at para 66 a similar comment specifically with respect to the ILC draft articles.

56 ILC Draft Articles (n 52), Article 7, General obligation to cooperate.

57 Ibid Article 8, Regular exchange of data and information.

58 Ibid Article 15, Planned activities.

59 Ibid Article 15(2) and paragraph (4) of the commentary.

60 Ibid Article 15(3).

61 Ibid Article 15, Monitoring.

62 Ibid Articles 4 & 5.

63 Ibid Article 10, 'Protection and preservation of ecosystems', Article 12, 'Prevention, reduction and control of pollution'

64 Ibid Article 5.

cooperation, appropriate bilateral and regional agreements, joint monitoring activities, and a joint management mechanism.[65]

In addition to the general duties to cooperate with respect to shared resources as well as the more specific rules in relation to transboundary aquifers, UNCLOS also contains a number of provisions enjoining the parties to cooperate including the duty to cooperate in relation to the conservation and management of living resources, the duty to co-operate with respect to enclosed and semi-enclosed seas and the duty to cooperate with respect to the protection of the marine environment.[66] There are rather more specific provisions requiring the relevant parties to pursue provisional arrangements in areas of overlapping maritime entitlements.[67] These provisions are the subject of more detailed examination below in section 3.3.

3.2 *The Treatment of Sub-Seabed Disposal of CO_2 in Shared Formations in the Work of Protocol to the London Dumping Convention*

The one treaty forum that has addressed storage in shared geological structures in marine areas is the Protocol to the London Dumping Convention (PLDC).[68] It will be recalled that the parties to the PLDC agreed to amend that instrument to permit geological sequestration of pure streams of CO_2.[69] The parties also agreed that the injection of CO_2 for enhanced oil recovery purposes does not engage the Protocol.[70] That amendment and the understanding with respect to CO_2 for EOR purposes did not resolve all issues relating to the potential use of the seabed for geological sequestration because Article 6 of the PLDC prohibits the export of wastes. The Parties identified two potential implications of this prohibition for geological sequestration of CO_2: export and migration (or 'migration-as-export'). 'Export' refers to the export of CO_2 from one country for disposal in the jurisdiction of another country. It was agreed that this activity was subject to the prohibition of Article 6, that the problem could not be resolved by way of an agreed interpretation, and that it could only be resolved by means of an amendment. An amendment to create an exception for the

65 Draft Articles, Articles 7, 9, 13, & 14.

66 UNCLOS (n 3) Articles 61(2), 64, 65, 66, 117, 118, 123, 194, & 197–201. Ong (n 50) at 781–785, Davenport (n 50) at 107.

67 UNCLOS (n 3) Articles 74 & 83.

68 PLDC (n 2).

69 LC 28/15.6 adopted November 2006 entered into force 10 February 2007; discussed in Dixon et al (n 12) at 433–436.

70 See International Maritime Organization (IMO) webpage, Carbon Capture and Sequestration <http://www.imo.org/en/OurWork/Environment/LCLP/EmergingIssues/CCS/Pages/default.aspx>

export of CO_2 for geological sequestration was adopted in 2009 but is still a long way from entering into force.[71]

The migration-as-export concern is particularly relevant for this chapter since it concerns the scenario in which CO_2 is injected into a shared geological structure in State A and the CO_2 plume migrates (either intentionally or unintentionally) from State A to State B (or to an area beyond national jurisdiction). The Parties concluded that such a migration does *not* constitute an export within the meaning of Article 6 (and adopted a resolution to that effect).[72] It is therefore a form of disposal which is permitted under the newly adopted amendment. However, the parties considered that it required special consideration under the terms of the Specific Guidelines for the Assessment of Carbon Dioxide for Disposal into Sub-seabed Geological Formations. The Guidelines were first adopted in 2007; the amended version which addresses the migration scenario was adopted in November 2012.[73] The Specific Guidelines (2012) address geological waste disposal of CO_2 generally but also contain a number of more specific provisions dealing with disposal into geological formations where there is potential for transboundary movement of injected CO_2. The general provisions address such matters as waste prevention audits, consideration of waste management options, the chemical and physical properties of the waste stream (i.e. proper characterization of the waste stream), an action list (a screening tool to determine whether a material is considered acceptable for dumping), site selection and characterization, assessment of potential effects (a risk assessment and an impact hypothesis), monitoring and risk management, and permit and permit conditions.[74]

The main provision dealing with disposal into geological formations where there is potential for transboundary movement of injected CO_2 is an overarching provision which precedes the general provisions outlined above and provides as follows:[75]

---

71 31 August 2018 only 5 parties had ratified the amendment: Norway, Netherlands, UK, Iran and Finland, LC 40/2, 31 August 2018.

72 Resolution LP.3(4), 30 October 2009, Recital 12.

73 Adopted 2 November 2012, LC 34/15, annex 8 (Specific Guidelines, 2012).

74 For more detailed discussion see Dixon et al (n 12) at 434. See also IEAGHG, Review of Project Permits under the London Protocol – An Assessment of the Proposed P18–4 CO2 Storage Site, Report: 2016/TR4, May 2016. The project involves injection and storage into the P18–4 field which is a near-depleted natural gas field located approximately 20 km off the Dutch Coast. The report involves an analysis of the Netherlands permitting exercise to test compliance with the Specific Guidelines. It does not contain an assessment of potential transboundary issues presumably because the storage complex was a confined rather than an open complex.

75 Specific Guidelines (n 73) para 1.10 – bolded letters and numbers added.

> [A] In the case of [1] transboundary sub-seabed geological formations that could be used by more than one country or [2] where sub-seabed geological formations are located in areas where there is the potential for transboundary movement of CO_2 streams after injection [3] the Contracting Party where the injection occurs should be responsible for the implementation of these Specific Guidelines. [B] Consent should be sought for the use of the sub-seabed geological formation from all countries with jurisdiction over this sub-seabed geological formation, without prejudice to international law including as reflected in the relevant provisions of UNCLOS. [C] The Contracting Party where the injection occurs should cooperate with other relevant Contracting Parties, other States and other relevant entities, to ensure adequate sharing of information as needed and in accordance with international law, including by way of arrangement or agreement to ensure that these Specific Guidelines are implemented effectively.

There are a number of key ideas embedded in this paragraph.

First, sentence [A] serves to both identify the circumstances under which the clause shall apply but also to allocate compliance responsibility. Thus, the Guidelines apply where, [A1] a sub-seabed geological formation could be used by more than one country, *or* [A2] there is a formation which is located in an area where there is *potential* for transboundary migration post-injection. This first sentence establishes low alternative thresholds for its applicability. The first alternative, threshold [A1], is satisfied by the mere existence of a geological formation that *could* be used by more than one country. The second alternative, threshold [A2], is satisfied by the *potential* for transboundary migration. The allocation of compliance responsibility is dealt with in the last part of the first sentence [A3] which resolves that it is the State where the injection occurs that is responsible for the implementation of the Guidelines.

Second, sentence [B] establishes the threshold for proceeding. This requires that consent should be sought from all countries with jurisdiction over the shared sub-seabed geological formation. This requirement is said to be without prejudice to international law generally and UNCLOS specifically. The consent requirement is broadly framed. It is not confined to seeking to obtain consent from states to which the plume may migrate, or states that may suffer a (significant) adverse effect from use of the shared formation; it extends to any state with whom the formation may be shared. The consent is framed in the passive voice. It does not specify that the consent is to be obtained by the Contracting Party (rather than, say, the operator of the project) but the section as a whole is very state-oriented and thus it seems reasonable to think that consent should be obtained at the state to state level. The clause does not prescribe the form that the consent must take, such as whether it might take the form of a treaty or an agreement between the respective regulatory authorities of the parties, or an exchange of diplomatic notes. It is difficult to divine the precise purpose of the 'without prejudice' clause but presumably it is intended to indicate that any consent that is obtained may deal with matters as between the parties to the consent but cannot discharge either party of obligations that either may owe to others, or to the international community more generally. Finally, the clause stipulates that 'consent should be sought'. The clause does not use the mandatory 'shall' but the softer imperative 'should'; nor does the clause stipulate the consequences of failing to obtain the consent of all formation states and certainly does not use the formulation 'should be sought *and obtained*'.

Third, in addition to the consents that 'should be sought', sentence [C] calls upon the Contracting Party where the injection is occurring, to cooperate with: (1) other relevant Contracting Parties, (2) other States, and (3) other relevant entities. It is evident that this obligation is not confined to formation states (or presumably this sentence would have used the same terminology as the consent clause) although one would think, given the scope of the consent clause, that a reservoir State would always qualify as a relevant Contracting Party or 'other State'. It is not clear what the term 'other relevant entities' is designed to capture but presumably it might include global and regional treaty organizations such as the Protocol secretariat and conference of the parties, OSPAR, and, where relevant, a body like the EU. The cooperation is designed for two ends: (1) to ensure adequate sharing of information, and (2) to ensure that these Guidelines are implemented effectively. These objectives should be the measure of the information that must be shared which must also be 'in accordance with international law'. One could anticipate that an exchange that is adequate to ensure effective implementation of these Guidelines will need to be both detailed and extensive and indeed address all of the topics covered by the substantive requirements of the Guidelines. The clause provides some guidance as to the means of effecting cooperation insofar as it references 'arrangement or agreement' but clearly these are just possible modes of attaining the objective.

The Guidelines contain two other references to transboundary movement both in the section of the Guidelines dealing with the characterization of the sub-seabed geological formation.[76] Thus, section 6.2 instructs that in selecting a site for geological storage of CO_2 the geological assessment should take account of potential migration and leakage pathways over time (including

76 Ibid, ss.6.2.5 and 6.5.

transboundary movement) and potential effects to the marine environment of leakage of CO_2. A second reference in the section largely follows part of the previously quoted and discussed section 1.10.[77]

While Dixon et al take the view that Guidelines under the Protocol are for guidance only, and not legally binding (unlike decisions adopted under OSPAR), these Guidelines offer the most detailed prescriptions for sub-seabed disposal of CO_2 into shared geological formations.[78] It is also possible that the Guidelines could attain additional normative weight through referential incorporation in other instruments which might, for example, require states to apply generally accepted standards or guidelines.[79]

3.3 *The Law Governing Seabed Operations in Areas of Overlapping Maritime Claims Pending Delimitation*

There are overlapping maritime claims in many areas of the world. These may include overlapping claims to a territorial sea, an EEZ and a continental shelf. Whereas Article 15 of UNCLOS establishes a default equidistance entitlement in the case of overlapping claims to a full 12 nm territorial sea,[80] there is no default entitlement rule in the case of overlapping EEZs and continental shelves, only a duty to negotiate a delimitation by agreement with a view to reaching an equitable solution, failing which the Parties shall resort to dispute settlement under Part XV.[81] Part XV (Article 298(1)(a)) permits a state to opt out of binding compulsory dispute settlement with respect to Articles 15, 74 and

77 Ibid s.6.5.

78 Dixon et al (n 12) at 436.

79 See generally Catherine Redgwell, 'Mind the Gap in the GAIRS: The Role of Other Instruments in LOSC Regime Implementation in the Offshore Energy Sector' in Nigel Bankes and Seline Trevisanut (eds), *Energy from the Sea: An International Law Perspective on Ocean Energy*, Brill Nijhoff (2014), and Angelica Bonfanti and Francesca Romanin Jacur, 'Energy from the Sea and the Protection of the Marine Environment: Treaty-Based Regimes and Ocean Corporate Social Responsibility' *ibid.* More specifically, UNCLOS (n 3) Article 208 requires coastal states to adopt laws and regulations to prevent, reduce and control pollution arising from sea-bed activities. Such domestic laws shall be 'no less effective than international rules, standards and recommended practices and procedures.'

80 UNCLOS (n 3) Article 15; absent agreement neither State is entitled to extend its territorial sea beyond the median line (subject to exceptions dealing with historic title or other special circumstances. See David Anderson and Youri van Logchem, 'Rights and obligations in areas of overlapping maritime claims' in Jayakumar et al (n 50) pp. 121–228 at 196 & 197 who observe that 'This rule is cast in unusual terms. It amounts to a restriction on the entitlement of each neighbouring State' and 'seeks to prevent overlapping claims to sovereignty by qualifying the entitlements of coastal States ... Hence, whilst ... the boundary remains disputed, and there are no special circumstances or historic titles, neither of the claimant States is '*prima facie*' allowed to exercise sovereignty beyond the equidistance boundary line.' (reference omitted).

81 UNCLOS Articles 74(1) and 83(1).

83 but in return such a party must accept the possibility of compulsory, but non-binding, conciliation.[82] This section focuses on overlapping EEZ and continental shelf claims. Where the shelf and EEZ have been delimited, the coastal state's rights are exclusive. Absent an agreed delimitation 'no such exclusivity can exist'.[83]

Articles 74 (EEZ) and 83 (continental shelf) impose procedural and substantive obligations on adjacent and opposite states pending agreement on delimitation.[84] In particular, common paragraph three provides that:[85]

> 3. Pending agreement as provided for in paragraph 1, the States concerned, in a spirit of understanding and cooperation, shall make every effort to enter into provisional arrangements of a practical nature and, during this transitional period, not to jeopardize or hamper the reaching of the final agreement. Such arrangements shall be without prejudice to the final delimitation.

This provision does not 'limit *de iure* the powers of each State in a contested area ... these powers thus remain'. Provided that the claim is a reasonable one powers 'generally attributed to the coastal State by the relevant UNCLOS provisions' may be exercised but the paragraph does impose 'a double condition on the exercise of those rights.'[86]

---

82 For a recent example of recent successful conciliation see Report and Recommendations of the Compulsory Conciliation Commission between Timor-Leste and Australia on the Timor Sea, 9 May 2018, Registry, Permanent Court of Arbitration. The Commission was able to assist the parties in negotiating the Treaty Between the Democratic Republic of Timor-Lese and Australia Establishing their Maritime Boundaries in the Timor Sea, New York, 6 March 2018. The treaty includes Annex B which is a joint development scheme for the Greater Sunrise field.

83 Anderson and van Logchem (n 80) at 198.

84 There is no similar provision in Article 15.

85 The leading decision interpreting the scope of these obligations is *Guyana v Suriname*, Arbitral Award, 17 September 2007, esp. at paras 453–486 available at <http://www.pca-cpa.org/showpage5751.html?pag_id=1147>. Useful assessments of both the case and the language of Articles 74 and 83 of LOSC include Enrico Milano and Irini Papanicolopula, 'State Responsibility in Disputed Areas on Land and at Sea' (2011) 71 ZaöRV 587 esp at 606–621, Anderson & van Logchem, (n 80) and Davenport (n 50). On the background to articles 73 and 84 see Rainer Lagoni, 'Interim Measures Pending Maritime Delimitation Agreements' (1984), 78 AJIL 345. Most recently the British Institute of International and Comparative Law (n 1), has offered a comprehensive review of state practice under Articles 74(3) and 83(3) of UNCLOS.

86 Milano and Papanicolopulu (ibid) at 612. A reasonable claim is one that accords with the settled jurisprudence of the ICJ, ITLOS and Annex VII tribunals on maritime boundary delimitations and in particular the *Black Sea Case*; an unreasonable claim would be one that is manifestly unjustifiable when measured against that standard (id at 630).

Rainer Lagoni has provided an authoritative account of the evolution of paragraph 3 of these two articles.[87] Lagoni recalls that the paragraph represented a compromise and indeed a compilation of two different ideas that had been introduced in proposed language for this article. On the one hand, those states committed to the preeminence of the median or equidistance line favoured a formulation that prohibited a state from extending its shelf or EEZ (or in some proposals engaging in any exploration or exploitation activities beyond the equidistance line) absent an agreement between the states concerned.[88] On the other hand, those states that favoured delimitation in accordance with equitable principles argued for a formulation that emphasized the importance of provisional arrangements.[89] As can be observed, the final text incorporated elements of both of these ideas insofar as it refers to both the importance of making provisional arrangements as well as the commitment not to do anything that would jeopardize or hamper conclusion of an agreement,[90] a point emphasized in the Award of the Arbitral Tribunal in *Guyana v Suriname.*[91] That Award also emphasized that the purpose of paragraph 3 is to permit utilization of disputed areas pending delimitation, thereby avoiding the suspension of economic development.[92] The history of this paragraph fully justifies Lagoni's assertion that, at least at the time the paragraph was negotiated, it was 'in no way a codification of customary international law but represents an example of its progressive development.'[93]

Paragraph 3 was subject to detailed analysis in the Award in *Guyana v Suriname* and the Tribunal helpfully distinguished the two core and complementary obligations within the paragraph, first, the duty to use every effort to agree on provisional arrangements of a practical nature, and second, the duty not to hamper or jeopardise a final agreement.[94]

87 Lagoni (n 85).

88 Ibid at 350–351.

89 Ibid.

90 Ibid at 349, and see also at 353 the provision 'combines two basically different views … attempts either to promote certain measures or to restrict others'.

91 (n 85).

92 Ibid at para. 460.

93 Lagoni (n 85) at 354.

94 For further analysis of the decision see Milano and Papanicolopula (n 85) Anderson & van Logchem (n 80) Davenport (n 50) and BIICL (n 3). See also ITLOS, Dispute Concerning Delimitation of the Maritime Boundary between Ghana and Côte d'Ivoire in the Atlantic Ocean, Request for the prescription of provisional measures, Order 25 April 2015. For comment see Yoshifumi Tanaka, 'Unilateral Exploration and Exploitation of Natural Resources in Disputed Areas: A Note on the *Ghana/Côte d'Ivoire* Order of 25 April 2015 before the Special Chamber of ITLOS' (2015), 46 ODIL 315. See also Judgment, 23 September 2017, at paras 604–634. The Chamber (at para 604) emphasized the importance of the duty to negotiate in good faith especially 'where neighbouring States conduct maritime activities

3.3.1 Application to CCS Activities

What then is the position in relation to proposed CCS related activities in an area of overlapping claims to the EEZ and continental shelf at least where both countries are parties to UNCLOS?[95] The main conclusion must be that the twin duties to make every good faith effort to enter into provisional arrangements and the duty not to take any actions that might jeopardize or hamper reaching a final agreement are just as applicable to CCS operations as they are to conventional oil and gas operations. The first branch would require a party proposing to authorize CCS evaluation activities in the disputed area to notify the other state(s) concerned, to provide relevant information, to seek cooperation and to commit to share any information obtained. Both parties should be flexible in their efforts to negotiate provisional arrangements. Such arrangements might include a joint development agreement or arrangement. The second branch would require the proposing state not to authorize drilling activities for any CCS related purposes such as for taking core samples from target formations to assess porosity, permeability and injectivity and for assessing the characteristics of sealing formations. Under the present state of the law, the proposing state cannot carry out these activities unilaterally but requires the agreement of the other state(s) concerned.

---

in close proximity.' The same would be true with respect to efforts to negotiate a 'provisional arrangement' but there could be breach of that duty (at para 628) absent a request for such an arrangement. Ultimately the Chamber found that Ghana was not in breach of either paragraph 3 but at least some of the Chamber's reasoning (at para 633) seems to turn on the specific and limited language of Côte d'Ivoire's request for relief.

95 As noted above, at the time of its adoption, paragraph 3 represented a progressive development of international law rather than its codification. It is still not clear whether both elements of paragraph 3 could be considered to be customary international law. Milano and Papanicolopulu (n 86) are not completely consistent on the point. Thus at 604 and 607 they suggest that while the duty not to hamper a final settlement might be grounded in the general principle to settle disputes peacefully and in good faith the same cannot be said of the positive duty to enter into arrangements of a practical nature. On the other hand, at 611, the authors suggest (without accompanying references or discussion) that paragraph 3 does reflect customary law. Anderson and van Logchem (n 80) at 208 seem content to conclude that this is 'a difficult question'. The BIICL Report (n 3) is equally cautious concluding (at para 401) that it is 'difficult to draw any general tends from the practice collated regarding the content of the obligations of Article 74(3) and 83(3) or any applicable customary international law obligations of restraint.'

4 The Treatment of Shared Resources in Maritime Delimitation Agreements – Is There Space for CCS Operations?

Having examined the rules that apply where States (or at least those States that are party to UNCLOS) have not entered into a delimitation agreement, we can now examine how States might deal with storage operations in shared geological formation in their delimitation agreements. There is an extensive state practice and literature on this issue in the context of hydrocarbon deposits.[96] The options[97] for dealing with this issue include: (1) silence, (2) the inclusion of a basic straddling deposit or unity of deposit clause, and variations on such a clause, (3) so-called framework agreements, and (4) joint development agreements. The following sections examine each of these options with a view to determining the extent to which these arrangements address the use of shared geological formations for CCS operations or CO_2\EOR operations.[98]

4.1 *Silence*

The parties may elect to effect a delimitation without addressing the issue of shared geological formations at all. This gives rise to the question as to what the default rules would be in such a case, either with respect to petroleum operations or CO_2 storage and disposal operations. It is fairly clear that Article 74 and 83 UNCLOS and, in particular, common paragraph 3 are no longer relevant – they are spent. The subject matter of the two articles is that of delimitation. Once the parties have made their delimitation agreement(s), those agreements assume priority – at least with respect to the issue of delimitation.[99] Furthermore, paragraph 3 is certainly no longer applicable. After all,

96 See, for example, Ong (n 50); Vasco Becker-Weinberg, *Joint Development of Hydrocarbon Deposits in the Law of the Sea*, Hamburg Studies on Maritime Affairs, vol. 30 (Springer, 2014); William T Onorato, 'Apportionment of an International Common Petroleum Deposit' (1977) 26 ICLQ 324–337, Rainer Lagoni, 'Oil and Gas Deposits Across National Frontiers' (1979) 73 AJIL 215–243 and Peter Cameron, 'The Rules of Engagement: Developing Cross-Border Petroleum Deposits in the North Sea and the Caribbean' (2006) 55 ICLQ 559–586.

97 I have applied this categorization previously in Nigel Bankes, 'Recent Framework Agreements for the Recognition and Development of Transboundary Hydrocarbon Resources' (2014), 19 International Journal of Marine and Coastal Law 666–690 and Nigel Bankes, 'The regime for transboundary hydrocarbon deposits in the maritime delimitation treaties and other related agreements of Arctic states" (2016) 47(2) Ocean Development and International Law 141–164.

98 See also UNFCCC, Transboundary CCS Technical paper (n 6) at para 110 noting that 'Existing agreements for the management and shared natural resources (e.g. on the use of the aquifer or reservoir) might apply even if they do not expressly contemplate CCS.'

99 See paragraph 4 of each Article.

paragraph 3 commences with the words 'Pending agreement'. The agreement that is referenced here is the agreement on delimitation referred to in paragraph 1. Thus, the twin obligations of that paragraph no longer apply. One should recall in particular that the duty 'not to jeopardize or hamper' is a duty not to jeopardize or hamper the reaching of a final agreement. The delimitation agreement is that final agreement. Lagoni agrees remarking at the end of a section headed 'Application *ratione temporis*' that 'The obligation [to pursue good faith negotiations for a provisional agreement] ceases when the final delimitation agreement is reached.'[100]

As a result, neither the two Articles, nor the interpretive case law, are directly applicable as treaty obligations post the conclusion of a delimitation agreement. However, the general duty to cooperate with respect to shared resources continues as does the parallel duty to cooperate as well as the general obligations under UNCLOS, as outlined above.[101] In addition, a party to the PLDC will need to observe the relevant provisions of the Specific Guidelines to the extent that it is contemplating disposal operations in a shared geological structure.[102]

4.2 *The Basic Clause and Its Variants*

Agreements delimiting the continental shelf and\or the EEZ frequently make provision for the discovery of hydrocarbon deposits that straddle the agreed delimitation line. An early and oft-cited example is the 1965 Agreement between the United Kingdom (UK) and Norway[103] which provides that:

> If any single geological petroleum structure or petroleum field, or any single geological structure or field of any other mineral deposit, including sand or gravel, extends across the dividing line and the part of such structure or field which is situated on one side of the dividing line is exploitable, wholly or in part, from the other side of the dividing line, the Contracting Parties shall, in consultation with the licensees, if any, seek to reach agreement as to the manner in which the structure or field

100 Lagoni, Interim Measures (n 85).

101 See also Eritrea v Yemen, Second Phase, December 17, 1999, <http://www.pca-cpa.org/showpage.asp?pag_id=1160> esp at paras 85–86. Reference might also be had to the general duty of good faith under both customary law and LOSC Article 300 as well as more general obligations of restraint based on the obligation not to take steps that would aggravate or extend a dispute. See BIICL Report (n 3) at paras 65–66 and also South China Sea Award, 12 July 2016, Registry, Permanent Court of Arbitration, at paras 1166–1173.

102 See section 3.2 above.

103 Agreement between the Government of the United Kingdom of Great Britain and Northern Ireland and the Government of the Kingdom of Norway relating to the delimitation of the continental shelf between the two countries, 10 March 1965, 551 UNTS 213, Article 3.

shall be most effectively exploited and the manner in which the proceeds deriving therefrom shall be apportioned.

In some cases the discovery of such straddling deposits and the resulting consultation has led to the conclusion of specific treaties elaborating on the rules for exploiting such a deposit, beginning with the Frigg Agreement in 1976.[104] These rules are typically based on the concept of unitization (which involves treating the entire pool as a single ownership unit)[105] with apportionment of costs and production based on the distribution of the underlying hydrocarbon reserves.[106] Project operations are governed by the terms of unitization and operating agreements concluded between the licensees on either side of the line, typically subject to the agreement or approval of both governments. The unit operator is appointed by agreement between the licensees subject again to the approval of the governments.[107]

The question for present purposes is whether a clause such as that contained in the UK/Norway Agreement as well as the subsequent pool-specific development agreements should be read as extending to CCS operations and CO_2\EOR operations. It is perhaps easiest to begin with CO_2\EOR operations. These operations are strictly ancillary to the principal purpose of exploiting a hydrocarbon reservoir – they are designed to enhance production. It seems clear therefore that such operations would fall within the language of Article 3 of the Norway/UK Agreement. Such operations would also likely fall within

---

104 See Agreement relating to the exploitation of the Frigg Field Reservoirs and the transmission of gas therefrom to the United Kingdom (London, 10 May 1976) 1098 UNTS 3. The Frigg Field was abandoned in 2004, see Norwegian Petroleum Directorate, *CO_2 Storage Atlas, Norwegian North Sea* (2012) at 10, <http://www.npd.no/Global/Norsk/3-Publikasjoner/Rapporter/PDF/CO2-ATLAS-lav.pdf>. Other reservoir-specific agreements include, Agreement relating to the exploitation of the Statfjord Field Reservoirs and the offtake of petroleum therefrom (Oslo, 16 October 1979) 1254 UNTS 379, Agreement relating to the exploitation of the Murchison Field Reservoir and the offtake of petroleum therefrom (Oslo, 16 October 1979) 1249 UNTS 173 and (between the UK and the Netherlands) the Agreement relating to the exploitation of the Markham Field Reservoirs and the offtake of petroleum therefrom (The Hague, 26 May 1992) 1731 UNTS 30235.

105 See Frigg Agreement, ibid, Article 1(1) 'The gas in the Frigg Field Reservoir and the hydrocarbons produced with or from the gas (excluding oil underlying the gas and gas and oil in other horizons), referred to in this Agreement as "Frigg Gas", shall be exploited as a single unit ...'.

106 Ibid, Article 2(2): 'The two Governments shall consult with a view to agreeing a determination of the limits and estimated total reserves of the Frigg Field Reservoir and an apportionment of the reserves therein as between the Continental Shelf appertaining to the United Kingdom and the Continental Shelf appertaining to the Kingdom of Norway. For this purpose the licensees shall be required to submit to the Governments a proposal for such determinations.'

107 Ibid Article 5.

the language of most pool-specific development agreements,[108] although the institution of such operations would likely require the consent of the two governments under the terms of these specific treaties.[109] Plans to commence CO_2\EOR operations may also lead to a restatement of recoverable reserves; however, all of these pool-specific development agreements provide for this eventuality and therefore would apply whether a request for a restatement is triggered by additional drilling which causes the licensees of the governments to revise their understanding of the parameters of the pool, or whether it is triggered by the introduction of EOR operations.[110]

But what is the position with respect to a shared saline formation? What is the position where a project in a conventional hydrocarbon reservoir moves from a CO_2\EOR operation to a pure disposal operation? And what is the position where a straddling reservoir has been abandoned and is now being considered for CCS purposes (the Frigg Gas field is a case in point).[111] The Norway\UK version of the straddling deposit clause addresses two subjects: (1) geological petroleum structures and petroleum fields; and, (2) single geological structures or fields for any other mineral deposit, including sand or gravel. Furthermore, the clause is concerned with the exploitation of these resources for commercial gain and presumably by means of extraction rather than exploitation by injection. There may be arguments both ways as to whether such clauses can be read as embracing the operations referred to above i.e. exploitation by injection of a saline reservoir, cross-over operations from EOR to pure disposal, and the use of abandoned petroleum reservoirs. Perhaps the arguments are strongest in relation to the use of abandoned petroleum reservoirs since CCS can be seen as the continued exploitation of such a reservoir. The arguments may be weaker with respect to a shared saline aquifer since it

108 I say 'most' because the Frigg Agreement, Article 1(1) only applies to 'Frigg Gas' and specifically does not apply to 'oil underlying the gas'.

109 See, e.g. Frigg Agreement (n 104), Article 4(1) calling for government approval of a proposed scheme for the conservation of gas; Article 11 of each of the Murchison and Statjford Agreements refers to the need for government approval of amendments to the exploitation programme which programme should have as its objective the exploitation of the reservoir 'in such a manner as to prevent the waste of petroleum and minimizes losses of reservoir energy' A CO_2 \EOR recovery programme would fall within this objective.

110 Frigg Agreement (n 104) Article 3.

111 See Norway's North Sea Storage Atlas, (n 104) at 59–60 which evaluates both a pure disposal operation and an enhanced gas recovery operation in the abandoned Frigg field. The operation would take advantage of 'a huge [saline] aquifer that is connected to the field'. The Report acknowledges that a prerequisite for any injection operation would be integrity studies of abandoned wells that penetrate the formation.

may be more difficult to describe such a structure as a 'geological structure or field *of any other* [i.e. non-petroleum] *mineral* deposit' (emphasis added).

The UK\Norway Agreement offers a relatively simple form of a straddling deposit clause. While delimitation agreements continue to be concluded with this type of clause, other agreements contain more sophisticated versions which might address such matters as the exchange of drilling information and perhaps dispute settlement.[112] In general, it seems fair to say that the more detailed the clause the more likely that it can be interpreted as referring only to hydrocarbon reservoirs and the exploitation of hydrocarbons. However, other forms of the clause use more general language that does not refer to hydrocarbon resources or petroleum reservoirs. For example, the France\Spain delimitation agreement of 1974 (which also creates a joint development zone (JDZ)) refers to the exploitation of a deposit of a natural resource.[113] It does not seem unreasonable to think that injection of CO_2 into a shared saline aquifer might qualify as the exploitation of that shared resource. But equally, other similar clauses are explicitly confined to petroleum and\or natural gas[114] or to the exploitation of deposits of mineral resources[115] and would therefore likely not apply to operations in saline aquifers.

112 See for example Article 2 of the Agreement between the Government of the Kingdom of Denmark together with the Home Government of the Faroe Islands and the Government of the United Kingdom of Great Britain and Northern Ireland relating to the Maritime Delimitation in the area between the Faroe Islands and the United Kingdom, Torshavn, 18 May 1999, 4 IMB 2956. IMB refers to the series *International Maritime Boundaries* (various editors, 1993 – date).

113 2 IMB 1727. See also Agreement between Italy and Spain Relating to the Delimitation of the Continental Shelf between the Two Countries, 19 February 1974, 2 IMB 1601, Article 2 applies to a 'deposit of natural resources'.

114 Agreement between the Governments of Malaysia and Indonesia on the Delimitation of the Continental Shelves between the Two Countries, 27, October 1969, 1 IMB 1019, Article 4; Maritime Delimitation Treaty between Columbia and Honduras, 2 August 1986, 1 IMB 503, Article 3, 'hydrocarbons or natural gas deposits or fields'; Agreement between the Government of the Kingdom of the Netherlands and the Government of the United Kingdom of Great Britain and Northern Ireland relating to the exploitation of single geological structures extending across the dividing line on the continental shelf under the North Sea, 6 October 1965, Cmd 3254, Article 1, 'geological mineral oil or natural gas field or structure' (this agreement is unusual since it was a stand-alone agreement dealing with straddling deposits and executed contemporaneously with a separate delimitation agreement).

115 Estonia\Russia (2005), 6 IMB 4582, Article 4; Trinidad and Tobago\Grenada (2010), 7 IMB 4705, Article VII straddling 'oil and gas structure or field or any other mineral deposit'; Venezuela\Netherlands (Antilles) (1978), 1 IMB 615, Article 6 'a single geological structure or mineral field containing oil or natural gas';

As for the field-specific unit operations agreements referred to above, it seems fairly clear that such agreements would not apply to either CO_2\EOR operations crossing over to pure disposal or to storage and disposal operations in an abandoned field (the Frigg situation). While such agreements may not terminate automatically with the cessation of production and abandonment,[116] it seems clear that all these agreements are concerned with production of petroleum and not with pure CO_2 disposal operations and are therefore inapplicable.

In sum, the general conclusion is that while delimitation agreements that deal with straddling deposits as well as subsequent field-specific agreements will likely also cover CO_2\EOR operations, it is much more difficult to conclude that such agreements were intended to address the use of shared saline aquifers for CO_2 disposal purposes. More specific drafting directed at the used of shared geological structures for storage or disposal purposes would be more appropriate. A simple version of such clause might be as follows:

> If any geological structure that has the potential to be used for geological storage or disposal purposes extends across the dividing line, and the part of such structure or field situated on one side of the dividing line is exploitable, wholly or in part, from the other side of the dividing line, the Contracting Parties shall, in consultation with the licensees, if any, seek to reach agreement as to the exploitation of that structure for storage or disposal purposes.

In addition, and in light of the discussion above of the London Protocol's Sub-Seabed Geological Formations Guidelines (2012), contracting parties to the Protocol might wish to add further text such as the following: 'Any such agreement should take account of any guidelines for the assessment of geological formations for storage or disposal purposes developed by any relevant international organization.' The suggested language would be broad enough to embrace not only the work within the London Protocol but also any future work within OSPAR or perhaps work by the IPCC.

4.3 *Framework Agreements*

Framework agreements dealing with shared hydrocarbon resources are bilateral agreements which aim to provide a detailed framework for the treatment of straddling petroleum deposits thereby obviating the need to negotiate field

116 The agreements provide for amendments or termination by agreement at any time: Frigg Agreement, (n 104) Article 31; Statfjord (n 104), Article 24, Murchison (n 104), Article 24.

specific treaties.[117] While most such agreements are free-standing agreements (e.g. the UK\Norway Framework Agreement), in some cases the framework provisions may be included in or appended to a delimitation agreement. Such is the case, for example, with the Norway\Russia Barents Sea Treaty. While specific answers to the above questions will of necessity turn on their precise language, these agreements generally apply only to 'petroleum activities' or some such similar term[118] or to hydrocarbon deposits.[119] It follows from this that while such agreements will therefore cover CO_2\EOR operations[120] they will not cover cross-over operations, saline aquifer disposal activities or disposal activities into abandoned reservoirs, except perhaps where the CO_2 is

117 The agreements include (1) Framework Agreement between the United Kingdom and Norway concerning Cross-Boundary Petroleum Co-operation, 4 April 2005, 7 IMB 5214, Report No. 9–15(6), (2) Agreement between Canada and the French Republic Relating to the Exploration and Exploitation of Transboundary Hydrocarbon Fields, 17 May 2005, unpublished, on file with the author, (the agreement has been ratified by France but not by Canada), (3) Framework Treaty relating to the unitization of hydrocarbon reservoirs that extend across the delimitation line between the Republic of Trinidad and Tobago and the Bolivarian Republic of Venezuela, 20 March 2007, 7 IMB 4649, Report No. 2–13(4), (4) Agreement between Iceland and Norway concerning transboundary hydrocarbon deposits, 3 November 2008, 7 IMB 5123, Report No. 5123, (5) Treaty between Norway and the Russian Federation concerning Maritime Delimitation and Cooperation in the Barents Sea and Arctic Ocean, 15 September 2010, 7 IMB 5167 and its attached Annex II dealing with Transboundary Hydrocarbon Deposits, Report No. 9–6(3), (6) Framework Agreement between the Republic of Cyprus and the Arab Republic of Egypt Concerning the Development of Cross-Median Line Hydrocarbon Resources, 21 January 2012 available on FAOLEX here http://faolex.fao.org/docs/pdf/bi-110369.pdf and (7) Agreement between the United States and the United Mexican States Concerning Transboundary Hydrocarbon Reservoirs in the Gulf of Mexico, 20 February 2012, 7 IMB 4613, Report No. 1–5(3). For discussion of some of these agreements see Cameron (n 96) and Bankes (n 97).

118 Article 1.1 of the UK\Norway Framework Agreement stipulates that the Agreement applies to cross-boundary cooperation between the two states with respect to petroleum activities. While the scope is arguably broader (since the substantive provisions also cover cross-boundary projects) other provisions make it clear that the agreement is confined to oil and gas activities and not disposal projects. Thus cross-boundary projects are defined in terms of transboundary petroleum reservoirs, pipelines and associated facilities. The Iceland\Norway Agreement refers only to the exploitation of transboundary hydrocarbon deposits; similarly the Canada\France Agreement is clearly confined to transboundary hydrocarbon fields and activities.

119 Norway\Russia, Article 5 and Annex II; Iceland\Norway; Cyprus\Egypt, Article 1, scope.

120 The UK\Norway Framework Agreement makes this clear insofar as it defines (Article 1.2) 'exploitation' as including injection, reinjection or storage of any substance used for or derived from the appraisal, production, treatment and processing of gas or liquids from a reservoir.

derived from the processing of produced substances (as is the case for example with the Norwegian Sleipner and Snøhvit projects).[121]

In sum, such framework agreements would require comprehensive redrafting to include pure CCS disposal operations. Alternatively, parties could negotiate a parallel framework agreement expressly addressing CCS operations in the full range of shared geological structures.

4.4 *Joint Development Agreements*

The literature refers to two main types of joint development agreements (JDAs).[122] The first category comprises JDAs which are put in place pending agreement on delimitation (provisional arrangements[123]) whereas the second category of JDA may be created as part of a delimitation agreement and is intended to be permanent. The distinction between these categories of agreement is likely of little significance for present purposes, since, as with the above arrangements, the question that concerns us is whether these agreements, whether temporary or permanent, whether associated with an agreement on delimitation or not, are intended to, or can be read to apply to, CCS related activities as well as petroleum activities.[124]

---

121 This exception follows from the definition of exploitation quoted, *id.* There is no doubt room to argue that the reinjection activities covered by the agreement are limited to injection in to hydrocarbon formations. The One North Sea Report (n 20) does however suggest (at 86) that the ideas underlying the Framework Agreement might inform a potential parallel agreement dealing with CCS.

122 See Masahiro Miyoshi, 'The Joint Development of Offshore Oil and Gas in Relation to Maritime Boundary Delimitation' (1999), 2(5) Maritime Briefing 1–51.

123 UNCLOS (n 3) Articles 74(3) & 83(3) and see discussion above in section 3.3.

124 The JDAs examined for the purposes of this chapter are as follows: (1) Agreement between the Government of Bahrain and the Kingdom of Saudi Arabia, 22 February 1958, 2 IMB 1489, Report No. 7–3; (2) Convention between France and Spain Concerning on the Delimitation of the Continental Shelf in the Bay of Biscay, 29 January 1974, IMB 1719, Report No. 9–2; (3) Agreement between Japan and the Republic of Korea Concerning the Establishment of Boundary in the Northern Part of the Continental Shelf Adjacent to the Two Countries, 30 January 1974, 1 IMB 1057, Report No. 5–15; (4) Norway\Iceland (Jan Mayen) (1980) 2 IMB 1755; (5) Memorandum of Understanding between Malaysia and the Kingdom of Thailand on the Establishment of a Joint Authority for the Exploitation of the Resources of the Sea-Bed in a defined Area of the Continental Shelf of the Two Countries in the Gulf of Thailand, 13 May 1990, 1 IMB 1099, Report No. 5–13(2); (6) Memorandum of Understanding between Malaysia and the Socialist Republic of Vietnam for the Exploration and Exploitation of Petroleum in a Defined Area of the Continental Shelf involving the Two Countries, 5 June 1992, 3 IMB 2335, Report No. 5–19; (7) Maritime Delimitation Treaty between Jamaica and the Republic of Colombia, 12 January 1993, 3 IMB 2179, Report No. 2–1, (8) Agreement between the Republic of Kazakhstan and the Russian Federation on Demarcation of the Seabed in the Northern Caspian Sea for the Purpose of Exercising Sovereign Rights to the Use of Subsoil

While some of these agreements are expressly confined to petroleum or mineral extraction activities[125] most are broader in their application. For example, the JDZ established by the France\Spain agreement of 1974 refers more generally to the prospecting for, and exploitation of, the resources of the JDZ[126] while the 1979 Agreement between Malaysia and Thailand refers generally to the 'exploration and exploitation of the non-living natural resources

---

Resources, 6 July 1998, and the Protocol, 13 May 2002, 5 IMB 4013, Report No. 11–1; (9) Memorandum of Understanding between the Royal Government of Cambodia and the Royal Thai Government regarding the Area of their Overlapping Maritime Claims to the Continental Shelf, 18 June 2001, 5 IMB 3735, Report No. 5–24, (an agreement to establish a joint development treaty); (10) Treaty between the Federal Republic of Nigeria and the Democratic Republic of São Tomé and Príncipe on the Joint Development of Petroleum and other Resources in Respect of Areas of the Exclusive Economic Zone of the Two States, 21 February 2001, 5 IMB 3638, Report No. 4–10; (11) Australia\Timor-Leste (2003), 5 IMB 3806, Report No. 6–20 (1) & (2); (12) Treaty Between the Republic of Guyana and the State of Barbados Concerning the Exercise of Jurisdiction in their Exclusive Economic Zones in the Area of Bilateral Overlap within each of their Outer Limits and beyond the Outer Limits of the Exclusive Economic Zone of Other States, 2 December 2003 5 IMB 3578, Report No. 2–27; (13) Agreement on the Exploration and Production of Hydrocarbons in the Common Interest Maritime Zone between the Democratic Republic of the Congo and the Government of the Republic of Angola, 30 July 2007, 6 IMB 4270, Report No. 4–15, and Protocol of Agreement… Approving Offshore Unitization Prospects, 10 September 2001, 6 IMB 4281, Report No. 4–16; (14) Mauritius\Seychelles, (2012) 6 IMB 4391. Mention might also be made to the Denmark\Faroes and UK agreement (n 112) This Agreement establishes a Special Area rather than a joint development zone. The Special Area is principally an area of shared rights and responsibilities with respect to fisheries matters but the Agreement also contemplates that special provisions should also apply to oil and gas operators working within the Special Area. Arguably these provisions and restrictions would also apply to CCS related operations. Finally, the chapter does not take account of the most recent Timor-Leste Australia (and its annex dealing with the Great Sunrise project) resulting from the compulsory conciliation proceedings initiated by Timor-Leste (n 82).

125 The Japan\Korea agreement applies to natural resources which is defined as petroleum and other resources produced in association with petroleum; the Malaysia\Vietnam JDZ applies to the exploration and exploitation of petroleum; the Congo\Angola agreement creates a 'common interest maritime petroleum zone"; the Australia\Timor-Leste agreement applies to the petroleum resources of the Joint Petroleum Development Area and the Norway\Iceland (Jan Mayen Agreement) applies to "co-operation [within the JDZ] … in connection with the exploration for and exploitation of hydrocarbon resources …'. Finally, while some of the language of the Protocol to the Kazakhstan\Russia Agreement refers more generally to economic activity in relation to straddling geological structures these provisions should likely be interpreted in light of the dominant concern of the agreement which is clearly the hydrocarbon resources of the Northern Caspian Sea.

126 France\Spain Agreement, Annex II. Paragraph 2 of the Annex refers to mining regulations but many jurisdictions licence CCS activities through the mineral or petroleum legislation. Other provisions in the Annex such as paragraph 7 which refers to products obtained from the JDZ may suggest a narrower interpretation that does not admit of the

of the sea-bed and subsoil' of the overlapping area of claims.[127] The treaty between Barbados and Guyana similarly applies to the 'jurisdiction, control, management, development, exploration and exploitation of living and non-living resources, as well as all other rights and duties established in the [Law of the Sea] Convention' within the Co-operation Zone.[128] The JDZ treaty between Nigeria and Sao Tome and Principe is also far-reaching.[129] This agreement applies to 'development activities' within the zone.[130] Such activities are defined as 'any economic activity in or concerning the Zone, including petroleum activity, fishing activity, all other activities for the development or exploitation of other mineral or living resources of the Zone and all forms of exploration and research related to any of the foregoing.' Furthermore, Article 32 provides that the Zone Plan (to be developed by the Joint Authority) may make provision for non-petroleum development activities within the Zone. Equally notable is the 2012 treaty between Mauritius and the Seychelles establishing the extended continental shelf beyond the respective EEZs of the two countries as a joint management area.[131] Within that area, the Contracting Parties, through a Designated Authority, 'shall jointly control, manage and facilitate the exploration of the continental shelf … and the conservation, development and exploitation of its natural resources'.[132]

It follows from the above that while all of these agreements would cover CO_2\EOR operations many of these agreements are also broad enough to cover cross-over operations from CO_2\EOR to pure disposal operations in depleted reservoirs, disposal operations in a previously abandoned reservoir[133] as well as to pure disposal operations in a saline aquifer. This would no doubt

application of the provisions to disposal activities but the general tenor of the treaty and the annex does admit of a broader application.

127 Malaysia\Thailand, MOU, Annex I, Article III. Furthermore Annex III (which establishes the joint authority) provides that the purpose of the joint authority 'exploration and exploitation of the non-living natural resources of the sea-bed and subsoil, in particular petroleum' which of course makes it crystal clear that the parties were contemplating the exploitation of other resources within the JDZ. See also the Guinea-Bissau\Senegal JDZ Agreement and accompanying Protocol of Agreement establishing a joint Agency and other institutions. These arrangements apply to fisheries and petroleum resources but also refers to other resources of the Zone.

128 Barbados\Guyana, Article 1.

129 Nigeria\Sao Tome and Principe (2001).

130 Ibid, Article 3.

131 Mauritius\Seychelles.

132 Ibid, Article 3.

133 See however Congo\Angola Agreement and the accompanying Protocol (which are concerned solely with a 'common interest maritime petroleum zone'). Article 10 of the Agreement contemplates that it will expire "upon completion of abandonment work of the installations of the final reservoir under exploitation. Article 16 of the annexed

require the joint authority (where provided for) to develop specific rules in relation to this activity.

5 Other Possible Arrangements

The analysis above shows that while delimitation treaties between adjacent and opposite states will frequently deal with the possibility of shared petroleum deposits they do not deal expressly with the use of shared geological structures for the purposes of CO_2 disposal or storage. While such agreements may certainly in some cases be read to include CO_2\EOR operations, it would be very difficult to read them as addressing other forms of CCS operations. Many JDAs are framed more broadly and thus might embrace CCS saline operations but that hardly addresses the question of whether they are really fit for the purpose. Accordingly, this last section canvasses the limited literature discussing what a bespoke agreement for CCS operations in a shared geological structure might look like.

The International Energy Agency has developed a Model Regulatory Framework for CCS activities.[134] While the principal purpose of the document is to provide advice on the design of a domestic regulatory framework, the Model also acknowledges the need to take account of a number of transboundary issues. These include not only export and transit issues but also the issues addressed in this chapter which the report lists as:[135]

- Unintended migration or leakage of injected CO_2 in the sub-surface across jurisdictional borders;
- Use of storage complexes that span jurisdictional borders;
- Secondary effects from storage activities occurring across jurisdictional borders (*e.g.* due to a sub-surface pressure front or displacement of sub-surface fluids across borders).

In response to these issues, the report identifies the need for regulatory coordination and harmonization and either the selection of a single competent authority for a CCS project or 'joint regulatory responsibility'.[136] Harmonization and coordination might extend to such things as coordinated bidding rounds for storage exploration and efforts to optimize the use of pipeline and other

---

unitization protocol contemplates that it will remain in force 'as long as the production period in the Unitization zone lasts'.

134 International Energy Agency, *Carbon Capture and Storage: model regulatory framework*, 2010. <https://www.iea.org/publications/freepublications/publication/model_framework.pdf>

135 Ibid at 30.

136 Ibid at 31.

infrastructure; application of market rules for access to storage; and agreement on liability and GHG accounting rules.[137]

In sum, the Model Framework provides some useful ideas but it is far from a systematic account as to the contents of either a state-to-state agreement or an operating agreement as between the private parties who might actually be charged with the responsibility for a CCS project.

Much the same can be said of the Technical paper prepared by the UNFCCC Secretariat in response to a request from the Subsidiary Body for Scientific and Technical Advice (SBSTA) in connection with qualifying CCS projects as CDM eligible.[138] That paper also recommends drawing upon our experiences with water (including aquifers) and petroleum resources[139] and briefly canvasses a broad range of relevant agreements. It notes that cooperation agreements in oil and gas already include common principles providing for *inter alia*: exchange of information; effective exploitation; equitable sharing of the resources; the protection of the environment including sometimes in joint management structures; appointment of contractors and operators; financial provision; and dispute resolution.[140] The report also notes that JDAs are frequently more integrated and institutionally richer than other bilateral arrangements.[141] The submissions to the Secretariat also emphasized that some guidance might be found in Decision 10/CMP.7 (and its Annex and Appendices), as to the topics that might be addressed in any bilateral agreement.[142]

Another source from which parties might draw are the unitization agreements typically used in the oil and gas industry to deal with production from straddling deposits of hydrocarbons.[143] Unitization will typically be achieved by agreement between those commercial parties who have an interest as a licensee (i.e. as a working interest owner) from either government to a

137 id at 31 & 32. The starting point for developing appropriate GHG accounting rules will be the default rules developed by the IPCC in its 2006 IPCC Guidelines for National Greenhouse Gas Inventories, available <https://www.ipcc-nggip.iges.or.jp/public/2006gl/>.

138 UNFCCC, Transboundary CCS Technical paper (n 6 and related text).

139 Ibid at para 66.

140 Ibid at para 70.

141 Ibid at para 158.

142 This is the decision of the meeting of the parties to the Kyoto Protocol accepting CCS project as eligible within the CDM. FCCC/SBSTA/2012/MISC.12 and Add.1 (submissions from Australia and Cyprus on behalf of the EU).

143 Jacqueline Lang Weaver and David F Asmus, 'Unitizing oil and gas fields around the world: a comparative analysis of national laws and private contracts' (2006), 28 Houston Journal of International Law 1–190; Ana E Bastida et al, 'Cross-Border Unitization and Joint Development Agreements: An International Law Perspective' (2007) 29 Houston Journal of International Law 355. See also the One North Sea Report (n 20) at 86, suggesting that relevant principles and ideas could be drawn from the UK\Norway Framework Agreement (n 117).

geographical area that forms part of the deposit. The unitization agreement will provide for the apportionment of production and expenses in accordance with the best scientific understanding as to the distribution of reserves underlying either side of the boundary and each of the different licence areas (or tracts). Thus each tract will be assigned a tract participation factor. The agreement will also make provision for future operations.[144] Unitization agreements are only necessary where there are multiple tracts and multiple parties. If there is a single tract, or if there is a single owner of the different tracts that form the reservoir (or storage unit), then unitization is arguably unnecessary. This would be an unusual scenario in an oil and gas context which is an industry which thrives on competition[145] but is perhaps a more likely scenario in a disposal context since operations are likely to be less competitive and more coordinated from the outset. Indeed, one of the goals of regulatory coordination between national authorities should likely be to try and ensure unified ownership of a target storage\disposal complex through coordinated offerings of storage acreage.

6 Conclusions

This chapter began with discussion of the storage part of the CCS chain considering CO_2\EOR operations as well as storage in a saline formation. A key conclusion is that injection in saline formations will likely affect a larger geographical area than injection in a depleted reservoir and that the effects of injection will include a pressure front in addition to the physical CO_2 plume.

Section 3 examined general rules pertaining to the duty to cooperate with respect to shared resources as well as work conducted under the auspices of the Protocol to the London Dumping Convention to develop a set of Guidelines for the Assessment of Carbon Dioxide for Disposal into Sub-seabed Geological Formations including one key provision dealing with cases where there is the potential for the transboundary movement of injected CO_2. A crucial point to

144 There may be one agreement dealing with unitization (the unitization agreement) and another agreement dealing with how to make decisions about subsequent operations (the unit operating agreement) or the two may be rolled together as a 'Unitization and Unit Operating Agreement'. The agreements will typically be confidential but standard forms are available and see also Weaver and Asmus, ibid, especially the Appendix commencing at 99 and containing sample clause from both unitization agreements and relevant domestic laws. One commonly used form is AIPN's (Association of International Petroleum Negotiators) 2006 Unitization and Unit Operating Agreement available on AIPN's website at <http://www.aipn.org/>.

145 Terence Daintith, *Finders Keepers? How the Law of Capture Shaped the World Oil Industry*, (Resources for the Future, 2010).

emphasise is that the Guidelines suggest that injection should only proceed in such a case where consent has been sought from all countries with jurisdiction over such a storage formation. Section 3 also examined the implications of Articles 74(3) and 83(3) UNCLOS for the situation in which one party may be proposing to engage in CCS operations in an area of overlapping maritime entitlements and in the absence of an agreed delimitation. The analysis suggested that any storage operations in the area of overlapping entitlements, including any evaluation operations that required drilling operations, could only be carried out with the consent of all parties with an overlapping entitlement. Unilateral operations would be a breach of the obligations of restraint under Articles 74(3) and 83(3). It would also seem to follow that injection operations conducted outside the area of overlapping entitlement which might affect that area (either the physical plume or the pressure front) would also be unlawful insofar as such effects might prejudice the sovereign rights of the other party (depending on the outcome of any agreement or decision on entitlement).

Section 4 examined existing boundary delimitation agreements and joint development agreements to see to what extent they may be read as being applicable to CCS and CO_2\EOR operations. The general conclusion is that most delimitation agreements, even those with straddling deposit clauses likely do not extend to cover CCS saline operations; however most if not all would cover CO_2\EOR operations on the basis that such operations are incidental to petroleum recovery. It should also be noted that such operations would likely only be conducted where there was already in place a treaty-based unitization arrangement with respect to the field and that that treaty might well address the introduction of EOR operations either directly or indirectly (by triggering a re-assessment of recoverable reserves). While some JDAs only address petroleum operations, others are broadly framed. Such agreements would certainly cover CO_2\EOR operations but some may be read more broadly to cover CCS operations in a saline formation. Whether they are fit for purpose however is another question.

Section 5 considered other possible arrangements for CCS operations in storage complexes that cross a maritime boundary. While some work has been done on this topic both in the context of the PLDC and the Kyoto Protocol (CCS eligibility under the CDM), this work falls far short of a model bilateral (or broader) treaty arrangement. Finally, while there is some suggestion that oil and gas unitization agreements might provide a good model for such projects, this claim may be doubted insofar as such arrangements are designed to deal with the situation of multiple owners and multiple tracts. Such situations are common in an oil and gas context but in the case of storage operations, neighbouring states should take a more coordinated approach from the outset with a view to avoiding the situation of multiple owners.

CHAPTER 18

Decommissioning of Offshore Installations: a Fragmented and Ineffective International Regulatory Framework

Seline Trevisanut

1 Introduction*

After decades of exploitation of offshore resources the problem of managing and removing platforms and installations constructed for this purpose urgently presents itself. The number of installations which are reaching the end of their life-cycle are increasing. The issue of removing outdated platforms and installations has become prominent since the 1980s. Due to technological advances in the oil and gas industries, the exploration of resources has extended to new depths and distances, increasing the number of structures in the sea and the impacts on the seabed. For instance, in 1950 there were only two offshore drilling rigs in the world. Yet, by 1988 the number of installations rose to 750,[1] and currently, around 7000 offshore platforms around the world are engaged in the exploitation of hydrocarbons.[2]

* The author would like to thank Nina Mileva for her research, Nikolaos Giannopoulos for his comments on previous drafts of this chapter, and Vivian Aiyedogbon for her editing work. The research for this paper has been conducted thanks to the financial support of the European Research Council under the European Union's Horizon 2020 research and innovation programme (Grant Agreement No 639070 - SUSTAINABLEOCEAN).

1 Z. Gao, 'International Law on Offshore Abandonment: Recent Developments, Current Issues and Future Directives', in Z. Gao (ed.), *Environmental Regulation of Oil and Gas*, London-The Hague-Boston, 1998, at 144.

2 Möller indicates that '[t]here are around 6.500 installations worldwide, about 4.000 of which are in the US Gulf of Mexico, 950 in Asia, 700 in the Middle East and 400 in Europe'; L. Möller, 'UN law on decommissioning offshore installations', M. Hammerson (ed.), *Oil and Gas Decommissioning*, Global Business Publishing, London, 2013, at 33. Lyons suggests however that there are more than '7000 offshore platforms installed globally for hydrocarbon exploration and production'; Y. Lyons, 'The New Offshore Oil and Gas Installations Abandonment Wave and the International Rules on Removal and Dumping', 29 *The International Journal of Marine and Coastal Law* (2014) 480–520, at 480.

Abandoned and decommissioned installations pose two main problems: firstly, they can consist of a threat to the safety of navigation, and secondly, they can be harmful to the marine environment. Already in 1958, the Geneva Convention on the Continental Shelf required the complete removal of abandoned and decommissioned installations,[3] because the exploitation of resources in the continental shelf should not result in any 'unjustifiable interference' with navigation, fishing, or the conservation of the living resources of the sea.[4] This rule however, was designed to mainly address unjustifiable interferences with other legitimate uses of the sea and the seabed.[5] The environmental aspect was not the main concern of the drafters in the early developments of the law of the sea, and this approach was subsequently confirmed with the adoption of the 1982 United Nations Convention on the Law of the Sea (UNCLOS).[6]

Nevertheless, installations of the latest generation have been designed and built considering the protection of the environment and the relevant legal framework. First generation structures, built in the period between 1950 and 1960, were indeed designed without any consideration for the potential need for their removal or abandonment. The issue of dealing with old structures became a point of concern for states only in the late 1980s. Actually, even the agreements on joint management and exploitation of offshore resources concluded in that period do not contain any specific provisions in this regard. The provisions concerning the prevention or reduction of pollution only referred to activities of extraction and transport of oil and gas.[7]

The international community however considered the issue at an earlier stage. The first steps towards managing the problem at the international level can be traced back to the Declaration of the United Nations Conference on

---

3 Art. 5, par. 5, of the 1958 Geneva Convention on the Continental Shelf, made in Geneva on 29 April 1958, entered into force on 10 June 1964.

4 Art. 5, par. 1, of the 1958 Geneva Convention on the Continental Shelf.

5 T. Treves, *Codification du droit international et pratique des Etats in le droit de la mer*, in *Recueil des Cours de l'Académie de droit international de la Haye*, 1990-IV, at 204.

6 Signed at Montego Bay on 10 December 1982, entered into force on 16 November 1994. Regarding the negotiations preceding the adoption of UNCLOS in 1982, Treves emphasized that 'le respect généralisé dont font l'objet les activités en matière pétrolière de ces temps de crise de l'énergie (années 1970) explique (…) le moins d'attention et de réglementation qu'a reçu jusqu'ici la pollution dérivant de ces activités'; cf. T. Treves, *La pollution résultant de l'exploration et de l'exploitation des fonds marins en droit international*, in *Annuaire Français de Droit International*, 1978, at 828.

7 M. Bathurst (ed.), *Joint Development of Offshore Oil and Gas, A Model Agreement for States for Joint Development with Explanatory Commentary*, London, 1989, at 355 ff.

the Human Environment adopted in 1972 in Stockholm.[8] Principle 21 of this Declaration contains a reminder that 'States have [...] the sovereign right to exploit their own resources pursuant to their own environmental policies', specifically pointing out their '*responsibility to ensure that activities within their jurisdiction or control do not cause damage to the environment of other States or of areas beyond the limits of national jurisdiction*' [emphasis added]. The reference to offshore activities is quite explicit, and addresses their development in the 1970s. Twenty years later, in 1992, Chapter 17 of the Agenda 21, concerning the 'Protection of the Oceans', details the obligations on states with respect to the prevention and fight against 'the degradation of the marine environment due to offshore activities', and provides in particular:

> States, acting individually, bilaterally, regionally or multilaterally and within the framework of IMO (International Marine Organization) and other relevant international organizations, [...] should assess the need for additional measures to address degradation of the marine environment [...] (c) From offshore oil and gas platforms, by assessing existing regulatory measures to address discharges, emissions and safety and assessing the need for additional measures. (para. 17.30).

This text, adopted prior to the entry into force of UNCLOS, highlighted a gap in the regime related to offshore installations and the need for a regulatory effort in this direction.[9]

More than twenty years after the adoption of Agenda 21, it may be noted that some grey areas persist, particularly with respect to the regime concerning abandoned or decommissioned platforms and installations. One can wonder whether there exists a general obligation to remove outdated structures, and if it does, what its content is. Furthermore, it can be discussed whether the removal of abandoned and disused structures and platforms is in the best interest of the marine environment. In fact, several examples in practice show

8 *Declaration of the United Nations Conference on the Human Environment*, Stockholm, 5–15 June 1972, available at <http://www.unep.org>.

9 The accident of the Deepwater Horizon drilling rig owned by the oil company British Petroleum (BP) in the Gulf of Mexico in April 2010 and its aftermath highlighted the gray areas that exist in this area of activity and related regulations. See, among others, R. Abeyratne, 'The Deepwater Horizon Disaster – Some Liability Issues,' in *Tulane Maritime Law Journal*, 2010, at 125 ff.; C. Chatterjee, Anna Lefcovitch, *The Gulf of Mexico Oil Disaster : some Legal Issues*, in *Amicus curiae : journal of the Society for Advanced Legal Studies*, 2010, at 17 ff.

the potential of former industrial structures to become artificial reefs offering habitats to certain species of fish.[10]

The present chapter analyses the international legal framework applicable to the decommissioning of offshore installations and highlights its ambiguities. The chapter also stresses how states still enjoy a broad discretion on how to deal with the issue at hand, specifically on whether to remove (partially or integrally) or not disused installations. This situation coupled with the fragmentation of the international regulatory framework is especially detrimental to the protection of the marine environment.

The analysis unfolds in four steps. First, the legal regime incorporated into the UNCLOS is sketched. The content of the obligation of partial removal is discussed and particular attention is given to the rules of reference contained in the relevant provisions (2). Then, the chapter focuses on the global generally accepted international rules and standards (GAIRS) and the instruments elaborated by the IMO (3). Thereafter, the focus moves to the GAIRS elaborated at the regional level in order to then present (4), in the following section, some considerations on the interaction between the universal GAIRS and regional instruments (5). The concluding remarks then stress how the lack of a binding instrument at the universal level and the geographical fragmentation of the existing legal regime act as an obstacle to the development of clear global GAIRS and ergo to a more coherent (and effective) legal framework for the decommissioning and removal of offshore installations (6).

2 The Law of the Sea Convention Framework

2.1 *The Obligation of Partial Removal*

With the 'offshore-isation'[11] of the exploitation of oil and mineral resources, the UNCLOS entered a new cycle of expansion of the rights of coastal states.[12] Technological advances and the increasing need for energy have added value to offshore seabed areas rich in natural resources, and over which coastal

10 M.J. Kaiser, Y. Yu, B. Snyder, 'Economic feasibility of using offshore oil and gas structures in the Gulf of Mexico for platform-based aquaculture,' in *Marine Policy* (2010) 699 ff.

11 T. DAHOU, 'La politique des espaces maritimes en Afrique. Louvoyer entre local et global,' in *Politique africaine*, n. 116, December 2009, at 10.

12 *Idem* (requests for extension of the continental shelf beyond 200 nautical miles).

states exercise sovereignty.[13] Therefore, in its exclusive economic zone (EEZ)[14] and continental shelf,[15] a coastal State has the right to build or authorize the construction of artificial islands, installations and structures for the exercise of licensed activities (Article 60(1); Article 80 UNCLOS[16]). These activities fall exclusively in its jurisdiction (Article 60(2)). The question of management of abandoned and decommissioned structures and platforms is addressed directly in Article 60 para. 3, which provides:

> Due notice must be given of the construction of such artificial islands, installations or structures, and permanent means for giving warning of their presence must be maintained. Any installations or structures which are abandoned or disused shall be removed *to ensure safety of navigation*, taking into account any generally accepted international standards established in this regard by the competent international organization. Such removal shall also have *due regard to fishing, the protection of the marine environment and the rights and duties of other States*. Appropriate publicity shall be given to the depth, position and dimensions of any *installations or structures not entirely removed.* [emphasis added]

From a strictly textual reading of the provision, it can be deduced that Article 60(3) does not impose an absolute obligation to remove offshore installations. While the general rule seems to be that of removal, the second part

13 Article 193 UNCLOS provides: 'States have the sovereign right to exploit their natural resources pursuant to their environmental policies and in accordance with their duty to protect and preserve the marine environment'.

14 The rights and obligations of coastal States in their exclusive economic zone are defined in Article 56 of UNCLOS. The exclusive economic zone is defined in Article 55 as: 'an area beyond and adjacent to the territorial sea, subject to the specific legal regime established in this Part, under which the rights and jurisdiction of the coastal State and the rights and freedoms of other States are governed by the relevant provisions of this Convention'.

15 The rights of the coastal State with respect to its continental shelf are defined in Article 77 of UNCLOS. The continental shelf is defined in Article 76(1) as : 'the seabed and subsoil of the submarine areas that extend beyond its territorial sea throughout the natural prolongation of its land territory to the outer edge of the continental margin, or to a distance of 200 nautical miles from the baselines from which the breadth of the territorial sea is measured where the outer edge of the continental margin does not extend up to that distance'.

16 Article 80 of UNCLOS provides that Article 60 of UNCLOS 'applies, mutatis mutandis, to artificial islands, installations and structures on the continental shelf.' The analysis and considerations regarding Article 60 of UNCLOS which will be detailed later in this study are also valid for structures on the continental shelf.

of the provision focuses on the safety of navigation, which weakens the absolute nature of the obligation. The removal seems to be mandatory only when the safety of navigation is at stake; thus it could be inferred that if the safety of navigation is not endangered, the structures could be left in place. Other legitimate uses of the sea (fishing) and the protection of the environment should be given 'due regard' in the decision making process, but do not consist of self-standing grounds for justifying the removal of the structure. Moreover, what clearly emerges is that entire installations and structures cannot be abandoned. The last part of Article 60(3) rather explicitly provides for an obligation to publicize 'the depth, position and dimensions of any installations or structures *not entirely removed*'. Nonetheless, Article 60(3) of UNCLOS only contains an obligation of partial removal, in order to ensure the safety of navigation.

This provision has raised much criticism[17] particularly since the wording of Article 60(3) differs substantially from that of the precedent Article 5(5) of the 1958 Geneva Convention on the Continental Shelf. The latter called on States to completely remove abandoned and disused installations.[18] Nevertheless, some authors read Article 60(3) as imposing a general obligation of full removal in order to ensure the safety of navigation, and saw abandonment or partial removal as exceptions subject to specific justifications.[19]

Articles 60 and 80 UNCLOS only apply to installations in the EEZ and continental shelf. There is no specific provision concerning installations in the territorial sea except that the coastal state can condition the exercise of the right of innocent passage to their protection (Article 19.2(k)); and 'shall give appropriate publicity to any danger to navigation, of which it has knowledge, within its territorial sea,' as for instance to offshore installations in activity or disused (Article 24.2). The coastal state is then bound by a series of obligations

---

17 See, among others, Z. Gao, 'Current Issues of International Law on Offshore Abandonment, with Special Reference to the United Kingdom,' in *Ocean Development and International Law*, 1997, at 59 ff. ; G.C. Kasoulides, 'Removal of Offshore Platforms and the Development of International Standards,' in *Marine Policy*, 1989, at 249 ff. ; AT McDade, 'The Removal of Offshore Installations and Conflicting Treaty Obligations as a Result of the Emergence of the New Law of the Sea: A Case Study,' in *San Diego Law Review*, 1987, at 645 ff.; A. Reglat-Boireau, 'La désaffectation des installations en mer,' in *Annuaire français de droit international*, 1982, at 873 ff.

18 Article 5(5) of the Convention on the Continental Shelf reads: Due notice must be given of the construction of any such installations, and permanent means for giving warning of their presence must be maintained. *Any installations which are abandoned or disused must be entirely removed* [emphasis added].

19 R. Wolfrum, N. Matz, *Conflicts in International Environmental Law*, Berlin-Heidelberg-New York, 2003, pat 110–111.

for the protection of the marine environment pursuant to Part XII UNCLOS, which also applies to the territorial sea.[20]

While exercising their sovereign rights pursuant to Article 193, a state shall comply with 'the obligation to protect and preserve the marine environment,' set in Article 192; and take the necessary measures to prevent, reduce and control pollution (Article 194).[21] Coastal states have a duty to ensure that pollution arising from activities within their jurisdiction and control does not extend beyond those areas (Article 194(2)).[22] If there is an imminent risk of damage to the marine environment, the coastal state has an obligation to notify any other state, which might be affected by such damage (Article 198).

Article 208 UNCLOS regulates the prevention, reduction and control of the pollution from seabed activities subject to national jurisdiction. This is, for instance, the pollution generated by the offshore oil and gas industry located in the EEZ or continental shelf. Coastal states are required to adopt laws, regulations and measures that 'shall be no less effective than *international rules, standards and recommended practices and procedures*' (Article 208(3), emphasis added).[23] Moreover, states, 'acting especially through *competent*

---

20 R. Beckman, 'Global Legal Regime on the Decommissioning of Offshore Installations and Structures', in M. Nordquist *et al.*, *The Regulation of Continental Shelf Development, Rethinking International Standards*, Martinus Nijhoff, Leiden, 2013, 259–280, at 279.

21 For a recent interpretation of those articles in relation to the construction of installations and artificial islands, see Permanent Court of Arbitration, *The South China Sea Arbitration*, Award of 12 July 2016, *spec.* paras 983 ff.

22 This UNCLOS provision re-affirms the well-established customary norm prohibiting transboundary pollution. First affirmed in the *Trail Smelter* case (United States v. Canada, 11 March 1941, RIAA 1941 vol. III, 1905), the International Court of Justice (ICJ) provided that 'The existence of the general obligation of States to ensure that activities within their jurisdiction and control respect the environment of other States or of areas beyond national control is now part of the corpus of international law relating to the environment' (ICJ, *Advisory opinion on the legality of the threat or use of nuclear weapons*, 8 July 1996, para. 29). See also Y. Lyons, 'Transboundary Pollution from Offshore Oil and Gas Activities in the Seas of the Southeast Asia', in R. Warner, S. Marsden (eds.), *Transboundary Environmental Governance. Inland, Coastland and Marine Perspective* (Ashgate, 2012) 167.

23 UNCLOS does not define concepts such as 'international rules', 'standards' and 'recommended practices and procedures', which remain vague also in practice. See R.R. Churchill and A.V. Lowe, *The Law of the* Sea, 3rd edition (Manchester University Press, Manchester, 1999) at 346; B. Oxman, 'The duty to Respect Generally Accepted International Standards', *New York University Journal of International Law and Politics*, 1991–92, vol. 24, pp. 109 ff.; S. Trevisanut, 'La Convention des Nations Unies sur le droit de la mer et le droit de l'environnement : développement intrasystémique et renvoi intersystémique', in H. Ruiz Fabri and L. Gradoni (eds.) *La circulation des concepts juridiques: le droit*

international organizations or diplomatic conference, shall establish global and regional rules, standards and recommended practices and procedures to prevent, reduce and control pollution of the marine environment referred to in paragraph 1' (Article 208(5), emphasis added). This provision builds an important 'bridge' between the UNCLOS and other relevant normative sources, which do not necessarily generate from treaties, but derive from the work of relevant actors. Such includes, international organizations; diplomatic conferences; and professional associations which develop recommendations and best practices in a specific economic sector.[24]

A similar rule of reference[25] is also contained in Article 210 UNCLOS concerning the prevention, reduction and control of pollution by dumping. Dumping consists of the placement for disposal at sea of wastes and other materials, and of 'vessels, aircraft, platforms or other man-made structures at sea' (Article 1(5)). The abandonment of installations or parts of the installations can be considered dumping[26] and thus falls under the relevant legal regime. The substantive rules on dumping and on seabed activities have been (or should have been developed) outside the UNCLOS, and within the framework that the convention creates through reference to generally accepted rules and standards (GAIRS) and the duty to respect them.

2.2 *The Duty to Respect Generally Accepted Rules and Standards*

UNCLOS was born as a framework convention and a living instrument. In order to avoid becoming outdated (or even obsolete), the drafters resorted in using rules of reference in many of its provisions,[27] in particular in those creating

international de l'environnement entre mondialisation et fragmentation (Société de législation comparée, 2009) 416. The issue is further discussed below (see Section 2.2.)

24 C. Redgwell, 'Mind the Gap in the GAIRS: The Role of Other Instruments in LOSC Regime Implementation in the Offshore Energy Sector', in N. Bankes, S. Trevisanut (eds), *Energy from the Sea: An International Law Perspective on Ocean Energy*, Brill Nijhoff, 2015, at 40.

25 Roach highlights how the rules of reference in art. 208 is however slightly different that the one contained in other articles regulating pollution: 'It should be recognized that this firm requirement on States ("shall adopt") contrast sharply with the much looser standard of "shall endeavour" to establish rules in article 207 (pollution of the marine environment from land-based sources), in article 210 (by dumping), and in article 212 (from and through atmosphere);' A. Roach, 'International Standards for Offshore Drilling', in M.H. Nordquist et al. (eds), *The Regulation of Continental Shelf Development, Rethinking International Standards* (Martinus Nijhoff, Leiden, 2013) 105.

26 For a discussion on whether the abandonment of installation qualifies as dumping, see *inter alia* Y. Lyons, (n2), 484–491.

27 On the drafting history of the rules of reference see Oxman, (n23), at 121 ff.

obligations in the field of environmental protection. Article 60(3) provides that the removal of any installations should be performed 'taking into account any *generally accepted international standards established in this regard by the competent international organization*' (emphasis added). Moreover, Article 208 provides that, states shall not only adopt laws, regulations and measures 'no less effective than *international* rules, standards and recommended practices and procedures' (para. 3, emphasis added), but they shall also elaborate those '*global and regional* rules, standards and recommended practices and procedures to prevent, reduce and control pollution of the marine environment' from seabed activities (para. 5, emphasis added). Additionally Article 210(4), concerning pollution by dumping, encourages states, 'acting especially through competent international organizations or diplomatic conference, (...) to establish *global and regional* rules, standards and recommended practices and procedures to prevent, reduce and control such pollution.'

The IMO has taken the lead concerning both the treatment of disused installations and pollution by dumping. It has elaborated some guidelines for the removal of installations and is the depositary of the London Dumping Convention, both discussed below. However, its role in the development of the regulatory framework in relation to the offshore oil and gas industry has raised some criticism.[28] The IMO mandate, as amended, includes 'the general adoption of the highest practicable standards in matters concerning *maritime safety* (...) and prevention and control of *marine pollution from ships*'.[29] Notwithstanding, the IMO Legal Committee has pointed out: 'while pollution directly arising from exploration/exploitation is however not of direct concern of IMO, the Organization may contribute to the establishment of international regulations'.[30] The IMO Legal Committee in particular supports the development of guidance for states in their effort to conclude arrangements at the bilateral and regional level on liability and compensation issues connected to transboundary pollution damage, resulting from offshore oil exploration and

28 Some states and some commentators have raised several points of criticism about the role of the IMO in the development of rules and standards concerning offshore installations. The details of such debate are beyond the scope of the present chapter. For a critical voice, refer to J.A. Roach, 'International Standards for Offshore Drilling,' in M.H. Nordquist *et al.* (eds) *The Regulation of Continental Shelf Development, Rethinking International Standards*, Martinus Nijhoff, Leiden, 2013, at 107.

29 Convention on the Intergovernmental Maritime Consultative Organization (Geneva, 6 March 1948, in force 17 March 1958) 289 UNTS 3, emphasis added.

30 See 'Implications of the United Nations Convention on the Law of the Sea for the International Maritime Organization', IMO Doc. LEG/Misc.7 Annex, 19 January 2012, at 18.

exploitation.[31] Not all IMO contracting parties, however, have support for what they perceive as an unjustified extension of the IMO mandate.[32] This undermines the 'generally accepted' character of the rules and standards which the organisation elaborates.

General acceptance needs to be assessed then on the basis of the subsequent practice of states. Consequently, the 1989 IMO *Guidelines and Standards for the Removal of Offshore Installations and Structures on the Continental Shelf and in the Exclusive Economic Zone* (hereinafter, 1989 IMO Guidelines)[33] are considered as GAIRS.[34] Within the London Dumping system, Lyons for instance suggests that, unlike the 1972 London Convention, its 1996 Protocol does not yet qualify as global rule under UNCLOS because it has gained general acceptance only in some regions of the world.[35] This also suggests that the 1996 Protocol may be regarded as GAIRS in some areas of the world where the majority of relevant states has ratified it, and hence as regional GAIRS and not global rules.

This interpretation can find further support in the joint reading of Articles 208 and 210 with Article 237 UNCLOS. Article 237 consists of a specific compatibility clause for Part XII of the UNCLOS. The provision in particular affirms that the provisions of Part XII are 'without prejudice (...) to agreements which may be concluded in furtherance of the general principles set forth in this Convention' (para. 1). This provision facilitates the application of UNCLOS in the relevant normative context and of environmental law instruments in the context of marine environment protection. It enables the application of the 1996 Protocol, within the UNCLOS, for states that are parties to it, and thus recognises their practice as relevant for the interpretation and application of the relevant UNCLOS provisions (Articles 208 and 210).

31 See IMO doc. LEG 99/14, 24 April 2012, at para. 13.16.

32 A. Roach, 'International Standards for Offshore Drilling', at 105. See also see A. Chircop, 'The International Maritime Organisation,' in D.R. Rothwell *et al.*, *The Oxford Handbook of The Law of the Sea*, Oxford University Press, Oxford, 2015, at 429.

33 IMO, Resolution A.672 (16), of 19 October 1989.

34 Y. Lyons, 'The New Offshore Oil and Gas Installations Abandonment Wave,' n_, at 495. Contra, see A. Proelss, United Convention on the Law of the Sea, A Commentary, Beck/ Hart, Oxford, 2017, at pp. 474–475.

35 Ibid., at 506, 510.

3 Generally Accepted International (Global) Rules and Standards

3.1 *IMO Guidelines*

The obligation of partial removal, as affirmed in the UNCLOS, is justified by financial considerations surrounding the operation, and confirmed by the 1989 IMO Guidelines.[36] These Guidelines were developed by the IMO pursuant to its competences in the field of maritime safety[37] and are thus primarily concerned with the safety of navigation. They deal to a lesser degree with fisheries and the environment, as the mandate of the IMO does not include general jurisdiction over environmental matters.

The Guidelines provide that the processing of installations is performed pursuant to a case by case determination by the coastal State. They are therefore characterized by a high degree of flexibility, which contributes to their success with coastal states.[38] Furthermore, the Guidelines provide for the removal of installations which are located at a depth of less than 75 meters, and weigh less than 4000 tons. Nonetheless, the possibility of partial removal or abandonment of the structures remains in cases when their removal is not feasible from a technical point of view (first generation structures); when the removal is excessively costly; or when the removal might pose an unacceptable risk for people or for the environment. Additionally, the Guidelines accommodate the possibility of re-using installation, in particular as an artificial reef for the development of fisheries. This is in line with the general criteria for removal provided by Article 60(3), namely that the removal procedure has due regard,

36 IMO, Resolution A.672 (16), of 19 October 1989, paras 3.6, 3.8, 3.12. In this regard see R. Beckman, 'Global Legal Regime on the Decommissioning of Offshore Installations and Structures', in M. Nordquist *et al.*, *The Regulation of Continental Shelf Development, Rethinking International Standards*, Martinus Nijhoff, Leiden, 2013, at 266.

37 S. Rosenne, 'The International Maritime Organization Interface with the Law of the Sea Convention,' in M.H. Nordquist, J.N. Moore (eds.), *Current Maritime Issues and the International Maritime Organization*, The Hague-Boston-London, 1999, at 260. The author recalls that UNCLOS requires the collaboration of the IMO, as the competent international organization in 72 sectors, 19 of which are not shared with other organization (such as FAO, ILO, or ICAO). The development of standards for the removal of offshore stuctures is featured among the 19 sectors over which the IMO has jurisdiction. On the role of the IMO in the development of international rules and standards, see A. Chircop, (n32), at 429; S.N. Nandan, S. Rosenne (eds.), *United Nations Convention on the Law of the Sea 1982, A Commentary*, vol. II, Dordrecht-Boston-London, 1993, at 587; T. Treves, (n5) at 206.

38 T. Treves, (n5), at 205: 'The success of the IMO initiative seems to confirm that the requirement to introduce a measure of flexibility in the rule on the obligation to remove is widespread in the international community'.

not only to the security of navigation, but also to 'fishing, the protection of the marine environment and the rights and duties of other States'.[39]

As illustrated by the 1989 IMO Guidelines, the objective of environmental protection sometimes clashes with the obligation to remove an installation when the latter has become an artificial habitat for natural resources, mainly for some fish stocks.[40] An abandoned or decommissioned offshore installation which has become an artificial habitat for certain species may still however be considered a form of pollution when it is an obstacle to fishing and other legitimate uses of the sea. States consequently find themselves balancing between environmental interests, especially the obligation to protect living resources, on the one hand, and the prevention and control of pollution on the other hand. More specifically, states find themselves stranded between, first, Articles 208 and 214 concerning the pollution resulting from seabed activities, and second Articles 56, 61, 117 and 118 concerning the protection of living resources. The Guidelines do not provide any further guidance on how to strike such a balance, leaving an important margin of discretion to coastal states.

3.2 *London Dumping Convention and Its 1996 Protocol*

As mentioned above, the remains of or disused structures can also be abandoned in accordance with the 1972 London Convention for the Prevention of Marine Pollution by Dumping of Wastes and Other Matter (hereinafter, 1972 London Convention).[41] Article III of the 1972 London Convention (replaced

39 Some authors claim that paragraph 3 of Article 60 does not impose any absolute obligation as to the removal of facilities but rather suggests general criteria for determining the removal procedures in specific cases; see S.N. Nandan, S. Rosenne (eds.), (n37) at 585.

40 There are even mentions of a "beneficial" environmental effect of certain structures which develop into habitats for commercially relevant species; see K. Bangert, 'Environment Protection – Offshore, Oil and Gas Production', in U. Karpen (ed.), *Maritime Safety – Current Problems of Use of the Baltic Sea*, Baden-Baden, 2005, 88; P Macreadie. A. Fowler, B. Booth, 'Rigs to Reef: will the Deep Sea Benefit from Artificial Habitats?,' in *Frontiers in Ecology and the Environment* (2011) 455. According to Beckman, '[t]here are three methods by which installations and structures could be used for the creation of artificial reefs. First, the top part could be removed and the bottom of the structure left in-situ, with sufficient clearance so that it does not pose a threat to navigation. Second, the structure could be toppled on site with sufficient clearance so that it does not pose a threat to navigation. Third, the structure could be cut up and parts of it moved to specific locations to attract marine life'; R. Beckman, 'Global Legal Regime on the Decommissioning of Offshore Installations and Structures', in M. Nordquist *et al.*, *The Regulation of Continental Shelf Development, Rethinking International Standards*, Martinus Nijhoff, Leiden, 2013, at 275.

41 Signed in London on 29 December 1972, entered into force on 30 August 1975. The 1972 Convention was substituted by the 1996 Protocol. *1996 Protocol to the Convention on the Prevention of Marine Pollution by Dumping of Wastes and Other Matter*, signed in London on 7 November 1996, entered into force on 24 March 2006.

by the 1996 Protocol) provides that 'dumping' means 'any deliberate disposal at sea of vessels, aircraft, platforms or other manmade structures at sea'. The 1972 London Convention does not contain a comprehensive ban on dumping,[42] and neither does the 1996 revised text of the Convention provide for an absolute prohibition of dumping of platforms and installations.

Through its amending Protocol of 1996, the 1972 London Convention integrates the precautionary principle. This reverses the logic of the previous system by prohibiting the dumping of seriously harmful substances[43] and the incineration of waste at sea. The preamble of the Protocol also considers relevant international instruments, including in particular UNCLOS, the Rio Declaration and Agenda 21 of 1992. Article 210 (5) and (6) of UNCLOS concerning pollution through dumping provides:

> (5) Dumping within the territorial sea and the exclusive economic zone or onto the continental shelf shall not be carried out *without the express prior approval of the coastal State*, which has the right to permit, regulate and control such dumping after due consideration of the matter with other States which by reason of their geographical situation may be adversely affected thereby. (6) National laws, regulations and measures shall be no less effective in preventing, reducing and controlling pollution than *the global rules and standards*. [emphasis added]

The UNCLOS here refers to the 1972 London Convention,[44] which preceded the UNCLOS and was taken into account in the drafting process. UNCLOS suggests the possibility of normative developments leading to a stricter regime. Indeed,

42 Z. Gao, (n1) at 164.

43 Pursuant to Article IV of the 1972 Convention, before its amendment in 1996, substances whose dumping is completely prohibited are inserted into the 'black list' of Annex I to the Convention; then there is the 'gray list' of Annex II, which contains the names of substances whose dumping requires special permission; and finally Annex III lists the substances and wastes the dumping of which may be allowed in general. Art. 4 of 1996 on the other hand prohibits all dumping except for substances listed in Annex 1, the dumping of which is subject to obtaining an authorization under the criteria set out in Annex 2.

44 C. Redgwell, 'From Permission to Prohibition: The 1982 Convention on the Law of the Sea and the Protection of the Marine Environment', in D. Freestone, R. Barnes, D.M. Ong (eds.), *The Law of the Sea, Progress and Prospects*, Oxford, 2006, at 190 : 'Far from being self-contained regimes, the [UNCLOS] and the London Convention of 1972 can be defined as 'Reinforcing integrated schemes', insofar as they contaminate each other by absorbing the changes that occur on one side and another, but also outside, as in the grounds of the IMO'; see also S. Trevisanut, 'La Convention des Nations Unies sur le droit de la mer et le droit de l'environnement: développement intrasystémique et renvoi intersystémique', in H. Ruiz Fabri, L. Gradoni (eds.), *La circulation des concepts juridiques : le droit international de l'environment entre mondialisation et fragmentation*, Paris 2009, at 411.

the regime created by the 1972 London Convention, as modified by the 1996 Protocol, may be seen as being slightly stricter than the one created by the UNCLOS. The dumping of offshore platforms and installations is mentioned in Annex 1 of the 1996 Protocol, in reference to wastes or other matter that may be considered for dumping (para. 1.4, Annex 1). The text of the new agreement allows states to adopt stricter national measures banning the dumping of wastes and other materials specified in Annex 1 of the Protocol (Article 4(2) of the 1996 Protocol).

Wastes and other materials may be considered for dumping provided such material that creates floating debris or otherwise contributes to the pollution of the marine environment has been removed; and only when it has been ascertained that the material dumped poses no serious obstacle to fishing or navigation (para. 2, Annex 1). Article 4(1)(2) of the 1996 Protocol indicates that particular attention shall be paid to opportunities to avoid dumping in favour of environmentally preferable alternatives. The national authority responsible for granting permission for dumping must assess the existence of less environmentally harmful alternatives; and the owner of the structure seeking permission to perform dumping must prove that there is no other more preferable alternative.[45] The regime created by the 1972 London Convention as amended by the 1996 Protocol is slightly stricter than the one created by the UNCLOS, in the sense that full removal is general standard, and dumping is a possible alternative only if there is no evidence to the contrary.

The London Convention system thus creates a presumption in favour of the removal of an abandoned or decommissioned structure. Hypothetically, this preference may be reversed if the structure becomes an artificial reef, namely when the abandonment takes place for environmental reasons, thus for a purpose other than the mere disposal of the installation. Unlike the UNCLOS, the London Convention does not only address coastal States, but also owners of structures who bear the burden of proof with respect to the inevitability of the dumping. These owners (private persons) thus assume an important role in the decision-making process concerning the treatment of disused structures. This might pose a problem with respect to the availability of means, as the two concerned entities – the national authority responsible for the granting of authorizations and the owner of the structure – may not have the same

45 IMO 2000 *Specific Guidelines for Assessment of Platforms or Other Man-Made Structures at Sea*, para. 5, available at: <https://cil.nus.edu.sg/2000-specific-guidelines-for-assessment-of-platforms-or-other-man-made-structures-at-sea/>. See also H. Esmaeili, *The Legal Regime of Offshore Oil Rigs in International Law*, Ashgate, Dartmouth, 2001, at 201.

technical and economic means at their disposal. It is sufficient to think of examples of developing countries where the *major* oil companies are most active. This imbalance might be mitigated in the future by the work the London Convention/Protocol TC Trust Fund[46] within the Technical Co-operation Committee (TC) of the IMO. London Convention/Protocol TC Trust Fund is responsible for managing donations for cooperation and technical assistance. The work of the fund is particularly important in regions where no specific instrument regulates the protection of the marine environment, generally, and dumping, specifically.

4 International (Regional) Rules, Standards and Recommended Practices and Procedures

Both Article 208.5 and 210(4) UNCLOS refer to the need to elaborate instruments at the regional level in order to regulate, respectively, pollution from seabed activities and dumping. Four regional seas instruments have either specific provisions or specific instruments regulating the decommissioning of offshore installations and/or their dumping. The content of each regional system is briefly described and analysed in order of strength of their provisions concerning the dumping of installations.

4.1 *The Offshore Protocol to the Kuwait concerning Marine Pollution Resulting from Exploration and Exploitation of the Continental Shelf*

The 1989 Offshore Protocol to the Kuwait Regional Convention for Co-operation on the Protection of the Marine Environment from Pollution,[47] permits partial removal in the interests of safety of navigation and fishing (Article XIII(1)(b)). Contracting states shall also have regard to any guidelines issued by the regional organization. The Marine Emergency Mutual Aid Centre (MEMAC)[48] has however not yet elaborated such guidelines concerning the decommissioning of installations. It has nonetheless adopted *Guidelines on Requirements for the*

---

46 IMO, doc. LC-LP.1/Circ.33, 19 January 2010.

47 *Protocol Concerning Marine Pollution Resulting from Exploration and Exploitation of the Continental Shelf*, 2065 UNTS 68, available at <https://treaties.un.org/doc/Publication/UNTS/Volume%202065/v2065.pdf>.

48 The Marine Emergency Mutual Aid Centre (MEMAC) was created in 1982 by the Protocol concerning Regional Co-operation in Combating Pollution by Oil and other Harmful Substances in Case of Emergency to the Kuwait Convention. For more information about this regional organisation, see: <http://memac-rsa.org/en/home>.

Conduct of Environmental Impact Survey and the Production of Environmental Impact System (hereinafter, *Guidelines on Requirements for the Conduct of Environmental Impact Survey*).[49] Those guidelines are particularly important because the Protocol does not contain a clear and general obligation of environmental impact assessment, (EIA) as it allows contracting parties not to require such an assessment before a new activity starts. When the contracting state decides not to request an EIA, it 'shall consider' performing a survey of the marine environment (Article IV.2). Contracting parties then have a wide margin of discretion under this instrument, which is, however, limited by the global legal framework analysed above.

The *Guidelines on Requirements for the Conduct of Environmental Impact Survey* mention in the terms of reference of the assessment that those terms require the consideration of '[p]lans for removal of any plant or equipment, and any reclamation, restoration or clearing of the site after cessation of operations' (4.2(j)). The same guideline however recognises the wide margin of appreciation to the authority requiring the assessment as it can strike out any term of reference which is considered unnecessary by the objectives of the survey (4.2.). In respect of decommissioning of offshore installations, the Kuwait Convention thus does not provide for clear obligations to complement the general legal framework sketched above.

4.2 *The Dumping and Offshore Protocols to the Barcelona Convention for the Protection of the Mediterranean Sea against Pollution*

The 1976 Barcelona Convention for the Protection of the Marine Environment and the Coastal Region of the Mediterranean contains two instruments which are relevant for the issue at hand: the Protocol for the Prevention of Pollution in the Mediterranean Sea by Dumping from Ships and Aircraft (hereinafter, Barcelona Dumping Protocol, not yet into force)[50] and the Protocol for the Protection of the Mediterranean Sea against Pollution Resulting from Exploration and Exploitation of the Continental Shelf and the Seabed and its Subsoil (hereinafter, Barcelona Offshore Protocol).[51]

The Barcelona Dumping Protocol applies to platforms and other structures (Article 3.1) and follows the same rational as the London Convention system,

49 Available here: <http://memac-rsa.org/assets/fileManager/1_GUIDELINES_Impact_Assessment_1.pdf>.

50 UNEP(OCA)/MED IG.6/7; text available at < http://web.unep.org/unepmap/who-we-are/legal-framework>.

51 UNEP(OCA)/MED IG.4/4; text available at <https://wedocs.unep.org/rest/bitstreams/2336/retrieve>.

namely dumping is prohibited except when it is permitted. The Barcelona Offshore Protocol, on the other hand, consists of a quite detailed instrument which aims at covering the complete life-cycle of an offshore operation. In setting a number of mandatory requirements for the authorization procedure (Articles 4–7), the protocol focuses on the role of both authorizing states and the industry, i.e. the operator, in assessing the environmental impact of a planned activity, in monitoring it and in reacting to possible emergencies.[52] Article 5 in particular requires that the project submitted by the candidate operator also includes '(g) The plans for removal of installations as specified in Article 20.' Article 20 extensively reaffirms Article 60.3 UNCLOS and adds elements concerning the role of private actors. The article provides that:

> The *operator shall be required by the competent authority to remove* any installation which is abandoned or disused, in order to ensure safety of navigation, taking into account the guidelines and standards adopted by the competent international organization. Such removal shall also have due regard to other legitimate uses of the sea, in particular fishing, the protection of the marine environment and the rights and duties of other Contracting Parties. Prior to such removal, *the operator under its responsibility shall take all necessary measures to prevent spillage or leakage from the site of the activities* (Article 20(1), emphasis added).

Moreover, Article 20 also creates an obligation upon coastal state to act in the event in which the operator 'fails to comply with the provisions of this Article.' '[T]he competent authority shall undertake, at the operator's expense' any necessary action, including the actual removal of the installation.

The Mediterranean Action Plan,[53] adopted in the framework of the Offshore Protocol, aims in particular at encouraging the adoption of further safety measures at the regional level, in the time frame of 2016–2024. These measures should include: '[s]etting-up a governance framework to support the implementation of the Action Plan and the adoption, enforcement and monitoring of regional standards, procedures and rules;' [d]efining commonly agreed regional offshore standards and guidelines'; and develop and adopt '[c]ommon

52 S. Trevisanut, 'The Role of Private Actors in the Offshore Energy Industry,' 29 *International Journal of Marine and Coastal Law* (2014) 645.

53 Mediterranean Offshore Action Plan in the framework of the Protocol for the Protection of the Mediterranean Sea against Pollution resulting from Exploration and Exploitation of the Continental Shelf and the Seabed and its Subsoil, UNEP(DEPI)/MED IG.22/28, available at: <https://wedocs.unep.org/rest/bitstreams/8381/retrieve>.

criteria, rules and procedures for the removal of installations and the related financial aspects adopted'.[54] The Action Plan also provides the creation of two sub-groups within the Barcelona Convention Offshore Oil and Gas Group (BARCO OFOG) which will be in charge of the removal of installation, namely the OFOG Sub-Group on environmental impact and the OFOG Sub-Group on health and safety.[55]

In relation to the development of regional standards and guidelines, the document emphasizes the need of Regional Guidelines on removal of installations and the related financial aspects,[56] highlighting in this way the shortcomings of the global and regional existing legal framework and the crucial importance of clarification and harmonization.

4.3 *The Helsinki Convention on the Protection of the Marine Environment of the Baltic Sea Area*

The Helsinki Convention for the protection of the Marine Environment of the Baltic Sea Area[57] contains Annex VI on the prevention of pollution from offshore activities. Pursuant to Regulation 8 of Annex VI, concerning disused offshore units, the 'Contracting Parties shall ensure that abandoned, disused, and accidentally wrecked offshore units are *entirely removed* and brought ashore *under the responsibility of the owner'* (emphasis added). This regulation restates Article 5(5) of the 1958 Geneva Convention[58] in stricter terms, and lays the primary responsibility of removal on the private operator (the owner). Differently to the Barcelona Offshore Protocol, the Helsinki Convention does not specify what actions the State should take if the owner of the structure refuses to comply with his responsibilities.

Annex VI applies to all offshore units defined in Regulation 1 which includes: 'any fixed or floating offshore installation or structure engaged in gas or oil exploration, exploitation or production activities, or loading or unloading of oil'. This definition thus limits the scope of application to structures engaged in activities related to the exploration and exploitation of hydrocarbons. Hence, the structures engaged in the production of renewable energy are not included.

54 Ibid, at 214, 220.

55 Ibid, at 215–216.

56 Ibid, at 220.

57 Adopted in Helsinki on 24 September 1992, entered into force on 17 January 2000. The text replaces the previous 1992 Convention for the Protection of the Marine Environment of the Baltic Sea for 1974. The agreed text of 1992 was amended four times (2000, 2002, 2004, 2007). The articles which are referred to in this study correspond to the text as amended by the entry into force in 2008 of the last amendments, available at <http://www.helcom.fi>.

58 K. Bangert, (n40) at 88.

The Helsinki Convention then prohibits pollution by dumping. This includes 'any deliberate disposal at sea or into the seabed of wastes or other matter from ships, other man-made structures at sea or aircraft' (Article 2(4)(a)(ii)). Offshore installations, including those which are not engaged in oil and gas sector, should not be dumped; thus, we can assume that there exists a presumption in favour of the removal of such installations. The prohibition of dumping is however not absolute. Contracting parties can issue a permit for the dumping of certain dredged material, provided that the criteria set out in Annex V are respected (Article 11(2)). Moreover, Article 11(3) provides that:

> The provisions of this Article shall not apply (…) in any case which constitutes a danger to human life, if dumping appears to be the only way of averting the threat and *if there is every probability that the damage consequent upon such dumping will be less than would otherwise occur*. Such dumping shall be so conducted as to minimize the likelihood of damage to human or marine life. [emphasis added]

It may be observed that this provision contains the same criteria as the 1996 London Dumping Protocol. Namely, dumping should take place only when no other alternative solution provides comparable guarantees. However, the Helsinki Convention does not provide detailed guidelines on the burden of proof regarding the inevitability of dumping as the London Convention. Given the role assigned to the owner of the offshore installation by Annex VI it may, however, be assumed that the owner is also called to play a role in the determination of the burden of proof for dumping, at least with respect to facilities used by the oil and gas industries.

4.4 *The OSPAR Convention for the Protection of the Marine Environment of the North-East Atlantic*

The North-East Atlantic is filled with offshore structures involved in the exploration and exploitation of non-living resources. Already in the 1970s, coastal states[59] established a system for the protection of the environment by adopting the 1972 Oslo Convention for the Prevention of Marine Pollution by Dumping.

59 The OSPAR system has 14 States Parties and the European Commission, representing the European Community, now the European Union. The States Parties are: Belgium, Denmark, Finland, France, Germany, Iceland, Ireland, Luxembourg, the Netherlands, Norway, Portugal, Spain, Sweden, Switzerland, and the UK. Not all Contracting Parties are coastal states. An example of this is Finland, whose participation is justified by the fact that part of its rivers flow into the Barents Sea; Luxembourg and Switzerland are contracting parties as they are located in the Rhine catchment area. For further information see <http://www.ospar.org>.

In 1974, the Oslo Convention was joined with the Paris Convention for the Prevention of Marine Pollution from Land-Based Sources, thus becoming the OSPAR Convention for the Protection of the Marine Environment of the North-East Atlantic in 1992.[60]

Annex III of the OSPAR Convention concerning the prevention and elimination of pollution arising from offshore activities contains an absolute prohibition on dumping of wastes and other material (Article 3(1) Annex III) derived from offshore installations. The definition of dumping includes 'any deliberate disposal in the maritime area of wastes or other matter from offshore installations' (Article 1(f)(i)(2) OSPAR Convention) but excludes the full or partial maintenance of the facilities in place, provided that this is compatible with the Convention (Article 1(g)(iii) OSPAR Convent). Article 5(1) of Annex III then specifies that no disused offshore installation shall be left wholly or partly in place in the maritime area without a permit issued by the competent authority on a case-by-case basis. If the disused offshore installation contains substances which may result in hazards for 'human health, harm to living resources and marine ecosystems, (…) or interference with other legitimate uses of the sea', no such permit should be issued (Article 5(2) Annex III). Any such permit should contain the relevant justifications and, since 1st January 1998, need to be communicated to the other OSPAR Contracting Parties through the OSPAR Commission, in order to make consultation possible (Article 5(3) Annex III).

This kind of obligation already existed in the Oslo conventional system. Such is why in February of 1995 the UK Department of Trade and Industry informed the Oslo Commission (the predecessor of the OSPAR Commission) of its intention to grant permission to Shell UK, to dump the remains of the Brent Spar Shell platform which was decommissioned in 1991. In April of the same year, multiple protests against the project were organized by environmental groups, and particularly by the non-governmental organisation *Greenpeace*. Without going into too much detail concerning the events,[61] these protests led to a debate during the Meeting of Ministers of the Oslo commission in

---

60 Concluded in Paris on 22 September, 1992, and entered into force on 25 March, 1998.

61 Many authors have dealt with this subject. See for example: L. de La Fayette, 'New Developments in the Disposal of Offshore Installations,' in *International Journal of Marine and Coastal Law*, 1999, at 523 ff. ; Z. Gao, (n1), at 59 ff. ; E.A. Kirk, 'The 1996 Protocol to the London Dumping Convention and the Brent Spar,' in *International and Comparative Law Quarterly*, 1997, at 957 ff.; S. Mankabady, 'Decommissioning of Offshore Installations,' in *Journal of Maritime Law and Commerce*, 1997, at 603 ff. ; J. Woodliffe, 'Decommissioning of Offshore Oil and Gas Installations in Europe Waters: The End of a Decade of Indecision?,' in *International Journal of Marine and Coastal Law* (1999) at 101 ff.

June 1995, which adopted a moratorium on the dumping of offshore installations. In January of 1998, Shell UK announced a new project for the disposal on-shore of the structure.

The immediate consequence of this episode was the adoption of the Decision OSPAR 98/3 for the Disposal of Disused Offshore Installation by the OSPAR Commission in July of 1998.[62] Pursuant to this text, which entered into force in February of 1999, '[t]he dumping, and the leaving wholly or partly in place, of disused offshore installations within the maritime area is prohibited'. However, the text states that where 'there are significant reasons why an alternative disposal mentioned below is preferable to reuse or recycling or final disposal on land' the competent authority may issue a permit for exemption to the prohibition. Thus, it may authorise the dumping or abandonment of parts of or the entire structure.

The OSPAR Commission has had a central role in the development of a procedure for the decommissioning and removal of offshore installations. Decision 98/3 with its ban on dumping establishes a presumption in favour of an obligation to remove a disused structure. Moreover, the mandatory system of consultation in case of derogation, provided in Annex 3 of Decision 98/3, guarantees the existence of control, international monitoring, and thus better protection of common interests, such as the protection of the marine environment.

5 Interaction between Regional and Global Rules: Complementarity and Fragmentation

The regulation of offshore oil and gas operations has developed following a liberal model of regionalism.[63] There is no attempt in the UNCLOS 'either to impose a uniform global standard comparable to that for ships, or even a minimum standard comparable to that for dumping at sea.'[64] The UNCLOS has thus only partially contributed to the coherent development of the legal framework in this field, mainly through the general obligations for the protection of the marine environment and for partial removal.

---

62 Available at < https://www.ospar.org/work-areas/oic/installations>.

63 A. Boyle, 'Globalism and Regionalism in the Protection of the Marine Environment,' in D. Vidas, *Protecting the Polar Marine Environment, Law and Policy for Pollution Prevention*, Cambridge University Press, Cambridge, 2000, at 24.

64 Ibid.

The existence of specialized regional instruments is a welcome phenomenon which, on the one hand, implements the obligation set in Article 208 UNCLOS and, on the other hand, develops state practice in this field. However, as highlighted in the analysis on the existing instruments, regional approaches seem to diverge on certain key issues, such as, absolute or relative prohibition of dumping and criteria for issuing dumping permits. Moreover, the level of regional integration varies significantly from one regional sea to the other. The OSPAR and Helsinki systems rely on a developed institutional framework which includes supervisory and monitoring bodies, the OSPAR Commission and the Helsinki Commission, respectively. Under the Barcelona Convention, compliance control lies with the Meeting of the Parties and is conducted through the revision of periodical reports submitted by the contracting parties (Articles 18.2 and 27 of the Barcelona Convention; Article 30 of the Barcelona Offshore Protocol). A similar system exists under the Kuwait Convention where the Council of the Regional Organisation for the Protection of the Marine Environment receives and evaluates the reports sent by the contracting parties (Article XVII). These institutional differences impact the level of implementation of the existing obligations and the progressive development of rules and standards at the regional level.

Boyle insightfully remarks that: '[R]egional cooperation may fragment the possibilities for, and the effectiveness of international supervision of compliance with environmental standards. (…) Without an overarching global scheme comparable to the London [Dumping] Convention, there is (…) no alternative supervisory mechanism and no accountability.'[65] This is not a call against regionalism. It is more a reminder that regionalism by itself, and in particular weak regionalism, without an international/global framework, does not solve all problems. The global framework is not enough developed at the normative and institutional level in order to guarantee the effectiveness of regional cooperation systems or to offer safeguards against their failures.

6 Concluding Remarks

Partial or complete removal of offshore installations from the seabed is the generally accepted rule under both global and regional instruments. The conditions under which the removal is supposed to be performed and the criteria for the exceptions however highly diverge from one treaty system to another.

65 Ibid., at 33.

This remains highly problematic in light of the considerable number of installations reaching the end of their life-cycle, and of the potential environmental harm they represent.

The lack of an agreed institutional framework should not necessarily consist of an obstacle to the development of a more coherent and comprehensive legal framework. As pointed out by Beckman, '[i]f the IMO is not considered to be the competent international organization for this purpose, then interested States should draft a global convention and convene a global diplomatic conference to consider its adoption.'[66] A specialized framework treaty could be an important step for the offshore industry in general and for the treatment of decommissioned or disused installations in particular. However, Scovazzi aptly highlights that, 'there is little appetite for a global regime,'[67] notwithstanding, the strict regional model[68] has poignantly shown its limits.

66 R. Beckman, 'Global Legal Regime on the Decommissioning of Offshore Installations and Structures', in M. Nordquist *et al.*, *The Regulation of Continental Shelf Development, Rethinking International Standards*, Martinus Nijhoff, Leiden, 2013, at 280.

67 T. Scovazzi, 'Maritime Accidents with Particular Emphasis on Liability and Compensation for Damage from the Exploitation of Mineral Resources of the Seabed, in A. De Guttry *et al* (eds), *International Disaster Response Law*, T.M.C. Asser Press, The Hague, 2012, at 457.

68 Tanaka distinguished four models on interaction between global and regional legal frameworks. Under the first model, the regional model, 'the role of the global treaty is very limited and marine pollution it to be regulated promarily by regional treaties.' The other three models are: the global-single regional model; the global-multiple regional model; and the global model. See Y. Tanaka, 'Four Models on Interaction between Global and Regional Legal Frameworks on Environmental Protection against Marine Pollution: The Case of the Marine Arctic,' *Ocean Yearbook* (2016) at 346.

CHAPTER 19

Re-using (Nearly) Depleted Oil and Gas Fields in the North Sea for CO_2 Storage: Seizing or Missing a Window of Opportunity?

Martha M. Roggenkamp

1 Introduction*

Carbon Capture and Storage (CCS) involves the capture of carbon dioxide (CO_2), the transport and injection and subsequent permanent storage into suitable geological formations.[1] According to the European Commission, CCS is a major instrument in view of the transition to a full low-carbon economy, as CO_2 storage can balance conflicting interests of increasing energy demands and thus continued reliance on fossil fuels on the hand and the need to reduce greenhouse gas emissions into the atmosphere on the other hand.[2]

---

* This chapter has been written as part of the CATO CCUS programme. The main part of the research was conducted by Daniëlle M. Hanema who was a junior researcher at the Groningen Centre of Energy Law, University of Groningen, The Netherlands until 1 November 2015. The main part of the research was concluded in 2015 but updates have been made where necessary.

1 Intergovernmental Panel on Climate Change, *IPCC Special Report on Carbon Dioxide Capture and Storage. Prepared by Working Group III of the Intergovernmental Panel on Climate Change* [B. Metz and others eds] (Cambridge University Press 2005) 54 <www.ipcc.ch/pdf/special-reports/srccs/srccs_wholereport.pdf>.

2 European Commission, 'Communication from the Commission to the European Parliament, the Council, the European Economic and Social Committee and the Committee of the Regions on the Future of Carbon Capture and Storage in Europe', Brussels, 27.03.2013, COM (2013) 180 final, 3; European Commission, Communication from the Commission to the European Parliament and others, 'A Clean Planet for all A European strategic long-term vision for a prosperous, modern, competitive and climate neutral economy', Brussels, 28.11.2018 COM(2018) 773 final, 15; L. Helman, G. Parchomovsky and E. Stavang, 'Dynamic Regulation and Technological Competition: A New Legal Approach to Carbon Capture and Storage' in D.N. Zillman and others (eds), *The Law of Energy Underground: Understanding New Developments in Subsurface Production, Transmission, and Storage* (Oxford University Press 2014) 296.

DOI:10.1163/9789004391567_021

CO_2 can be stored in different kinds of reservoirs such as aquifers and depleted oil and gas fields.[3] This chapter looks into the re-use of depleted oil and gas fields for the purpose of CO_2 storage.[4] Since CO_2 storage onshore is met by fierce public opposition,[5] offshore storage is currently considered as the main option in Europe. Below the focus will therefore be on CO_2 storage in (nearly) depleted reservoirs offshore. We will in particular concentrate on the North Sea as this area has the largest storage capacity in Europe and a suitable geology for the purpose of CO_2 storage.[6] Because this offshore area is characterized by large scale oil and gas exploitation,[7] it also has the potential for reusing these fields for CO_2 storage. Depending on the size and location of these reservoirs, it may even be necessary to decide that CO_2 storage will require a coordinated approach and turn several depleted reservoirs into CO_2 storage facilities at the same time.[8] CO_2 storage on the Dutch continental shelf may, for example, rely on clustering several reservoirs given their relatively small size.[9] In order to form an adequate cluster, these reservoirs have to be

3 See: IPCC (n 1) 94. Depleted offshore oil and gas fields are important for CO_2 storage due to their geophysical features, the years of experience in exploiting them and the presence of suitable infrastructure that could be reused: IPCC (n 1) 215.

4 This reuse involves the transition of an exploitation permit to a storage permit and scaling up in terms of infrastructure: EBN & Gasunie, 'CO_2 transport- en opslagstrategie' (EBN/Gasunie Advies 2010) 11 <www.ebn.nl/ebn-publicaties/overige-publicaties> accessed 17 October 2015 (Dutch only).

Storage sites can only be operated provided there is a storage permit: Directive 2009/31/EC of the European Parliament and of the Council of 23 April 2009 on the geological storage of carbon dioxide and amending Council Directive 85/337/EEC, European Parliament and Council Directives 2000/60/EC, 2001/80/EC, 2004/35/EC, 2006/12/EC, 2008/1/EC (CCS Directive) [2009] OJ L140/114, art 6(1).

5 See for example: COM (2013) 180 final (n 2) 18. Also: M. Pragnell, 'Communications for Carbon Capture and Storage: Identifying the benefits, managing risks and maintaining the trust of stakeholders' (supported by Global CCS Institute 2013) 12–13 <http://hub.globalccsinstitute.com/sites/default/files/publications/92266/communications-carbon-capture-storage.pdf>.

6 Scottish Carbon Capture & Storage (SCCS), *SCCS Recommendations and Conference 2013 Report. Unlocking North Sea CO_2 Storage for Europe: Practical actions for the next five years* (SCCS 2013) 10–11 < www.sccs.org.uk/images/expertise/reports/unlocking/downloads/SCCSConference2013Report.pdf >.

According to the European Commission, storage capacity in the North Sea has been estimated at over 200 gigatonnes of CO_2: COM (2013) 180 final (n 2) 18.

7 <http://euoag.jrc.ec.europa.eu/node/63>.

8 See also chapter 17 of this book, N. Bankes, 'The use of sub-seabed transboundary geological formations for the disposal of carbon dioxide'

9 DHV and TNO (commissioned by NOGEPA and Dutch Ministry of Economic Affairs), 'Potential for CO_2 storage in depleted gas fields on the Dutch Continental Shelf. Phase 1:

situated in close proximity to each other.[10] Even if these reservoirs are located in close proximity to each other it does not necessarily mean that they will reach a state of near depletion at the same time. This raises issues with regard to existing abandonment and removal obligations. Is it possible to keep offshore installations in place instead or removing them with the aim of re-using the reservoirs and installations for CO_2 storage on a later stage?

In order to answer the question whether infrastructure can be kept in place with the aim of reusing it at a later stage for CO_2 storage, it needs to be assessed whether States can allow for offshore installations not being removed after production has ceased. For this purpose, we will first consider in Section 2 the position of coastal States under international law and in particular the Law of the Sea as the latter provides for the legal basis for developing energy activities offshore. In Section 3, we will then examine how some coastal States have implemented these principles of international law in their national jurisdictions. We will particularly examine the regimes in the Netherlands, the United Kingdom (UK) and Norway. Subsequently Section 4 will discuss some specific challenges and possible solutions with regard to the reuse of depleted and abandoned fields and infrastructure. Finally, Section 5 will provide some concluding remarks.

2 The Exercise of State Jurisdiction Offshore

2.1 *Introduction*

Historically, the use of the sea was unrestricted and open to all mankind.[11] The jurisdiction of States and the rights to issue and enforce legislation was in principle limited to the territory of States. This situation changed when due to technical developments it became possible to produce oil offshore.[12] In order to deal with offshore claims the 1958 United Nations Convention on the Continental Shelf provided for the concept of a continental shelf (CS) and the subsequent right of coastal states' functional jurisdiction with regard to the exploration and exploitation of oil and gas on the continental shelf.[13] The

technical assessment' (June 2008) 31, 44 <www.nlog.nl/resources/StorageCO2/DHV_TNO_CCS_offshore_%20final_30_06_08.pdf>.

10 DHV and TNO (n 9) 44.

11 See the concept of 'Mare Librum' as issued by Hugo de Groot in 1609.

12 See Truman Proclamation.

13 United Nations, Convention on the Continental Shelf. Done at Geneva, on 29 April 1958, no. 7302. United Nations –Treaty Series 1964 Vol. 499, 311–354 <https://treaties.un.org/doc/Publication/UNTS/Volume%20499/volume-499-I-7302-English.pdf>.

1982 United Nations Conventions on the Law of the Sea (UNCLOS)[14] provided coastal States with the possibility to claim an additional maritime zone – an exclusive economic zone or EEZ – and thus additional rights as regards the production of energy from other sources like waves and other economic activities.[15]

Whereas States have sovereignty and thus full jurisdiction on their territory (including the territorial sea),[16] their jurisdiction is limited on the CS and in the EEZ to economic activities such as the exploration for and the production of oil and gas. Coastal States may thus issue legislation governing these activities offshore. Although UNCLOS does not specifically refer to carbon storage it is generally assumed that this activity also is part of a coastal State's functional jurisdiction, either on the basis of the regime governing the continental shelf or the regime governing the EEZ.[17]

2.2 *Jurisdiction under UNCLOS*

It follows from the above that coastal States have sovereign rights in the EEZ[18] and on the CS[19] and thus a functional jurisdiction for the purpose of the exploration and exploitation of natural resources.[20] To be able to exercise these rights, coastal States have the right to make use of the seabed and to establish

---

14 United Nations, Convention on the Law of the Sea. Concluded at Montego Bay on 10 December 1982, no. 31363. United Nations – Treaty Series 1994 Vol. 1833, 396–581. <https://treaties.un.org/doc/Publication/UNTS/Volume%201833/volume-1833-A-31363-English.pdf>.

15 The 1958 UN Convention on the Continental Shelf and UNCLOS have been signed and ratified by the North Sea States. Currently UNCLOS is the main legal basis for developing activities offshore. See also M. Brus, 'Challenging Complexities of CCS in Public International Law' in M.M. Roggenkamp and E. Woerdman (eds), *Legal Design of Carbon Capture and Storage. Developments in the Netherlands from an International and EU Perspective* (Intersentia 2009) 29.

16 UNCLOS (n 13), art 2(1), (2) and art 3.

17 Brus (n 15) 29.

18 The EEZ is an area beyond and adjacent to the territorial sea, which shall not extend beyond 200 nautical miles from the baselines from which the breadth of the territorial sea is measured: UNCLOS (n 14), arts 55, 57. An EEZ has to be declared by a coastal State.

19 UNCLOS (n 14), art 76(1): The CS of a coastal State comprises the seabed and subsoil of the submarine areas that extend beyond its territorial sea throughout the natural prolongation of its land territory to the outer edge of the continental margin, or to a distance of 200 nautical miles from the baselines from which the breadth of the territorial sea is measured where the outer edge of the continental margin does not extend up to that distance.

20 UNCLOS (n 14), art 56(1)(a) (EEZ) and art 77(1) (CS). Natural resources include the mineral and other non-living resources of the seabed and the subsoil: UNCLOS (n 14), art 56(1)(a) (EEZ) and art 77(4) (CS).

as well as use offshore installations.[21] It entails that coastal States have the exclusive right to construct, authorize and regulate the construction, operation and use of all installations necessary for the exploration and production of energy offshore.[22] The functional jurisdiction is extended to those cables and pipelines constructed or used in connection with exploitation of resources.[23]

The functional jurisdiction[24] needs to be balanced with another main principle of UNCLOS: the freedom of fishing, navigation as well as the laying of cables and pipelines. The freedom to use the high sea cannot be restricted more than really necessary. Consequently, when an oil or gas production ceases, installations necessary for its production need to be removed. The extent to which these installations have to be removed has been a matter of discussion since the 1980s. Below we will examine the removal obligations applying to offshore installations – e.g. reservoirs, wells, platforms and subsea facilities – and offshore pipelines as these obligations are crucial when assessing a possible reuse for the purpose of CO_2 storage.[25]

2.3 *Removal of Offshore Installations*

2.3.1 International Legal Framework

Under the regime of the 1958 Convention on the Continental Shelf abandoned or disused offshore installations located on the CS had to be removed entirely.[26] The introduction of UNCLOS in 1982 has limited this obligation as it now generally provides for the removal of abandoned or disused installations without requiring complete removal.[27] It therefore facilitates a partial removal of installations, and 'appropriate publicity shall be given to the depth, position and dimensions of any installations not entirely removed'.[28] UNCLOS does

---

21 UNCLOS (n 14), art 56(1)(b)(i), 60(2) (EEZ).

22 UNCLOS (n 14), arts 60(1)(b), 56(1)(a) (EEZ) and arts 80, 60(1)(b), 56(1)(a) (CS).

23 UNCLOS (n 14), arts 56(3), 79(4) (EEZ) and art 79(4) (CS).

24 H.K. Müller and M.M. Roggenkamp, 'Regulating Offshore Energy Sources in the North Sea- Reinventing the Wheel or a Need for More Coordination?' [2014] The International Journal of Marine and Coastal Law 716, 718.

25 See further IEAGHG, 'Re-use of Oil & Gas Facilities for CO_2 Transport and Storage', 2018/06, July 2018.

26 United Nations, Convention on the Continental Shelf. Done at Geneva, on 29 April 1958, no. 7302. United Nations –Treaty Series 1964 Vol. 499, 311–354, art 5(5) <https://treaties.un.org/doc/Publication/UNTS/Volume%20499/volume-499-I-7302-English.pdf >.

27 UNCLOS (n 14), arts 80, 60(3) (CS).

28 Also: R. Beckman, 'Global Legal Regime on the Decommissioning of Offshore Installations and Structures' in M.H. Nordquist (ed), *The Regulation of Continental Shelf Development. Rethinking International Standards* (Nijhoff 2013) 263; C. Redgwell, 'International Regulation of Energy Activities' in M.M. Roggenkamp and others (eds), *Energy Law in Europe. National, EU and International Regulation* (Oxford University Press 2007) 65.

not provide any detailed rules as to when and how these installations have to be removed. It merely states that coastal States have to take into account 'any generally accepted international standards established by the competent international organization',[29] which in this case is the International Maritime Organization (IMO).[30] In 1989, the Assembly of the IMO adopted Guidelines and Standards for the removal of offshore installations on the CS and in the EEZ (IMO Guidelines).[31]

Starting-point of the IMO Guidelines is the obligation to remove abandoned and/or disused installations.[32] Entire removal is explicitly required if the offshore installations are located in shallow waters.[33] However, the IMO Guidelines also provide possibilities to allow for non- or partial removal of installations,[34] in which case the IMO has to be notified of the decision.[35] An exemption from the removal obligation would, for example, be possible if the existing installation (or part thereof) will serve a new purpose.[36] At first glance, this possibility could be relevant for reusing installations for the purpose of CO_2 storage. However, the IMO Guidelines are not legally binding[37] and a decision not to remove or partially remove abandoned or unused installations needs to be balanced against the duties stemming from UNCLOS, most particularly the freedom of navigation and fishery. Although coastal States basically are bound by a duty to remove unused offshore installations and thus need to

---

29 UNCLOS (n 14), art 60(3) (EEZ) and arts 80, 60(3) (CS).

30 'Competent or relevant international organizations under the United Nations Convention on the Law of the Sea' Law of the Sea Bulletin No. 31, 82. <www.un.org/depts/los/doalos_publications/LOSBulletins/bulletinpdf/bulletinE31.pdf>. On the relationship between UNCLOS and the IMO: International Maritime Organization, 'Implications of the United Nations Convention on the Law of the Sea for the International Maritime Organization' (LEG./MISC.7), IMO 19 January 2012). <www.imo.org/OurWork/Legal/Documents/Implications%20of%20UNCLOS%20for%20IMO.pdf>.

31 IMO, Resolution A. 672(16), adopted on 19 October 1989 (Annex: Guidelines and Standards for the Removal of Offshore Installations and Structures on the Continental Shelf and in the Exclusive Economic Zone). <www.imo.org/blast/mainframe.asp?topic_id=1026>.

32 IMO Guidelines (n 31), para 1.1.

33 This concerns offshore installations which stand in less than 75 meters of water (100 meters if it has been emplaced on the sea-bed on or after 1 January 1998) and weighs less than 4,000 tonnes in air (IMO Guidelines (n 31), paras 3.1 and 3.2).

34 The non- (or partial) removal has to be consistent with the IMO Guidelines: IMO Guidelines (n 31), para 1.1.

35 IMO Guidelines (n 31), para 1.3.

36 IMO Guidelines (n 31), para 3.4(1).

37 IMO, Resolution A. 672(16) (n 29), point 2. At the same time, the IMO Guidelines need to be taken into consideration by Member Governments: UNCLOS (n 14), art 60(3). Also: Redgwell (n 28) 66.

ensure that removal actually takes place,[38] UNCLOS and the IMO Guidelines also provide some flexibility and as decisions with regard to (partial) removal and the timing of removal are based on a case-by-case evaluation.[39]

2.3.2 What Is an Installation?

According to UNCLOS the removal obligation applies to abandoned and disused installations and structures in order to ensure safety of navigation and will take into account international standards (e.g. IMO Guidelines) and have due regard to fishing, the protection of the marine environment and the rights and duties of other states.[40] It is generally accepted that this obligation applies to all exploration and production platforms. It is not clear whether the removal obligation also applies to abandoned or unused pipelines on the CS or in the EEZ.[41] The wording of UNCLOS indicates that the removal obligation could apply if a pipeline qualifies as an installation or other structure.[42] In the absence of a clear definition in UNCLOS of the terms 'installation' and 'structure', it is generally accepted that a pipeline is not considered as an installation because most provisions of UNCLOS regarding installations do not apply to submarine pipelines. This is, for example, the case with the requirement that coastal states have to establish a safety zone around installations. Such a requirement does not apply to pipelines (and would be difficult to establish). Moreover, UNCLOS explicitly links the need for removal to safety of navigation. Unused pipelines will usually not affect safety of navigation.

Although generally speaking pipelines are not considered as an installation in its own right, it is still possible that some pipelines (and cables) are considered as being part of a (production) installation. In such case the removal obligation applying to installations may extend to some subsea pipelines. We therefore conclude that by virtue of the absence of a directly applicable international legal norm, it is within the competence of the coastal State to determine the legal regime governing the removal of offshore pipelines and

---

38 IMO Guidelines (n 31), para 1.2.

39 IMO Guidelines (n 31), para 2.1. For the specific conditions: see inter alia IMO Guidelines (n 31), para 2.4. The coastal State also has obligations with regard to the installation, as it should ensure that the legal title to the installation is unambiguous and that responsibility for maintenance and the financial ability to assume liability for future damage are clearly established: IMO Guidelines (n 31), para 3.11.

40 UNCLOS (n 14), art 60(3).

41 M.M. Roggenkamp, *Het juridisch kader van pijpleidingen in de olie- en gasindustrie* (Intersentia 1999) 409.

42 UNCLOS (n 14), art 60(3).

that a distinction needs to be made between different types of pipelines, i.e. pipelines being part of the installation and field-to-shore pipelines.

2.3.3 The Timing of Removal

The next question relevant for our research is when exactly offshore installations have to be removed. The removal obligation applies if an installation is abandoned or disused. UNCLOS does not provide a clear guidance for this purpose. Generally speaking an installation is disused if production has ceased or if a licence to produce has lapsed. However, not any termination of the production activities implies that the installation is disused. Although the IMO Guidelines merely are recommendations,[43] they provide some further guidance in this respect as they refer to 'permanent disuse'. In other words, they seem to indicate that a distinction needs to be made between disuse and permanent disuse.

UNCLOS does neither provide for a specific time frame for any removal of unused installations. Again the IMO Guidelines seem to provide some guidance as they state that the removal should be performed 'as soon as reasonably practicable after abandonment or permanent disuse'.[44] In practice these norms may still prove to be rather vague as different opinions may be held about the term 'reasonable practicable'. When is something reasonable practicable and who decides whether it is reasonable?

This phrasing of UNCLOS therefore seems to leave the possibility for leaving an installation in place after production has ceased with the aim of reusing it on a later stage for carbon storage as in such scenario the permanent disuse of the installation will be postponed after the injection of CO_2 into the reservoir has been completed. Last but not least, disuse of an installation does not mean that the installation is abandoned. As long as an installation is being maintained and an operator is in charge there is no abandonment taking place.

More problematic is the situation when an installation is kept in place for future use (and thus not permanently disused) but at the same time is abandoned as the operator in charge of the production activities is no longer in charge. What is the status of the installation in the period of (temporary disuse) and reuse? To what extent should this situation be considered as constituting illicit disposal under international law, especially in view of the duty of coastal States to adopt laws to prevent, reduce and control pollution of the

43 IMO, Resolution A. 672(16) (n 31), point 2.

44 IMO Guidelines (n 31), para 1.2.

marine environment caused by any source,[45] including dumping.[46] The term 'dumping' refers to any deliberate disposal of offshore installations[47] but does not include 'the leaving wholly or partly in place of a disused offshore installation, provided that any such operation takes place in accordance with any relevant provision of the Convention and with other relevant international law'.[48] A situation where an unused installation is kept in place for future use would therefore not constitute a form of dumping. The dumping conventions applying to the North Sea area do subsequently not apply and will not be discussed in this chapter.

2.4 *The Reuse of Installations*

The above has shown that UNCLOS requires that abandoned or disused installations or structures shall be removed to ensure safety of navigation, taking into account international standards and having due regard to fishing, the protection of the marine environment and the rights and duties of other states. This obligation applies to offshore (production) installations and to those pipelines that are considered part of the installation. As regards the field-to-shore pipelines UNCLOS does not require any removal. Coastal States may thus decide to leave those offshore pipelines that are not part of an installation in situ.

Neither UNCLOS nor the IMO Guidelines impose a clear time frame on the coastal State with regard to the removal of installations. Moreover, the IMO Guidelines seem to imply that coastal States have some discretion as to the timing of the removal and to the possibility of temporarily leaving the installations in situ for purposes of CO_2 storage. Another matter is how coastal States have interpreted these norms. This issue will be discussed below.

3 Functional Jurisdiction and Abandonment Approaches in the North Sea

3.1 *Introduction*

It follows from the above that coastal States have a functional jurisdiction with regard to the exploration for and exploitation of natural resources on the CS and are thus entitled to regulate oil and gas production and subsequently also

---

45 UNCLOS (n 14), arts 194(1), 194(3).

46 UNCLOS (n 14), arts 194(1), 194(3) and 210(1).

47 Convention for the Protection of the Marine Environment of the North-East Atlantic. Done at Paris, on 22 September 1992, art 1(f)(ii)(i). <http://www.ospar.org/convention/text> accessed 28 October 2015.

48 OSPAR Convention (n 47), art 1(g)(iii)(i).

the construction and removal of all necessary installations. Similarly, we noted that coastal States have generally assumed that their functional jurisdiction on the CS and/or in the EEZ also applies to carbon storage. In this section we will discuss how these principles of jurisdiction have been applied in the North Sea area. Whilst doing so it has to be noted that international law addresses States and States usually are not directly involved in energy production – and carbon storage – but rather transfer those rights to individual companies.

Most coastal States surrounding the North Sea are members of the European Union (EU). Norway is not a member of the EU but party to the European Economic Area (EEA). Based on the EEA Agreement, Norway may be required to implement all relevant EU directives and regulations.[49] Consequently, Norway has implemented both the Hydrocarbons Licensing Directive (Directive 94/22/EC) and the CO_2 Storage Directive (2009/31/EC). These directives provide general rules on EU level for governing exploration and production of hydrocarbons and the possibility of carbon storage. Whereas the Hydrocarbons Licensing Directive does not explicitly refer to the continental shelf and thus assumes that EU Member States have functional jurisdiction, the European legislator apparently wanted to avoid any doubt about the extent of coastal States' jurisdiction as it has included in Directive 2009/31/EC an express provision allowing for CO_2 storage on the continental shelf and in the exclusive economic zone within the meaning of UNCLOS.[50]

The Hydrocarbons Licensing Directive and the CO_2 Storage Directive are both based on an authorization regime. EU Member States may have the right to issue exclusive authorizations for the exploration and production of oil and gas[51] but may also issue exploration permits for selecting suitable storage sites and permits for storing carbon dioxide.[52] It should be noted, however, that the Hydrocarbon Licensing Directive does not include any provisions regarding the closure of reservoirs, the removal of installations and pipelines, nor regarding the possibility of reusing depleted oil and gas fields for storage of carbon dioxide.[53] Neither of these directives nor any other EU legislation specifically applies to the abandonment and removal of offshore oil and gas installations

49 Agreement on the European Economic Area [1994] OJ L 1/3, art 7(b). <www.efta.int/legal-texts/eea>.

50 CCS Directive (n 4), art 2(1) and preamble, recital 18.

51 Directive 94/22/EC, arts 1(3), 3.

52 CCS Directive (n 4), art 1 and 2(1).

53 Also: M.M. Roggenkamp and D.M. Hanema, 'New Uses of the Underground in the Netherlands: How to Manage a Crowded Subsoil?' in D.N. Zillman and others (eds), *The Law of Energy Underground: Understanding New Developments in Subsurface Production, Transmission, and Storage* (Oxford University Press 2014) 375.

and the possible re-use of subsoil areas and installations. In the absence of any guidance at EU level we will therefore focus on the way in which coastal States have dealt with these issues. Below we will thus discuss the national regimes in three North Sea States – the Netherlands, the United Kingdom (UK) and Norway. These States have been mainly selected on account of their significance in offshore oil/gas production[54] and their potential for CO_2 storage.[55] Moreover, all three coastal states also are faced with the need to decommission offshore installations. Despite these similarities there are also some major differences. One of these differences relates to the size of the fields and thus the extent to which it may be necessary to cluster depleted fields in order to develop economically viable CO_2 storage.

Below we will briefly discuss the rules applying to the exploration and production of oil and gas and the storage of carbon dioxide. The main focus will, however, be on the rules governing abandonment and removal of installations and the possibility for re-using these installations for storing CO_2. We will in particular examine the Dutch Mining Act of 2003 (MA 2003)[56] and its

54 In 2013, the primary production of natural gas in the EU amounted to 131.755 kilotonne of oil equivalent (ktoe) and the primary production of crude oil and other hydrocarbons (together) amounted to 72.041 ktoe. In the same year, the Dutch primary production accounted for 61.767 ktoe (natural gas) and 2.192 ktoe (crude oil and other hydrocarbons), whereas the UK primary production was 32.870 ktoe (natural gas) and 41.921 ktoe (crude oil and other hydrocarbons). In Norway, the primary production in 2013 accounted for 95.602 ktoe (natural gas) and 84 932 (crude oil and other hydrocarbons). Eurostat, *Energy balance sheets 2013 data* (Eurostat 2015) 9 (EU), 49 (Netherlands), 67 (UK), Norway (69) <http://ec.europa.eu/eurostat/web/products-statistical-books/-/KS-EN-15-001>. General information on offshore oil and gas production in the European Union/North Sea is available through: <http://euoag.jrc.ec.europa.eu/node/63>.

55 The Netherlands: see DHV and TNO (n 9). The UK: see UK Department of Energy and Climate Change (DECC), 'CCS Roadmap. Storage strategy' (URN 12D/016d, DECC April 2012) <www.gov.uk/government/uploads/system/uploads/attachment_data/file/48320/4904-ccs-roadmap–storage-strategy.pdf>. Norway has extensive experience with regard to CO_2 storage: Norwegian Petroleum Directorate, 'CO_2 Storage Atlas Norwegian North Sea' (2011) 6 <www.npd.no/Global/Norsk/3-Publikasjoner/Rapporter/PDF/CO2-ATLAS-lav.pdf>.

56 Mining Act 2003. Dutch: *Wet van 31 oktober 2002, houdende regels met betrekking tot het onderzoek naar en het winnen van delfstoffen en met betrekking tot met de mijnbouw verwante activiteiten* (Act of 31 October 2002, regarding regulations concerning the exploration for and the production of minerals and concerning mining related activities), *Staatsblad* 14 November 2002, number 542 (Dutch). A (non-authoritative) translation of this Act in English: <http://nlog.nl/en/legal/legislation.html>.

accompanying Mining Decree[57] and Mining Regulation,[58] the UK Petroleum Act 1998 (PA 1998)[59] and the Guidance Notes[60] as well as the Norwegian Petroleum Act 1996 (PA 1996) and the accompanying Petroleum Regulations 1997 (PR 1997).[61]

3.2 *The Netherlands*

In the Netherlands the Minister of Economic Affairs and Climate (MEA) is entitled to award oil and gas exploration and/or production licences as well as CO_2 storage licences on a competitive basis.[62] Exploration and production licences are usually applied for by and awarded to joint ventures for a specific period of time. The joint venture parties appoint an operator that needs to be approved by the MEA.[63] The State participant – EBN – will usually participate financially via an agreement of cooperation.[64] When awarding a licence the MEA may require the licensee for a financial security in order to ensure that abandonment and removal costs can be paid for.[65] So far the MEA has not made use of this possibility.

---

57 Mining Decree. Dutch: *Besluit van 6 december 2002, houdende regels ter uitvoering van de Mijnbouwwet* (Decree of 6 December 2002, regarding regulations to implement the Mining Act), *Staatsblad* 2002, number 604 (Dutch). A (non-authoritative) translation of the Mining Decree in English: <http://nlog.nl/en/legal/legislation.html>.

58 Mining Regulation. Dutch: *Mijnbouwregeling, Staatscourant* 16 December 2002, number 245 (Dutch). A (non-authoritative) translation of this Regulation in English: <http://nlog.nl/en/legal/legislation.html>.

59 Petroleum Act 1998, c 17. An Act to consolidate certain enactments about petroleum, offshore installations and submarine pipelines <www.legislation.gov.uk/ukpga/1998/17/contents>.

60 Department of Energy and Climate Change, 'Guidance Notes. Decommissioning of Offshore Oil and Gas Installations and Pipelines under the Petroleum Act 1998' (DECC Guidance Notes), <www.gov.uk/oil-and-gas-decommissioning-of-offshore-installations-and-pipelines>.

61 Petroleum Act 1996. Norwegian: *Lov om petroleumsvirksomhet* (Petroleum Act), <https://lovdata.no/dokument/NL/lov/1996-11-29-72>. A non-authoritative translation of the Act in English can be consulted at: <www.npd.no/en/Regulations/Acts/Petroleum-activities-act/>. Petroleum Regulations 1997. Norwegian: Norwegian: *Forskrift til lov om petroleumsvirksomhet* (Petroleum Regulations). Sections 10 till 18 NO PA 1996 provide the legal bases of these Regulations. <https://lovdata.no/dokument/SF/forskrift/1997-06-27-653>. A non-authoritative translation of the Regulations in English can be consulted at: <www.npd.no/en/Regulations/Regulations/Petroleum-activities/>.

62 MA 2003 (n 56), arts 6 and 31b–32. So far, the Ministry has awarded one CO_2 storage permit.

63 M.M. Roggenkamp 'Energy Law in the Netherlands' in M.M. Roggenkamp and others (eds), *Energy Law in Europe. National, EU and International Regulation* (Oxford University Press, forthcoming in 2016).

64 M.M. Roggenkamp 'Energy Law in the Netherlands' (n 63).

65 MA 2003 (n 56), art 47.

The MA 2003 explicitly provides that 'a mining installation that is no longer in use has to be removed'.[66] As the Dutch CS has shallow waters with an average water depth of 35 meters, rising to well over 60 meters in the northern parts,[67] these installations will in accordance with the IMO Guidelines be removed entirely.[68] So far, approximately 2,000 wells (both onshore and offshore) have been abandoned, 23 platforms removed – with some 150 platforms remaining – and some 200 km pipelines decommissioned.[69]

The closure, abandonment and removal of offshore installations are subject to an abandonment and removal plan, which is issued by the operator.[70] This plan needs at least to describe the way in which the installation, debris and other matter will be removed[71] as well as their final destination,[72] and the timing of the involved activities.[73] The removal plan has to be send to the MEA for approval.[74] The MEA can only refuse his consent in case of risk of potential damages.[75] Although neither the MA nor the Mining Decree provides for a time limit for actual removal, the Minister has the right to set such a time limit.[76] This provision allows the MEA to force the (former) licensee[77] to remove the installation.[78] By doing so the State passes on to the (former) licensees the obligation to remove unused offshore installations.

66 MA 2003 (n 56), art 44(1). MA 2003 (n 56), art 1(o) refers to the term mining installations being 'a mining work anchored in or present above the soil of surface water'.

67 'Ontwerp- Beleidsnota Noordzee 2016–2021', 5 (Dutch). This policy document is an appendix to the 'Ontwerp Nationaal Waterplan 2016–2021' and was presented to Dutch parliament in December 2014, see: Kamerstukken II 2014/15, 31 710, number 35 (Dutch).

68 Section 2.3. (this article). M.M. Roggenkamp 'Energy Law in the Netherlands' (n 61).

69 EBN c.s., 'Netherlands masterplan for decommissioning and re-use', 2017. See https://kennisbank.ebn.nl/masterplan-decommissioning-and-re-use/.

70 Mining Decree (n 57), art 60(1).

71 Mining Decree (n 57), art 61(1)(a), (1)(c).

72 Mining Decree (n 57), art 61(1)(d).

73 Mining Decree (n 57), art 61(2).

74 Mining Decree (n 57), art 60(2).

75 Mining Decree (n 57), art 60(3).

76 MA 2003 (n 56), art 44(4).

77 MA 2003 (n 56), arts 47(2) in combination with art 41(4). If there is more than one party to the licence, the operator will be responsible instead: TK 1998–1999, 26 219, number 3, 27.

78 TK 1998–1999, 26 219, number 3, 27. According to the MEA, the same provision could create an opportunity for postponement of removal if the platform is not used anymore for production processes, but is still a link in the transportation system. This appears to be relevant if multiple fields share their infrastructure and there is/are still some producing field(s).

Following the main principles of international law, offshore pipelines will usually remain in place unless the MEA prescribes such removal.[79] A decision to remove unused pipelines will be made on the basis of comparing the costs and benefits for society with the environmental consequences and safety issues.[80] If a pipeline remains in situ, the MEA may require that the pipeline meets certain requirements as regards its condition[81] and that the operator will regularly inspect the abandoned pipeline.[82] This provision seems rather contradictory as it can be assumed that the company operating a pipeline will terminate its activities when the pipeline is being disused. It is therefore not surprising that this provision will be reassessed.[83]

It follows from the above that by contrast to submarine pipelines, unused installations have to be completely removed on the basis of an abandonment and removal plan that has to be approved by the MEA. Such approval can only be rejected if the plan would result in dangerous situations. If not being rejected on those grounds, the Minister has to grant its approval and is not entitled to deviate from the plan submitted by the operator. Hence, the MEA is not in the position to facilitate that an unused and abandoned installations remains in place if a licensee/operators submits a removal plan and initiates a removal of an unused installation within a specific time frame, the MEA is neither entitled to reject the plan in order to facilitate future CO_2 storage. Hence, the legal framework does not really facilitate re-use of installations.

However, the approach towards re-use of installations is gradually changing since EBN jointly with the organisations representing the oil and gas operators (NOGEPA) and service industry (IRO) presented in November 2016 a Masterplan for Decommissioning & Re-use in the Netherlands. This led to the establishment of a National Platform for Re-use & Decommissioning (Nexstep). In response to the discussions on potential re-use options, the MEA is proposing an amendment of the MA to enable the re-use of offshore assets.[84]

79 MA 2003 (n 56), arts 49(1),(5) with reference to Mining Decree (n 57), art 103. In case the MEA has commanded removal of an offshore pipeline, the same regime as regards the removal of offshore installations will apply: MA 2003 (n 56), art 45(1).

80 'Ontwerp- Beleidsnota Noordzee 2016–2021' (n 65), 35.

81 MA 2003 (n 56), arts 49(1),(5) with reference to Mining Decree (n 57), art 104(2).

82 MA 2003 (n 56), arts 49(1),(5) with reference to Mining Decree (n 57), art 104(3). If necessary, the MEA may stipulate remedial action.

83 An investigation will be carried out between 2016 and 2021 in order to establish whether it would be possible to leave the pipelines in situ without inspection, whether the duty of inspection should be transferred to the government or to consider the possibility of removing the pipeline. See 'Ontwerp- Beleidsnota Noordzee 2016–2021' (n 67), 35.

84 TK 2017–2018 (Duurzame ontwikkeling en beleid), 30196, nr. G, 2.

3.3 *The United Kingdom*

Whereas exploration for and production of oil and gas are based on licensing pursuant to the Petroleum Act 1998 (PA 1998) as amended, permits to store carbon dioxide are governed by the 2008 Energy Act.[85] By contrast to the oil and gas exploration and production licensing regime on the UK continental shelf, section 18(3) of the 2008 Energy Act also requires a lease from the Crown Estate for storing CO_2 offshore.[86] An offshore hydrocarbons licence covers all phases of the development, i.e. exploratory drilling, appraisal and production, but is split into three terms and may last for approximately 25 to 30 years. The licence may be awarded upon condition that the licensee provides for financial security[87] by way of a trust or other arrangements.[88] The licence is usually awarded to an unincorporated joint venture who appoint amongst themselves an operator.[89] By contrast to the Netherlands (and Norway) the State does not participate.

The CS of the United Kingdom (UKCS) comprises (mostly) of deep waters and although it is presumed that unused installations have to be entirely removed, the PA 1998, consistent with OSPAR Decision 98/3, allows for the possibility of partial removal.[90] Rules regarding abandonment and removal can be found in the PA 1998, which provides that the Secretary of State (SoS) may require the submission of an abandonment programme for the abandonment of an offshore installation[91] or a submarine pipeline.[92] The power in the hands

85 Energy Act 2008, c 32. 'An Act to make provision relating to gas importation and storage' (…….) <www.legislation.gov.uk/ukpga/2008/32/contents> accessed 13 October 2015. See: A. McHarg and M. Poustie, 'Risk, Regulation, and Carbon Capture and Storage: The United Kingdom Experience' in D.N. Zillman and others (eds), *The Law of Energy Underground: Understanding New Developments in Subsurface Production, Transmission, and Storage* (Oxford University Press 2014) 250.

86 See also EnA2008, Explanatory Notes, para 55.

87 PA 1998 (n 59), ss 38A, 38B.

88 PA 1998 (n 59), s 38A(1).

89 G. Gordon, A. McHarg and J. Paterson 'Energy Law in the United Kingdom' in M.M. Roggenkamp and others (eds), *Energy Law in Europe. National, EU and International Regulation* (Oxford University Press forthcoming 2016).

90 PA 1998 (n 59), s 29(4)(c).

91 Offshore installation means any installation which is or has been maintained, or is intended to be established, for the carrying on of – inter alia – the exploitation of mineral resources in or under the shore or bed of relevant waters. (PA 1998 (n 59), ss 45, 44(1)- 44(4)). Relevant waters include the Continental Shelf of the UK (PA 1998 (n 59), s 44(4)(b) PA 1998 (n 59), Continental Shelf Act 1964, s 1(7) and the Continental Shelf (Designation of Areas) Order 2013).

92 Submarine pipeline means a pipeline, i.e. a pipe or system of pipes (excluding a drain or sewer) for the conveyance of anything, together with all apparatus, works and services associated with its operation, which is (or is intended to be established) in, under or over

of the Secretary of State to make regulations for decommissioning[93] have not been used, but Guidance Notes have been issued and regularly updated by the relevant Department[94] and it is by means of these that the UK specifies in detail how it will go about implementing its obligations under international law with regard to decommissioning.[95] Usually the abandonment programme will be submitted to the SoS by the operator, but the SoS may serve a notice requiring submission of such a programme might to a wide range of parties.[96] In case of any unused pipelines, a notice can only be served upon the owner of the pipeline, any person who owns a significant interest in the pipeline or associated companies with this person.[97] An abandonment programme will be subjected to a stakeholder consultation process[98] and shall in any case contain an estimate of the costs of the measures proposed in it and shall either specify the times at/within the proposed measures are to be taken or shall make provision as to determine these times.[99]

The SoS may reject the programme or approve it conditionally or unconditionally. If an abandonment programme has been approved by the SoS, the person who has submitted the programme has to secure that it will be carried out and that any abandonment and/or removal conditions are complied with.[100] This implies that the operator – who will in most cases be the designated person[101] – will be responsible for the monitoring and the maintenance

waters in the territorial sea and the CS. (PA 1998, ss 45 and 26(1), Continental Shelf Act 1964, s 1(7), and the Continental Shelf (Designation of Areas) Order 2013).

PA 1998 (n 59), s 29(1).

Even though the Act refers to an abandonment programme, the preferred and generally accepted term is decommissioning programme: DECC Guidance Notes (n 58), para 2.1.

93 PA 1998 (n 59), s39.

94 Currently the Department for Business, Energy and Industrial Strategy (commonly referred to by the abbreviation BEIS).

95 The most recent Guidance Notes were issued in November 2018 and are available at: https://assets.publishing.service.gov.uk/government/uploads/system/uploads/attachment_data/file/760560/Decom_Guidance_Notes_November_2018.pdf.

96 PA 1998 (n 59), ss 29,30. Also: G. Gordon, A. McHarg and J. Paterson 'Energy Law in the United Kingdom' in M.M. Roggenkamp and others (eds), *Energy Law in Europe. National, EU and International Regulation,* 3d edition, Oxford University Press, 2016.

97 PA 1998 (n 59), ss 29, 30.

98 See G. Gordon, A. McHarg and J. Paterson 'Energy Law in the United Kingdom' in M.M. Roggenkamp and others (eds), *Energy Law in Europe. National, EU and International Regulation.* 3d. edition, Oxford University Press, 2016.

99 PA 1998 (n 59) 9.

100 PA 1998 (n 59), s 36.

101 PA 1998 (n 59), ss 29, 30.

activities that are being part of the programme. Consequently, the operator will also bear the costs associated with these activities.

In the event of non-compliance, the SoS may serve a default notice. Failure to comply may give rise to criminal penalties and will permit the SoS to undertake any remedial action necessary, at the defaulting party's expense.[102] In case the designated person is no longer able to carry out the approved abandonment programme, the SoS may decide to instruct another company to carry out the work instead.[103] However, only companies that were previously in receipt of a notice or persons on whom notices could have been served are eligible to take over responsibilities.[104] One has to note that this competence should only be used as a measure of last resort.[105] Moreover, the SoS may in some few instances have the power to prepare an abandonment plan itself, for example if an abandonment plan is rejected or a notice to submit an abandonment programme has not been complied with.[106] In such a case, the SoS shall inform the persons to whom notice was given in the first place of the terms of the abandonment programme.[107]

About 500 installations and 3000 pipelines are placed on the UKCS. In 2017 just 7% of this infrastructure has been decommissioned. However, the rate of decommissioning is expected to accelerate rapidly in the years ahead.[108] When considering the possible re-use of installations, it is important to note that the UK government recognizes that disused facilities – including pipelines – may represent important infrastructure for its CS and therefore should provide the means for new developments such as geological storage of carbon dioxide.[109] If a specific opportunity for re-use has been identified, a deferral of the duty of abandonment might be considered.[110] In that case the proposed abandonment programme should include detailed provisions as to any continuing use and maintenance of the installation.[111] Moreover, the operator is required to

102 PA 1998 (n 59), s 37.
103 PA 1998 (n 59), s 34(1)(b). DECC Guidance Notes (n 60), para 16.1.
104 PA 1998 (n 59), s 34(2). DECC Guidance Notes (n 60), para 16.1.
105 DECC Guidance Notes (n 60), para 16.1.
106 PA 1998 (n 59), s 33(1).
107 PA 1998 (n 59), s 33(7). In that case, the provisions of PA 1998 (n 59), part IV shall have effect as if the abandonment programme had been submitted by the notified persons and approved by the SoS.
108 Bureau Veritas, Decommissioning on the UK Continental Shelf – an overview of regulations, May 2017.
109 DECC Guidance Notes (n 60), para 5.18.
110 DECC Guidance Notes (n 60), para 5.18.
111 PA 1998 (n 59), s 29(4)(c).

implement arrangements for monitoring, maintenance and management of the decommissioned site and any remaining parts of the offshore infrastructure.[112]

The question of reuse is now more acute since the arrival of the Oil and Gas Authority (OGA) in 2015.[113] Tasked with implementing the UK's Maximising Economic Recovery Strategy,[114] which includes a concern to avoid stranded reserves in the event that infrastructure is decommissioned prematurely, the OGA is the body to whom an operator must apply in relation to cessation of production prior to decommissioning. The OGA advises the Secretary of State on "alternatives to abandoning or decommissioning the installation or pipeline, such as reusing or preserving it" and on "how to ensure (whether by means of the timing of the measures proposed, the inclusion of provision for collaboration with other persons, or otherwise) that the cost of carrying out the (decommissioning) programme is kept to the minimum that is reasonably practicable in the circumstances".[115] It is noteworthy also that in the exercise of its functions, the OGA must have regard, inter alia, to the "The development and use of facilities for the storage of carbon dioxide, and of anything else (including, in particular, pipelines) needed in connection with the development and use of such facilities".[116]

3.4 *Norway*

The Ministry of Petroleum and Energy (MPE) is entitled to issue an exploration, production and pipeline licence on the basis of the 1996 Petroleum Act and a CO_2 storage license on the basis of a Regulation governing the storage of carbon dioxide on the Norwegian CS.[117]

---

112 DECC Guidance Notes (n 60), para 5.17, also 86.

113 The OGA is established as an independent Government Company by the Energy Act 2016, s1.

114 Available at: https://www.ogauthority.co.uk/media/3229/mer-uk-strategy.pdf.

115 BEIS Guidance Notes, para. 2.7.

116 Energy Act 2016, s8.

117 CO2 Storage Regulation 2014 (*Forskrift om utnyttelse av undersjøiske reservoarer på kontinentalsokkelen til lagring av* CO_2 *og om transport av* CO_2 *på kontinentalsokkelen*) implements Directive 2009/31/EC and entered into force on 20 January 2015. The Regulation is based on Act no. 12 of 21 June 1963 governing the exploitation of other offshore natural resources (*Lov om vitenskapelig utforskning og undersøkelse etter og utnyttelse av andre undersjøiske natureforekomster enn petroleumsforekomster*) The Act and the Regulation can be found at <https://lovdata.no/dokument/SF/forskrift/2014-12-05-1517>. A non-authoritative translation of the Act in English can be consulted at: www.npd.no/en/Regulations/Acts/Scientific-research-act/.

Exploration and production licenses are awarded on the basis of licensing rounds and are usually awarded to joint ventures established by the MPE. Licences are awarded for a specific period of time and the State participates via the State company Petoro in all licences.[118] The State can provide for financial security ensuring that unused installations will be removed.[119]

The Norwegian CS comprises of deep (and very deep) waters, which means that installations can be partially removed. The holder of a production licence[120] and/or a pipeline licence[121] has the duty to submit a decommissioning plan to the MPE.[122] The obligation to submit a decommissioning plan depends upon two main events: the permanent disuse of the installations or the expiration of the licence, whichever event occurs first.[123] The decommissioning plan has to either include a proposal for continued production or a proposal for shutdown of production and disposal of facilities. Disposal may, inter alia, constitute other uses or partial (or complete) removal.[124] As a general rule, pipelines may be left in place when they do not obstruct, or present a safety risk for fishing.[125]

More detailed provisions are provided in the PR 1997, which state that the decommissioning plan shall consist of two parts: a disposal plan describing the disposal alternatives and an impact assessment.[126] The disposal plan has to present all disposal alternatives relevant to the specific case, which may range from complete or partial removal to continued use for other purposes. The latter also includes the mere abandonment of an installation in combination with continued maintenance and inspections.[127] The licensee has to describe the technical, safety, environmental and economic aspects and the relationship to other users of the sea for each alternative and has to recommend one of them.[128] The MPE will select one of the alternatives presented by the licensee – which is

118 F. Arnesen and others, 'Energy Law in Norway' in M.M. Roggenkamp and others (eds), *Energy Law in Europe. National, EU and International Regulation*, 3rd edition, Oxford University Press, 2016.

119 PA 1996 (n 61), ss 1–4, 1–6(l).

120 PA 1996 (n 61), s 3–3.

121 PA 1996 (n 61), s 4–3. Also: Arnesen and others (n 107).

122 PA 1996 (n 61), s 5–1. The MPE may waive the requirement to submit a decommissioning plan.

123 PA 1996 (n 61), s 5–1.

124 PA 1996 (n 61), s 5–1.

125 Arnesen and others (n 118).

126 PR 1997 (n 61), s 43.

127 Arnesen and others (n 118).

128 PR 1997 (n 61), s 44.

not necessarily the recommended one – and decides accordingly.[129] About 500 installations are placed on the Norwegian CS. In 2018, around 20 decommissioning plans have been processed and approved by the authorities. In most instances the final decision has been to partially remove disused facilities and transport them to shore.[130]

In case the MPE agrees that an unused installation should be abandoned but not removed, the licensee or owner will be liable for any damages or inconveniences caused wilfully or inadvertently in connection with the abandoned facility, unless the MPE decides otherwise.[131] The PA 1996 also provides for a situation that the licensees/owners and the State conclude an agreement as a result of which the maintenance, responsibility and liability will be taken over by the State on the basis of an agreed financial compensation.[132] The State may also take over a fixed facility when its use has been terminated permanently or when a licence is expired, surrendered or revoked.[133] This possibility has to be regarded in view of the Norwegian government's aim to continue the use of reservoirs and the installations, which are considered beneficial for the government.[134]

3.5 *Approaches in the North Sea Area*

It follows from the above that the approach taken in Norway and the UK differs from the approach in the Netherlands. This is partly due to the fact that the installations on the Dutch continental shelf are situated in shallow waters and thus subject to a requirement of complete removal. In Norway and the UK decisions regarding the removal of unused installations are based on a case-by-case approach and will usually result in a decision to partially remove the installation. More important, however, is the fact that currently the Dutch MEA has no discretionary powers when assessing an abandonment plan. This contradicts with the Norwegian and UK approaches that facilitate future uses of offshore infrastructure. However, all three countries have identified re-use

---

129 PA 1996 (n 61), s 5–3. According to Arnesen and others, this provision leaves a very wide margin of discretion to the MPE: Arnesen and others (n 118).

130 Norwegian Ministry of Petroleum and Energy and Norwegian Petroleum Directorate, <http://npd.no/en/Publications/Facts/>.

131 PA 1996 (61), s 5–4. Civil liability will exist as long as the installation is physically able to cause damage, there is no preclusion in this respect: H.J. Bull and K. Kaasen, 'Abandonment and Reclamation of Energy Sites and Facilities: Norway' [1992] Journal of Energy and Natural Resources 37, 45.

132 PA 1996 (n 61), s 5–4.

133 PA 1996 (n 61), s 5–6.

134 Roggenkamp 'Juridisch kader' (n 41) 504–505.

as an important issue that may affect abandonment and decommissioning policies.

4 Challenges and Possible Solutions with regard to the Re-use of Installations

4.1 *Introduction*

In the above we have examined how international law and national laws regulate the abandonment and removal of installations that have been used to produce oil and/or gas from fields situated on the CS. Now we will discuss in more detail the possible re-use of depleted oil and gas fields for permanently storing CO_2 and how this may have an impact on the abandonment and removal obligations. Given our focus on the re-use of reservoirs and related infrastructure, we will not discuss CO_2 storage in offshore aquifers despite its potential for large-scale storage.

When examining the re-use of depleted offshore oil/gas fields for the purpose of CO_2 storage we distinguish between several scenarios. First, we will discuss the option that the holder of a hydrocarbons production licence wishes to re-use the reservoir for CO_2 storage. Thereafter we will discuss the possibility that the holder of the production licence is not interested in such a development and wants to terminate its activities and abandon the installations. To which extent can the national authorities intervene and decide that the installations will not be removed? Who will in such a scenario be responsible for the installation, its maintenance and future removal?

4.2 *Re-use by an Existing Licensee*

Re-use by an existing holder of a production license is the most straightforward option. In such a scenario the competent authority will probably not require an exploration permit as sufficient information is available following the hydrocarbons production activities. It is therefore most likely that the licensee can apply directly for a CO_2 storage permit. However, following the CO_2 Storage Directive sus a permit has to be awarded in competition. This means that the application will be assessed on the basis of its technical and financial merits. If the applicant is financially sound the assessment will mainly focus on the technical capability of the applicant. Given the knowledge of the reservoir following the prior production activities it can be assumed that this particular applicant will be awarded the storage permit. This will become even more likely if the holder of the production license has made use of the possibility to inject CO_2 for the purpose of enhanced hydrocarbons recovery.

Another issue involves the abandonment and removal plan. What is the status of this plan if the holder of the production licence will be awarded the storage licence? As the UK and the Norwegian authorities are aiming at facilitating future uses of existing infrastructure, the holder of a production license under UK and Norwegian law may submit an abandonment plan that takes future use into account. Although not explicitly stated under Dutch law, it seems that such an approach would be possible as well. The abandonment and removal plan should then clearly explain the situation and how removal will be dealt with in the future. Although the MEA still has the right to set a time limit for removal it is not very likely that he will do so if a storage permit has been awarded or is pending. Moreover, under current Dutch law there is no time frame within which an abandonment and removal plan has to be submitted. As a consequence the licensee and the MEA may agree to postpone submission of such a plan.

Turning a production into a storage facility will also have some financial consequences. A joint venture holding a production licence will usually have made some arrangements in the Joint Operating Agreement with regard to abandonment costs. The government will be keen to ensure that any financial arrangements involving removal costs will be transferred to the new licensee. In case not all companies of the joint venture will be part of the storage venture, it may be necessary to rely on provisions that normally are included in farm-in and farm-out agreements. Similarly account has to be taken of the State participant. If a State participates in the production of hydrocarbons it will also be responsible of paying a share of all costs, including abandonment costs. However, if the State is not participating in the storage activities arrangements have to be made with regard to the postponed payment of any removal costs. Given the more stringent abandonment provisions in the CO_2 Storage Directive it is likely that the costs of abandonment and removal will be higher in the future. In order to avoid disputes about the payment of removal costs in the future, it seems necessary that all parties involved ensure that there is a clear understanding of future obligations.

4.3 *Re-use by Another Party*

It is possible that the holder of a production license is not interested in re-using the reservoir and the existing installations for storing CO_2 in the depleted reservoir. It is, nevertheless, possible that the government is of the opinion that the reservoir is suitable as a storage site. After the area has been returned to the government, all interested parties can apply for a storage permit. Depending on the national regimes, such a permit can be awarded on the basis of a permitting round or via an open door approach, i.e. inviting others to submit a competitive application after an application has been made.

As in the above scenario, it will be necessary to consider the procedures regarding the requirement to submit an abandonment and removal plan. What happens if the government wishes to re-use the field and related installations but the holder of a production plans submits an abandonment and removal plan aiming at a complete or partial removal of the installations and closure of the field? In case an abandonment and removal plan has been submitted, the Dutch MEA cannot intervene and can only refuse to give his consent in case of risks of potential damages. As the MA does not provide for a clear time line for submitting abandonment and removal plans, the only solution seems to be that the MEA and holder of the production license enter into a prior agreement on the approach towards the abandonment and removal and the payment of costs involved. The UK and Norway seem to have a different approach. If in the UK a specific opportunity has been identified, the SoS may consider a deferral of the duty of abandonment and removal.[135] Norwegian law opens up for the possibility that the State will take-over the installation if a licence has been expired.[136]

4.4 *Future Re-use of Depleted Fields*

A third scenario involves the situation that depleted fields are suitable for CO_2 storage but no parties can be found to apply for a storage permit or the use of one or more depleted fields for storage purposes only will materialise on the longer term, for example, due to the need to cluster several fields in order to develop transport facilities and/or to make it economically viable. Such a scenario requires that the State will be actively involved in order to avoid that fields are closed down and installations removed but also that the State will be able to ensure that the period between the cessation of the production activities and/or the termination of the production licence and the commencement of CO_2 storage on the longer term will be bridged in terms of monitoring and maintenance of the fields and the existing infrastructure.[137]

135 See above section 3.3.

136 See above section 3.4.

137 Also: M.M. Roggenkamp 'Regulating Underground Storage of CO_2' in M.M. Roggenkamp and E. Woerdman (eds), *Legal Design of Carbon Capture and Storage. Developments in the Netherlands from an International and EU Perspective* (Intersentia 2009) 225.

As discussed above, international law requires States to ensure that unused installations are removed but does not provide for clear guidance as to the timing of removal. Consequently, States have to ensure that no installations become 'orphans' as a result of which there are no funds available for abandonment and removal in the future. In the absence of a licensee to maintain unused infrastructure, it is easy to point to the State as the party responsible for keeping the infrastructure in place. This follows from international law and is also the approach taken by Norway. However, who or what is the State? The State itself is not conducting such activities and will have to appoint a person or entity that will be able to take care of the abandoned infrastructure on its behalf. This could be a State oil and gas company but in the North Sea area the previous State oil and gas companies have been mostly privatised and any State participation is currently restricted to a financial participation. It is therefore not very likely that State participants like EBN and Petoro will be charged with the task to keep installations in place for future re-use. Apart from a financial participation the State is also involved as an independent health and safety authority. However, being able to supervise exploration and production activities and check whether or not installations are maintained and safe does not put these authorities into a position as a temporary operator. It therefore is most likely that the State has to appoint an oil and gas producing company to maintain these installations on its behalf as a temporary operator since these companies have the necessary expertise and technical skills.

Which company can be appointed or act as a temporary operator and how could this be arranged? In the absence of a State oil and gas company, a possible solution would be to appoint an 'operator of last resort' in order to take care of the abandoned installations on behalf of the State until a storage permit has been awarded. This 'operator of last resort' would be somewhat comparable to the 'supplier of last resort' in the downstream energy sector, i.e. a situation where an energy company is taking over the supply of energy to household consumers in a case where another supplier fails to deliver due to a bankruptcy or withdrawal of a licence.[138] The appointment of an 'operator of last resort' should be based on a competitive regime or based on a prior selection of companies active on the CS of the coastal States involved. Such a temporary operator or an 'operator of last resort' will be responsible for maintaining the

138 Directive 2009/73/EC of the European Parliament and of the Council of 13 July 2009 concerning common rules for the internal market in natural gas and repealing Directive 2003/55/EC [2009] OJ L211/94, art 3(3).

infrastructure and transfer the operations and responsibility to the holder of the CO_2 storage permit as soon as it has been awarded.[139]

The appointment of an 'operator of last resort' has several consequences. As regards the abandonment obligations it can be assumed that the operator as a temporary operator will not be interested to take the full abandonment obligation without further arrangements. Hence, either the abandonment and removal obligations resting on the holder of the (previous) production licence will be transferred temporarily to the State or to the 'operator of last resort' until a CO_2 storage permit has been granted. Such transfer needs to be accompanied with all financial consequences as apply in any transfer of a licence. The holder of the (previous) production license will thus not be relieved from paying abandonment and removal costs and the 'operator of last resort' will not be charged with abandonment and removal costs, which is fair as this temporary operator of the installation will not have any income or profits from the position it has achieved. In order to act as a temporary operator and depending on the applicable tax regimes it may be necessary to consider some tax exemptions for 'operators of last resort'. Moreover, it may also be necessary to consider some financial incentives to compensate the 'operators of last resort' for their efforts. The easiest and most straightforward solution would be to require the State to pay the operator of last resort a fair compensation. Such financial compensation could potentially be considered as a State aid. However, such compensation could escape classification as a State aid if the undertaking ('operator of last resort') is selected on the basis of a tendering procedure and if it involves a service in the general interest (public service obligation), the compensation is established in a transparent way and does not exceed what is necessary to cover the costs incurred and neither gives the recipient undertaking an advantage over competing undertakings.[140] These criteria seem to apply to the 'operator of last resort' if the undertaking is selected via a transparent tendering procedure and the compensation is limited to actual costs made for maintenance. Moreover, the re-use of installations would be in the general interest as it is the State who is subject to international and EU obligations regarding CO_2 greenhouse gas emissions reduction and as such has an interest keeping these installations in place for re-use on the long(er) term.

---

139 M.M. Roggenkamp 'Regulating Underground Storage of CO_2' (n 137) 225.

140 Cf. the criteria developed in the Altmark Trans case (C-280/00).

5 Concluding Remarks

CCS is being perceived as a possibility for governments to mitigate climate change and to be able to fulfil the still increasing demands for fossil fuels at the same time, by way of storing CO_2. The purpose of this chapter was to investigate – from a legal perspective – how to enhance the deployment of CO_2 storage in (nearly) depleted oil and gas fields in the North Sea area. The chapter focuses in particular on legal aspects governing the reuse of existing but redundant infrastructure on the seabed of the continental shelf.

Any re-use of depleted oil and/or gas fields requires that installations are kept in place when production ceases and the reservoir is not shut down. As offshore installations potentially may obstruct other uses of the sea such as navigation and fishing, UNCLOS provides a clear obligation that unused installations have to be removed. The IMO Guidelines limit this obligation to a situation where installations are permanently disused. Moreover, neither UNCLOS nor the IMO Guidelines contain a strict time frame for removal. Coastal States may thus have more room to manoeuvre than expected at first sight. This is illustrated by the abandonment and removal plans, which are applied in the Netherlands, the UK and Norway. The regimes in the UK and Norway are rather similar as they facilitate partial removal of installations and for the possibility to keep existing infrastructure in place if relevant for future uses. In such case, the abandonment plan should contain provisions regarding continuing inspection and maintenance, for which the licensee/operator would be responsible. In Norway, the State may even decide to step in and take over responsibilities and assets. The situation in the Netherlands differs as the MA requires a complete removal after production has ceased and does not provide the MEA with any powers to amend and intervene in abandonment and removal plan after it has been submitted. The current MA does not provide any strict time frame for submitting an abandonment and removal plan and in practice a solution can be found by postponing the submission of such a plan. However, this situation may change in the near future if the proposal to amend the MA in order to facilitate re-use is accepted.

The interest in re-using unused offshore facilities is increasing in the North Sea area. However, when considering the options for re-using depleted oil and gas fields and all relevant installations for carbon storage, we note that some uncertainties exist for the transitional period between production ceases and a CO_2 storage permit is awarded. Is there a guarantee that the holder of a production license will be awarded a storage permit? If so, what kind of arrangements will be made for transferring abandonment and removal obligations to the holder of the storage permit and how does this impact future removal? How can we ensure that fields are not abandoned and installations not removed without a storage permit has been awarded? Is it possible to provide for a temporary solution by appointing an 'operator of last resort'? These questions have been discussed in this chapter and an attempt is made to provide some answers. However, regulatory certainty can only be provided by the legislators. We therefore conclude that from a legal perspective, storage of CO_2 in depleted oil and gas fields is surrounded by some obstacles that need to be resolved. Solutions can feature on the level of international, EU and national law and may range from amending the IMO Guidelines, the CO_2 Storage Directive and national legislation. As the exploitation of oil and gas fields in the North Sea area is ageing and maturing and although first steps have been made to facilitate re-use of installations, it is time for legislators and policy makers to act and seize this window of opportunity.

CHAPTER 20

International Investment Law and the Regulation of the Seabed

James Harrison

1 Introduction*

Modern technology has increased the possibilities for greater parts of the oceans and ocean floor to be opened up for exploitation. Oil and gas extraction is still one of the most prevalent seabed activities, with almost a third of oil consumed in the world coming from seabed drilling[1] and new fields continue to be discovered in ever deeper waters. Innovations in marine technology also mean that many other uses of the seabed are now anticipated. One of the fastest growing offshore activities is renewable energy generation, which involves the attachment of structures, platforms or devices to the seabed in order to capture energy from wind, wave, or tidal power.[2] Many companies are also turning to the deep seabed as a source of valuable minerals, such as manganese, cobalt, copper and others. Significant deposits of these metals are thought to be located on many parts of the ocean floor, both within and beyond national jurisdiction.[3] These examples illustrate the diversity and fast-growing volume of the types of 'seabed investments' that will be addressed in this chapter.

Many of the seabed activities with which we are concerned in this chapter are undertaken by private actors, operating under a licence or concession from the coastal state. Coastal states have a significant degree of control over what seabed activities may take place in the waters adjacent to their territory. The

* The author would like to thank participants in the workshop held in Oslo for their feedback, as well as Filippo Fontanelli for his comments on a previous draft of this chapter.

1 J. Rochette, *Towards an International Regulation of Offshore Oil Exploitation*, IDDRI Working Paper No. 15, July 2012, 5. See also Global Oceans Commission, *From Decline to Recovery – A Rescue Package for the Global Ocean*, Summary Report 2014, 35.

2 For an overview of the relevant technology, see D. Leary and M. Esteban, 'Climate Change and Renewable Energy from the Ocean and Tides' (2009) 24 I.J.M.C.L. 617–651. The world's first full-scale floating windfarm was announced off the coast of Scotland in 2017 and new projects are emerging. See BBC, 'World's first floating wind farm emerges off coast of Scotland,' 23 July 2017 <http://www.bbc.co.uk/news/business-40699979>.

3 See P.A. Rona, 'The Changing Vision of Marine Minerals' (2008) 33 *Ore Geology Rev.* 618.

1982 United Nations Convention on the Law of the Sea (UNCLOS) confers 'sovereign rights' on the coastal state over almost all seabed activities taking place in waters up to 200 nautical miles from their coast,[4] as well as on the seabed up to the edge of the continental margin.[5] The rights of the coastal state are inherent and they do not depend upon proclamation or occupation.[6] They are also exclusive, so that 'if the coastal state does not explore the continental shelf or exploit its natural resources, no one may undertake these activities without the express consent of the coastal state.'[7] Operators will therefore have to comply with the laws and regulations established by the coastal state in relation to the exploitation of the seabed.

Most states will want to encourage broad participation in offshore activities, whilst also ensuring that companies exploiting seabed resources respect the marine environment and make some contribution to the national economy, through taxes, royalties, or other benefit-sharing arrangements. At the same time, the discretion of the coastal state to design its domestic legal framework will also be constrained by international law. Many states have entered into treaties, which seek to protect the interests of foreign investors, by imposing minimum standards of treatment. These instruments also allow companies to bring claims directly against states, without having to rely upon diplomatic protection from their own state. This is a burgeoning field of international law, with an increasing number of disputes being initiated.[8] There have been only a handful of disputes concerning seabed activities to date, but it is worth noting that land-based energy and mineral extraction projects are currently the biggest users of investor-state arbitration.[9]

This chapter explores the extent to which the investment treaty framework can be applied to seabed investments and, if so, the scope of protection that may be expected by investors. It focuses on seabed investments made within the jurisdiction of coastal states. The chapter is divided into three parts. Firstly, there will be a brief introduction to the international investment treaty regime

4 This includes sovereign rights for the purpose of exploring and exploiting the natural resources of the seabed, as well as jurisdiction over the placement of installations and structures for the production of energy from the water, currents and winds; see 1982 United Nations Convention on the Law of the Sea (UNCLOS), Articles 56–57, 60.

5 UNCLOS, Articles 76–77.

6 UNCLOS, Article 77(3).

7 UNCLOS, Article 77(2).

8 2017 saw at least 65 investment claims being initiated, bringing the total number of known claims to 855; see UNCTAD, *World Investment Report 2018*, 91.

9 The International Centre for the Settlement of Investment Disputes (ICSID) reports that oil gas and mining projects are the subject of 26% of claims, whilst electric power & other energy are the subject of 17% of claims; *The ICSID Caseload – Statistics*, Issue 2016–1, 12.

as it has evolved since its inception in the late 1950s. Secondly, the chapter addresses the geographical scope of international investment treaties, considering the extent to which they can be applied to offshore investments. This seemingly simple question has received very little attention in the literature to date and the discussion in this part will reveal a number of complexities that arise both from the ambiguous nature of the relevant treaty provisions, and also from the various scenarios in which seabed investments can take place. Thirdly, the chapter considers the substantive protections that are available if a seabed investment is found to fall within the scope of an investment treaty. This is an issue that raises complex questions about the balance between the rights of states to regulate activity falling under their jurisdiction and the limits that are imposed on those rights for the benefit of economic actors. The analysis demands an understanding of how international investment law interacts with relevant rules in the international law of the sea or international environmental law. Given the intensification in seabed activity around the world, this analysis will provide an important contribution to clarifying key issues that arise in the interpretation and application of investment standards in this context. Ultimately, the chapter aims to propose a framework of principles to reconcile these two areas of law.

2 The International Investment Law Framework

International investment law can trace its origins to the rules of customary international law relating to the protection of aliens, but international investment law emerged in its current form at the end of the 1950s, when states started to negotiate treaties for the protection of their nationals investing overseas.[10] These treaties aimed to both encourage the expansion of foreign investment between the parties and to create 'favourable conditions for nationals and companies of either State in the territory of the other State.'[11] This objective was achieved through the elaboration of substantive standards of protection. Another important feature of most investment treaties is that they confer a direct right on investors to bring a claim against a host state through arbitral proceedings. Indeed, it has been suggested that 'the availability of arbitration

10 The first modern Bilateral Investment Treaty (BIT) was concluded between Germany and Pakistan on 25 November 1959.

11 Germany-Pakistan BIT, preamble. Similar language is used in the preamble of many other BITS.

is probably the most important part of the "treatment" the foreign investor is looking for.'[12]

Over time, the number of investment treaties has increased incrementally to the point where there are 3,322 agreements in existence today.[13] Coverage of investment treaties is still far from universal, although most states in the world are party to at least one investment agreement.[14] Most of these treaties are bilateral in nature,[15] although there is a recent trend towards the conclusion of so-called mega-regional economic agreements that bring together larger groupings of states within a region into a closer economic relationship. Examples include the Convention on Protection of the Rights of the Investor,[16] the North American Free Trade Agreement (NAFTA),[17] the China/Korea/Japan Trilateral Investment Agreement,[18] and the Trans-Pacific Partnership (TPP).[19]

Another important agreement in the current context is the Energy Charter Treaty (ECT),[20] which contains a chapter dedicated to investment protection in the energy sector.[21] Its scope includes all 'economic activity concerning the exploration, extraction, refining, production, storage, land transport, transmission, distribution, trade, marketing or sale of energy materials or products.'[22] This includes prospecting and exploration for oil and gas, as well as construction and operation of power generation facilities, including those powered

12 *Impregillo v Argentina*, ICSID Case No. ARB/07/17, Award, 21 June 2011, Dissenting Opinion of Bridgette Stern, para. 32.

13 UNCTAD (n8) 88. The Report notes that the total fell for the first time in 2017, due to several IIAs being terminated, but that an increasing number of IIAs take the form of regional agreements which create many treaty relationships between the contracting parties. The text of most BITs and other investment agreements cited in this paper can be found on the UNCTAD website: <http://investmentpolicyhub.unctad.org/IIA>.

14 See <http://investmentpolicyhub.unctad.org/IIA>.

15 At the last count, there were 2946 BITs; UNCTAD (n8) 101.

16 1997 Convention on Protection of the Rights of the Investor, concluded by Armenia, Belarus, Moldova, Tajikistan, Kazakhstan, Kyrgyzstan, and Russia. Russia withdrew from the Convention in 2007.

17 The NAFTA was concluded in 1992 between the United States, Canada and Mexico. It entered into force in 1994. At the time of writing, major reforms of NAFTA were expected to be finalized in the near future, which would include renaming the instrument as the United States, Mexico and Canada Agreement.

18 The Trilateral Investment Agreement was concluded in 2012 and entered into force in 2014.

19 The TPP was negotiated between the United States, Australia, Brunei, Canada, Chile, Malaysia, Mexico, New Zealand, Peru, Singapore, Vietnam and Japan. The US pulled out of the TPP in January 2017.

20 The ECT was concluded in 1994. It entered into force on 1998. It has 52 parties.

21 ECT, Articles 10–17.

22 ECT, Article 1(5).

by wind and other renewable resources.[23] This instrument is also important because it has become the most frequently invoked investment agreement in terms of investor-state arbitration.[24]

What is important to remember is that all of these instruments have been individually negotiated and therefore no two treaties are identical. With that in mind, the following analysis will attempt to identify major trends in investment treaty drafting and interpretation in order to discover the extent to which investment treaties can be applied to seabed activities and what types of protection may be afforded to seabed investments. At the same time, it must be borne in mind that individual treaties may differ in their precise terms, which may influence the outcome in a particular dispute.

3 The Application of Investment Treaties to Seabed Activities

3.1 *The Application of Investment Treaties to Maritime Zones*

To identify the geographical scope of investment treaties, we must look firstly to the terms of the treaties themselves.[25] Most early investment treaties simply apply to investments 'in [the] territory'[26] or 'situated in the territory of' a contracting party.[27] For example, such language is found in the first investment treaty concluded between the United Kingdom and Egypt, which refers to the protection of investments within the territory of the two countries, without any explanation of what is meant by territory.[28] The first French BITs evidence a similar level of ambiguity, simply referring to 'investissements ... situé sur le territoire de l'autre Etat ...'[29] Even the 1992 NAFTA, seen by many as a watershed for a new style of investment agreement,[30] simply provides that the investment protection chapter applies to 'investments of investors of

23 See Understanding with respect to Article 1(5) of the ECT.

24 UNCTAD (n8) 93, which attributes 113 cases as being initiated under the ECT.

25 Note that there may be other provisions which define the scope of the treaty, potentially excluding some seabed investments from protection; see e.g. the discussion of sectoral exclusions in S. Trevisanut and N. Giannopoulos, 'Investment Protection in Offshore Energy Production: Bright Sides of Regime Interaction' (2018) 19 *Journal of World Investment and Trade* 789, 794–795.

26 E.g. 1991 Austria – Korea BIT, Article 3(2); 1959 Germany-Pakistan BIT, Article 3.

27 E.g. 1970 Belgium – Indonesia BIT, Article 5.

28 1975 UK-Egypt BIT.

29 See e.g. 1975 France-Congo BIT, Article 2.

30 N.W. Ranieri, 'NAFTA: An Overview', in L. Trakman and N.W. Ranieri (eds), *Regionalism in International Investment Law* (OUP, 2013) 90.

another Party in the territory of the Party.'[31] Such ambiguities are still found today. The TPP is a case in point, as it simply defines a covered investment as 'an investment in its territory' without defining territory.[32]

The language of these treaties creates some ambiguity about their precise geographical scope. Their application to seabed activities will depend upon the interpretation of the phrase 'territory.' This is in fact a question that has received little attention in either the jurisprudence or the literature. Whilst it clearly applies to investments made within the land territory of a state, does it also extend to the maritime zones of a coastal state? If so, does it include both the 'territorial sea' over which the coastal state exercises sovereignty[33] and other maritime zones in which the coastal state has sovereign rights for the purpose of exploring and exploiting natural resources?[34]

The inclusion of territorial sea within the scope of territory is a relatively unproblematic proposition. It is a general principle of treaty law that the application of treaties extends to 'the entire territory of each party'[35] and when drafting the rules on this subject, the International Law Commission noted that the reference to the 'entire territory' is 'a comprehensive term designed to embrace all the land and appurtenant territorial waters and air space that constitute the territory of the State.'[36] This phrase is understood as the internal waters and territorial sea of a state.[37] At the same time, the Commission did not explicitly mention the continental shelf as being part of the territory of a state, despite the fact that the Commission had shortly beforehand been involved in codifying rules on coastal state jurisdiction over the seabed.[38] Not too much significance should be attributed to this omission, however, as the question does not appear to have been considered in detail by the Commission.[39]

Indeed, there are other reasons to believe that the term 'territory' should be interpreted to extend to seabed activities on the continental shelf. It must be remembered that the very notion of the continental shelf is premised upon

31 NAFTA, Article 1101(1)(b).

32 TPP, Article 9.1.

33 UNCLOS, Article 2.

34 UNCLOS, Articles 56 and 77.

35 1969 Vienna Convention on the Law of Treaties, Article 29.

36 International Law Commission, 'Commentary to Article 25 of the Draft Articles on the Law of Treaties' (1966 - II) *Yearbook of the International Law Commission* 213.

37 See O. Doerr and K. Schmalenbach, *Vienna Convention on the Law of Treaties: A Commentary* (Springer, 2012) 497.

38 The Commission initiated discussion on the continental shelf in 1950 and it produced draft articles, which were used as the basis for the subsequent 1958 Convention on the Continental Shelf.

39 Doerr and Schmalenbach (n37) 496.

the natural prolongation of a coastal state's land territory into the seabed. In the words of the International Court of Justice (ICJ) in one of its seminal judgments on the continental shelf, 'the submarine areas concerned may be deemed to be actually part of the territory over which the coastal State already has dominion – in the sense that, although covered with water, they are a prolongation or continuation of the territory, an extension of it under the sea.'[40] Whilst the Court was speaking in general terms and it was not concerned with providing a legal interpretation of the term 'territory', its dicta in this respect nevertheless would support a broad understanding of the concept of territory for the purpose of international investment law.[41]

A teleological interpretation may also support a broad understanding of 'territory' for the purposes of international investment law.[42] One of the principal objectives of investment treaties is the protection of investment and it has thus been argued that an interpretation that offers more protection to investors should be preferred in cases of doubt.[43] From this perspective, an interpretation of a treaty that excludes investments that undoubtedly fall within a coastal state's jurisdiction as a matter of general international law could be considered as an unreasonable result that undermines the objectives of the treaty. There are obviously limits to this approach and the object and purpose cannot be used as a factor to override other more compelling arguments that suggest a different interpretation, for example an explicit exclusion of continental shelf investments by the contracting parties.[44] Nevertheless, it would support an assumption in favour of extending protection to seabed investments under the jurisdiction of the coastal state.

---

40 *North Sea Continental Shelf Cases* (1969) ICJ Reports 3, para. 43.

41 See also the *Aegean Sea Continental Shelf Case*, although the question in this case was not whether the continental shelf was a form of territorial right, but rather whether the dispute relating to the continental shelf boundary between Greece and Turkey was one which 'relates to the territorial status' of Greece; *Aegean Sea Continental Shelf Case* (1977) ICJ Reports 3, para. 81.

42 A similar argument has been made in the context of extending the protection of investment treaties to occupied territories; see R. Happ and S. Wuschka, 'Horror Vacui: Or why investment treaties should apply to illegally annexed territories' (2016) 33 *Journal of International Arbitration* 245, 261.

43 *SGS v Philippines*, ICSID Case No. ARB/02/6, Decision on Jurisdiction, 29 January 2004, para. 116; *Sanum v Laos*, PCA Case No. 2013–13, Award of Jurisdiction, 13 December 2013, para. 240.

44 Indeed, as Happ and Wuschka point out, the award in *Sanum v Laos* was ultimately annulled because it was contradicted by diplomatic exchanges by the parties to the treaty; Happ and Wuschka (n42) 261.

In practice, it would appear that this issue is not controversial. *Mobil and Murphy Oil v Canada* involved a claim by two oil companies in relation to their investments on the continental shelf of Canada.[45] In this case, the geographical scope of the NAFTA was not even raised by the parties and it appears to have been assumed by the tribunal that the concept of 'territory' in the agreement extended to this type of investment.

More recent developments in treaty drafting also confirm a broader geographical application of international investment law to maritime zones beyond the territorial sea. Practice on this issue appears to have taken a turn in the mid-1990s, when we see the question of the application of investment treaties to maritime zones being addressed in more explicit terms. For example, the Portugal-Korea BIT, concluded in May 1995, explicitly defines the term territory as 'the territory of either of the Contracting Parties, as defined by their respective laws, including the territorial sea, and any other zone over which the Contracting Party concerned exercises, in accordance with international law, sovereignty, sovereign rights or jurisdiction.'[46] This is also the practice that has been adopted by the European Union (EU) in its negotiation of investment provisions in its recent Free Trade Agreements[47] and it is a practice that is adopted in many modern model BITs.[48] This widespread practice supports the inclusion of maritime zones within the concept of 'territory' for the purposes of international investment law.[49]

Finally, it is worthwhile noting that investment treaties generally cannot be interpreted to apply to seabed investments in areas beyond national jurisdiction. Such areas are clearly not within the territory or maritime zones of a coastal state[50] and therefore investors in these areas would have to rely upon the sui generis provisions of Part XI of UNCLOS if they were to seek protection of their investments.[51]

45 *Mobil and Murphy Oil v Canada*, ICSID Case No. ARB(AF)/07/4, Decision on Liability and Principles of Quantum, 22 May 2012.

46 Portugal-Korea BIT, Article 1(4). See also Paraguay-Switzerland BIT, Article 1(4).

47 EU – Viet Nam Free Trade Agreement, Article 801.

48 UK Model BIT, Article 1(e); Italy Model BIT, Article I(6); India Model BIT, Article 1(f); Canada Model BIT, Article 1; Germany Model BIT, Article 1(4); US Model BIT, Article 1.

49 Not all treaties necessarily include maritime zones within the concept of 'territory.' Some treaties explicitly distinguish territory and maritime areas, but nevertheless include the latter within their scope of application. See e.g. 2007 France-Seychelles BIT, Article 1(4); ECT, Article 10(1).

50 UNCLOS, Article 137(1).

51 See particularly UNCLOS, Article 152(1), Annex III, Articles 17–19. See also Part XI Agreement, Annex, Section 1, para. 13.

3.2 *The Application of Investment Treaties to Joint Exploitation Zones and Disputed Maritime Areas*

The previous discussion was based on the assumption that an investment was made within the established maritime boundaries of a coastal state. Yet, in some parts of the ocean, states have been unable to agree upon maritime boundaries. Two scenarios may present themselves in this context.

3.2.1 Application to Joint Exploitation Zones

In the first scenario, coastal states may have chosen to pursue joint development of seabed resources in an area in which they cannot agree on the appropriate maritime boundary. Such arrangements are explicitly encouraged by UNCLOS.[52] There are various types of arrangement that have been entered into for this purpose.[53] One common mechanism is the establishment of joint commission, which is granted powers to authorize and regulate seabed activities within a joint development area. An example is provided by the Thailand-Malaysia Memorandum of Understanding, which identifies an area of overlapping claims and establishes a Joint Authority composed of representatives of both countries.[54] This body has legal personality[55] and it assumes 'all rights and responsibilities on behalf of both parties for the exploration and exploitation of the non-living natural resources of the seabed and subsoil in the overlapping area ... and also for the development, control and administration of [said] area.'[56] The Joint Authority is empowered to enter into contracts with oil and gas companies wishing to carry out operations in the

52 UNCLOS, Articles 74(3) and 83(3). See also chapter 8 of this book, E. Røsæg, 'Framework legislation for commercial activities in the Area'; J. Dingwall, 'International Investment Protection in Deep Seabed Mining Beyond National Jurisdiction' (2018) 19 *Journal of World Investment and Trade* 890–929.

53 For a useful overview of arrangements, see M. Miyoshi, 'The Joint Development of Offshore Oil and Gas in Relation to Maritime Boundary Delimitation', Maritime Briefing, Volume 2, Number 5, 1999. See also P. Tzeng, 'Investment Protection in Disputed Maritime Areas' (2018) 19 *Journal of World Trade and Investment* 828, 855–858.

54 1979 Memorandum of Understanding between Malaysia and Thailand on the Establishment of the Joint Authority for the Exploitation of the Resources of the Seabed in a Defined Area of the Continental Shelf of the Two Countries in the Gulf of Thailand (1979 MOU). For another example, see the 2001 Treaty between the Federal Republic of Nigeria and the Democratic Republic of São Tomé e Príncipe on the Joint Development of Petroleum and other Resources.

55 1990 Agreement between the Government of Malaysia and the Government of Thailand on the Constitution and other Matters relating to the Establishment of the Malaysia-Thailand Joint Authority (1990 Agreement), Article 1.

56 1979 MOU, Article III(2).

Joint Development Area.[57] It is also the Joint Authority that is responsible for formulating policies for seabed activities in the Joint Development Area.[58]

For present purposes, the question is how such arrangements would be addressed from the perspective of international investment law. Would such a Joint Development Area be classified as being within 'the territory' of a coastal state for the purposes of international investment law? This is a difficult question given that the whole purpose of such arrangements is to avoid the determination of which state has sovereignty or sovereign rights over a contested area. Thus, it may be better to consider a Joint Development Area as a tertium genus.[59]

Another problem that arises for investors in such a scenario is that the arrangement may confer the power to regulate seabed activities on an autonomous international organization with distinct legal personality. If so, any regulatory measures adopted by the organization arguably could not be attributed to the state that is party to the investment treaty.[60] Instead, it raises issues of the responsibility of the organization itself.[61] It follows that the investor will not be able to claim protection under the treaty and its remedies will be limited to those that are available on the basis of its contractual relationship with the organization. Thus, it is important that the investor ensures that a contract contains relevant provisions to safeguard its interests against interference from the organization, such as the inclusion of minimum standards of treatment or a stabilization clause,[62] as well as an appropriate dispute settlement mechanism.[63]

3.2.2 Application to Disputed Maritime Areas

In the second scenario, states may not be able to reach agreement on provisional arrangements to apply to a disputed area, but they will both continue

57 1990 Agreement, Article 7(2)(e).

58 1990 Agreement, Article 7(1).

59 For other uses of this concept in the law of the sea, see *Fisheries Jurisdiction Case* (1974) ICJ Reports 3, para. 54.

60 See discussion in the International Law Commission, 'Commentary to Chapter IV of the Draft Articles on State Responsibility' (2001 - II) *Yearbook of the International Law Commission* 64.

61 By way of example, albeit in the context of international human rights law, see *Behrami and Behrami v France*, Judgment of the European Court of Human Rights, 2 May 2007.

62 See e.g. L Cotula, 'Regulatory Takings, Stabilization Clauses and Sustainable Development', paper presented at the OECD Global Forum on International Investment, 27–28 March 2008. Available at: http://www.oecd.org/investment/globalforum/40311122.pdf.

63 For discussion of contractual claims in this context, see Tzeng (n53) 850–852.

to claim authority therein. There are many maritime boundaries around the world that remain un-delimited[64] and several areas in which more than one coastal state claims rights.

It is generally accepted that international law does not prevent states from carrying out unilateral prospecting of seabed resources in a disputed area, although they may be obliged to refrain from activities that might affect the rights of other states.[65] The fact that investments may be made in disputed areas is demonstrated by the maritime boundary case between Cote d'Ivoire and Ghana, in which Cote d'Ivoire alleged that Ghana had proceeded to authorize seabed activities in an area that Cote d'Ivoire argued belongs to it.[66]

The question that arises for present purposes is if the investor would have a claim against the state that had authorized it to proceed, if the investor is later forced to stop its activity because the area is found to be under the sovereignty of another state.[67] Strictly speaking, in this situation, the investment is not within the maritime zone of the authorizing state and therefore an investment treaty would not prima facie be applicable. Yet, it seems unfair to allow a state to evade responsibility when it had itself led an investor to believe that it did have jurisdiction over the disputed area. This situation may thus call for the application of the principle of ex factis jus oritur, so that the exercise of de facto sovereign rights in a contested area could be sufficient to establish that an investment was in the maritime zone of the state.[68] An alternative approach

64 See C. Schofield and C. Carleton, 'Key Technical Issues in the Law of the Sea Dispute Resolution and the Role of the Technical Expert', in A.G. Oude Elferink and D. Rothwell (eds), *Oceans Management in the 21st Century: Institutional Frameworks and Responses* (Martinus Nijhoff, 2004) 254.

65 See Miyoshi (n53) 4. See also *Aegean Sea Continental Shelf Case* (*Provisional Measures*) (1976) ICJ Reports 3, paras 28–33; BIICL, *Report on the Obligations of States under Articles 74(3) and 83(3) of UNCLOS in respect of Undelimited Maritime Areas* (2016).

66 Ghana raised the impacts on investors in its pleadings concerning the imposition of provisional measures pending the settlement of the dispute; see *Dispute concerning Delimitation of the Maritime Boundary between Ghana and Cote d'Ivoire in the Atlantic Ocean*, ITLOS Case No. 23, Provisional Measures Order, 25 April 2015, para. 85. Ultimately, the Special Chamber ordered Ghana to 'take all the necessary steps to ensure that no new drilling either by Ghana or under its control takes place in the disputed area' (para. 102), although the order did not apply to 'ongoing activities … in respect of which drilling has already taken place.' (para. 99) A final judgment was delivered by the Special Chamber on 23 September 2017 in which it absolved Ghana from any responsibility for carrying out unlawful activities in the disputed area.

67 A more complicated set of legal questions arise if an investment claim is made whilst a sovereignty dispute is ongoing; see Tzeng (n53) 839–848.

68 Similar arguments have been made in relation to unlawfully annexed territories; Happ & Wuschka (n42) 264.

would be to apply the doctrine of estoppel to prevent a state from pleading a lack of jurisdiction in circumstances in which it had previously asserted its jurisdiction to the detriment of the investor.[69] Whichever approach is preferred, the result is to extend the application of investment treaties to disputed maritime zones.

4 Standards of Investment Protection under Investment Treaties and Implications for the Regulation of Seabed Investments

4.1 *Investment Protection Standards*

Having considered the geographical scope of investment treaties, it is now necessary to examine what type of protection may be granted to an investment thereunder. In the following section, particular attention is given to the manner in which investment agreements may constrain coastal states in exercising their regulatory powers in relation to seabed activities.

4.1.1 Pre-investment Protection

Generally speaking, states have a broad degree of discretion as to which investments they allow to take place within their territory and most investment treaties leave this discretion unfettered, as protection only kicks in once the investment has been made. However, a number of more modern investment treaties do impose some obligations on states at this initial decision-making stage. In particular, it is increasingly common to find a requirement that states grant most-favoured nation treatment and national treatment to investors with respect to the establishment, acquisition or expansion of their investments.[70] Such requirements would not restrict the content of any decision, but they would demand that decisions are made in a non-discriminatory manner.

---

69 For a discussion of estoppel in international law, see I.C. MacGibbon, 'Estoppel in International Law' (1958) 7 I.C.L.Q. 468–513; C. Brown, 'A Comparative and Critical Assessment of Estoppel in International Law' (1996) *U. Miami L. Rev.* 369. However, see Tzeng (n53) 847 who argues that this approach 'is arguably at odds with the tribunal's arguably inherent duty to examine proprio motu its own jurisdiction.'

70 See e.g. TPP, Articles 9.4 and 9.5 applying MFN treatment and National Treatment to the establishment of investments. ECT, Article 10(1) requires states to 'create stable, equitable, favourable and transparent conditions for Investors of other Contracting Parties to make Investments in its Area.' Article 10(2) of the ECT, however, only requires states to 'endeavour' to accord MFN and National Treatment to investors. Often, such pre-establishment rights are also subject to exceptions, particularly when it comes to 'subsidies' or 'procurement' decisions; see e.g. *Mesa Power Group v Canada*, PCA Case No. 2012–17, Award, 24 March 2016, paras 378–507.

Some investment treaties go even further in limiting the discretion of host states to impose so-called performance requirements on investors. Performance requirements involve the imposition of conditions on companies relating to various aspects of their operations and they are often used by host states to 'help materialize expected spill-over effects from foreign investment.'[71] Investments treaties do not prohibit all forms of performance requirements, rather they tend to target those that create significant economic inequalities, such as conditions requiring the purchase of local goods or services, conditions relating to the achievement of certain levels of exports or foreign exchange earnings, or conditions mandating the transfer of technology.[72] Investment treaties usually address performance requirements relating to both the authorization or continuing operation of an investment, as well as similar conditions attached to advantages, such as loans, grants or subsidies.

The interpretation of the performance requirements provision in the NAFTA was raised in the case of *Mobil and Murphy Oil v Canada*, in which the two oil companies challenged the revision of certain requirements concerning expenditure on research and training activities. Both companies had accepted certain expenditure requirements at the time when they made the investment. However, the Canadian authorities introduced revised guidelines in 2004, which the investors claimed significantly increased the burden imposed upon them to carry out these sorts of activities. In particular, the new guidelines required that expenditure should meet a certain level during specified periods. The companies argued that the revisions constituted a breach of, inter alia, the NAFTA provisions on performance requirements. The principal issue in dispute was whether the changes could be considered as falling within the prohibition on performance requirements 'to purchase, use or accord a preference to ... services provided in its territory' under Article 1106(1)(c) of the NAFTA. Research and training were both held to be services for the purposes of the NAFTA.[73] Yet, Canada argued that the 2004 guidelines did not expressly compel the investors to use particular local services in order to meet their commitments. In rejecting the Canadian defence, the Tribunal looked beyond the actual text of the guidelines to determine their de facto effect, concluding that '[they] are designed to ensure that expenditures for [research and training] services occur in the Province, and thereby implying a legal requirement for

71 See 2015 Investment Policy Framework for Sustainable Development, 'Policy Options for International Investment Agreements,' example of Standards of treatment and protection. Available at: <http://investmentpolicyhub.unctad.org/ipfsd/section/3/toc-item-details/277>.

72 See e.g. TPP, Article 9.10; NAFTA, Article 1106.

73 *Mobil* (n45) para. 221.

the purposes of Article 1106.'[74] The majority of the tribunal also took a restrictive interpretation of the reservations to this obligation.[75] This case serves as a reminder of the important constraints that may be imposed by investment agreements.

4.1.2 Post-investment Protection

Once an investment has been made, one of the greatest risks that can face an investor is having their property or assets taken by the host state without justification. To counter such threats, most investment agreements contain provisions prohibiting a state from nationalizing or expropriating an investment without due process and without payment of compensation.[76] Many agreements also prohibit a host state from restricting or delaying the transfer of profits or proceeds of an investment.[77] Claims based upon such provisions are rare, as states do not often have recourse to such heavy-handed tactics. In contrast, it is much more common for international investment law to be used to challenge more subtle changes to the legal framework applicable to investors.

One of the most difficult issues in investment arbitration is the extent to which investment protection standards restrict the ability of host states to increase taxes or strengthen regulations after an investment has been made.

Early case law suggested that standards of protection could operate in such a way to significantly constrain states in their exercise of regulatory powers and there was a fear that investment law may have a 'chilling effect' on regulation.[78] This is particularly the case when tribunals emphasized the need for stability when interpreting investment protection standards in light of the legitimate expectations of investors.[79]

---

74 Ibid, para. 242. See also para. 238.

75 This decision was subject to a dissenting opinion by Professor Philippe Sands QC.

76 E.g. ECT, Article 13; NAFTA, Article 1110.

77 E.g. ECT, Article 14.

78 See e.g. J. Bonnitcha, 'Outline of Normative Framework for Evaluating Interpretation of Investment Treaty Protections, in C. Brown and K. Miles (eds), *Evolution in Investment Treaty Law and Arbitration* (CUP, 2011) 133–140; see also K. Tienhaara, 'Regulatory Chill and the Threat of Arbitration', in C. Brown and K. Miles (eds), *Evolution in Investment Treaty Law and Arbitration* (CUP, 2011) 606–627.

79 Legitimate expectations has featured in the interpretation of both the indirect expropriation standard (*Tecmed v Mexico*, ICSID Case No. ARB(AF)/00/2, Award, 29 May 2003, para. 122) and the fair and equitable treatment standard (*Biwater Gauff v Tanzania*, ICSID Case No. ARB/05/22, Award, 24 July 2008, para. 602). See discussion in Y. Levashova, 'Fair and Equitable Treatment and the Protection of the Environment: Recent Trends in Investment Treaties and Investment Cases', in Y. Levashova, T. Lambooy and I. Dekker (eds), *Bridging the Gap between International Investment Law and the Environment* (Eleven International Publishing, 2016) 64.

Yet, more recent developments in investment case law would seem to suggest that the standards should not be interpreted in a restrictive manner and states possess a degree of flexibility when deciding how to regulate investments, particularly when 'defining the issues that affect its public policy or the interests of society as a whole, as well as the actions that will be implemented to protect such values.'[80] On the basis of this case law, it is clear that international investment law cannot be relied upon to shield investors from any changes to the legal framework. This was made explicit in *Mobil and Murphy Oil v Canada* when the Tribunal held that the fair and equitable treatment standard 'was never intended to amount to a guarantee against regulatory change, or to reflect a requirement that an investor is entitled to expect no material changes to the regulatory framework within which an investment is made.'[81] In another recent case involving the modification of the regulatory framework for the renewable energy sector, the Tribunal similarly held that 'in the absence of a specific commitment toward stability, an investor cannot have a legitimate expectation that a regulatory framework … is not to be modified at any time to adapt to the needs of the market and to the public interest.'[82] Moreover, it is also now generally accepted that 'non-discriminatory regulatory actions by a Party that are designed and applied to protect legitimate public welfare objectives, such as public health, safety and the environment, do not constitute indirect expropriations, except in rare circumstances.'[83]

Similar considerations apply to the taxation framework that is applied to investors. For example, in *Paushok v Mongolia*, the Tribunal held that 'significant modification of taxation levels represents a serious risk, especially when investing in a country at an early stage of economic and institutional development' and the imposition of a windfall tax imposed by Mongolia could not by itself constitute a breach of the fair and equitable treatment standard.[84] The Tribunal went on to note that an investor could seek to protect itself by negotiating a stability agreement with the government, but '[i]n the absence of such a stability agreement, Claimants have not succeeded in establishing that they had legitimate expectations that they would not be exposed to significant

80 *Tecmed* (n79) para. 122.

81 Ibid, para. 153. See also *Saluka Investments v Czech Republic*, UNCITRAL Arbitration, Partial Award, 17 March 2006, para. 306; *Perenco v Ecuador*, ICSID Case No. ARB/08/6, Decision on Remaining Issues of Jurisdiction and Liability, 12 September 2014, para. 560.

82 *Charanne Construction Investments v Spain*, Stockholm Chamber of Commerce Arbitration No. 062/2012, Award, 21 January 2016, para. 510 (unofficial translation by Mena Chambers). The majority of the Tribunal also rejected the proposition that the regulatory framework in that case involved an implicit stabilization clause. See however the dissenting opinion of Prof. Guido Santiago Tawil.

83 TPP, Annex 9B, para. 3(b).

84 *Paushok v Mongolia*, UNCITRAL Arbitration, Award, 28 April 2011, para. 302.

tax increases in the future.'[85] Even a 99% windfall tax applied to oil industry profits was held not to violate the expropriation standard in *Perenco v Ecuador* because the tax did not affect the operation and management of the company, nor did it come close to extinguishing the business.[86]

4.2 *Balancing Investment Protection and Environmental Concerns in the Context of Seabed Investments*

4.2.1 Need for Flexibility in Interpreting Investment Standards

The preceding analysis suggests that the requirements of investment protection must be balanced against the interests of the host state to protect certain public interests. The protection of the environment clearly counts as an interest that can be promoted by the host state in exercising its regulatory powers. Indeed, it must be appreciated that international law places obligations on the coastal state to 'adopt laws and regulations to prevent, reduce and control pollution of the marine environment arising from or in connection with seabed activities subject to their jurisdiction and from artificial islands, installations and structures under their jurisdiction.'[87] The precise content of these laws and regulations will be determined in the first instance by the coastal state, albeit subject to an overarching obligation to exercise due diligence to minimize the risk of significant harm to the marine environment.[88] It has also been argued that, in light of the uncertainties concerning the impact of new technologies on the marine environment, emerging seabed industries require a precautionary approach to regulation,[89] meaning that states should adopt measures to prevent serious and irreversible harm, even if there is not yet proof that harm may occur.[90] Meeting the due diligence obligation imposed by international law may also require the rules and standards to be adapted over time as perceptions of risks change in response to developments in our understanding of the oceans.[91]

---

85 Ibid., para. 302.

86 *Perenco* (n81) paras 680–687.

87 UNCLOS, Article 208(1). This provision must also be read in light of Article 194(3)(c).

88 On the due diligence standard, see *South China Sea Arbitration*, PCA Case No 2013–19, Merits Award, 12 July 2016, para. 944. See also J. Harrison, *Saving the Oceans through Law: The International Legal Framework for the Protection of the Marine Environment* (OUP, 2017) 209–225.

89 See e.g. The World Bank, *Precautionary Management of Deep Sea Mining Potential in Pacific Island Countries* (2016).

90 See 1992 Rio Declaration on the Environment and Development, Principle 15.

91 See Advisory Opinion on *Responsibilities and Obligations of Sponsoring States with Respect to Activities in the Area* (2011) ITLOS Reports 10, para. 117. On the importance of adaptive

There is little doubt that these international obligations relating to the protection of the marine environment provide relevant context for informing the scope and content of international investment law. As noted by in one recent investment award, 'consideration of a host state's international obligations may well be relevant in the application of [an investment treaty standard] to particular circumstances.'[92] Yet, it is not to say that these environmental obligations will automatically take priority. Rather, one should pursue 'mutual supportiveness'[93] of economic and environmental objectives and, if possible, it is for the Tribunal to identify an interpretation that reconciles both sets of rules.

The need for flexibility in the interpretation of investment rules has been expressly recognised in relation to industries where there are known environmental concerns.[94] This is an area in which states possess a 'high level of deference ... [and] it is not for an investor-state tribunal to second-guess the substantive correctness of the reasons which an administration were to put forward in its decisions, or to question the importance assigned by the administration to certain policy objectives over others.'[95] However, there are certain principles that host states must nevertheless respect when developing environmental regulations, as discussed in the following section.

4.2.2 Guiding Principles When Developing National Environmental Regulations in Order to Satisfy International Investment Law

Firstly, tribunals have stressed the need for states to take a cooperative approach when introducing changes to the legal framework,[96] particularly where procedures for consultation or negotiation are themselves recognized in the applicable laws.[97] At a minimum, states should avoid conduct that demonstrates 'a lack of due process leading to an outcome which offends judicial propriety – as might be the case with a manifest failure of natural justice in

---

management as a regulatory approach to environmental protection, see Addis Ababa Principles and Guidelines for the Sustainable Use of Biodiversity, Practical Principle 4.

92 *Peter Allard v Barbados*, UNCITRAL Arbitration, Award, 27 June 2016, para. 244. In this case, the Tribunal was talking about the full protection and security standard.

93 See e.g. R. Pavoni, 'Mutual Supportiveness as a Principle of Interpretation and Law-Making' (2010) 21 E.J.I.L. 649–679.

94 *Methanex v United States of America*, UNCITRAL Arbitration, Final Award, 8 March 2005, Part IV, Chapter D, para. 9.

95 *Crystallex International Corporation v Venezuela*, ICSID Case No. ARB(AF)/11/2, Award, 4 April 2016, para. 583. See also Mesa Power (n70) para. 505.

96 *Suez* and others *v Argentina*, ICSID Case No. ARB/03/19, Decision on Liability, 30 July 2010, para. 223.

97 Ibid, para. 237.

judicial proceedings or a complete lack of transparency and candour in an administrative process.'[98] The threshold for a breach of relevant investment standards is, however, a high one, which requires something more than a 'perceived unfairness.'[99]

Secondly, international investment law would prevent states from reneging on clear and explicit representations that were made by the state in order to induce the investment and relied upon by the investor.[100] Even in this scenario, the investor may not be successful in their claim unless they can show that they have relied upon the representation to their detriment and the breach of the representation was in some way arbitrary or unreasonable.[101]

Thirdly, states must take the least restrictive option that is available to them to achieve their environmental objectives. In other words, 'where a state can achieve its chosen level of environmental protection through a variety of equally effective and reasonable means, it is obliged to adopt the alternative that is most consistent with open trade.'[102] This is an important consideration in the field of marine environmental protection, where the due diligence obligations under UNCLOS would appear to leave a degree of discretion to states as to how they comply.

Fourthly, there may be a potential breach of relevant investment standards if the state can offer no reasonable explanation for a proposed change in the regulatory framework. In other words, there must be *some* evidential basis for a measure, if it is not to be considered as 'arbitrary'[103] or 'disproportionate.'[104] Thus, the denial of a permit was found to be a violation of international investment law in *Crystallex v Venezuela* because it was based upon reasons that had not been raised in the exchanges between the company and the state, whilst ignoring the scientific evidence that had been submitted as part of the application.[105]

That is not to say that there must always be clear proof of environmental harm before a host state may take action. Article 201 of UNCLOS encourages

98 *Waste Management v Mexico*, ICSID Case No. ARB(AF)/00/3, Award, 30 April 2004, para. 98. See also *Glamis Gold v United States of America*, UNCITRAL Arbitration, Award of 8 June 2009, para. 627; *Mobil* (n45) para. 152.

99 *Perenco* (n81) para. 559. See also *Bilcon v Canada*, UNCITRAL Arbitration, Award, 17 March 2015, para. 444.

100 *Mobil* (n45) para 156. See also *Glamis Gold* (n98) para. 621; *Suez* (n96) para. 226.

101 See e.g. *Mesa Power* (n70) para. 502. See also *Allard* (n92) paras 200–204.

102 *SD Myers v Canada*, UNCITRAL Arbitration, Partial Award, 13 November 2000, para. 221.

103 Waste Management (n94) para. 98. See also *Glamis Gold* (n94) para. 627; *Mobil* (n43) para. 152.

104 See *Tecmed* (n75) para. 122.

105 *Crystallex* (n95) paras 591–597.

'appropriate scientific criteria for the formulation and elaboration of rules, standards and recommended practices and procedures for the prevention, reduction and control of pollution of the marine environment,'[106] but it does not specify how much scientific evidence is necessary and it has been suggested that 'the [precautionary approach] will have a role to play in the operation of Article 201 as the establishment of appropriate scientific criteria will need to acknowledge potential risks even if they are not fully known.'[107] The precautionary approach is also accepted as being implicit in the obligation of due diligence.[108] It follows that the precautionary approach may be invoked to justify action for the protection of the environment, even when there is no conclusive proof that environmental harm will occur.[109] At the same time, there are limits inherent in the precautionary approach itself, which is apparent from recent case law on the subject.

Windstream Energy v Canada is one of the few investment cases in which the precautionary approach has been raised in proceedings. The case involved an American investor challenging a moratorium on offshore wind farms in Lake Ontario in Canada. The company claimed that it had invested in a project for offshore wind development based upon encouragement from the Government of Ontario, which later changed its mind and decided to halt any further development until further scientific studies had been carried out. Canada justified its decision to take 'a cautious approach and develop a comprehensive regulatory framework before allowing any offshore wind energy facilities to be built' as a reasonable use of its regulatory powers.[110] This position was supported by the Tribunal, which accepted that Ontario's policy was 'at least in part driven by a genuine policy concern that there was not sufficient scientific support for establishing an appropriate ... exclusion zone for offshore wind projects.'[111] The reference to a 'genuine' policy concern is important because it emphasizes that there is a minimum threshold of evidence that must be met in order to invoke

106 UNCLOS, Article 201.

107 T Stephens, 'Article 201', in A Proelß (ed.), *United Nations Convention on the Law of the Sea: A Commentary* (Beck 2017) 1345.

108 *Sponsoring States Advisory Opinion* (n91) para. 131.

109 On the relationship between the precautionary approach and the standard of proof and sources of evidence, see J. Harrison, 'Addressing the Procedural Challenges of Environmental Litigation in the context of Investor-State Arbitration, in Y. Levashova et al (eds), *Bridging the Gap between International Investment Law and the Environment* (Eleven International Publishing, 2016) 99–104.

110 See e.g. *Windstream Energy v Canada*, Government of Canada's amended Response to the Notice of Arbitration, 5 December 2013.

111 *Windstream Energy v Canada*, UNCITRAL Arbitration, Award, 27 September 2016, para. 376.

the precautionary approach. Thus, as explained by another international tribunal, even a precautionary approach requires that 'there are plausible indications of potential risks.'[112] Such an understanding ensures that environmentalism is not used as a pretext for measures that substantially interfere with investment activity.

Indeed, another restriction is that the precautionary approach only justifies 'cost-effective' measures.[113] In other words, a tribunal could still evaluate the effects of a measure on an investor, even if it accepts that some sort of precautionary measure would be appropriate. In some cases, complete restriction may be justified, but this outcome should not be assumed without some sort of assessment of alternative measures. Furthermore, the Tribunal in *Windstream Energy* explained that, even if a moratorium would be appropriate, the precautionary approach does not give a state freedom to postpone a project for as long as it wants – the Tribunal ultimately held that Canada had violated the fair and equitable treatment standard because it had done 'relatively little to address the scientific uncertainty surrounding offshore wind that it had relied upon as the main publicly cited reason for the moratorium.'[114] This would seem to suggest that the precautionary approach may justify provisional restrictions on an investment, but a state should take further steps to gather the scientific evidence that may be necessary to make a full assessment of the environmental effects of an activity. This interpretation of the precautionary approach follows the rules agreed in other economic treaty regimes, where states are permitted to provisionally adopt restrictive measures where there is scientific uncertainty, but states must 'seek to obtain the additional information necessary for a more objective assessment of risk and review the [provisional] measures accordingly within a reasonable period of time.'[115]

Ultimately, it is important to acknowledge that there may be situations in which investment rules will require protection to be afforded to an investor, despite the environmental concerns of a state. This may be particularly true in the case of regulatory measures which '[render] the value to be derived from their activities so marginal or unprofitable as to effectively deprive them of

112 *Sponsoring States Advisory Opinion* (n91) para. 131.

113 See 1992 Rio Declaration on Environment and Development: 'Where there are threats of serious or irreversible damage, lack of full scientific certainty shall not be used as a reason for postponing cost-effective measures to prevent environmental degradation.' The requirement of cost-effectiveness was emphasized in *Sponsoring States Advisory Opinion* (n91) para. 128.

114 *Windstream Energy* (n111) para. 378.

115 WTO Agreement on the Application of Sanitary and Phytosanitary Measures, Article 5(7).

their character as an investment.'[116] This is a very high threshold, but it might be met, for example, in the case of a declaration of a marine protected area (MPA), in which all activity is prohibited. Such findings have been made in cases concerning the establishment of terrestrial protected areas, for example in *Metalclad v Mexico*, where the Tribunal held that an Ecological Decree designating a cacti reserve in the area of the investment constituted an act tantamount to expropriation because it severely restricted the carrying out of any activity within the area.[117] The fact that the international community has repeatedly endorsed the development of MPA networks would not prevent such a finding.[118] After all, states have a large amount of discretion as to which areas they designate as MPAs, as well as what activities they restrict therein. Moreover, it must be understood that international investment law would not prevent the establishment of a MPA, but the state would be required to pay compensation to investors that are prevented from continuing activities that they had been previously authorized to conduct in that area. In this respect, it has been noted that '[w]hile an expropriation or taking for environmental reasons may be classified as a taking for a public purpose, and thus may be legitimate, the fact that the [p]roperty was taken for this reason does not affect either the nature or the measure of the compensation to be paid for the taking.'[119] This is a reminder that coastal states will have to take into account international investment law when carrying out their regulatory mandate in relation to seabed activities and the establishment of MPAs. It also suggests that, in order to avoid conflicts, coastal states should, as far as possible, try to identify ecosystems that require protection *prior* to authorizing investments in the relevant areas.[120]

5 Conclusion

It is likely that uses of the seabed will continue to increase in the coming years, both in terms of an expansion of existing activities and the emergence of new ones. In the exercise of their sovereignty and sovereign rights over the seabed,

116 *EnCana Corporation v Ecuador*, LCIA Case No. UN 3481, Award, 3 February 2006, para. 174.

117 See *Metalclad v Mexico*, ICSID Case No. ARB(AF)/97/1, Award, 30 August 2000, paras 109–112.

118 See e.g. Sustainable Development Goal 14.5.

119 *Santa Elena Development Company v Costa Rica*, ICSID Case No. ARB/96/1, Award, 17 February 2000, para. 71.

120 To this effect, see UK House of Commons Energy and Climate Change Committee, *The Future of Marine Renewables in the United Kingdom* (2012) para. 92.

coastal states have a broad degree of power to ensure that these activities are carried out in such a way that both provide benefits to local communities and do not cause serious or irreparable harm to the marine environment. This paper has explored the extent to which international investment law applies to the measures adopted by coastal states in relation to seabed activities and what constraints may be imposed on coastal states thereby.

It has been shown that there are clear arguments that investment treaties can be interpreted to apply to investments in all maritime zones of coastal states, even if they are not explicit in this respect.

The chapter also identified some situations in which the application of international investment law becomes more complex, notably seabed activities in joint development areas or disputed areas. Investors should be more cautious in these situations, as they may not necessarily benefit from the protections of international investment law. Thus, they may need to enter into special arrangements in order to receive the assurances that they may want before they invest.[121] It is also important to remember that international investment law also does not extend to investments in areas beyond national jurisdiction.

When it comes to the substantive protection that may be enjoyed by an investor, it was argued that international investment law must be interpreted and applied in light of state's international commitments in relation to the protection of the marine environment, with a view to achieving mutual supportiveness between these two areas of law. The paper proposed a framework of principles to guide states in the exercise of their regulatory powers, in order to minimize any conflict with investment protection standards. In doing so, the paper identified the opportunities for applying a precautionary approach to seabed regulation, whilst also pointing out the limits of this principle as a defence to investment claims.

Ultimately, international investment law offers a large degree of discretion to states in exercising regulatory powers, although the precise limitations will depend upon the nature of the obligations they have undertaken in their investment treaties and the restrictiveness of measures that are adopted.

121 In this context, see the discussion in Tzeng (n53) of contractual measures and political risk insurance.

CHAPTER 21

Navigating Legal Barriers to Mortgaging Energy Installations at Sea – the Case of the North Sea and the Netherlands

Jaap J.A. Waverijn

1 Introduction

Annual investments in the global energy sector exceed $1.5 trillion.[1] Tens of billions per year thereof are invested in oil, gas and electricity infrastructure situated above, on or under the seabed of the North Sea. In the second half of the 20th century, the investments in the North Sea energy sector predominantly concerned the exploration and exploitation of oil and gas fields. Since the turn of the century, the number of wind farms constructed on the North Sea seabed has increased drastically. The energy sector can be characterized as highly capital intensive. This is particularly true for the offshore energy sector. Hundreds of millions or billions of euros are invested in individual offshore oil and gas and offshore wind projects.[2] It is common business practice that debt is attracted to finance such operations.[3] The companies active in the oil and gas sector in the last century had no issues attracting the required capital, in particular when oil and gas prices were high, as they could use their extensive assets and oil and gas reserves as collateral.[4] This is different in the current sector as oil and gas prices are relatively low, the discovered fields are smaller and many fields are nearing depletion. New companies with fewer assets have

1 'World Energy Investment 2018' (OECD/International Energy Agency 2018). In 2017, total investments in the energy sector, including electricity generation and supply, oil and gas supply, energy efficiency and coal supply, amounted to 1.8 trillion USD. In 2015 and 2016 global investments in the energy sector were higher.

2 Construction and operation of individual wind farms often exceeds €1 billion, 'Offshore Wind in Europe – Key Trends and Statistics 2018' (*WindEurope*, 2019) 21 <windeurope.org/about-wind/statistics/offshore>.

3 The following explanation receives more attention in C.G. Verburg & J.J.A. Waverijn, 'Liberalizing the Global Supply Chain of Renewable Energy Technology: The Role of International Investment Law in Facilitating Flows of Foreign Direct Investment and Trade' (2019) Brill Open Law and my forthcoming PhD thesis.

4 'Economic Report 2016' (*Oil and Gas UK*, 2017) 34 <http://oilandgasuk.co.uk/wp-content/uploads/2016/09/Economic-Report-2016-Oil-Gas-UK.pdf>.

entered the market, for example specializing in the offshore wind sector, ocean energy, the production of mature oil and gas fields or the re-use of oil and gas infrastructure. These companies require different types of collateral since reserves are not present.

In order to hedge their risks, lenders require the establishment of a comprehensive security package to cover the hundreds of millions of euros provided.[5] The current possibilities to offer financial security could be improved in many jurisdictions such as the Netherlands, where mortgaging installations on the seabed beyond the territorial sea (22.2km) is impossible.[6] Strengthening the security package with such a mortgage could decrease the cost of capital in certain jurisdictions. Moreover, if the result is that a less complex security package can be established and legal opinions have to be provided, transaction costs would decrease.

In the civil law jurisdictions bordering the North Sea, the regulations governing security rights are arranged for by law. This chapter discusses a rule of private international law which legislators could adopt and which other laws they should take into account to allow for the establishment of mortgage against installations at sea. Relevant rules of the law of the sea, private international law and national property law will be discussed. At the level of national law, the Netherlands will be used as the example as it is currently impossible to establish such a mortgage under Dutch law and the legal questions and discussions concerning current and possible future Dutch legislation may be exemplary for other states.

This chapter starts by setting out the extent to which coastal states can regulate the construction and operation of oil, gas, electricity and other energy infrastructure above, on or below the seabed according to the Law of the Sea (Section 2). This discussion includes the question whether coastal states enjoy the right to adopt property laws concerning installations constructed on the seabed. Subsequently, the practical potential for mortgage to be established against offshore installations will be illustrated by discussing investments in the production of oil, gas and electricity in the North Sea area and in how far costs are carried by private actors and states (Section 3). The North Sea area will be used as an example in this part, as some countries bordering the North Sea have a long history regulating oil and gas production and increasing experience with the regulation of electricity production at sea. Subsequently, the

5 See, for example, Philip Benger and Patrick Holmes, 'Ancillary Finance Documentation' in John Dewar (ed), *International Project Finance Law and Practice* (2nd ed, OUP 2015) 463.

6 There is a range of legal questions and uncertainties connected to different security rights established concerning main assets of projects, such as the license and subsidies. Questions pertaining in particular to the Netherlands are discussed in my PhD on this topic and further publications thereon are forthcoming.

focus lies on the legal questions which the legislator in the Netherlands faces to introduce the possibility to mortgage installations on the seabed beyond the territorial sea (Section 4).

2 Legal Basis in International Law

The limits of coastal state jurisdiction over oil, gas and electricity production at sea are determined by the law of the sea. In light of the focus on the mortgaging of installations at sea, particular attention will be given to whether the right to adopt property laws falls within coastal state jurisdiction.

The UN Convention on the Law of the Sea (UNCLOS) sets out coastal states' rights and duties in the different maritime zones.[7] To begin with, a distinction has to be made between the territorial sea on the one hand and the continental shelf and Exclusive Economic Zone (EEZ) on the other.[8] The reason therefore is that the territorial sovereignty of the coastal state extends to its territorial sea.[9] As a result, legislation applicable to the territory of a coastal state applies to its territorial sea as well, unless it contains a provision explicitly providing otherwise.[10] The property laws of the coastal state thus govern, *inter alia*, installations in its territorial sea, unless the laws explicitly exclude application within the territorial sea.

Beyond their territorial sea, coastal states enjoy sovereign rights for the purpose of exploring and exploiting the natural resources on their continental shelf[11] and can claim sovereign rights for these purposes and for the production of energy from the water, currents and winds in the EEZ.[12] This includes

7 The parties to UNCLOS 1982 currently include 167 states and the European Union. Important provisions of the treaty became customary international law, especially those granting states rights rather than obligations, as they were applied by both parties and non-parties, Robin Churchill and Vaughan Lowe, *The Law of the Sea* (3rd edition, Manchester University Press 1998) 19.

8 Following UNCLOS 1982, arts 3, 57 and 76; the maximum breath of the territorial sea is 12nm, of the EEZ is 200nm and the breath of the continental shelf can exceed 200nm under certain circumstances.

9 UNCLOS 1982, arts 2, 56 and 77. See, for example, Peter Malanczuk, *Akehurst's modern introduction to international law*, (Routledge 1997) 76; Bernard H. Oxman, 'Jurisdiction of States' (2007) MPEPIL, para 13 <opil.ouplaw.com/home/EPIL>; Ian Brownlie, *Principles of International Law* (7th ed, OUP 2008) 105.

10 See, for example, Malcolm Evans, 'Law of the Sea' in Malcolm Evans, *International Law* (4th ed, OUP 2014) 657; Brownlie (n 9) 105.

11 1958 Continental Shelf Convention, art 3; UNCLOS 1982, art 77. The ICJ ruled that customary law provides that coastal states have a continental shelf *ab inito* and *ipso jure*, North Sea Continental Shelf Cases [1969] ICJ Rep 3.

12 UNCLOS 1982, art 56(1)(a).

sovereign rights concerning the cables and pipelines used for the aforementioned purposes. The entire seabed and subsoil of the North Sea falls under the continental shelf regime of UNCLOS. In the 1960s, the United Kingdom, Belgium, the Netherlands, Germany, Denmark and Norway have entered into agreements dividing the vast majority of the seabed and subsoil of the North Sea between them.[13] The states bordering the North Sea have all claimed an EEZ and the associated sovereign rights, which allows them to regulate the production of oil and gas as well as electricity from the wind, waves and tides at sea.[14]

In the areas where a coastal state enjoys sovereign rights, its jurisdiction is not full but limited to the rights and area defined in UNCLOS. UNCLOS does not explicitly mention that these sovereign rights allow the coastal states to adopt property laws in this area. Instead, Article 60(1) UNCLOS provides that on their continental shelf and in their EEZ, coastal states have the exclusive right to regulate the construction, operation and use of installations and structures for the purposes of exploring and exploiting the natural resources and the production of electricity from the water, currents and winds. Article 60(2) UNCLOS provides that coastal states have exclusive jurisdiction over such artificial islands, installations and structures, including jurisdiction with regard to customs, fiscal, health, safety and immigration laws and regulations.[15] There are convincing arguments supporting that Article 60 UNCLOS also grants coastal states the right to adopt property laws governing these installations.[16] These include that with the adoption of Article 60 UNCLOS a clear choice was made for coastal state jurisdiction, rejecting the suggestion made during the UNCLOS negotiations to grant jurisdiction to the state constructing or operating the installation.[17] Also as a result of this choice, the interpretation that coastal

13 Most of these bilateral treaties were signed between 1964 and 1966. However, in some cases reaching agreement on the exact delimitation of the continental shelf proved to be more difficult. Most famously, the International Court of Justice delivered judgment on continental shelf delimitation in the *North Sea Continental Shelf* cases concerning the borders of the German, Dutch and Danish continental shelves; North Sea Continental Shelf Cases [1969] ICJ Rep 3.

14 Act of 22 April 1999 relating to Belgian's exclusive economic zone in the North Sea, Belgian Official Journal of 10 July 1999; The UK Exclusive Economic Zone Order 2013, SI 2013/3161; Exclusive Economic Zone (Establishment) Act (1999) Netherlands Official Journal 281; Act No. 91 of 17 December 1976 relating to the Economic Zone of Norway; Danish Act No. 411 of 22 May 1996 on Exclusive Economic Zones.

15 Pursuant to UNCLOS 1982, art 80, art 60 also applies *mutatis mutandis* to artificial islands, installations and constructions on the continental shelf.

16 In agreement: Maria Gavouneli, *Functional Jurisdiction in the Law of the Sea* (Martinus Nijhoff Publishers 2009) 11; Christopher Staker, 'Jurisdiction' in Malcolm Evans (ed), *International Law* (OUP 2014) 316.

17 Barbara Kwiatkowska, *The 200 Mile Exclusive Economic Zone in the New Law of the Sea* (Martinus Nijhoff 1989) 112.

states can adopt property laws for these installations does not conflict with the rights of other states and is in line with the rest of the treaty: the treaty grants coastal states exclusive jurisdiction concerning these installations.[18]

3 Energy Activities in the North Sea and Their Financing

The states bordering the North Sea have adopted laws and regulations to govern the construction, use and removal of installations, cables and pipelines on the seabed of their territorial sea and EEZ. In some respects distinctly different approaches were used, while in others the approaches are similar, at times because of EU regulation. In the following, I will firstly provide some background on the development and regulation of the production of oil and gas, wind and ocean energy and transmission infrastructure present on and above the seabed of different countries bordering the North Sea. This includes discussion of investments in these sectors and by whom these investments are carried, the private sector or the state, be it through state-owned companies or direct state participation. Finally, paragraph 3.3 will reflect on different forms of finance.

3.1 *Production*

3.1.1 Offshore Oil and Gas

In the continental shelves of Denmark and the Netherlands and in particular Norway and the United Kingdom, significant oil and gas deposits were found. The total oil and gas expenditures on the Norwegian[19] and UK[20] continental shelves are, since 2011, annually between €15 and €23 billion. In addition, expenses must be made for the removal of infrastructure at the end of the lifetime of oil and gas fields, or alternatively for the re-use of production platforms and pipelines, as there, for example, is ample potential for CO_2 storage in reservoirs beneath the North Sea seabed.[21]

Companies investing in the offshore oil and gas sector are confronted with high costs for the exploration of the seabed and its subsoil in search of

18 The argumentation supporting coastal state jurisdiction to adopt property laws regarding these installations and structures is discussed in more detail in the fourth chapter of my PhD.

19 'Investments in oil and gas, manufacturing, mining and electricity supply', (*Statistics Norway*, 21 February 2019) <ssb.no/en/energi-og-industri/statistikker/kis>.

20 Oil and Gas UK (n 4), 33–34.

21 See chapter 19 of this book, M. Roggenkamp,' Re-using (nearly) depleted Oil and Gas Fields in the North Sea for CO_2 Storage: Seizing or Missing a Window of Opportunity?', discussing the re-use of depleted oil and gas fields in the North Sea for CO_2 storage.

hydrocarbon deposits. The cost of drilling a single exploration well ranges from millions of euros in shallow waters to more than 100 million euros for a single deep-water well. Over the life-time from exploration until decommissioning, capital expenditures form more than half of the costs for installations on the seabed of the continental shelf in Norway and the United Kingdom.[22] Operating costs form almost forty percent of the total costs made in the United Kingdom, but a significantly smaller part in Norway.[23]

The governments of the countries bordering the North Sea provide licenses to private companies who wish to undertake exploration or exploitation activities. In accordance with the Hydrocarbons Licensing Directive, these licenses are provided on a competitive basis.[24] As regards state participation, the United Kingdom does not, while the Netherlands, Denmark and Norway do know a system in which the state, through a state-owned company, participates in all or certain oil and gas exploration and exploitation licenses.[25] The Danish state-owned company Nordsøfonden becomes a licensee, while the Dutch EBN does not. In Norway, the State's Direct Financial Interest is managed by Petoro, which is a licensee in the vast majority of licenses, but in a few instances its participation is limited to a right to a share of possible profit.

The state-owned companies in these countries have not always been, and in other countries still are not always, responsible for their share of the investments and costs. For example, until 1988 the Norwegian state participant did not financially contribute to costs incurred for exploration, but contributed only to production.[26] Following developments such as low oil and gas prices and the discovery of fewer large fields, the Norwegian, Dutch and Danish state participants currently do financially contribute to exploration activities in order to stimulate investments. The share the state-owned company obtains is in the Netherlands by law set at forty percent, is in Denmark usually twenty

---

22 See, for example, figures provided by the Norwegian Petroleum Directorate, www.norskpetroleum.no/en/economy/investments-operating-costs and graphics.wsj.com/oil-barrel-breakdown.

23 Ibid.

24 Directive 94/22/EC of the European Parliament and of the Council of 30 May 1994 on the conditions for granting and using authorizations for the prospection, exploration and production of hydrocarbons [1994] OJ L164/3.

25 See, *inter alia*, section 3–6 and Chapter 11 Norwegian Petroleum Act 1996; articles 81 to 97 Dutch Mining Act; section 8 Act on the Danish Subsoil and the Act on the Danish North Sea Fund; the United Kingdom used to know a system of state participation, but this was abolished in the 1980s.

26 Ernst Nordtveit, 'Regulation of the Norwegian upstream petroleum sector', in Tina Hunter (ed), *Regulation of the Upstream Petroleum Sector: A Comparative Study of Licensing and Concession Systems* (Edward Elgar 2015) 145.

percent and is for current projects in Norway between five and 63 percent.[27] Finally, state-participants are expected to provide their part of the required investments in decommissioning.[28] The private companies holding the oil and gas exploration and exploitation licenses are thus not responsible to carry all investments and costs in a number of jurisdictions. In case the state does not participate, the private companies are fully responsible for capital costs, but they can still rely on existing laws decreasing their tax burden.

3.1.2 Offshore Wind

Whereas wind farms in the North Sea were scarce in number in the early 2000s, during the past decade the installed capacity has increased exponentially.[29] The driving forces behind this increase are the commitments states have made regarding the consumption of electricity from renewable energy sources.[30] For several states, reaching these targets by constructing onshore capacity proved to be difficult, *inter alia*, because of public opposition against onshore wind and spatial constraints.[31] Therefore, within the boundaries of the Electricity Directive and the Renewable Energy Directive,[32] states bordering the North Sea have decided to introduce varying legislation on licensing and support schemes which allow for the realization of offshore wind projects,

27 Articles 88 and 94 Dutch Mining Act. EBN has a fifty percent share in production activities for which the exploration license was granted under a previous regime. In two licenses, the Danish state-owned company Nordsøfonden has a 36.36 percent rather than a twenty percent share, see <nordsoefonden.dk>.

28 See, for example, J.J.A. Waverijn & L. Baljon, 'Verslag van European Energy Law Seminar 2018' (2018) 4 *Nederlands Tijdschrift voor Energierecht* 133–147.

29 Offshore Wind in Europe (n 2) 12. The Norwegian offshore wind sector is currently limited to one, floating, wind mill both because the already installed hydroelectric capacity in Norway limits the need for offshore wind farms and because the water depths off the Norwegian coast provide technical and thereby financial challenges which other North Sea states do not face.

30 The states bordering the North Sea have committed themselves to national renewable electricity production targets and to targets at EU level through Directive 2009/28/EC of the European Parliament and of the Council of 23 April 2009 on the promotion of the use of energy from renewable sources [2009] OJ L 140/16 and Directive 2018/2001/EU of the European Parliament and of the Council of 11 December 2018 on the promotion of the use of energy from renewable sources [2018] OJ L 328/82 (the RES Directive).

31 See, for example, Anita Rønne, 'Opposition to Wind Farms and the Possible Responses of the Legal System', in Lila Barrera-Hernández and others (eds), *Sharing the Costs and Benefits of Energy and Resource Activity: Legal Change and Impact on Communities* (OUP 2016) 176.

32 Currently, Directive 2009/72/EC of the European Parliament and of the Council of 13 July 2009 concerning common rules for the internal market in electricity [2009] OJ L 211/55 and the RES Directive (n 30).

giving the sector an opportunity to mature and try to reduce costs.[33] Furthermore, the states bordering the North Sea have signed various agreements and understandings to further the development of offshore wind in the North Sea and identified the North Sea as a 'power house' which may deliver as much as 8 percent of the energy supply of Europe by 2030.[34] As will be elaborated on in the following paragraph, these agreements also concern increased electricity transport through cables laid on the seabed of the North Sea.

The developers of offshore wind farms in the North Sea are private companies. These include (partially) state owned power companies such as Ørsted, Equinor and Vattenfall, which jointly represent around thirty percent of the current market.[35] As the subsidy schemes currently in place for offshore wind can only be relied upon once production starts, developers have to fully provide the capital required to construct the wind farms.[36] For large scale offshore wind farms in the North Sea, this usually concerns €1–2 billion, while operational expenditures are relatively low.[37] As regards the European offshore wind sector, yearly investments grew from €8 billion in 2010 to €18 billion in 2016.[38] Between 2017 and 2021, the annual capacity of permitted and planned projects is comparable to or greater than the capacity installed during the year 2015.[39] On the longer term, both national governments and the EU are aiming for ambitious targets regarding the consumption of electricity produced from renewable energy sources by 2030 and 2050 and offshore wind is expected to play an important role in reaching these targets.[40]

33 For an overview in the development of and support schemes used for offshore wind in the UK, Germany, the Netherlands and Denmark see Hannah Katharina Müller, *A Legal Framework for a Transnational Offshore Grid in the North Sea* (Intersentia 2015) 145, 163, 174, 190.

34 See, in particular, Council of the EU, 'Political declaration on energy cooperation between the North Seas Countries' 8673/16 of 13 May 2016, which echoes the 2016 Manifesto Northern Seas as the Power House of North Western Europe which was signed by twenty Members of the European Parliament from countries neighbouring the North Seas.

35 Offshore Wind in Europe (n 2) 27–28.

36 Previously and currently used support schemes include renewable obligation certificates, contracts for difference, feed-in premiums and feed-in tariffs. See also, n 33.

37 Offshore Wind in Europe (n 2) 21.

38 'The European Offshore Wind Industry – Key Trends and Statistics 2016' (*WindEurope*, 2017) 33 <windeurope.org/about-wind/statistics/offshore>.

39 Ibid, 24–25.

40 See, for example, Commission, 'Energy Roadmap 2050' (Communication) COM (2011) 885 final.

3.1.3 Ocean Energy

Electricity can be produced from an array of ocean sources, of which the tides and waves are currently regarded to be most promising.[41] The majority of the installed capacity is situated within territorial seas. The expectation is that production will extend to the EEZ where facilities, such as network connections, may be shared with offshore wind farms.[42]

The main arguments in favour of developing ocean energy technologies are that the sources are renewable, add diversity to national generation, and – unlike electricity production from wind and sun – the output is predictable. Thus, electricity production from these sources is favourable in terms of balancing and the achievement of renewable electricity goals. For these reasons, the European Commission is active in stimulating the development of ocean energy technologies within the EU and have provided around €150 million in support in the last ten years.[43] The potential worldwide market for electricity production from ocean energy sources has been estimated at €535 billion between 2010 and 2050.[44] A significant part thereof may develop in the North Sea, as the EU hosts around half of the world's tidal energy developers, sixty percent of all wave energy developers and seventy percent of the ocean energy research and testing infrastructure.[45] Between 2007 and 2015, €2.6 billion was invested in ocean energy in the EU of which 75 percent by the private sector.[46] Until 2030, a further €2.8 to €9.4 billion will be invested in the sector.

3.2 *Transport Infrastructure*

To transport electricity, oil and gas from either the production site to shore or from one country to another, a large number of electricity cables and oil and gas pipelines have been laid down on or under the seabed of the North Sea. Producers constructed oil and mainly gas pipelines connecting the production platforms to shore and were responsible for these investments. As a

---

41 Commission, 'Renewable Energy: a Major Player in the European Energy Market' (Staff Working Document) SWD (2012) 0164 final.

42 See, for example, Eric D. Stoutenburg, 'Integrating Wind and Wave Power in California' (PhD thesis, University of Stanford 2012); Sharay Astariz and Gregorio Iglesias, 'Enhancing Wave Energy Competitiveness through Co-Located Wind and Wave Energy Farms. A Review on the Shadow Effect' (2015) 8 Energies 7344.

43 See, for example, Davide Magagna and Andreas Uihlein, '2014 JRC Ocean Energy Status Report' (Scientific and Technical Research Reports) EUR (2015) 26983; The 2018 Annual Economic Report on EU Blue Economy (European Union, Joint Research Centre 2018) 73.

44 Commission, 'Blue Energy – Action needed to deliver on the potential of ocean energy in European seas and oceans by 2020 and beyond' (Communication) COM (2014) 08 final.

45 The 2018 Annual Economic Report on EU Blue Economy (n 43) 70.

46 Ibid, 73.

result of efficiency considerations, offshore pipeline systems have been constructed which connect multiple production platforms to each other and to shore, of which the most elaborate network lies on the Norwegian continental shelf. Initially, the companies in control of these upstream pipelines were owned by the producers and the state participated. For example, almost all of the Norwegian upstream gas pipeline network is owned by Gassled. Gassled is a joint venture of which the state indirectly holds 51 percent of the shares.[47] In the Netherlands, the Dutch state participant EBN has a share in five offshore upstream gas pipelines, among which a 45 percent stake in NOGAT.

The majority of electricity cables on the seabed in the North Sea are radial connections.[48] The national regimes concerning offshore wind farm connection to shore differ. For example, in the United Kingdom, the wind farm operators are responsible for financing the construction but choose whether they or an offshore transmission system operator (OFTO) construct the connection.[49] Until 2016, developers in the Netherlands were obliged to fund and construct and owned the connection. However, the Dutch national TSO TenneT has been appointed to construct joint connections from multiple wind farms, funded through government support, with the aim to reduce costs.[50] This brings the Netherlands in line with Denmark and Germany where, respectively, Energinet and TenneT are responsible for both construction and funding.

The other main category of offshore pipelines and electricity cables are those which cross borders, i.e. interconnectors. These pipelines and electricity cables cross the North Sea, connecting for example the Netherlands to Norway or the United Kingdom to Germany. The construction costs of a number of interconnectors exceeded €500 million.[51] The vast majority of interconnectors are owned, operated and funded by state-owned TSOs.[52]

---

47 The 51 percent of the state involvement consists of 46,6 percent of Gassled shares held by Petero and 5 percent by Equinor. The state has a controlling interest in Equinor as it holds 67 percent of the shares.

48 A radial connection directly connects electricity production facilities to shore.

49 Electricity (Competitive Tenders for Offshore Transmission Licences) Regulations 2013, SI 2013/175.

50 The Minister of Economic Affairs has appointed TenneT as offshore TSO in September 2016 by means of a decision taken on the basis of the Act of 23 March 2016 Amending Electricity Act 1998 (tijdig realiseren doelstellingen Energieakkoord), article III.

51 The NorNed electricity cable between the Netherlands and Norway cost at least €600 million, Commission, 'The European Investment Bank finances NorNed (Press Release) BEI (2007)118. The BBL Pipeline – natural gas – between the Netherlands and the UK cost €500 million, see <bblcompany.com>.

52 See, for example, Electricity Interconnectors (Ofgem) <ofgem.gov.uk/electricity/transmission-networks/electricity-interconnectors> and Commission, 'Exemption Decisions and Pending Notifications of National Exemption Decisions for Gas and Electricity' <ec.europa.eu/energy/sites/ener/files/documents/exemption_decisions2017.pdf>. In

The aforementioned cooperation agreements also concern the potential construction of a cross-border offshore grid, connecting offshore wind farms and the electricity networks of coastal states bordering the North Sea.[53] The envisaged benefits of an offshore grid would include further interconnection of European energy markets, which should reduce electricity prices, construction costs and should contribute to decarbonisation and the share of consumed renewable energy. Realization of such infrastructures is estimated to cost between €59 and €107 billion.[54] Before such a grid can be built, technical, legal and financial hurdles will have to be overcome.[55]

3.3 *Finance*

The developers of these energy exploration, production and transmission activities at sea use different methods to raise the required debt or equity to finance their activities. This section touches upon corporate finance and project finance and a few significant differences between the two. Regardless of the method used, it is common that the assets and liabilities connected to a single activity are housed in a separate legal entity. This is always the case when project finance is used.[56] The assets and liabilities held by this special purpose vehicle (SPV) include the permits, the off-take agreement and construction contracts.[57] In case of project finance, equity and debt are used to fund the SPV. The defining feature of project finance is that lenders of the debt only have recourse against the assets of the SPV and not against any other assets of the shareholders of the SPV.[58] Conversely, financial constructions wherein the recourse of the lenders is broader, for example when the parent company

---

Norway, until recently, only the fully state owned national TSO could own and operate interconnectors. This has been changed, but the authorities are now considering reversing the decision.

53 See, for example, Commission, 'Second Strategic Energy Review: an EU energy security and solidarity action plan' (Communication) SEC (2008) 2870, 2871, 2872; Memorandum of Understanding on the North Seas Countries' Offshore Grid Initiative (NSCOGI) of 3 December 2010; Council of the EU (n 34).

54 Commission, 'Study of the benefits of a meshed offshore grid in northern seas region' (2014).

55 See, for example, Müller (n 33); Ceciel Nieuwenhout, 'Offshore Hybrid Grid Infrastructures: The Kriegers Flak Combined Grid Solution', in Martha M. Roggenkamp & Catherine Banet (eds), *European Energy Law Report* XII (Intersentia 2019) 95–112.

56 In the upstream petroleum sector, unincorporated joint ventures are most commonly used. When attracting debt, lenders often require incorporating the company, Aled Davies and James Orme, 'International Projects – Sector Focus Section A – Oil and Gas' in John Dewar (ed), *International Project Finance* (2nd ed, OUP 2015) 152.

57 Stefano Gatti, *Project Finance in Theory and Practice* (Elsevier Academic Press 2013) 271.

58 Robert Clews, *Project Finance for the International Petroleum Industry* (Elsevier Academic Press 2016) 7.

provides a guarantee to lenders, fall outside the scope of project finance and are referred to as corporate finance.[59]

Around ten percent of oil and gas funding is provided through project finance.[60] In the European offshore wind sector, the share of project finance in total investments is significantly higher with 40 percent in 2015 and 2016, out of total investments of €31.5bn.[61] In 2018, €8 out of €10.3 billion invested in new projects was raised using project finance.[62] On top of that, another €8.5 billion of debt was refinanced in 2018.[63]

The average debt raised for individual North Sea project finance deals is around one billion euro. This debt is provided by commercial banks, insurance companies, pension funds and public and semi-public financial institutions.[64] The public and semi-public financial institutions are instrumental to the success of these deals and include development banks and export credit agencies, such as the European Investment Bank, the German development bank KfW and the Denmark's export credit agency EKF, which *inter alia* can provide debt, guarantees and insurance.[65] As commercial banks rarely offer risk commitments exceeding €200mln to individual projects, a consortium of commercial banks is involved in these deals.[66] In addition, a growing number of institutional investors, such as pension funds and insurance companies, are involved in financing offshore wind. This can be by providing debt, but also equity as, for example, some utilities sell off minority stakes in their wind farm to free up capital to invest in other projects. In this model, the utility remains in control over the wind farm while an institutional investor such as a pension fund, seeking equity returns, becomes a shareholder.[67]

Debt is attracted for almost all energy activities carried out by private parties, regardless of whether this is done through corporate finance or project finance. Further distinctions include that the cost of capital for debt attracted through corporate finance is often higher: The lenders only have recourse against the assets of the SPV and therefore may face higher risks which translate

59 Definitions along these lines are broadly, but not exclusively, used. Referring to sources used in this contribution in particular: 'Funding challenges in the oil and gas sector' (EY 2014) 2 <ey.com/oilandgas>; The European Offshore Wind Industry (n 38) 20.

60 Funding challenges in the oil and gas sector (n 59) 2.

61 The European Offshore Wind Industry (n 38) 33–34.

62 Offshore Wind in Europe (n 2) 35.

63 Offshore Wind in Europe (n 2) 36.

64 Cathy Marsh and Andrew Pendleton, 'Project Participants and Structures' in John Dewar (ed), *International Project Finance* (OUP 2011) 35–36.

65 Davies and Orme (n 56) 155; The European Offshore Wind Industry (n 38) 32.

66 Mark Plenderleith, 'Sources of Funding' in John Dewar (ed), *International Project Finance* (OUP 2011) 61.

67 Marsh and Pendleton (n 64) 34–35.

to higher interest rates. In case of project finance, it is thus of greater importance that the security which can be provided in light of the individual project is as extensive as it can be, as this decreases risks and thus may lead to lower interest rates and thus a lower cost of capital.[68] The cost of capital can be a major part of the life-time expenses of a project, as they can for example total a billion euros over the lifetime of a two-billion-euro loan. The cost of capital can thus be prohibitive and result in projects not being developed. This is currently an issue in the ocean energy sector. Developers of ocean energy have particular trouble attracting sufficient funds to scale up to commercial levels as the projects are too capital intensive to attract venture capitalists, too risky to attract private equity and borrowing from banks is too expensive because of the risks and subsequent cost of capital.[69] Governments will thus have to provide financial support and legislative support. One manner in which governments can provide legislative support is by ensuring that the law does not contain unnecessary barriers, for example by allowing for mortgage of installations on the seabed as will be discussed in the following paragraph.

4 Mortgaging Property on the Seabed

4.1 *Introduction*

We have established that the investments required for the construction and operation of offshore installations, cables and pipelines are significant and that a large part of the costs involved are carried by private companies. In order to hedge their risks, lenders ensure to be granted a security rights. In case of project finance this concerns a comprehensive security package usually concerning all assets of the SPV including the shares of the SPV itself, even if a certain security or a security regarding certain assets is doubtful or limited in efficacy.[70] This includes the license, the income streams from the power purchase agreement, a possible subsidy agreement, the bank accounts of the SPV or the physical installations. Developers not using project finance also mortgage and pledge their assets. Practice in the Norwegian offshore oil and gas sector is an excellent example thereof. The Norwegian Petroleum Act contains a chapter dedicated to the mortgage of licenses.[71] The Ministry regularly allows licensees

68 Shannon Pratt & Roger Grabowski, 'Relationship between Risk and the Cost of Capital', in Shannon Pratt & Roger J. Grabowski (eds.), *Cost of Capital* (5th ed, Wiley 2014) 70–87.

69 Commission, 'Market Study on Ocean Energy' (Publications Office of the European Union, 2018) 45.

70 Joanne Robertson and Patrick Holmes, 'Ancillary Finance Documentation' in John Dewar (ed), *International Project Finance* (OUP 2011) 319.

71 Norwegian Petroleum Act 1996, ch 6.

to mortgage an offshore license to finance activities carried out under another licence than the one which has been mortgaged, which means this mortgage does not concern project finance as that would be limited to the licensed activity.[72] In the context of specific laws addressing security rights of offshore energy activities, Denmark knows laws specifically allowing for mortgaging physical wind farms at sea.[73]

In the Netherlands, such specific legislation has not been adopted. The possibility to mortgage energy installations on the seabed beyond the territorial sea would provide lenders with additional security, however, which could result in a reduction in the cost of capital. The benefits of mortgaging offshore energy installations were recognized by the Netherlands State Commission for Private International Law (the State Commission), but the Dutch government did not follow their initial recommendations.[74] The following paragraphs discuss the advice issued by the State Commission and a subsequent publication by the chairman of the State Commission. Through this discussion, a number of legal issues are identified which have to be resolved in order to allow for mortgaging of offshore installations in conformity with Dutch law. The discussion of the Dutch regime illustrates the questions of property law which other governments could also be confronted with.

4.2 *Property Law and Installations on the Seabed of the Continental Shelf*

Upon request of the Dutch government, in 1990 the State Commission issued an advice concerning national legislative jurisdiction on the Dutch continental shelf beyond the territorial sea.[75] The main area of attention identified by the State Commission related to property laws and installations. Private international law dictates which national laws apply to installations on the continental shelf beyond the territorial sea.[76] Problems arise as the *lex rei sitae* – the place where the property is situated – determines which property laws apply.[77] As the continental shelf has no internal civil law, the State

72 Pursuant to Section 6–2 Norwegian Petroleum Act 1996; Olav Nordli, 'Pantsettelse av utvinningstillatelser' (2011) 3 Tidsskrift for Forretningsjus 131.

73 Danish Promotion of Renewable Energy Act, s 25(4).

74 The State Commission, whose members include judges, academics and practicing lawyers, provides the Dutch government with requested and unrequested advice concerning international private law.

75 Staatscommissie voor het Internationaal Privaatrecht, 'Advies van 12 december 1990' in E.N. Frohn, E. Hennis (eds), *Geselecteerde adviezen – naar een afgewogen IPR* (T.M.C. Asser Instituut 1995). Even though it is not included in the main text, the Dutch EEZ overlaps with the Dutch continental shelf beyond the territorial sea.

76 Staatscommissie voor het Internationaal Privaatrecht (n 75) 247.

77 Ibid; DCC, art 10:127(1).

Commission concluded that the *lex rei sitae* refers to a *locus sine lege* – a place without law – and that a legal vacuum concerning property laws exists.[78]

In order to increase legal certainty for both investors and developers, the State Commission advised the Dutch government to propose a new rule of private international law, along the following lines:

> Installations on the continental shelf of the coastal state are treated as installations within that coastal state.[79]

The Minister of Internal Affairs decided against pursuing the introduction of this rule. The reasons provided were that financial practice in the offshore oil and gas sector would not require such a rule and that according to the State Commission there may already be an unwritten rule of referral to Dutch property law concerning the Dutch continental shelf, which would allegedly resolve the issue when encountered in practice.[80] In an advice issued in 1996, the State Commission itself advised against the adoption of a comparable rule.[81] The State Commission argued that the rule would lead to confusion as regards which laws should and which should not be applied. It also used the argument put forward by the Minister, that the legal structures used in the practice of financing petroleum installations on the continental shelf did not require a change in law.[82]

Even though the State Commission may have been correct in asserting that the financial structures used in the past half century in the petroleum industry do not seem to have hampered investments in the Netherlands, the landscape has changed radically after the advice was issued and the need for legal certainty and financial security has increased, as previously mentioned. Allowing for the mortgaging of installations at sea would in fact strengthen the security package and thereby increase legal certainty for investors in offshore energy activities.[83]

While the appetite for investments in offshore wind in the North Sea area is currently high, this increase in legal certainty could prove vital for investors

78 Ibid.

79 Ibid 247–248. This is a translation and interpretation of the much more elaborate rule drafted by the State Commission in Dutch.

80 Netherlands Parliamentary Papers II (1990–91) 22 390, nr 1.

81 Staatscommissie voor het Internationaal Privaatrecht, *Rapport aan de Minister van Justitie – Internationaal Goederenrecht* (1 November 1998) 8.

82 Ibid.

83 In agreement: Frank M.J. Verstijlen, 'Eigendom van Delfstoffen en in de ondergrond opgeslagen stoffen' in: *Preadvies Nederlandse Vereniging voor Energierecht: Energie en Eigendom* (Intersentia 2011) 19; X.E. Kramer & H.L.E. Verhagen, *Asser 10-II Internationaal Vermogensrecht* (Wolters Kluwer 2015) 413.

in the development of ocean energy, storage potential or other activities in the North Sea in the future, which may involve new technologies and thus greater risks and therefore costs. Moreover, considering that the *lex rei sitae* is used to identify which property laws apply in the vast majority of countries,[84] comparable issues as those concerning the Dutch continental shelf are expected to exist in a significant number of other coastal states, many of which have a comparatively high cost of capital. In the coastal states with potential for offshore production of electricity from renewable energy sources and storage, cost reductions will be determinant in realizing projects.

4.3 *National Law*

The adoption of a rule prescribing that installations on the continental shelf are regarded to be situated within the territory of the coastal state would resolve certain but not all barriers to mortgaging installations on the seabed of the Netherlands continental shelf. To illustrate this, it is necessary to briefly discuss mortgage (*hypotheekrecht*) under Dutch law.

> A mortgage is established by a notarial instrument drawn up between the parties in which the grantor grants a mortgage to the mortgagee over registered property, followed by the entry of the instrument, in the appropriate public registers provided for that purpose.[85]

The following questions must be answered to ascertain whether installations at sea can be mortgaged or pledged: What is the definition of registered property? Can installations at sea, inside and outside the territorial sea, be defined as such? The definition of registered property is enshrined in Article 3:10 of the Dutch Civil Code (DCC):

> Registered property is property the transfer or creation of which requires entry in the appropriate public registers.[86]

The most common category of property the transfer of which requires registration in the public register is immovable property.[87] Immovable are the land, not yet mined minerals, plants connected to the land, and buildings and

84 Malanczuk (n 9) 73.

85 DCC, art 3:260. Hans Warendorf, Richard Thomas and Ian Curry-Summer, *The Civil Code of the Netherlands* (1st ed, Wolters Kluwer 2009) 498.

86 Hans Warendorf (n 85) 434.

87 DCC, art 3:89. Warendorf (n 85) has been used for this and the following references to the DCC.

constructions permanently attached to the land, either directly or through connection with other buildings or constructions.[88] All tangible property which is not immovable is movable property.[89] Immovable property is not the only category of property requiring registration to transfer or establish. The most prominent other example is ships. While a ship is movable property, Dutch law provides that ships are registered property.[90]

The result is that both movable property for which law provides that transfer or creation requires registration and immovable property can be mortgaged. The mortgage has to be established by notarial mortgage deed which must be entered into the public registers. Against movable property outside this category, *inter alia*, a non-possessory security right comparable to pledge can be established (*pandrecht*).[91] The qualification of property as immovable or movable therefore has an important influence on which type of security right can be established.

4.4 *Installations at Sea, Immovable or Movable Property?*

As Dutch law does not contain a rule that transfer of installations in the EEZ requires registration in a public register, the question is whether installations at sea are movable or immovable property. This is discussed assuming that Dutch property law applies pursuant to a comparable rule as proposed by the State Commission.

The response of the Dutch Minister of Economic Affairs to questions from members of the Lower House provides an example of how unclear the legal qualification of offshore installations currently is. The minister stated that windmills in the EEZ are immovable property because they qualify as such pursuant to the Valuation of Immovable Property Act (VIPA).[92] This act concerns municipal taxes, while the limits of jurisdiction of Dutch municipalities is one kilometer offshore.[93] Without providing further supporting arguments, the minister added that the fact that this law does not apply within the EEZ did not affect his conclusion.[94] Solely supporting this argumentation with a law which

88 DCC, art 3:3(1).

89 DCC, art 3:3 (2).

90 DCC, art 8:199(1).

91 DCC, art 3:237(1).

92 Letter of the Minister of Economic Affairs to the Lower House as regards questions and amendments for the bill 'Electricity- and Gas law (32 199), 2 October 2015, DGETM-EI / 15138937.

93 Act of 2 November 1990, concerning provincial and municipality borders along the North Sea coast 1984 (1990) Netherlands Official Journal 553.

94 Letter of the Minister of Economic Affairs (n 92).

does not apply is clearly insufficient. Considering the aforementioned conclusion that Dutch property laws do not apply in the Dutch EEZ in particular, there is no reason to assume that the VIPA would influence the property law status of installations in the Dutch EEZ.

The then chair of the State Commission and Advocate-General to the Netherlands Supreme Court Roelvink, addressed the question whether installations beyond territorial waters are movable or immovable in a publication following the advice discussed above.[95] In this publication, Roelvink argued for the adoption of a rule of private international law. Coastal state property law should apply to installations within the continental shelf and EEZ in so far the installations are used for the purpose of the exploration and exploitation of natural resources.[96] This rule would create an unfortunate situation where the material property laws of the coastal state provide both that these installations are immovable and that for the transfer and mortgaging of immovable property registration in the public registers is required, but where the public registers do not extend to the relevant part of the sea.[97] Roelvink added that Dutch public registers do not extend beyond territorial borders. He further observed that other commentators have argued that matters would become worse if the seabed beyond the territorial sea is to be considered a *res nullius* and if national law is considered to provide that through accession the windmills legally accede to, and thus are part of, the seabed and therefore also are *res nullius*.[98] Roelvink concluded as follows:

> I have a different point of view, at least as regards Dutch property law. I defend the position that an installation, connected to the ownerless continental shelf, can qualify as and can be treated as movable property. I do not see good grounds to assume that ownership of property is lost because of connection to a *res nullius,* nor to assume that the absence of public registers makes transfer impossible.[99]

Roelvink takes a very pragmatic approach. His conclusion that these installations qualify as movable property is not based on the legal requirements

95 H.L.J. Roelvink, 'Het Continentaal plat als IPR-aanknopingspunt', in S.C.J.J. Kortmann and others (eds), *Op recht* (*lib. am. Struycken*) (W.E.J. Tjeenk Willink 1996).

96 National property law should not apply to anchored beacons for shipping, telephone cables, transmission islands or tunnels, according to Roelvink (n 95) 280.

97 Roelvink made note of that the Dutch public registers do not extend beyond the territorial sea, Roelvink (n 95) 820.

98 Roelvink (n 95) 280.

99 Translated from Roelvink (n 95) 280. Moving on to another part of his publication, Roelvink does not provide further insight on the legal reasoning supporting his views.

included in Dutch property law, specified in extensive case law. Instead, he attaches decisive significance to avoiding the legal issues created by the treatment of these installations as immovable property. While the issues caused by qualification as immovable property are undesirable, the present author is of the opinion that this cannot be considered a valid justification for deviation from Dutch property law, which is the law Roelvink explicitly refers to.

The applicability of Dutch property law to installations in the territorial sea is without doubt. The adoption of a rule in line with the proposal by the State Commission would ensure the same for installations on the continental shelf and EEZ. In the following, it will be assessed whether these installations qualify as movable or immovable property pursuant to contemporary Dutch law, and whether problems such as those mentioned by Roelvink arise.

Pursuant to Dutch law, immovable property includes constructions and buildings which are permanently attached to the land.[100] The *travaux préparatoires* reveal that the land should be interpreted as the outer layer of the earth's crust and the solid layers beneath the crust.[101] Considering that the seabed is the outer layer of the earth's crust, albeit covered by water, it qualifies as the land in the sense of this provision. Permanently attached is interpreted as requiring that a construction is designed and inherently intended to stay in place for a prolonged period of time.[102] This relies on the intention of the constructing party; insofar this intention is clear to third parties through the nature and design of the construction.[103] The question whether constructions at sea, such as wind farms and oil or gas production platforms are immovable thus requires a case-by-case assessment. Arguments supporting classification of installations such as wind mills and oil and gas production platforms as immovable property are their size,[104] and – where applicable – their concrete foundations, which are designed and constructed with the sole purpose of supporting the

100 DCC, art 3:3(1).

101 C.J. van Zeben, M.M. Olthof in cooperation with J.W. Du Pon (eds), *Parlementaire Geschiedenis van het Nieuwe Burgerlijk Wetboek. Boek 3. Vermogensrecht in het algemeen* (Kluwer 1981) 69. No different opinions have been supported since, according to *Groene Serie Vermogensrecht*, DCC, art 3:3, comment 16 (1-1-2017).

102 Netherlands Supreme Court, 13 June 1975, ECLI:NL:HR:1975:AC3080 (*Amercentrale*); Van Zeben, Olthof and Du Pon (n 101) 69.

103 Netherlands Supreme Court, 23 February 1994, ECLI:NL:HR:1994:ZC5591 (*Inzake Onroerende Windmolens*); Netherlands Supreme Court 13 May 2005, ECLI:NL:HR:2005:AT5469; Netherlands Supreme Court, 31 October 1997, ECLI:NL:HR:1997:ZC2478 (*Portacabin*) para 3.3.

104 Netherlands Supreme Court, 8 July 1997, ECLI:NL:HR:1997:AA2223 (*Rijdende Havenkranen I*); Netherlands Supreme Court 24 December 2010, ECLI:NL:HR:2010:BO3644 (*Rijdende Havenkranen II*); H.D. Ploeger, 'Een mobiele onroerende zaak?' (1998) 6321 WPNR 472.

installations.[105] Further arguments include that these installations have the sole purpose to produce electricity, oil or gas and that they are not constructed for temporary use. Indeed, many of these installations have been in place for a long period time which has been brought forward as evidence of the intention by Advocate-General to the Dutch Supreme Court Mok.[106] This sufficiently supports that offshore production platforms and wind farms are permanently attached to the land, as was also concluded in a case regarding onshore wind mills by Advocate-General Moltmaker.[107] As a result, both requirements are satisfied, and installations in the territorial sea, on the continental shelf and in the EEZ would qualify as immovable property and thereby also as registered property.

4.5 *Ownership and Registration*

As a consequence of the conclusion that production platforms and wind farms in the EEZ and continental shelf would be immovable property if Dutch law applied, questions of ownership and registration should be taken into consideration when discussing mortgage.

Firstly, ownership of the seabed should be considered. Pursuant to UNCLOS 1982, the seabed beneath the territorial sea is subject to national sovereignty, while the seabed beyond the territorial sea is not owned by the coastal state. Dutch law provides that the state is the owner of the seabed beneath the territorial sea.[108] It further provides that the state owns immovable property which has no other owner.[109] This would include the seabed beyond the territorial sea after adoption of a rule as proposed by the State Commission. The DCC also provides that the ownership of the land comprises buildings and constructions permanently attached to the soil.[110] The Dutch state thus is the owner of

105 Conclusion Advocate-General A.J. Moltmaker in *Inzake Onroerende Windmolens* (n 103).

106 Conclusion Advocate-General Mok, ECLI:NL:PHR:1980:AC1719, who interprets permanent as at least several years. Also referred to by Advocate-General Moltmaker in his conclusion in *Inzake Onroerende Windmolens* (n 103).

107 Conclusion Advocate-General A.J. Moltmaker in *Inzake Onroerende Windmolens* (n 103).

108 DCC, art 5:25.

109 DCC, art 5:24.

110 DCC, art 5:20. Many scholars support the view that pursuant to DCC, art 3:4 buildings and installations such as wind mills, in line with Roman law, accede to the land, Th.F. de Jong, *De structuur van het goederenrecht* (PhD thesis, University of Groningen 2006) 115 fn 196. The owner would thus also be the owner of the seabed. This rule of law also formed the basis of the argument mentioned by Roelvink that installations on the continental shelf beyond the territorial sea become *res nullius* through accession to the seabed. However, as the alternative is that no law applies (*locus sine lege*), this conclusion implies that Dutch property law applies which also contains the aforementioned DCC, art 5:24, which provides that the state would be the owner of the seabed rather than it being a *res nullius*.

the wind turbines, transformer stations and transmission cables in the territorial sea and would also be the owner of those on the Dutch continental shelf beyond the territorial sea were the suggested rule adopted. To avoid this, the state has to grant developers a right of 'long leasehold' (*erfpacht* bears some similarities to leasehold) for the relevant seabed area and rights of superficies (*opstalrecht*) regarding the installations and cables. This is already done for wind farms constructed on the seabed of the Dutch territorial sea.[111]

Secondly, the requirements of Dutch law regarding the registration of notarial deeds cause practical problems. Comparable to other jurisdictions, Dutch law requires the entry of deeds for the establishment and transfer of immovable property and property rights therein into the appropriate public registers. This includes the aforementioned rights of long leasehold and right of superficies. To allow for registration, the public registers should extend to the Dutch continental shelf beyond the territorial sea, which is currently not the case.

5 Conclusions

Capital expenditures in the offshore energy sector are high. The investments in offshore wind and electricity and gas transport infrastructure in particular are increasing. The investments in ocean energy and the re-use of installations, for example for storage purposes or power-to-gas, may also greatly increase in the coming decades motivated by targets for carbon emissions reduction and electricity consumption from renewable energy sources.

It is common practice that debt is raised to finance large energy and infrastructure projects. The greater the risks, the higher the cost of capital. The debt providers require a comprehensive security package to hedge their risks. Mortgage is commonly used and the strongest onshore security right in the Netherlands, but currently cannot be used in the EEZ. The first question is which laws regulate such a transaction. Analysis of public international law results in the conclusion that the coastal state can exercise full civil jurisdiction over artificial islands, installations and structures within its jurisdiction in the EEZ and on its continental shelf. Subsequently, the legal situation depends on whether the coastal state has exercised its jurisdiction. In jurisdictions where such laws are not in place and the *lex rei sitae* is used to determine which property laws apply, a *locus sine lege* or legal vacuum exist as regards property

111 Johan Dekkers, 'Rapportage Proces Vergunningverlening Offshore Windpark Egmond aan Zee' (Noordzeewind, August 2007) <www.noordzeewind.nl/wp-content/uploads/2012/02/OWEZ_R_192_20070820_vergunningen.pdf>.

laws. This vacuum may be resolved by the judiciary, but this does not provide investors with certainty.

Considering that most states worldwide use the *lex rei sitae* to determine which property laws apply and that practice in the countries bordering the North Sea shows that the adoption of property laws governing installations on the continental shelf and in the EEZ is not commonplace, it is expected that a property law vacuum exists in a significant number of coastal states. The impossibility to mortgage installations on the seabed could be resolved with the adoption of a rule of private international law stating that installations connected to the continental shelf should be treated as if they were situated within the territory of the coastal state. The public registers would have to be extended to the continental shelf and EEZ. Further rights may have to be granted by the state, depending on the applicable laws on accession and ownership. In line with Dutch law, the state may have to grant rights of long leasehold and superficies to transfer ownership of the installations.

The adoption of legislation allowing for the mortgaging of offshore installations can increase legal certainty and strengthen the security package offered by parties using debt to finance their business. In coastal states where the cost of capital is comparatively high, consequences can include a reduction in the cost of capital, increased investor appetite and the realization of additional projects at sea.

PART 7

Conflicting Uses or Coexistence, Resolving Mechanisms and Protection Regimes: Towards a More Integrated Approach

CHAPTER 22

Crossing the Sectoral Divide: Modern Environmental Law Tools for Addressing Conflicting Uses on the Seabed

Rosemary Rayfuse

1 Introduction

The oceans are becoming increasingly crowded. They are the venue for a vast range of human activities including merchant shipping, fishing, seabed mining, construction of artificial islands, undersea cables and pipelines, production of renewable energy, marine scientific research, and military uses. However, this complex, multidimensional and multi-faceted environment also provides a whole range of ecosystem services for humans, ranging from climate modulation and CO_2 absorption to provision of food stuffs and mineral resources. Because of the fluid nature of the ocean medium, activities on and under the seabed, such as seabed mining or cable laying, interact with the water column, while activities on or in the water column, such as bottom-trawling, may also interact with the seabed. In other words, all ocean uses interact with the marine environment. Thus, existing and emerging ocean uses can pose serious threats to the marine environment from, among other things, over-exploitation and pollution.

Seabed activities, in particular, pose complex threats due to the perturbation of marine biodiversity and the water column caused by construction, exploration or exploitation activities. Even apart from pollution emergencies such as that caused by the Deep-Water Horizon incident, seabed mining may destroy critical fish habitat or interfere with submarine cables. Bottom-trawling may interfere with submarine cables or destroy critical habitat for a range of sedentary species and other marine genetic resources. Exploitation of marine genetic resources may destroy critical marine biodiversity or interfere with ocean energy development and production. Increasingly, the varying demands on ocean space and resources are leading to conflict between both existing sectors such as shipping and fishing and between existing and emerging sectors including seabed mining, ecotourism and marine renewable energy. This increasing competition is also leading to increasing pressures on the marine environment.

Despite the overarching regime established by the Law of the Sea Convention (LOSC),[1] regulation of ocean activities is essentially a sectoral matter. In areas under national jurisdiction, where coastal states enjoy jurisdictional competence in respect of seabed uses, marine governance regimes have traditionally been largely characterised by high levels of sector-specific, uncoordinated institutional fragmentation.[2] The situation is even more fraught in areas beyond national jurisdiction (ABNJ) where sectoral fragmentation has been compounded by substantive inadequacy and regulatory ineffectiveness.[3] While a complex array of treaty regimes exists, governance, regulatory and substantive gaps hinder the ability of these regimes to adequately address both existing and emerging threats to the marine environment.[4] The potential for inter-sectoral conflict such as between shipping, fisheries, mineral exploration and exploitation, cable and pipeline operations and mining operations, only exacerbates the problem.

In recent years, consensus has emerged on the need to promote cross-sectoral cooperation and coordination in order to avoid inter-sectoral conflicts and to achieve sustainable environmental outcomes for the oceans. A number of general principles have emerged which provide the foundation for the development of normative frameworks and management approaches. These, in turn, provide the framework for the development of new management approaches and tools, designed to address conflicting uses on the seabed and conserve and protect the marine environment, that better reflect the holistic, fluid nature of the ocean and its uses. This chapter begins with a discussion of these principles (Section 2) and the broad normative frameworks (Section 3) and management approaches (Section 4) that have emerged. It then turns to a discussion of the key tools that have been developed to assist environmental management of marine and seabed uses (Section 5) and an examination of emerging frameworks for cross-sectoral management aimed specifically at avoiding, minimising or resolving conflicting uses on the seabed, particularly in ABNJ (Sections 6 and 7). It ends with some brief concluding remarks (Section 8).

---

1 United Nations Convention on the Law of the Sea, 10 December 1982, 1833 UNTS 3.

2 RK Craig, 'Marine Biodiversity, Climate Change and Governance of the Oceans' (2012) 4 *Diversity* 224, 231.

3 R Rayfuse, 'Climate Change, Marine Biodiversity and International Law' in M Bowman, P Davies and E Goodwin (eds) *Handbook on Biodiversity and Law* 123, 125 (Edward Elgar, 2016).

4 K Gjerde, H Dottinga, S Hart, E Molenaar, R Rayfuse, R Warner, *Regulatory and Governance Gaps in the International Regime for the Conservation and Sustainable Use of Marine Biodiversity in Areas Beyond National Jurisdiction*, IUCN Marine Law and Policy Paper No 1 (IUCN Gland, 2008).

2 Basic Principles Applicable to Uses of the Oceans and Seabed

A glance at any law of the sea textbook reveals a plethora of principles, some contested, some well accepted, that are applicable to ocean and seabed uses. In the context of addressing conflicting uses of the seabed for the purposes of protecting the marine environment, the basic principles articulated in the preamble of the LOSC include those of due regard for the sovereignty of all states, facilitation of international communication, promotion of peaceful uses, equitable and efficient utilization of marine resources, conservation of living marine resources, and protection and preservation of the marine environment. These principles play a valuable role in helping to understand and assess the operation and efficacy of the management approaches and tools that are examined in the following sections. They are therefore worthy of brief elaboration.

The principle of due regard reflects the notion of equality of states and the requirement that states refrain from any acts which might adversely affect the rights and duties of other states. This requirement evokes two principles relevant to determining issues of compatibility of uses: protection of life at sea, and the necessity to ensure that, to the greatest extent possible, activities co-exist rather than trump one another. It is envisaged that some accommodation or adjustments may be needed to the manner in which activities are carried out in order to ensure these principles are met. For example, navigation may be restricted temporarily in areas surrounding weapons tests or oil and gas installations, or permanently by the construction of deep-water ports. Where accommodation is not possible then states must refrain from activities that interfere with the exercise by other states of their rights under the law of the sea.[5]

The principle of facilitation of international communication is essentially a manifestation of the principle of the freedom of the seas and its corollary of exclusivity of flag state jurisdiction according to which no state may appropriate unto itself sovereignty over the high seas or interfere in any way with the vessels of another state. Of course, the LOSC itself allocates sovereignty and sovereign rights over ocean spaces to coastal states. However, while coastal states may exercise sovereignty over their territorial seas, this sovereignty is subject to the obligation to allow innocent passage to the vessels of other states.[6] Even less interference with foreign shipping is permitted in the exclusive economic

5 M Nordquist, (ed) United Nations Convention on the Law of the Sea: A Commentary, vol III, 86 (Martinus Nijhoff: Dordrecht, 1995).

6 LOSC Article 17.

zone (EEZ) where the coastal state is required to permit freedom of navigation and other states have the right to lay cables and pipelines, albeit subject to the coastal state's rights in respect of the continental shelf, exploitation of natural resources and prevention, reduction and control of pollution, and subject to the consent of the coastal state to the delineation of the course for the laying of such cables and pipelines.[7]

The principle of peaceful purposes is generally taken to refer to military activities and includes the general international law prohibition, articulated in Article 2(4) of the Charter of the United Nations, on the threat or use of force against the territorial integrity or political independence of other states.[8] Military activities, per se, are not prohibited, but are rather governed by other rules of international law outside the LOSC regime.[9] However, articulation of the peaceful purposes principle in the LOSC reflects the desire of the negotiators to ensure that activities such as marine scientific research and other seabed activities would be carried out exclusively for peaceful purposes.[10]

Equitable and efficient utilization of resources entitles all states to an equitable share of marine resources which are to be utilised in such a way as to avoid wastage. This does not mean that all states receive an equal share in any particular resource. This is clear for example, in the context of fisheries where the coastal state enjoys the sovereign right to exploit[11] and can exclude other states from fishing in its waters. Where the principle has its true effect is in areas beyond national jurisdiction. In the high seas, the freedom of fishing applies equally to all states, although not all states will share equally in either the physical practise of, or the economic benefit from, the catch. An even more important manifestation of this principle is reflected in the designation in Part XI of the LOSC of the non-living resources of the deep seabed beyond national jurisdiction as the common heritage of mankind and the establishment of a supranational management regime administered by the International Seabed Authority (ISA).[12]

The principle of conservation of marine living resources embodies the need to sustainably manage marine living resources, which include fish, marine mammals and all living marine genetic resources, for both current and future

7 LOSC Article 79.
8 LOSC Article 301.
9 RWG de Muralt, 'The Military Aspects of the UN Law of the Sea Convention' (1985) 32(1) *Netherlands International Law Review* 79.
10 LOSC Articles 141, 143(1), 147(2)(d) and 155(2).
11 LOSC Articles 61 and 62.
12 LOSC Articles 136 and 137(2).

generations.[13] The principle is reflected in the duty on all states to take measures, both unilaterally and collectively, for the conservation of marine living resources. Importantly, the principle also encompasses effects not only on target species but on dependent and associated species as well.[14] Thus it anticipates, at least in part, an ecosystem approach to the protection and preservation of the living elements of the world's oceans.

The principle of protection and preservation of the marine environment reflects the realisation of the importance of the oceans and their resources to human existence. This fundamental principle, embodied in Article 194 of the LOSC, imposes on all states the obligation to protect and preserve the marine environment from all sources of pollution, whether land, ship or air based, including the release of toxic, harmful or noxious substances, pollution from vessels, pollution from installations, intentional dumping, accidental discharges, and the introduction of alien species. Also required is the protection and preservation of rare or fragile ecosystems as well as the habitat of depleted, threatened or endangered species and other forms of marine life. This principle of environmental protection applies in respect of all ocean related activities, irrespective of where they are carried out. Thus, the principles above must also be read in light of this principle. The question is, however, how to operationalise the principle of protection and preservation of the marine environment, particularly in light of the multiplicity and variety of ocean uses.

3 The Normative Frameworks for Environmental Approaches to Oceans Management

Management of the oceans in a manner capable of protecting and preserving the marine environment is a complicated matter due to their dynamic spatial and temporal nature. In the context of environmental protection, two overarching approaches to the management of ocean activities have emerged: the precautionary approach and the ecosystem approach.

3.1 *The Precautionary Approach*

The precautionary approach – also referred to as the precautionary principle – is of relatively recent vintage. Post-dating the LOSC, the essential core of the

13 R Rayfuse, 'Article 119 – Conservation of the Living Resources of the High Seas' in A. Proelss (ed), *United Nations Convention on the Law of the Sea (UNCLOS): Commentary* 830–850 (Verlag CH Beck, 2017).

14 LOSC Article 119.

precautionary principle as articulated in Principle 15 of the Rio Declaration[15] is that in cases of possible serious or irreversible damage, lack of full scientific certainty shall not be used as a reason for postponing cost-effective measures to prevent environmental degradation. In such cases states are to apply the precautionary approach.[16] The approach thus aims to guide decision making while recognising the existence of sometimes radical uncertainty and ambiguous and contestable scientific knowledge.

Neither the application nor the effect of the precautionary approach is free from controversy. For some, the approach provides the basis for early action to address threatening environmental issues. For others, the application of a precautionary approach is said to result in over-regulation and unwarranted limitations on human activity. Conflicting interpretations of the approach range from the requirement merely to act carefully when taking decisions that may have an adverse impact on the environment, to the requirement to regulate and possibly even prohibit activities and substances which may be environmentally harmful even in the absence of conclusive proof of such likely harm, to the requirement that the person wishing to carry out a particular activity must affirmatively prove it will not cause environmental harm.[17] This latter interpretation, in particular, requires polluters to establish that their activities will not adversely affect the environment before they can be authorized to undertake the proposed activity, thus raising the connection between precaution and the requirements of environmental impact assessment (see section 5.1 below).

The precautionary approach is now required in an increasing number of environmental treaties, particularly treaties relating to the protection of the marine environment and the conservation and management of marine living resources. It is a key requirement in the 1995 United Nations Fish Stocks Agreement (FSA),[18] and is now widely adopted in fisheries management regimes at both the national and the international level. In the seabed mining context, the precautionary approach lies at the heart of the development by

15 Rio Declaration on Environment and Development, 13 June 1992, UN Doc. A/CONF.151/26; (1992) 31 *International Legal Materials* 874.

16 For comprehensive examinations of the precautionary principle in international law see, e.g., D Freestone, *The Precautionary Principle: The Challenge of Implementation* (Kluwer Law International, 1996) and A Trouwborst, *Evolution and Status of the Precautionary Principle in International Law* (Kluwer Law International, 2002).

17 P Sands and J Peel, Principles of International Environmental Law (3rd), (Cambridge University Press, 2012) 220.

18 Agreement for the Implementation of the Provisions of the United Nations Convention on the Law of the Sea of 10 December 1982, Relating to the Conservation and Management of Straddling Fish Stocks and Highly Migratory Fish Stocks, 4 August 1995, 2167 UNTS 3.

the International Seabed Authority (ISA) of the its Mining Code[19] and it has been recognised as a necessary element of the due diligence of states in regulating and conducting deep seabed mining.[20]

Given its wide acceptance in international and national law, it is generally accepted that the precautionary approach is trending towards becoming part of customary international law.[21] While opinion may still be divided as to its precise scope, meaning and effect, there can be no doubt as to its central importance to the law of the sea and the protection and preservation of the marine environment. The question is simply how to operationalise it. In this respect, it is important to note that the precautionary approach does not prevent activities with unknown effects from proceeding. Rather, it merely requires that if they do proceed they must do so with caution, with awareness of unknown potential impacts, and with appropriate checks and risk-minimizing controls in place. Precaution also involves seeking out and evaluating alternatives to the proposed action, including the possibility of taking no action. Thus, decisions as to the measures to be taken cannot appeal solely to scientific or technical justifications but must also align with social and cultural values about what harm may – or may not – be considered acceptable.[22] In other words, the concrete operationalisation of precaution is ultimately a task not only for law but also for the political realm. This is particularly relevant in the context of managing potentially conflicting seabed uses.

3.2 *The Ecosystem Approach*

Like the precautionary approach, the ecosystem approach is of relatively recent origin, having developed in response to recognition of a crisis in biodiversity conservation and the realisation of the inadequacies of single species

19 A Jaeckel, *The International Seabed Authority and the Precautionary Principle*, Brill 2017. See, as well, Chapter 8 in this book, H. Jessen, 'Advancing the Deep Seabed "Mining Code": Key Environmental Elements of the Regulatory Framework for the Commercial Exploitation of Mineral Resources'. The 'Mining Code' refers collectively to the rules, regulations and procedures adopted by the ISA regulations to regulate prospecting, exploration and exploitation of marine minerals in the international seabed area.

20 ITLOS, Seabed Disputes Chamber, *Responsibilities and Obligations of States Sponsoring Persons and Entities with Respect to Activities in the Area*, Case No. 17, Advisory Opinion of 1 February 2011, https://www.itlos.org/fileadmin/itlos/documents/cases/case_no_17/17_adv_op_010211_en.pdf.

21 Ibid, para. 135.

22 Rayfuse, 'Precaution and Climate Change: What Role for the Precautionary Principle in Addressing Global Warming' in T Hebeler, E Hoffmenn, A Proelss and P Reiff (eds) *Protecting the Environment for Future Generations* 61–78, 65 (Erich Schmidt Verlag, 2017).

management.[23] A seemingly intuitive approach, it recognises that the management and regulation of human activities that affect species, ecosystems and natural processes must be based on scientific knowledge of the wider systems in which such species, ecosystems or processes are situated, and that management measures must be designed and continuously adapted with consideration to the scales and dynamics (including the lack of full understanding) of ecosystem characteristics.

The scientific ideas on which the ecosystem approach is premised can be traced to the first half of the 20th century and elements of it can be discerned in the LOSC provisions requiring protection of species dependent on and related to targeted fish stocks,[24] protection and preservation of rare or fragile ecosystems and the habitat of depleted, threatened or endangered species and other forms of marine life in the pollution context.[25] The ideas underlying the approach can also be traced to the requirements to regulate for the prevention of interference with the ecological balance of the marine environment and of damage to its flora and fauna in the deep seabed mining context.[26]

While the ecosystem approach famously forms the core objective of the 1982 Convention on the Conservation of Antarctic Marine Living Resources,[27] it only gained general recognition in 1995 when the parties to the Convention of Biological Diversity (CBD)[28] adopted a 'common understanding' of the approach and called on all governments and international organisations to apply it.[29] According to the 'common understanding' the ecosystem approach 'is a strategy for the integrated management of land, water and living resources that promotes conservation and sustainable use in an equitable way …'. It 'requires adaptive management to deal with the complex and dynamic nature

23 RE Grumbine, 'What is ecosystem management' (1994) 8(1) *Conservation Biology* 27–38; KK Arkema, SC Abramson, BM Dewsbury, 'Marine ecosystem-based management: from characterisation to implementation' (2006) 4(10) *Frontiers in Ecological Environment* 525–532.

24 LOSC Articles 61(4) and 119(1)(b).

25 LOSC Article 194(5).

26 LOSC Article 145.

27 9 ATS 1982.

28 1760 UNTS 79.

29 On earlier expressions of the ecosystem approach or 'ecosystems thinking' in international law and policy, see A Trouwborst, 'The Precautionary Principle and the Ecosystem Approach in International Law: Differences, Similarities and Linkages' 18:1 RECIEL (2009) 26–37, at pp. 27–30.

of ecosystems and the absence of complete knowledge or understanding of their functioning'.[30]

The ecosystem approach has come to feature particularly strongly in the context of ocean management, having been endorsed, inter alia, by the parties to the Helsinki and OSPAR Conventions relating to the protection of the marine environment of the Baltic Sea and North East Atlantic, respectively[31] and, at the global level, in the FSA[32] and in the emerging requirements of the ISA's Mining Code. An ecosystem-based approach is also increasingly found in national law and policy relating to environmental and natural resource use and conservation. Nevertheless, despite its acceptance, in principle, there exist a wide variety of definitions of the ecosystem approach and an even larger number of opinions as to the way in which it can be implemented or operationalized.

Given the various, and varying, features and complexities of both ecosystems and the legal/administrative systems governing them, effective implementation of the ecosystem approach can be highly challenging and requires consideration of a vast number of interconnected – and inter-disciplinary – issues. This makes it difficult to articulate universally applicable rules for its effective operationalisation. Despite these challenges, however, the approach is now well established as a guiding principle (even if not necessarily a legal principle) in many contexts and its application is often seen as a prerequisite for the successful management of ecological systems (or rather the human activities affecting such systems). As with the precautionary approach, the question remains how to implement it.

4 Management Approaches for Operationalizing Precaution and Ecosystem Management

Implementing the precautionary and ecosystem approaches requires the development of management approaches capable of dealing with the ever-growing pressures on the marine environment from resource use and commercial activities. Increasingly, new approaches are being trialled. In the context of

30 UNEP/CBD, Fifth Meeting of the Conference of the Contracting Parties to the Convention on Biological Diversity (15–26 May 2000) Dec. V/6: Ecosystem Approach, U.N. Doc UNEP/CBD/COP/DEC/V/6 (16 May 2000).

31 Record of the First Joint Ministerial Meeting of the Helsinki and OSPAR Commissions (Bremen, 26 June 2003) (OSPAR/HELCOM statement), Annex 5 ('Towards an Ecosystem Approach to the Management of Human Activities').

32 FSA Preamble and Article 5.

resource management, these include approaches such as co-management, adaptive management, adaptive co-management, active adaptive management and, most notably, ecosystem-based management.

Co-management refers to a collaborative approach whereby responsibility for resource management is shared between government and user groups.[33] It is generally seen as an effective way of improving the legitimacy and efficiency of governance processes and management functions by decentralising resource management decisions, encouraging stakeholder participation and fostering dispute and conflict resolution. Inclusiveness in decision-making processes can both empower stakeholders and make them more accountable. It can also provide flexibility and the ability to adapt management regimes to new learned experiences.[34] However, co-management can also lead to unreasonable and unfulfilled expectations regarding process and result, exacerbate existing power imbalances, and provide room for the hijacking of environmental protection concerns and non-use values by extractive interests or the vested interests of other user groups.[35] In the absence of strong institutional structures it is generally considered that co-management arrangements can crumble under the weight of excessive and/or excessively diverse participation.[36] Nevertheless, it does provide a platform for the resolution of disputes and the negotiation of trade-offs.[37]

Adaptive management refers to management by feedback loop whereby lessons learned from management decisions are fed back into the following rounds of decision-making.[38] It ensures refinement and further development

---

33 S Sen, JR Nielsen, 'Fisheries co-management: a comparative analysis' (1996) 20 *Marine Policy* 405–418; S Singleton, 'Co-operation or capture? The paradox of co-management and community participation in natural resources management and environmental policy-making' (2000) 9(2) *Environmental Policy* 1–21.

34 E Pinkerton (ed) Co-operative Management of Local Fisheries: New Directions for Improved Management and Community Development (UBC Press, 1989); JR Nielsen, P Degnbol, KK Viswanathan, M Ahmed, M Hara, NMR Abdullah, 'Fisheries co-management and institutional innovation? Lessons from South East Asia and Southern Africa' (2000) 28(2) *Marine Policy* 151–160.

35 S Singleton, 'Co-operation or capture? The paradox of co-management and community participation in natural resource management and environmental policy-making' (2000) 9(2) *Environmental Politics* 1–21.

36 M Haward, 'Outstanding issues with regimes for ocean governance' in D Wilson and R. Sherwood (eds) *Oceans Governance and Maritime Strategy* 121–128 (Allen and Unwin, 2000).

37 L Carlsson and F Berkes, 'Co-management: concepts and methodological implications' (2005) 75 *Journal of Environmental Management* 65–76.

38 CR Allen, JJ Fontaine, J Pope, AS Garmestani, 'Adaptive management for a turbulent future' (2011) 92 *Journal of Environmental Management* 1339–1345.

as experience and knowledge increase. Adaptive management is said to address the challenge of operating with either impartial or incomplete knowledge thereby allowing progress in the absence of complete information.[39] In this respect, it can be seen as a manifestation of the precautionary approach. Importantly, it attempts to account for complexity by considering multiple sectors and policies,[40] thereby embracing complexity, variability and uncertainty. However, while theoretically an iterative process that can reduce uncertainty and deal with change through management of short and long-term impacts, the danger exists for passivity to replace learning and adaptation thereby pushing the system to a threshold at which abrupt, and unwelcome, change occurs.[41]

Active adaptive management and adaptive co-management are more responsive or highly developed versions of their namesakes. Active adaptive management is an iterative process of experimentation and re-experimentation designed to test hypotheses through the use of ecosystem scale holistic management experiments. However, this continued hypothesis-generation and testing can significantly impact the security and stability of the operating environment for both commercial operators and management authorities.[42] Nevertheless, this approach to management embraces ecosystem scale and system complexity in both human and ecological terms.[43] Adaptive co-management, for its part, links adaptive management with co-management and ecosystem dynamics. Seen as a means of studying or structuring increasingly coupled social and ecological systems it is considered more suited to developing adaptive capacity, social-ecological resilience, sustainable resource use and enhanced efficiency and effectiveness of management.[44] Its success

39 E Ogier, J Davidson, P Fidelman, M Haward, AJ Hobday, NJ Holbrook, E Hoshino and GT Peci, 'Fisheries management approaches as platforms for climate change adaptation: Comparing theory and practice in Australian fisheries' (2016) 71 *Marine Policy* 82–93, 86.

40 F Berkes, Implementing ecosystem-based management: evolution or revolution? (2012) 13 *Fish and Fisheries* 465–476.

41 J McDonald, MC Styles 'Legal strategies for adaptive management under climate change' (2014) 26(10) *Journal of Environmental Law* 25–53.

42 C Allen and A Curtis, 'Nipped in the bud: why regional scale adaptive management is not blooming' (2005) 36(3) *Environmental Management* 414–425.

43 Ogier et al, n 39.

44 P Olsen, C Folke and F Berkes, 'Adaptive co-management for building resilience in social-ecological systems' (2004) 34(1) *Environmental Management* 75–90.

is, however, highly dependent on the quality of decision-making and communication processes and on high levels of engagement by industry or other relevant stakeholders.[45]

The final management approach to be mentioned is that of ecosystem-based management which, as its names implies, is generally considered to be the most effective way of implementing the ecosystem approach because of its explicit consideration of all aspects of the main extrinsic forcers on ecosystem dynamics and their impacts on both human and non-human species and the environments in which they live.[46] Ecosystem-based management recognises the needs of an ecosystem as a whole, as opposed to those pertaining only to the particular target of management. In doing so it incorporates consideration of non-target dependent and related species as well as habitats and ecological communities and aims to sustain a broader range of ecosystem services. It also aims to integrate decision-making relating to all human activities that affect a particular ecosystem. In theory, by maximising ecosystem resources and services, it makes for their more efficient and sustainable use. As Warner puts it, ecosystem-based management has added 'a new dimension to marine environmental protection which has previously focused on prevention and control of marine pollution and the protection of single species'.[47] However, implementing ecosystem-based management is notoriously difficult, not least because of its complex data requirements, potentially prohibitive monitoring costs, and a general lack of scientific knowledge of ecosystem dynamics and/or their responses to human interventions. Robust precautionary environmental management tools are therefore required to support its implementation.

5 Modern Environmental Management Tools That Support Precautionary Ecosystem-Based Management

Key tools that have been developed to support precautionary ecosystem-based management include environmental impact assessments (EIAs) and area-based management tools, such as marine protected areas (MPAs).

45 F Berkes, 'Evolution of co-management: the role of knowledge generation, bridging organisational and social learning (2009) 90(5) *Journal of Environmental Management* 1692–1702; GP Kofinas, 'Adaptive co-management in Social-Ecological Governance, in FS Chapin III, GP Kofinas, C Folke (eds) *Principles of Ecosystem Stewardship: Resilience Based Natural Resources Management in a Changing World*, (Springer New York, 2009) 77–101.

46 Ogier et al, n 39.

47 R Warner, 'Conserving marine biodiversity in the global marine commons: co-evolution and interaction with the Law of the Sea' (2014) 1(6) *Frontiers in Marine Science* 1–23, 5.

5.1 *Environmental Impact Assessment*

EIA is generally defined as 'a process of evaluating the likely environmental impacts of a proposed project or development taking into account inter-related socio-economic, cultural and human health impacts, both beneficial and adverse'.[48] EIAs are particularly useful for determining and analysing the likely environmental impacts of human activities, developing mitigation measures, and identifying activities that should not be authorised to proceed because the impacts will either be too severe or too uncertain. Acknowledged in Principle 17 of the Rio Declaration[49] as a key element in the suite of tools for environmental protection, EIAs are a well-established practice in the domestic realm. The obligation to undertake EIAs in the transboundary context is also well recognised in international law.[50]

In ABNJ, however, the requirement to carry out EIAs is less well defined. Article 206 of the LOSC requires states to carry out environmental assessments 'when States have reasonable grounds for believing that planned activities under their jurisdiction or control may cause substantial pollution or significant and harmful changes to the marine environment'. This includes activities in ABNJ. However, the obligation itself is neither well-defined nor universally implemented.[51] While the general obligation has been made more specific in some other global agreements and in some sector specific and regional agreements, there remain significant gaps with respect to a number of activities that have the potential to cause significant impacts to the marine environment, including seabed activities other than seabed mining.[52]

With respect to seabed activities in ABNJ, no specific EIA requirements exist in relation to the laying of cables and pipelines, construction and operation of seabed installations, the conduct of marine scientific research, bioprospecting and sea-based tourism. Only in the contexts of deep seabed mining and bottom-fishing have detailed EIA requirements begun to emerge. With respect to the former, assessments of the environmental impacts of activities in the Area are

---

48 Voluntary Guidelines on biodiversity-inclusive impact assessment. CBD COP 8 Decision VIII/28 on Impact Assessment, Annex, sec. 5.

49 1992 Rio Declaration on Environment and Development, UN Doc. A/CONF.151/26, (1992) 31 ILM 874.

50 See, e.g., Convention on Environmental Impact Assessment in a Transboundary Context, 25 February 1991 (Espoo Convention); Pulp Mills on the River Uruguay (Argentina v Uruguay) Judgement, ICJ Reports 2010, para 204.

51 R Warner, 'Environmental assessment in marine areas beyond national jurisdiction' in R Rayfuse (ed), *Research Handbook on International Marine Environmental Law* (Edward Elgar, 2015) 291–312, 292.

52 E Druel, Environmental impact assessments in areas beyond national jurisdiction' Studies No 01/13 (IDDRI, Paris France, 2013).

mandated by the provisions of Part XI of the LOSC and its 1994 Implementing Agreement and have now been recognised as an obligation under customary international law.[53] Detailed regulations for the conduct of such assessments have been incorporated into the regulations relating to prospecting and exploration for deep seabed minerals adopted by the ISA as part of its Mining Code[54] and are the subject of intense discussion in the ongoing discussions on the development, by the ISA, of its exploitation regulations.[55] With respect to the latter, assessment of the adverse impacts of bottom fishing activities on vulnerable marine ecosystems is required by UN Resolution[56] and detailed EIA requirements have been developed in the FAO International Guidelines for the Management of Deep-Sea Fisheries in the High Seas and adopted by a number of regional fisheries management organisations (RFMOs).

Even where detailed requirements have been adopted, considerable inconsistency exists between regions and across sectors. In addition, not all sectors have developed legally binding requirements on EIAs in ABNJ. Moreover, no requirement exists to assess the cumulative impacts of human activities in ABNJ or to conduct cross- or inter-sectoral assessments.[57] The current negotiations on the possible development of an implementing agreement to the LOSC on marine biodiversity in ABNJ may result in the establishment of a default mechanism for the assessment and regulation of new and emerging activities as well as those not currently covered by EIA requirements and/or a standardised model for EIA requirements. As currently conceived, this would complement, rather than negate, the need for existing sectoral and regional EIA processes.[58]

5.2 *Area-Based Management Tools*

Area-based management tools have long been accepted in both the national and the international sphere. While there is no universally accepted definition of 'area-based management', it clearly refers to management on the basis of an area in which the regulation of human activity is more stringent than in the immediately surrounding area. A traditional single sector management approach, area-based measures include everything from areas closed to fishing

53 ITLOS, Advisory Opinion, n 20, para 145.

54 Nodules Regulations, reg 31(6), Sulphides and Crusts Regulations, reg 33(6).

55 A Jaeckel, *The International Seabed Authority and the Precautionary Principle: Balancing Deep Seabed Mining and Marine Environmental Protection* (Brill, Nijhoff, 2017) 230–251.

56 UNGA Resolution 61/105 of 8 December 2006.

57 Druel, n 52.

58 Warner, n 51, 307.

during spawning season to marine areas in which shipping or other human activities are regulated or restricted in some manner.

In ABNJ, traditional area-based management tools are often used by RFMOs to close fisheries either to specified gear types, or for seasonal purposes, or once quotas have been reached.[59] They are also manifest in the IMO's designations of Particularly Sensitive Sea Areas (PSSAs). In the seabed mining context in ABNJ, area-based measures are explicitly included in the ISA's Exploration Regulations which provide for the establishment of both preservation reference zones (PRZs) and impact reference zones (IRZs); the former consisting of pristine areas in which mining is not allowed, the latter consisting of areas in which mining takes place.[60] The purpose of these zones is to facilitate assessment of the effects of mining on the marine environment.[61] A more comprehensive area-based approach is taken in the Environmental Management Plan for the Clarion Clipperton Zone (EMP-CCZ), adopted by the ISA in 2012,[62] which sets out a spatial management plan for an area of deep seabed in the eastern central Pacific which is roughly equivalent in size to Europe.[63] A central feature of the plan is the designation of nine Areas of Particular Environmental Interest (APEIs) in which mining activities are prohibited. The purpose of these areas is both to protect representative habitat and to facilitate marine scientific research, although questions have been raised as to their utility given the adjustments to size and location that were made prior to adoption of the EMP-CCZ to accommodate pre-existing mining leases.[64] Nevertheless, the EMP-CCZ is generally considered a good example of both precautionary and ecosystem-based management.[65]

The tool that has received the most attention in recent years is that of marine protected areas (MPAs). Although similarly lacking a singular definition, MPAs include such things as marine sanctuaries, marine parks, wildlife refuges, fisheries closures, no-take MPAs, multiple use MPAs, marine reserves

---

59 R Rayfuse, 'Regional Fisheries Management Organisations' in D Rothwell, A Oude Elferink, K Scott and T Stephens, *Oxford Handbook of the Law of the Sea* 439–462 (Oxford University Press, 2015).

60 Nodules Regulations reg 31(6) and Sulphides and Crusts Regulations reg 33(6).

61 Jaeckel, n 55, 211–212.

62 ISA, ISBA/17/LTC/7 (13 July 2011), para 21.

63 M Lodge, et al, 'Seabed Mining: International Seabed Authority Environmental Management Plan for the Clarion-Clipperton Zone. A Partnership Approach' (2014) 49 *Marne Policy* 66–72, 72.

64 LM Wedding et al, 'Managing Mining of the Deep Seabed' (2105) 349 *Science* 144–145 and Jaeckel, n 55, 206.

65 M Lodge, 'Protecting the environment of the deep seabed' in R Rayfuse (ed), *Research Handbook on International Marine Environmental Law* (Edward Elgar, 2015), 151–169, 167.

and ecological reserves. Clearly, the ISA's PRZs and APEIs can also be seen as a form of protected area or MPA. However, it must be remembered that the ISA can only regulate activities relating to deep seabed mining and not other activities in these areas. They are thus not multi-sector MPAs.

The use of multi-sector MPAs as a management tool is increasingly seen as a central element of ecosystem-based management. In 2002 the World Summit on Sustainable Development called for the use of diverse approaches and tools to protect marine biodiversity including though the establishment of representative networks of MPAs by 2012.[66] The Aichi Biodiversity Targets, adopted by the parties to the CBD in 2010, call for 10 percent of coastal and marine areas to be conserved through MPAs by 2020,[67] although it must be admitted that, at this stage, achievement of that goal seems unlikely. Recent assessments show only 5.1 percent of areas under national jurisdiction and 0.17 percent of the high seas as being protected by MPAs.[68] While this latter number has increased with the adoption by the Commission on the Conservation of Antarctic Marine Living Resources of the Ross Sea MPA in 2016,[69] the prospects of meeting the global 2020 target seem slim. This is in large part due to lingering uncertainties as to the rationale for and efficacy of these measures, many of which are designated in remote areas,[70] lack management plans, allow many types of extractive activities,[71] and are not enforced or monitored,[72]

66 Plan of Implementation of the World Summit on Sustainable Development, para 32(c).

67 This was originally to have been achieved by 2012 however it became clear that target would not be achieved so the deadline was extended to 2020. CBD 2010 Decision X/2 UNEP/CBD/COP/10/27.

68 BH e Costa, J Claudet, G Franco, K Erzini, A Caro, EJ Gonçalvez, 'A regulation-based classification system for Marine Protected Areas (MPAs)' (2016) 72 *Marine Policy* 192–198, 192.

69 CCAMLR, Conservation Measure 91–05 (2016) 'Ross Sea region marine protected area', https://www.ccamlr.org/en/measure-91-05-2016.

70 PJS Jones and EM De Santo, 'Viewpoint – is the race for remote, very large marine protected areas (VLMPAs) taking us down the wrong track?' (2016) 73 *Marine Policy* 231–234.

71 MD Spalding, I Meliane, NJ Bennett, P Dearden, PG Patie and RD Brumbaugh, 'Building towards the marine conservation end-game: consolidating the role of MPAs in a future ocean' (2016) 26 (Suppl. 2) *Aquatic Conservation: Marine and Freshwater Ecosystems* 185–199.

72 R Devillers, RI Pressey, A Greech, JN Kittinger, GJ Edgar, T Ward, R Watson, 'Reinventing residual reserves in the sea: are we favouring ease of establishment over need for protection?' (2015) 25(4) *Aquatic Conservation: Marine and Freshwater Ecosystems* 48–504; P Leenhardt, B Cazalet, B Salvat, J Claudet, F. Feral, 'The rise of large-scale marine protected areas: conservation or geopolitics?' (2013) 85 *Oceans and Coastal Management* 112–118; J Lubochenko, K Grorud-Colvert 'Making waves; the science and politics of ocean protection (2015) 350(6259) *Science* 382–383; AN Rife, B Erisman, A Sanchez, O Aburto-Oropeza, 'When good intentions are not enough ... Insights on networks of 'paper parks' marine protected areas' (2013) 6(3) *Conservation Letters* 200–212.

as well as to their potential knock-on effects of excluding stakeholders either from their livelihoods entirely or from participation in the development and/or implementation of management measures.[73]

An additional complicating factor in the acceptability and efficacy of MPAs as a management tool arises where these are established on a sectoral basis thereby creating the potential for inter-sectoral conflict. This is relevant in the case of seabed uses where implementation and enforcement of an MPA designed to protect the water column (i.e. fisheries) may interfere with seabed uses (i.e., resource extraction, wind farms, shipping) and vice versa; or where a seabed MPA is adopted to protect from the adverse effects of seabed mining but not from other seabed uses such as cable laying, oil and gas extraction, wind farm construction or marine scientific research. Still greater levels of complication arise in the case of MPAs in ABNJ where no overarching management authority exists and individual treaty regimes are fragmented sectorally, substantively, geographically and in terms of participation.[74] Questions thus persist as to how MPAs (and other area-based measures) can be brought under an integrated protection scheme both within a particular sector and beyond and, more importantly, how cooperation and coherence, or harmonisation, between competent management authorities can be shaped in order to cross the sectoral divide.[75]

6 Crossing the Sectoral Divide: Marine Spatial Planning

When it comes to managing ocean uses, as noted above, traditional management structures have revolved around single sector or even, in the case of living resources, single species management. However, the dynamic nature of new and emerging ocean uses, particularly when coupled with the realisation of existing and emerging threats to the oceans from pollution, over-exploitation, habitat destruction, climate change and ocean acidification, call into question the effectiveness and flexibility of traditional, sectoral management approaches. In areas under national jurisdiction this has led to the development of new management approaches that cross the sectoral divide in order

73 SC Gall and LD Rodwell, 'Evaluating the social acceptability of Marine Protected Areas' (2016) 65 *Marine Policy* 30–38, 30.

74 K Gjerde, 'Marine protected areas beyond national jurisdiction: Some practical perspectives for moving ahead' (2012) 27(2) *International Journal of Marine and Coastal Law* 351–373.

75 P Drankier, 'Marine Protected Areas in Areas beyond National Jurisdiction' (2012) 27 (2) *International Journal of Marine and Coastal Law* 291–350.

to ensure the more orderly and sustainable use of ocean resources and to avoid or minimise conflict between sectors. Most notable amongst these is marine spatial planning (MSP).

The concept of spatial planning is nothing new; it is a commonly adopted approach to land-use planning. However, its application in the marine context is of relatively recent origin.[76] Although no agreed definition exists, MSP can generally be said to refer to 'a process of analysing and allocating parts of the three-dimensional marine spaces to specific uses to achieve ecological, economic and social objectives that are usually specified through the political process'.[77] In general, MSP is seen as a 'complicated but necessary process, to establish a more rational organisation of using marine space and the user interactions in order to protect the biological diversity of the marine environment, while taking into account social and economic values'.[78] It is a forward-looking, proactive planning process intended to result in a comprehensive plan or vision for a marine region that both allocates and reconciles the use of marine space on an integrated, cross-sectoral basis.[79] It is widely considered to be a useful tool for ensuring the equitable and efficient utilisation of resources and for avoiding, or at least managing, conflict both between various sectors and between human uses and the marine environment. Thus, MSP can also be utilised to resolve conflicts over uses of and activities on the seabed and to promote ecosystem-based management.[80]

MSP builds on the concept of integrated coastal zone management (ICZM) and other non-spatial integrated approaches to deliver a more spatial approach

---

76 T Potts, 'Marine spatial planning and various uses and interest relating to the marine environment' in N Soininen and D Hassan (eds) *Transboundary Marine Spatial Planning and International Law* (Routledge, 2015) 42–59. For reviews of recent MSP practice see S Jay, W Flannery, J Vince, W Liu, JG Xue, M Matczak, J Zaucha, H Janssen, J van Tatenhove, H Toonen, A Morf, E Olsen, JL Suarez de Vivero, JCR Mateos, H Calado, J Duff and H Dean, 'International Progress in Marine Spatial Planning' (2013) *Ocean Yearbook* 171–212 and K Scott, 'The Evolution of Marine Spatial Planning in New Zealand: Past, Present and Possible Future' (2016) 31(54) *International Journal of Marine and Coastal Law* 652–689.

77 C Ehler and F Douvere, 'Marine Spatial Planning: A Step-by-Step Approach Toward Ecosystem-Based management', *Intergovernmental Oceanographic Commission and Man and the Biosphere Programme, IOC Manual and Guides* No 53, IOCAM Dossier No 6 (UNESCO Paris 2009), 18.

78 F Maes, 'The international legal framework for marine spatial planning' (2008) 32 *Marine Policy* 797–810, 798.

79 N Soininen and D Hassan, 'Marine spatial planning as an instrument of sustainable ocean governance' in N Soininen and D Hassan (eds) *Transboundary Marine Spatial Planning and International Law* (Routledge, 2015) 3–12, 4; CN Ehler and F Douvere, 'An international perspective on marine spatial planning initiatives' (2010) 37(3) *Environments* 9–20.

80 Ehler and F Douvere, n 77; Maes, n 78 808.

which includes the zoning of marine spaces.[81] Similarly, it builds on the concept of other area-based measures, in particular MPAs, which are, in effect, small-scale models of ecosystem-based marine spatial plans.[82] However, it is important to distinguish between MSP and MPAs. MPAs are aimed at ensuring the protection of vulnerable marine ecosystems. MSP is aimed at ensuring the orderly and sustainable use of ocean spaces. Thus, MPAs are only one tool in the MSP tool-kit which is used to achieve more than mere area protection.[83] Of course MSP is not intended to exhaustively plan every inch of ocean space. Rather, it provides a process by which areas of strategic importance can be identified, cumulative impacts can be addressed, and conservation of ecosystem services can be maximised in an integrated, cross-sectoral management framework. The processes by which these ends are to be achieved are generally considered to include transparent and meaningful engagement and stakeholder participation, data collection, mapping and analysis of existing uses, identification of alternative use patterns, monitoring and enforcement.[84] To these can be added the need to identify and/or establish the authority responsible for the planning process and the requirement of adequate financing and resources.[85]

At its heart, MSP seeks to ensure integration of oceans management on sectoral, spatial and temporal scales. MSP is thus a response to the traditional fragmented, incoherent and uncertain approach to oceans management. Sectoral overlaps can be identified and resolved. International, regional, national and local regulatory scales can be identified, and their application assessed and coordinated. Social, economic and ecological interests can be identified and accounted for, although competition between these various interests may be fierce and strong political mechanisms will be needed to resolve them. In this respect, it is important to remember that MSP does not do away with the need for effective (and enforceable) sectoral ecosystem-based management. The 'mere act of delineating spaces' can never be 'sufficient to achieve management objectives'.[86] Rather, the idea of MSP is that it complements and coordinates sectoral approaches, taking a particular area or region

81 Potts, n 76, 43.

82 PJS Jones, LM Lieberknecht and W Qiu, 'Marine Spatial planning in reality: Introduction to case studies and discussion of findings' (2016) *Marine Policy* 256–264, 262.

83 Scott, n 76, 656.

84 Ehrle and Douvere, n 77.

85 Potts, n 76, 44.

86 J Duff, 'Trends on Ocean Zoning – Layers of Confusion and Approaches to Clarity' in AChircop, T McDorman and S Rolston (eds) *The Future of Ocean Regime Building: Essays in Tribute to Douglas M. Johnston* (Nijhoff Brill, 2009) 159–174, 162.

as a whole and integrating policies and objectives across different sectors in order to achieve holistic, integrated, coherent, rational and ecologically sustainable use of marine spaces and resources.[87] In this respect, MSP is said to provide the strategic context for sustainable development in the marine realm.[88]

In theory, MSP appeals as a process that recognises and integrates environmental and ecological interests with socio-economic ones.[89] In practice, reviews of MSP implementation in the national domain suggest that it is not necessarily delivering ecosystem-based management.[90] MSP is often focused on achieving specific sectoral objectives, such as the promotion of offshore wind energy,[91] and on promoting 'blue growth' activities.[92] Even in the Great Barrier Reef Marine Park, the quintessential 'poster child' for MSP, environmental interests are often compromised by economic ones such as terrestrial farming and port construction, with the Australian government having to fend off attempts to have the Great Barrier Reef placed on the World Heritage in Danger List. Clearly, as a tool for ecosystem-based management, MSP requires the integration of frameworks to assess ecosystem services and the implementation of precautionary and adaptive management approaches.[93] This requires convincing both new and existing sectors of the advantages and opportunities that genuine integrated, collaborative and ecosystem-based approaches can offer, a difficult task even at the national level.[94] In ABNJ, the effective application of MSP further requires a level of coherence and cooperation among and between global and regional agreements, institutions and national administrations that simply does not yet exist.[95] As discussed in the next section, however, some developments are occurring in this regard.

87 P Gilliland and D Laffoley, 'Key elements and steps in the process of developing ecosystem-based marine spatial planning' (2008) 32 *Marine Policy* 787–796.

88 Soininen and Hassan, n 79, 9.

89 F Douvere, 'The importance of marines spatial planning in advancing ecosystem-based sea use management' (2008) 32 *Marine Policy* 762–771.

90 E Domínguez-Tejo, G Metternicht, E Johnston and L Hedge, 'Marine Spatial Planning advancing the Ecosystem-Based Approach to coastal zone management: A review' (2016) 72 *Marine Policy* 115–130.

91 Potts, n 76, 53; Jones et al, n 82.

92 Jones et al, n 82, 262–263; M Young, 'Building the Blue Economy: The Role of Marine Spatial Planning in Facilitating Offshore Renewable Energy Development' (2015) 30 *International Journal of Marine and Coastal Law* 148–173.

93 Domínguez-Tejo, n 90.

94 Potts, n 76, 57.

95 JA Ardron, K Gjerde, S Pullen and V Tilot, 'Marine spatial planning in the high seas' (2008) 32 *Marine Policy* 832–839.

7 Crossing the Sectoral Divide in ABNJ

Regulation of ocean activities in ABNJ is characterised by high levels of governance, regulatory and substantive fragmentation. Absent a single centralised regulatory authority, a comprehensive integrated approach to ABNJ management such as that represented by MSP requires high degrees of international and inter-regime cooperation. While not as developed as in the domestic sphere, cross-sectoral initiatives are emerging which indicate possible ways forward for the integration of MSP-type management of seabed uses in ABNJ. Two examples discussed here relate to the protection of vulnerable benthic ecosystems in the North East Atlantic from the adverse effects of bottom-fishing, and the protection of undersea cables from deep seabed mining activities. These examples provide interesting studies in the possibilities, through coordinated management, of the implementation of the general principles articulated in the LOSC; the first exploring the balance between the principles of due regard, equitable and efficient use of resources and protection and preservation of the marine environment, while the second adds consideration of the principle of international communication to the mix.

7.1 *Conservation of Seabed Ecosystems vs Bottom-Fishing in the North East Atlantic*

Protection of vulnerable benthic ecosystems has its origins in the 1990s. It emerged from the recognition of the need for holistic ecosystem-based approaches to environmental protection coupled with the increasing realisation of the adverse effects of fishing activities on other parts of the marine ecosystem and on the protection of marine biodiversity.

In the North East Atlantic, two regional organisations are changed with protection of aspects of the marine environment: the OSPAR Commission for the Protection of the Marine Environment of the North-East Atlantic and the North-East Atlantic Fisheries Commission (NEAFC). While the two commissions cover the same geographic area, neither their membership nor their areas of application are the same. OSPAR has competence within both areas under and areas beyond national jurisdiction, while NEAFC's competence is limited to the high seas beyond national jurisdiction. In addition, their mandates, while complimentary, are completely distinct. NEAFC has competence over fisheries in ABNJ while OSPAR has competence in relation to the management of human activities (including those on the seabed) that impact the marine environment, including the protection and conservation of ecosystems and biodiversity in ABNJ. OSPAR's competence specifically excludes fisheries, shipping and seabed mining, although it does have a mandate to bring any such issues to the attention of the relevant international organisations.

In 2003 the OSPAR Commission formally adopted the ecosystem approach as the basis for its management activities and adopted a strategy to establish, by 2010, an ecologically coherent network of MPAs in the OSPAR Convention area, including in those parts of the area that constitute ABNJ.[96] Given their respective mandates, the only way in which OPSAR could ensure the coherence and protection of its MPAs was to enlist the support and cooperation of NEAFC. In the years that followed parallel, rather than coordinated, efforts were expended in both organisations leading to tensions, particularly on the part of NEAFC, which was concerned by what it saw as attempts by OSPAR to usurp its managerial competence. Initial efforts by OSPAR in 2002 to propose cooperation were soundly rejected by NEAFC. The prospects of a counter-productive 'turf war' were further heightened in 2004 when OSPAR, making good on its mandate to draw to the attention of NEAFC the need to protect some coral areas on the high seas from bottom fishing, suggested the start of a dialogue. NEAFC simply responded by saying that it was working on the issue.[97] Eventually, NEAFC realized that OSPAR was going to move ahead with MPAs which might impinge on its regulatory area. Thus, in 2006 the NEAFC parties amended the NEAFC Convention to remove any doubts as to NEAFC's ability to apply an ecosystem approach in order to protect vulnerable marine ecosystems (VMES) on the seafloor from the adverse impacts of bottom-trawling.[98] This allowed NEAFC to adopt the first of its precautionary bottom-trawling closures. These were expanded in 2008 and 2009 to closures on the Mid-Atlantic Ridge and the Charlie Gibbs Fracture Zone. Meanwhile, in 2007, OSPAR commenced consideration of its own MPAs in ABNJ, including areas on the mid-Atlantic Ridge. In 2010 OSPAR established six high seas MPAs, the world's first 'network' of high seas MPAs, to which a seventh was added in 2012.[99]

Taken together, the OSPAR and NEAFC closures have been said to represent an effective network of high seas MPAs in the North East Atlantic. In truth,

96 Bremen Joint Ministerial Statement on the Ecosystem Approach to the Management of Human Activities 'Towards an Ecosystem Approach to the Management of Human Activities", 26 June 2003. https://www.ospar.org/site/assets/files/1232/jmm_annex05_ecosystem_approach_statement.pdf.

97 I Kvalvik, 'Managing institutional overlap in the protection of marine ecosystems on the high seas. The case of the North east Atlantic' (2012) 56 *Ocean and Coastal Management* 35–43.

98 See 'Status of the 1980 Convention on Future Multilateral Cooperation in North-East Atlantic Fisheries' explaining the amendments, https://www.neafc.org/system/files/status-of-1980_convention-03.pdf.

99 BC O'Leary, RL Brown, DE Johnson, H von Nordheim, J Ardron, T Packeiser and CM Roberts 'The first network of marine protected areas (MPAs) in the high seas: The process, the challenges and where next' (2012) 36 *Marine Policy* 598–605.

neither the ecological coherence nor the efficacy of these MPAs has yet been assessed,[100] and upon closer examination a less sanguine picture emerges. Analysis of the geographical coordinates establishing the NEAFC closures and the OSPAR MPAs shows that the areas only partially overlap, potentially leading to a situation where activities undertaken under the regulation of one organisation may violate the protective measures adopted by the other. Moreover, only two of the OSPAR MPAS cover both the water column and the seabed (the Milne and Charlie Gibbs South MPAs). The others pertain to areas in which either Portugal or Iceland have claims to extended continental shelves. Until these states' submissions to the Commission on the Limits of the Continental Shelf have been dealt with it is unclear who has jurisdictional competence over what. This gives rise to very difficult issues relating to the scientific validity and ecosystem coherence of these MPAs, not to mention complicated questions of co-management as between OSPAR and the coastal states. In addition, the possibility exists that contracts could be issued by the ISA for mineral exploration and exploitation in areas currently within the OSPAR MPAs, in particular in the Charlie Gibbs Fracture Zone on the Mid-Atlantic Ridge thereby completely undermining (both literally and figuratively) the OSPAR MPAs. With the ISA considering adoption of an environmental management plan for the Mid-Atlantic Ridge, the further possibility exists for inconsistent environmental protections along the physical continuum of the Mid-Atlantic Ridge ecosystem and for conflict between the OSPAR and ISA regimes.

Cooperation between OSPAR and NEAFC has now been formalized beyond their initial Memorandum of Understanding[101] into the Collective Arrangement Between Competent International Organizations on Cooperation and Coordination Regarding Selected Area in Areas Beyond National Jurisdiction in the North East Atlantic,[102] which provides an informal framework for dialogue and the sharing information of mutual interest within the overall context of respecting each other's mandates and legal competencies.[103] However, the two main targets of the Collective Arrangement efforts, the ISA and the IMO, have, as yet, declined to join the Collective Arrangement. Admittedly, the ISA has made an effort to participate in the Collective Agreement meetings.

100 See OSPAR, Summary Record of the Meeting of the Biodiversity Committee, 29 February to 4 March 2016, BDC 16/9/1-E.

101 Available at https://www.ospar.org/about/international-cooperation/memoranda-of-understanding.

102 Established in 2014. See https://www.ospar.org/news/collective-arrangement.

103 For a history of the development of the Collective Arrangement, see, S Ásmundsson and E Corcoran, The Process of Forming a Cooperative Mechanism Between NEAFC and OSPAR, NEAFC and OSPAR (2015), https://www.ospar.org/documents?v=35111.

Nevertheless, a comprehensive, cooperative mechanism for coherent, coordinated and holistic ecosystem-based management of multiple and possibly conflicting human uses of the seabed in the OSPAR MPAS is still missing.

7.2 *Submarine Cables and Deep Seabed Mining*

Another example of emerging cross-sectoral planning relates to submarine cables and deep seabed mining.[104] Submarine cables are the backbone of the global economy carrying up to 98 percent of international telecommunications. This critical international infrastructure, which is almost entirely privately owned, criss-crosses the world's oceans over a total combined length of 1,576,481 kms. Cables are light weight, unprotected, and carefully laid to minimize the risk of fault or damage. Repairs to cables, particularly in ABNJ, are expensive, time consuming and highly technical.

Traditionally the greater water depths in ABNJ have protected cables from human interference, although they have always been subject to natural threats such as earthquakes, landslide and deep currents.[105] With the advent of deep seabed mining operations in ABNJ, however, the potential exists for cable laying and mining operations to come into conflict with each other. Two areas of particular current concern are the Eastern Pacific Ocean and the Indian Ocean where cables traverse areas over which the ISA has already issued exploration permits for polymetallic nodules.[106] The fear of the cable owners is that mining operations may damage the cables. The fear of the mining contractors is that cable laying and repair operations will interfere with or possibly cause damage to their mining operations. In both cases, economic losses could be substantial.

In contradistinction to seabed mining companies, whose operations and interests are the concern of the ISA, cable operators are not represented by any inter-governmental organisation, although their activities are carried out in pursuance of a number of international treaties. Rather, they are represented by the non-governmental International Cable Protection Committee (ICPC). In 2015, the ISA and the ICPC held a workshop with the purpose of identifying and addressing the issues involved in the potential risk of interference between submarine cables and deep seabed mining and finding 'practical solutions for

104 On the commercial aspects of submarine cables, see Chapter 26 in this book, L.O. Askheim 'Commercial arrangements and liability for crossing pipelines, power cables and telecommunication cables on the seabed'.

105 ISA, 'Submarine Cables and Deep Seabed Mining: Advancing Common Interests and Addressing UNCLOS 'Due Regard' Obligations' Technical Study No 14 (ISA, 2015), 17–19.

106 ISA, 'Submarine Cables and Deep Seabed Mining: Advancing Common Interests and Addressing UNCLOS 'Due Regard' Obligations', Briefing Paper 03/2105 (ISA, 2015), 3.

the peaceful coexistence of both uses in' ABNJ.[107] It will be recalled that the LOSC expressly authorises both activities however it equally expressly stipulates that these activities must be carried out with 'due regard' for each other (and for other users). In other words, both cable owners and mining contractors are required to exercise due regard for each other. Unfortunately, the LOSC does not define what constitutes 'due regard'. Thus, the workshop looked at practical procedures to reduce risks and avoid disputes between the two constituencies. Key to these was an understanding that, at the very least, 'due regard' requires first notice, either actual or constructive, and then consultation between the cable owners and contractors engaged in competing activities.[108] It was agreed, among other things, that the ISA and the ICPC should cooperate as points of contact, facilitators and information disseminators to ensure that their respective audiences would be able to identify solutions to cable location should exploitation occur in the future. It was further suggested that mapping of exploration areas under contract with the ISA and eventual cable locations would be useful, as would a joint code of conduct or the unilateral adoption by both organisations of appropriate recommendatory guidance on how to avoid conflicts with each other's operations.[109] Whether such recommendations will be adopted remains to be seen. Suffice it to say here that the need to reduce conflicts between seabed activities in ABNJ has, at least, been identified, and some concrete practical steps have been suggested as to how best to go about it.

What these two examples show is an increasing awareness of the need to cooperate across sectors in order to avoid, minimise or manage conflict, even at the international level. This is being achieved, to an as yet minor extent, through memoranda of understanding and other agreements or arrangements providing for cooperative activities such as joint workshops, work programs, and meetings, and participation in each other's meetings as observers. The success of these ventures remains, however, dependent on some common understanding of their potential benefits, such as certainty, prevention of duplication, increase in efficiency in achieving agreed goals and targets, and on the further development of significantly more, more complex, and more constructive processes for cooperation.[110]

107 ISA, n 104, 5.

108 Ibid, 27.

109 Ibid, 33–34.

110 JA Ardron, R Rayfuse, K Gjerde, R Warner, 'The sustainable use and conservation of biodiversity in ABNJ: What can be achieved using existing international agreements?" (2014) 49 *Marine Policy* 98–108.

8 Conclusion

This chapter has provided an overview of the key environmental approaches and tools that are being developed in order to reduce, minimise and avoid conflict between different uses and users of the seabed. It will be immediately apparent that a plethora of such approaches and tools exist, although not all are necessarily focused on precautionary and ecosystem-based management. In both areas under and areas beyond national jurisdiction certain key features of sound environmental management exist. First, it is necessary to know and understand what each sector is doing. Then the impacts that the activities of each sector can have or are likely to have on another sector must be assessed. Finally, cross-sectoral cooperation, aimed at ensuring each sector is interfered with only to the extent necessary and in a manner that does not compromise ecosystem integrity, is needed. This is not an easy task. However, the future orderly, peaceful and environmentally sustainable development and use of the seabed depends on it.

CHAPTER 23

Commercial Arrangements and Liability for Crossing Pipelines, Power Cables and Telecom Cables (Connectors) on the Seabed

Lars Olav Askheim

1 Introduction[1,2]

A direct consequence of the increasingly interconnected world and offshore energy production facilities – such as platforms producing petroleum and wind farms – is the growing number of crossings between subsea transport assets such as oil and gas pipelines, power cables of various voltages and telecom cables such as fiber-optic cables and any traditional telegraph and telephone cables. In particular in closed-basin seas such as the Mediterranean Sea, the Baltic Sea and the North Sea, the number of crossings has increased exponentially over the years.[3] As a general rule, each crossing is governed by an agreement, almost invariably called a 'crossing agreement', entered into by the owners.

This chapter focuses on offshore crossings, i.e. crossings occurring on the seabed outside the territorial seas. While the issues are much the same onshore and with a similar need for entering into crossing agreements, offshore agreements have some specificities. For example, the coastal state may have used its jurisdiction to affect the crossing arrangements, influencing the liability and compensation regime. Furthermore, there will most probably be additional types of pipelines, such as water and sewer pipelines on shore. The analysis

1 My colleague Mr. Knut Gjelsten, whose experience in this field exceeds my own, has kindly reviewed this article and provided numerous comments. His input is greatly appreciated, but as always, any errors and omissions are my responsibility.

2 The views expressed in this chapter are personal ones. They have not been approved by my employer Statnett SF, who as the Norwegian transmission system operator for electricity has many existing and future subsea power cables, which are subject to a substantial number of crossings.

3 By way of example, the NordLink power cable between Norway and Germany currently under construction, will have approximately 20 subsea crossings. The North Sea Link (NSL) power cable project between Norway and Great Britain, also under construction, will have approximately 30 subsea crossings.

 | DOI:10.1163/9789004391567_025

in this chapter is restricted to three types of 'connectors'[4] that are mentioned above: oil and gas pipelines, power cables and telecom cables.

Further clarifications in terms of terminology and scope are necessary. The 'intruding' installation will hereafter be referred to as the 'Crossing Connector', the 'incumbent' installation at the crossing point as the 'Affected Connector', and the owners as the 'Crossing Party' and the 'Affected Party' respectively. While the Affected Connector might be buried in the seabed, thereby not necessarily entailing any physical contact between it and the Crossing Connector, the situation will still be deemed a crossing. In a situation where the two connectors are adjacent to one other but without any actual crossing, the owners often enter into a 'proximity agreement' governing their relationship. This type of agreement will not be covered in this chapter. This chapter is based on experience from crossing agreements in the North Sea, most of which involve pipelines and power cables. There might be some regional specificities, including legal traditions, but the issues covered are universal and therefore relevant for other parts of the world. Finally, the chapter takes a practising-lawyer approach, focusing primarily on practical issues.

Project specificities constitute important starting points. The physical aspects of crossings are fascinating, particularly as work will be carried out on the seabed, possibly at a depth of hundreds of metres. The dark environment and substantial water pressure require the use of very specialized equipment, largely operated remotely from vessels on the surface. While telecom cables usually merely rest on the ocean floor, power cables are generally buried (below the seabed). For oil and gas pipelines, both approaches are adopted. Thus, the design of crossings will vary, depending on the types of connectors involved, their vulnerability and other factors. For each type of crossing scenario, fairly standardized solution concepts have been developed. If the Affected Connector has a free span (i.e. hanging unsupported in the sea) at the crossing point, the parties need to agree on a specific solution addressing the particular concerns free spans create.

This chapter starts by reviewing the relevant provisions under the United Nations Convention on Law of the Sea (UNCLOS) (Section 2). It continues by identifying the different interests of the parties involved (Section 3), before reviewing in detail the main features and provisions of crossing agreements

4 My choice of terminology is inspired by the term 'interconnector' used in the European Union (EU) regulatory framework for cross-border exchange of gas and power, designating gas pipelines and power cables between Member States (the Electricity Regulation and the Electricity Directive, as well as the Gas Regulation and the Gas Directive).

(Section 4). The next sections focus on the installation phase (Section 5), and importantly, the liability and indemnity regime applied by the crossing agreements (Section 6). The occurrence of 'new situations' after completion of the installation is envisaged (Section 7). The chapter ends with some concluding remarks (Section 8).

2 International Law Aspects: Relevant Provisions in UNCLOS

2.1 *Applicable Regime on the High Seas*

On the high seas (UNCLOS, Part VII), pursuant to Articles 87 and 112(1), all states have the right to lay cables and pipelines. This includes the right to lay new pipelines and cables that cross existing cables and pipelines, without the need for consent. The opposite alternative, in the view of this author my view, would give the Affected Party more power than necessary to adequately protect his interests. Further, the Crossing Party usually has very few practical alternatives to crossing the Affected Connector, implying that the Affected Party would have the power to veto beneficial projects. It would also be contrary to the principle of freedom of the seas if the Affected Party should be regarded as an incumbent with quasi-property rights to the seabed.

However, the Crossing Party will have to give due consideration to the Affected Connector, as provided for in Article 79(5), which also applies to the high seas (see Art. 112(2)):

> *When laying submarine cables or pipelines, States shall have due regard to cables or pipelines already in position. In particular, possibilities of repairing existing cables or pipelines shall not be prejudiced.*

Article 114 states some basic principles for liability:

> *Every State shall adopt the laws and regulations necessary to provide that, if persons subject to its jurisdiction who are the owners of a submarine cable or pipeline beneath the high seas, in laying or repairing that cable or pipeline, cause a break in or injury to another cable or pipeline, they shall bear the cost of the repairs.*

Several principles can be identified in this provision. The first principle is the one of strict liability, since there is no requirement for wilful misconduct or negligence. The second principle is that the provision is neutral between the

Crossing Party and the Affected Party, in the sense that it applies to both: if one of them causes damage to the other, it is liable. The third principle is that the provision pertains only to repair costs, which means that other losses, such as loss of revenue, are not protected.

2.2 *Applicable Regime in the Exclusive Economic Zone*

With respect to the exclusive economic zone (UNCLOS, Part V), the same rules for pipelines and cables apply as for the high seas (see Article 58 (1) and (2)). It must be noted that the interests of the coastal state have been given some protection in paragraph 3, however this does not seem very relevant in the context of crossing agreements:

> *In exercising their rights and performing their duties under this Convention in the exclusive economic zone, States shall have due regard to the rights and duties of the coastal State and shall comply with the laws and regulations adopted by the coastal State in accordance with the provisions of this Convention and other rules of international law in so far as they are not incompatible with this Part.*

2.3 *Applicable Regime on the Continental Shelf*

On the continental shelf (UNCLOS, Part VI), the liability regime established in Article 114 will still apply. However, there are significant changes with respect to the laying of pipelines and cables. First, the coastal state has broad jurisdiction over pipelines and cables placed in its territorial waters or used for the exploitation of the continental shelf (Article 79(5)). Second, while the principle of free laying of pipelines and cables is maintained for pipelines and cables crossing its continental shelf (Article 79(1)), the coastal state may protect its interests relating to exploitation of the continental shelf and the prevention of petroleum pollution (Article 79(2)), but may not impede laying or installation. Third, the course of the pipeline or cable is subject to consent from the coastal state (Article 79(3)).

3 The Interests of the Parties

Before going into the content of the crossing agreements, it is beneficial to review the interests of the parties.

The main concern of the Affected Party with respect to crossings is that the Affected Connector is covered for damage or exposure to a higher risk of damage resulting from the Crossing Connector or the installation work. Repair costs are substantial, particularly in terms of use of vessels and other marine operation costs. Repairs will consume considerable time, as they require comprehensive planning and the mobilization of specialized equipment/vessels, and may be heavily dependent on weather conditions, including seasonal 'weather windows'.

In addition to covering the repair costs, the Affected Party will also suffer a financial loss due to the Affected Connector being unavailable for its intended use. The amount of the loss will depend on the capacity of the Affected Connector and the economic value of that capacity. In comparison, the value of capacity in telecom cables will generally be substantially lower than the value of capacity in the connectors transporting energy (petroleum, power). The impact of the Affected Connector becoming unavailable will also depend on the redundancy, if any, in the networks with which the Affected Connector is affiliated. It would then be a question of whether the 'traffic' could be rerouted through other connectors. Due to the relatively high costs of subsea connectors, redundancy will tend to be much lower than for onshore connectors, so there will be a considerable exposure for the Affected Party, through loss of revenues, or compensation payable to purchasers of capacity. And even if the Affected Party succeeds in protecting himself against this, there will be a socio-economic loss. Finally, there may be a significant pollution risk relating to oil spillage from a damaged oil pipeline.

The interests of the Crossing Party mainly relate to the Crossing Connector. The costs and loss related to damage to it are much the same for the Affected Party as for the Affected Connector, as described in the previous paragraph, but the risk of causing damage to the Crossing Connector is obviously much smaller. However, the fact that the Affected Connector is situated in international waters does not mean that fundamental legal principles regarding the protection of property belonging to others do not apply. As the newcomer on the scene, the Crossing Party should respect the fact that the Affected Connector is already on site and will have to be given appropriate consideration. Furthermore, once the installation of the Crossing Connector has been completed, the Crossing Party will undoubtedly expect to benefit from the same crossing principles with respect to any subsequent newcomers that cross its own connector. Thus, the combination of those principles and obligations create a positive environment between the parties.

After the crossing has been completed, a new risk scenario emerges in the crossing area. For the Affected Party, access to the Affected Connector in the event of repair or maintenance work becomes more difficult, in particular directly underneath the Crossing Connector. For the Crossing Party, the existence of the Affected Connector will impede repair and maintenance work to some extent.

4 Crossing Agreements in a Nutshell

The parties to the crossing agreement will be the Affected Party and the Crossing Party, but several other companies will also have a direct interest in the content of the agreement, and particularly those affiliated with the Crossing Party. These interested third parties will typically be lenders, insurance providers, and the company that will carry out the installation of the Crossing Connector under contract to the Crossing Party.

4.1 *Main Features*

The main features and issues in a crossing agreement can be summarized as follows (issues that are not always addressed are marked with an asterisk):
- Identification of the crossing point;
- The crossing angle, i.e. the angle between the Crossing Connector and the Affected Connector at the crossing point;
- The Affected Party approves the crossing, either based on (i) an agreed technical solution in an appendix, or, more rarely, on (ii) a defined design process, the result being subject to final consent;
- Definition of a time slot for the installation;
- Steps before installation, i.e. sharing of data on the Affected Connector, design criteria and process, etc., notification prior to start of installation;
- The installation work itself, including representatives, notifications, standard of work, emergency procedures;
- Follow-up work, including preparation of as-installed documentation;
- Liability and indemnity, including basis for liability, extent of liability, liability caps;
- Insurance requirements*;
- Future repair and maintenance;
- Upgrading and reinvestment*;
- Crossing Connector projects with crossings*;
- Decommissioning*.

With respect to the contract period, this is often left open. It is assumed that the crossing agreement will apply as long as the crossing exists.

Naturally, there will also be 'boiler plate' type provisions dealing with issues such as confidentiality, waivers, amendments, notices, all of which are found in most commercial contracts.

4.2 *Model Agreements*

There is no universal, fixed format for crossing agreements. However, experience shows that it is fruitful to distinguish between different crossing scenarios, depending on the type of connector involved. As demonstrated in the table below, substantial efforts have been made in developing model agreements in the North Sea.

TABLE 23.1

		Affected Connector		
		Pipeline	**Telecom cable**	**Power cable**
New	**Pipeline**	IS / CS	?	? – ad hoc?
Connector	**Telecom cable**	?	CS	? – ad hoc?
.	**Power cable**	CS / based on IS	CS	? – ad hoc?

IS = Industry standard
CS = Company standard – might apply for the whole company or just for a single project
? = This author is unaware of any specific standards applying to this situation or whether standard from other crossing types are adopted.

In Great Britain and Norway, the industry associations (UK Oil & Gas Industry Association and Norwegian Oil and Gas) have produced model agreements for pipeline crossings.[5] Oil companies may also have established company standards. In the North Sea, the major telecom operators have produced their own company standard agreements; these are far from identical but are still quite well-aligned.

The company standards have tended to favour the interests of the Affected Party. They are generally drawn up by incumbents, some of whom own or operate a large number of connectors. However, the drafters of some of these standards have wanted to avoid a situation where each crossing has to be negotiated extensively. Given that the draft is frequently heavily in favour of the Affected Party, one wishes to avoid the Crossing Party disputing every deviation

5 For the UK, for Oil and Gas UK, see 'Pipeline Crossing Agreement & Proximity Agreement Pack October 2015' (OP115), available at: <http://oilandgasuk.co.uk/product/pipeline-crossing-agreement-proximity-agreement-pack-october-2015/>. For Norway, for Norwegian Oil and Gas, see 'Norwegian Oil and Gas recommended Model Agreements for use of Pipeline to pipeline, Tie-in, Crossing, Proximity, Confidentiality', available at: <https://www.norskoljeoggass.no/drift/standardkontrakter-og-modellavtaler/modellavtaler/>.

from the draft. Instead, they have prepared a model agreement that is not necessarily totally balanced but is largely within acceptable limits for the Crossing Party. With such a standard, it will be easier to insist that 'no variations are accepted' – or that variations may be acceptable but will require substantial internal processes over a lengthy and undetermined period of time.

The understanding of this author is that the development of industry standards in the petroleum sector has been based on the awareness that today's Crossing Party may become tomorrow's Affected Party in a new crossing, elsewhere at a later time. Thus, a more natural balance can be obtained in a longer run.

The UK standard pipeline crossing agreement is structured as follows:

- Main body of the agreement. It numbers approximately 20 pages, includes contractual conditions and signature pages. Interestingly, on some issues, the text provides alternative solutions.
- Schedule 1: Plan of the Route of the Crossing Pipeline
- Schedule 2:
  - Part (A) Scope of Work
  - Part (B) Plans, Specifications, Construction Procedures and Risk Assessments
  - Part (C) Specimen Composite As-built Survey Drawing
  - Part (D) Construction Operations and Location of the Representative
- Schedule 3: Facilities for the Representative Offshore

The Norwegian standard pipeline crossing agreement has a slightly different structure:

- Special Terms and conditions. This is a short, top-level document of 4 articles only. It contains consent to the crossing, list of the contract documents, a few basic definitions and signature page.
  - Article 1 – Definitions, Agreement Documents and Interpretation
  - Article 2 – Consent to Pipeline Crossing
  - Article 3 – Capacity of the Parties
  - Article 4 – Representatives and Notices
- Appendix A General Terms and Conditions for Pipeline Crossing (approximately 15 pages)
- Appendix B Planned Route of the Crossing Pipeline and Crossing Point(s)
- Appendix C Laying Operation

4.3 *The Parties' Motivation for Entering into Crossing Agreements*

As indicated above, the Crossing Party does not need consent from the Affected Party. More generally, there is no legal obligation to enter into a crossing agreement with regard to crossings in international waters.[6] In practice, most crossing situations entail the signing of a crossing agreement.

In the view of this author, the first major benefits from having a crossing agreement are clarity and predictability. The parties will know their legal situation, which is obviously an advantage for both. The Crossing Party can assess the legal risks relating to the crossing, and the Affected Party will have some influence over how and where the crossing shall take place. In addition, by having an agreement with a choice of law provisions, the parties avoid potentially difficult choice of law issues that may arise in the absence of an agreement. After all, *lex loci delicti*, which is the general principle for tort claims in private international law,[7] does not provide much guidance for crossings in international waters.

Secondly, a crossing agreement may establish solutions that differ from the ones provided in UNCLOS. A very important example in this respect is the introduction of maximum amounts of liability. There is no basis for such liability caps in UNCLOS, but there is a clear commercial need, not just for the Crossing Party, but also for his lenders and his insurance providers.[8]

A third advantage is that the crossing agreement can be made much more specific and detailed than the fairly short and general language in UNCLOS. The parties will still adhere to the principles found in UNCLOS, but they will define how these principles shall apply to the crossing at hand. The prime example of this is the often quite detailed body of provisions on cooperation before, during and after the construction work. These are much more operational and practically appropriate than the general language in Article 79(5) on the consideration to be given to the Affected Connector.

Closely related to this is the parties' opportunity to address issues that are outside the provisions of UNCLOS. The liability provisions of Article 114 only cover costs of repair to the Affected Connector and the Crossing Connector. The parties may also find it appropriate to provide solutions for damage to

6 There may be nationally based exceptions relating to owners of pipelines and cables which enter into territorial waters (and thus do not just cross the seabed outside these limits, or owners of pipelines and cables used for the exploitation of resources on the continental shelf.

7 The essence of this is that the tort claim is governed by the law of the place where the key elements in the tortious act occurred. There is substantial legal practice and theory here, but for reasons of space, I will not go further into these issues.

8 The construction of the New Connector will often be covered by project insurance (CAR = Construction All Risks) which will regularly include liability cover, including liability towards the Affected Party. For these reasons the providers of CAR insurance will attach great importance to the existence and contents of crossing agreements.

other assets belonging to them, damage to their personnel, as well as third party liability. Further, no insurance requirements exist in UNCLOS either.

All these advantages can be condensed into a single observation: the negotiations for a crossing agreement create an opportunity for the parties to find more effectual solutions than those that might have been adopted if the Crossing Party had simply proceeded on his own. The solutions can be based on the specifics of the crossing itself and the interests of the parties.

4.4 *Timing for Entering into a Crossing Agreement*

In projects for a new connector, it will be necessary at an early stage to clarify issues such as the type of crossings to be made, with appropriate details of the existing connectors and their owners. This will partly be a desktop exercise based on available information, supplemented by information from the owners of known or suspected connectors. Furthermore, data from the early seabed surveys will confirm the existence of a crossing situation and indicate any additional crossings. It may be a bit surprising, but not all existing connectors are properly documented, and some of them may no longer be known even to their owners.

At this stage, it is also beneficial to identify any 'dead' connectors, i.e. connectors that are no longer in use and for which there is no plan to make further use of them. The connector project will not aim to enter into crossing agreements for such connectors even if the historical owners were identified. Such connectors are usually disregarded for installation purposes and are regarded as having been abandoned by their owners, who thereby are deemed to have relinquished ownership.

When there is a real crossing situation, it is advantageous for a connector project to enter into crossing agreements at a fairly early stage, particularly if there are third parties with substantial interests in them. This is obviously the case when construction is to be financed on a project-finance basis. Under this approach, the crossing agreement will most likely only identify the crossing point, while the design of the crossing itself and the installation methods and procedures will be determined at a later stage.

However, in practice, many crossing agreements tend to be entered into fairly late. This allows for the involvement of the installation company selected by the Crossing Party and which will carry out much of the physical work. The insights and inputs from the installation company are obviously valuable for the successful and timely completion of the project. The detailed crossing design can be appended to the crossing agreement. It might also be difficult to convince the Affected Party to engage in quick negotiations if the laying of the crossing is several years in the future.

In some instances, there is a two-stage process, with an early quasi-agreement: once the contact has been negotiated between the parties, the Affected Party will issue a short 'letter of no objection'. This letter usually refers to the standard technical requirements of the Affected Party and its standard crossing agreements, both of which are attached to the letter. In the letter, the Affected Party states that, on the basis of compliance with those two documents, it has no objection to the crossing. It is unlikely, and not intended, that this letter is legally binding. However, it still serves a useful purpose by indicating the basis on which the Affected Party is ready to enter into a crossing agreement. It also provides some assurance to the Crossing Party and parties related to it, such as lenders and insurers.

5 Considerations Surrounding the Installation Phase

5.1 *Pre-installation Phase and Related Provisions*

The objective is here to facilitate the development of a good crossing solution, where the interests of the Affected Party are duly protected. Among the common topics covered are exchange of information about existing infrastructures and design of the crossing;

The content of the provisions concerning the pre-installation phase will depend on whether the parties have chosen an 'early' or 'late' crossing agreement, as discussed in Section 4.4. If the 'early' approach has been chosen, the Affected Party will need to provide information about the Affected Connector, usually covering existing documentation and without involving any liability for the correctness and completeness of the information submitted. Under the 'late' approach, the crossing agreement will usually only describe the detailed crossing design (prepared by the Crossing Party or a subcontractor, approved by the Affected Party), without any reference to information, if any, provided by the Affected Party.

The crossing agreement places the responsibility for the design of the crossing on the Crossing Party. This will also include installation methods and procedures. There is also almost invariably a requirement for a pre-installation survey to be carried out. The Crossing Party will have to produce appropriate drawings and progress schedules and may also be obliged to prepare risk assessments and/or other quality assurance/quality control documentation. The anchoring patterns for the installation vessel is a particular concern. The crossing agreement may contain references to 'good engineering practice' or similar expressions, although such requirements will usually apply even if they are not explicitly mentioned.

Following this, the Affected Party has the option of approving or disapproving the corpus of documents submitted. The Affected Party will be given a maximum time period for this, and it is usually stated that consent cannot be unreasonably withheld or delayed.

In some instances, the crossing solution entails that the Crossing Connector is installed underneath the Affected Connector. This usually requires cutting or lifting the Affected Connector so that the latter will be out of service for some time. The Affected Party might want to insist in the agreement that any work done directly on the Affected Connector be carried out by his contractor, or at least that there must be only one contractor commissioned to do all work, one who is qualified to perform work on both connectors.

To the extent approvals from public authorities are necessary, the Crossing Party will assume the task of obtaining them. In practice, it is usually stated that the party is obliged to ensure compliance with all applicable legislation.

The choice of contractor lies with the Crossing Party, but the crossing agreement might set some specific stipulations in terms of qualification and experience requirements.

5.2 *The Installation Work Itself*

The installation work is the obligation of the Crossing Party and is to be carried out at his cost and risk. Before installation starts, the Crossing Party must respect a minimum notification period.

The crossing agreement will often require, in a standard clause, that the Crossing Party conduct all installation work in compliance with general standards of care and reason. This concept can be expressed in various ways, such as 'good oil industry practice' or 'acting as reasonable and prudent operators' to give only two examples.

The installation of the Crossing Connector will be preceded by some preparatory works, including deepening the position of the Affected Connector in the seabed, as well as building trenches, foundations, ramps, and/or bridges for the Crossing Connector. When the Crossing Connector has been put in its proper place, the Crossing Party will have to carry out any protective measures agreed in advance, such as the placing of protective materials – mats, rock, other materials – over the Connectors.

Many crossing agreements will allow the Affected Party to have a representative on site during installation work. This representative is usually given observer status – expressly stated in the crossing agreement – and shall not be called upon to give approvals or act as a representative of the Affected Party. However, the representative may have the right to stop the installation work under certain circumstances, such as emergencies or risk of damage to the Affected Connector.

In the event the Affected Connector is damaged during installation, a situation arises for which it is difficult to make proper provisions in advance. The crossing agreement might specify that any repair work must be carried out by the Affected Connector or by a contractor chosen by him. The costs will be at the expense of the Crossing Party, as discussed in more detail in Section 6. The crossing agreement might also state that the protection and repair of the Affected Connector shall have priority over the installation work, if the Affected Party so requests. The Affected Party might be entitled to request assistance from the Crossing Party, for example to use the vessel and marine equipment the Crossing Party has on site. However, this cannot be strictly required. The suitability of the vessel, equipment and workmanship should fall under the provision stipulating appropriate tools and materials, and to a 'best efforts'-type of reservation.

The Crossing Party will be subject to reporting obligations on a cyclical and/or more or less current basis during the installation period.

5.3 *Follow-Up Activities*

After completion of the physical installation work, crossing agreements often provide for some additional obligations for the Crossing Party. While the content of those obligations will vary, the following issues seem to be most pertinent:

- An as-installed survey shall be carried out by the Crossing Party and made available to the Affected Party at no cost. Some crossing agreements introduce an approval mechanism here, implying that the Crossing Party may have to re-do the survey in full or in part. The purpose of the survey is both to document the current physical situation at the crossing point and in the area around it, and to verify that the Crossing Party has carried out the installation work in compliance with the terms of the crossing agreement.
- The as-installed survey will be included in the as-installed documentation, which shall also be provided to the Affected Party.
- Rectification of any deviations, whether identified through an as-installed survey or not, shall be carried out by the Crossing Party.
- A 'quarantine period' of some months will apply, during which all defects identified in the Affected Connector will be deemed to have been caused by the Crossing Party, entailing that any related financial loss will be borne by the Crossing Party. Such a period should be seen in conjunction with the liability and indemnity provisions of the crossing agreement, which will be reviewed below.

6 Liability and Indemnity

6.1 *Balancing Parties' Interests and Capabilities: General Matters*

The liability and indemnity provisions are often seen as the core of the crossing agreement in a legal perspective. They will address not only the direct tort liability of the parties between themselves, but also the extent to which a party can seek indemnification from the other party in the event of third-party claims.

Concerning liabilities between the parties, there are both the costs of repairing any damage to the Affected Connector and any follow-on losses such as loss of revenue, exposure under contracts with users of the Affected Connector, and in some cases also exposure under regulatory mechanisms relating to the unavailability of the Affected Connector.[9]

The number of potentially affected companies and individuals can be quite large. The first set of third parties is related to one of the parties in some manner, and includes inter alia affiliates of the parties, contractors and subcontractors, as along with employees of all these entities. These will be referred to as the 'Affected Group', and the equivalent group on the other side as the 'Crossing Group'.

The second set of third parties is occasionally referred to as 'true' third parties and is quite diverse. It can include owners of other facilities in the vicinity of the crossing point, shipowners, fishermen, users of the Affected Connector, etc. A special type of claims relates to pollution liability, which in practical terms is significant mainly for oil and gas pipelines.

The liability and indemnity provisions cannot be considered in isolation of the insurance arrangements. On a superficial level, the insurance arrangements can be seen as designed to cover the liabilities and indemnities assumed by a party in the crossing agreement. However, in the view of this author, it is more productive to see it the other way round: to the extent that an exposure is protected by insurance, it does not really matter which of the parties is responsible under the liability and indemnity provisions.[10] Consequently, the financial liability exposure should be assumed by the party who has the best access to insurance and can obtain the most favourable premiums and other conditions, including deductibles. The parties should attempt to avoid a situation where both carry insurance for the same potential event, as this is not efficient in terms of premiums, i.e. the Affected Party under his general

9 This exposure can arise for connectors that are part of an open access network with tariffs developed under public regulation.

10 The liability and indemnity provisions will still be relevant for any deductibles and exclusions from the insurance cover.

insurance, the Crossing Party under insurance relating to his project in general or to the crossing itself. 'Double insurance' can also create disputes as to which insurer shall pay.

Given UNCLOS Article 114 and the incumbent position of the Affected Party, it should not come as a surprise that in a crossing situation, as a general rule, the Crossing Party will carry the liability in the event of damage to the Affected Connector, howsoever caused. However, in practice this general rule is rarely applied to its full extent and exceptions are made. Based on the review of several crossing agreements, the impression given is that there is considerable variation on that point.

6.2 *Specific Liability and Indemnity Issues*

The first and most fundamental issue is the ***extent, in time and space***, of the special compensation regime the crossing agreements establish. Most of them relate this to the installation operation, either by simply looking at this as a period of time, or by requiring a link (with language like 'arising out of or in connection with' the work or the installation) between the installation and the damage.

The next issue is ***the damage the special compensation regime applies to.*** Damage to the Affected Connector is obviously included, but in practice the regime will also apply to other forms of damage to property and personnel in the belonging to the Affected Group, as well as claims from third parties.

Damage might not necessarily manifest itself during the installation period. It may also be that although the damage occurred during installation, it only comes to the attention of the parties at a later time. In order to give protection to the Affected Party in this respect, many crossing agreements establish what can be referred to as an '***extended liability period***'.

The effect of the extended liability period, in its purest form, is that damage discovered during the period will be deemed to relate to the installation, and thus be covered by the special regime, regardless of actual cause. In some instances, this is softened into a presumption that the Crossing Party can rebut if he can produce adequate evidence that the damage was not caused by him. In practice, this will mean that he will have to substantiate that a person specifically caused the damage, or that the damage is caused by a particular type of activity, for example trawling. The extended liability period can be quite long – 6 to 24 months is not unheard of. It can start when installation work is completed or can be linked to the delivery or approval of the as-installed documentation.

The next issue is the ***basis for liability***. In contract practice, the prevailing, sufficient basis for liability is that the Affected Connector is damaged. It is therefore not a necessary condition that the damage is caused by the Crossing

Party or any other individual or company for which he is responsible, such as contractors, affiliates or employees of such companies. However, in a few crossing agreements, causation is only a presumption that the Crossing Party may attempt to rebut, and the burden of proof will lie with him. It seems appropriate that the Crossing Party should be deemed liable in the event of inherent defects in the Affected Connector.

Is the principle of liability for the Crossing Party maintained also in the event that it can be substantiated that the damage was caused by the Affected Party or the Affected Group? This may be regarded as unreasonable, and many legal systems will not accept such a principle when there is serious misconduct. There are several ways to attempt to resolve this, and the proper solution depends on the governing law chosen by the parties. One element here is the gravity of the acts of the Affected Group, where an exception will be made in cases of wilful misconduct, possibly also extending to gross negligence. The other element is the identity of the persons in question: some crossing agreements limit the exclusion in this context to such actions taken at a management level.

It is customary to agree on a ***maximum amount of financial responsibility*** for the Crossing Party. The amount will obviously depend on several factors, including probability of damage, an expected range of loss and insurance costs. In Norway, in respect of crossings involving pipelines, a global cap amount of USD 100 million for the crossing is prevailing and is reflected in the industry standard. Still in Norway, the cap amount for crossings involving telecom cables tends to be significantly smaller, but here the cap often works at two levels: one cap for each incident, and a higher total cap.

Another important issue is whether ***indirect/consequential losses*** caused by damage to the Affected Pipeline are excluded. Here there is no generally prevailing solution. In the two oil industry standards referred to in Section 4.2, the Crossing Party is liable within the limits of the liability cap also for consequential damage. This is contrary to normal practice in most commercial contracts. For crossings involving telecom cables, the crossing agreements made available seem in general to take the opposite approach, excluding consequential loss.

Pollution liability, including both damages caused and clean-up costs, is particularly relevant for oil pipelines. Some crossing agreements involving oil pipelines exclude such liability from the special regime and will instead generally provide that financial losses due to spillage/leakage of oil from an oil pipeline, will be assumed by the owner of the pipeline in question. Exceptions might apply in the event of gross negligence or wilful misconduct.

7 Dealing with the New Situation in the Post-completion Phase

After completion of the installation, a new situation is created with two connectors in the same area and various protection arrangements in place. Both connectors are vulnerable with respect to the performance of work on the other connector. For the connector located underneath, there is an obvious additional challenge of obtaining access.

In the normal course of business, there should be little need for physical work in the crossing area, as the risks of defects in the connectors themselves are limited. Then, only a small part of the connector is located in the crossing area. Meanwhile, defects can occur, and one or both connectors could be damaged by outside forces. Some crossing agreements do not address this situation at all, which might be explained by a lack of foresight, deliberate omission after difficult negotiations, or just a business judgment that the risk is fairly small and can be assumed by the parties if it should arise. However, the consequences of defects can be far-reaching. Consequently, rules are needed that define the rights and obligations of the parties in such situations.

Most agreements do not distinguish the solutions depending on the parties, meaning that the provisions are symmetrical and that the Affected Party will have the same position as the Crossing Party. The Affected Party can thus not expect to avoid any subsequent costs deriving from the crossing merely due to the fact that he was the first in the area, even if his situation can be said to be worse than that of the Crossing Party. However, it may happen that the Affected Party obtains some concessions in the event of difficult access or temporary removal of the Crossing Connector, typically due to some particular circumstances.

An important element of the crossing agreements will relate to the physical delimitation of the area in which the post-completion provisions apply. This area is usually defined by a given radius around the crossing point. Outside that area, any obligation on the parties must be based on other legal grounds in the crossing agreement.

Some types of work do not involve actual physical contact with the seabed or the Connectors, such as surveys carried out by remotely operated vehicles (ROVs) (miniature, unmanned submarines controlled from the surface). The tendency is that each party can carry out intentional work, but potentially subject to a notification requirement.

Other types of work will involve physical presence, such as the use of anchors and other mooring equipment, as well as ploughs or other equipment affecting the seabed. For those physical works, several approaches are followed in practice. One approach is to make all such work conditional upon a separate agreement being entered into. Obviously, this approach will give the other party a strong negotiating position. An alternative approach is to retain the concept of consent but soften it by stating that consent shall not be unreasonably withheld. This language is not very precise, but it is hard to see that it is possible to improve it significantly. A third approach is to state that the parties shall enter into a new crossing agreement which must reflect the first crossing agreement. With this approach, the status of 'intruder' in the new crossing is determined independently of the status in the first crossing agreement. Thus, the Affected Party in the second crossing agreement could be the Crossing Party in the first crossing agreement. Regardless of the approach, it is convenient to have an exception for emergency situations, where there is insufficient time to obtain agreement or consent. The costs of such work are generally to be assumed by the party triggering the work, even if some of them relate to the connector of the other party.

Some crossing agreements foresee a particular need for changing the configuration of the crossing. For example, the Affected Connector which originally lies underneath the Crossing Connector may instead cross over it at a later stage. The contents of such change in crossing configuration clauses will vary. One approach is to modify the provisions on consent requirements to facilitate the changes but also clarify the cost consequences, e.g. that the Affected Party will assume all costs.

Finally, any work in the crossing area after completion of installation will generate liability and indemnity issues, governed either by the existing crossing agreement or in a new agreement to be entered into.

8 Concluding Remarks

Crossing agreements are commonly seen as practical arrangements, with limited interest from a commercial point of view. In companies with a significant portfolio of subsea connectors, crossing agreements tend to be dealt with by a small group of specialists who are highly experienced but with a strong preference for their established practices and model agreements/clauses.

For the reasons stated, agreeing on a crossing agreement can be a fairly long, but not necessarily a very complicated process. There is little general controversy and the industry shares some common practices along with the joint need to realise respective projects. However, if a Crossing Party faces many crossings in a project, this will create additional challenges in managing the entire process.

Crossing agreements constitute a fascinating example of the conflict between incumbents and intruders in relation to the use of the same seabed area. The basic question remains: how to strike the balance between the interests of the party who first moved into an open area and the interests of newcomers. UNCLOS provides the main principles, but as the review of contract practices has shown, many additional provisions are needed and different solutions must be found to solve possible conflicting situations. Therefore, there are significant benefits to be derived by both parties agreeing on a crossing agreement. Not least, the potential 'reversal of roles' in the post-completion phase regarding subsequent crossings/projects will make the parties more inclined to seek balanced solutions than if the crossing in question was seen as an isolated case.

CHAPTER 24

Balancing Competing Interests When Building Marine Energy Infrastructures: the Case of the Nord Stream Pipelines

David Langlet

1 Introduction

The age of large-scale hydrocarbon infrastructure projects may, partly as a consequence of climate change policy, be nearing its end. However, significant projects are still being planned and executed. Natural gas is also touted as a 'bridge' between more carbon intense coal-based energy production and carbon neutral, or almost neutral, renewable energy sources. This, together with new gas production technologies could further increase the need for gas transport infrastructure, significant parts of which are likely to be sea based.[1] Also, if carbon capture and storage (CCS) emerges as a large-scale climate change mitigation technology, which some see as imperative if climate change is to be tackled affectively,[2] that is likely to result in demand for submarine pipelines to transport carbon dioxide to offshore injection points.[3]

---

1 Offshore is estimated to account for almost two-thirds of the world's remaining conventional natural gas resources. International Energy Agency, Offshore Energy Outlook (OECD/IEA, 2018) 16.

2 See, inter alia, European Parliament Resolution of 14 January 2014 on Implementation Report 2013: Developing and Applying Carbon Capture and Storage Technology in Europe (2013/2079(INI)); and OECD/IEA, Technology Roadmap: Carbon Capture and Storage (OECD/IEA 2013) 5. However, there is no lack a critics of CCS. See e.g. E Rochon, 'False Hope: Why Carbon Capture and Storage Won't Save the Climate' (Greenpeace International, May 2008) <http://www.greenpeace.org/usa/research/false-hope-why-carbon-capture/> accessed 15 June 2014. The technology is associated with many problems, not least the lack of a viable business case for its employment in most jurisdictions. On the role of CCS and bioenergy with CCS (BECCS) in scenarios that would limit warming to 1.5°C above pre-industrial levels, see H de Coninck, A Revi, M Babiker, P Bertoldi, M Buckeridge, A Cartwright, W Dong, J Ford, S Fuss, J-C Hourcade, D Ley, R Mechler, P Newman, A Revokatova, S Schultz, L Steg, and T Sugiyama, 'Strengthening and Implementing the Global Response', in V Masson-Delmotte et al (eds), Global Warming of 1.5°C. An IPCC Special Report on the impacts of global warming of 1.5°C above pre-industrial levels and related global greenhouse gas emission pathways, in the

 DOI:10.1163/9789004391567_026

The construction and operation of submarine pipelines engage several parts of the law of the sea. Such activities can also affect many potentially competing interests, such as the freedom of transport, the utilization of resources in and on the seabed, protection of the marine environment, and access to secure energy supplies. With transit pipelines – i.e. those pipelines that pass over the continental shelf (CS) of one or more States without entering their territorial waters – potential conflicts between interests take on additional levels of complexity since the interests pertain to different States who may be quite differently affected by the pipelines. The States can also be in very different positions as to their ability to exercise control in relation to the pipelines and thereby influence the balancing of the interests concerned. If submarine energy pipelines are to be built and operated without unnecessarily interfering with other legitimate activities and interests, and vice versa, significant levels of coordination or cooperation between the affected States is often required.

The challenge of handling constructively and sustainably the many interests potentially affected by submarine pipelines is compounded by the relative vagueness of the applicable legal regime, not least with regard to the nature and extent of the jurisdiction that may be exercised by the States concerned.

This chapter aims to discuss, from an international- and partly EU law perspective, the challenges of managing conflicting interests associated with the use of the seabed for transport purposes. To do so it uses the Nord Stream pipeline project in the Baltic Sea, one of the most complex and contentious energy transport projects in Europe, as a case study. The focus is on issues pertaining to the physical presence and operation of the pipelines in a specific location and the interests that may prompt restrictions of the laying and operation of pipelines.

After a presentation of the Nord Stream pipeline project (Section 2), the chapter discusses which framework such sea based transport projects offer to manage competing interests, both in general terms (Section 3) and specifically

---

context of strengthening the global response to the threat of climate change, sustainable development, and efforts to eradicate poverty' (2018, in press).

3 In some regions offshore storage is the only or most abundant storage option while it in other cases may be preferred for policy reasons. Intergovernmental Panel on Climate Change (IPCC), IPCC Special Report on Carbon Dioxide Capture and Storage, prepared by Working Group II of the IPCC (Cambridge University Press, Cambridge, 2005) 223, 257 and S Teir, J Hetland, E Lindeberg, A Torvanger, K Buhr, T Koljonen, J Gode, K Onarheim, A Tjernshaugen, A Arasto, M Liljeberg, A Lehtili, L Kujanpaii and M Nieminen, Potential for Carbon Capture and Storage (CCS) in the Nordic Region (VTT Research Note 2556, 2010) 73.

in the case of Nord Stream (Section 4). The chapter ends with some concluding remarks (Section 5).

2 The Nord Stream Project

The Nord Stream project has been highly contentious and much debated. This has partly been due to the environmental risks associated with a large energy project being carried out in a unique and fragile marine area which is already subject to very high human pressures, e.g. in the form of eutrophying emissions and quite intense shipping activities. To a large extent, the debate has also been influenced by the mounting political tensions between Russia and several other European countries affected by the project.

The concerns raised by affected States vary from direct physical impacts on the environment or marine security, over economic implications of a sea-based gas transport route, to issues related to national security.[4] While acknowledging that this inevitably affected the political context in which the relevant legal provisions where construed and applied the main emphasis is here on the applicable law and on issues that may be of relevance also beyond the specific region and project.

Before engaging with the legal framework, the Nord Stream project as such will be introduced.

2.1 *Background, Facts and Figures*

The Nord Stream project in its current form comprises two 1,224 kilometres long parallel gas pipelines with the combined capacity to transport 55 billion cubic meters (bcm) of natural gas from Russia to Germany annually. The submarine pipelines originate near Vyborg in Russia, then pass through the exclusive economic zone (EEZ) and over the CS of Finland and Sweden, before entering the EEZ but also the territorial waters of Denmark, and finally landing in Lubmin, near Greifswald, in Germany.[5] The first of the two pipelines became operational in 2011, the second one about a year later. In 2017, the pipeline system

4 On the various objections raised and contentions associated with the project see B Solum Whist, 'Nord Stream: Not Just a Pipeline' FNI Report 15/2008 (2008) and D Langlet 'Nord Stream, the Environment and the Law: Disentangling a Multijurisdictional Energy Project' (2014) 59 ScStL 179.

5 On the notions 'exclusive economic zone' and 'continental shelf', see section 3.2.

operated at 93 per cent of its design capacity, delivering 51 billion cubic metres (bcm) of natural gas.[6]

Initial feasibility studies of different routing options were carried out in the late 1990s by North Transgas, a company whose major owners were the Russian OAO Gazprom (Gazprom) and the Finnish Fortum Oil and Gas Oy (Fortum). They found a submarine pipeline solution, similar to that eventually built, to be the most feasible option for connecting Russia's natural gas fields with the central European market.[7]

The pipelines were eventually built and are now operated by Nord Stream AG, a company founded in 2005 and currently owned by Gazprom, with 51 per cent of the shares, Wintershall Holding GmbH (a BASF subsidiary) and PEG Infrastruktur AG (an E.ON Beteiligungen subsidiary) with 15.5 per cent each; and N.V. Nederlandse Gasunie and ENGIE (formerly GDF SUEZ) with 9 per cent each.[8] Nord Stream AG is based in Zug, Switzerland, where also the control centre, from which the pipelines are monitored and operated, is situated.[9] Nord Stream AG is thus to be regarded as a national of Switzerland and subject to Swiss jurisdiction in accordance with the nationality principle.[10]

2.2 *Purposes and Developments*

While the purpose of the Nord Stream pipeline system is to transport gas from Russia to central and Western Europe the project is also very much about avoiding transporting gas via established land-based routes, thereby reducing the control of transit States, and completely avoiding certain States becoming transit States. As Nord Stream AG has itself explained, the Nord Stream pipeline system offers a natural gas connection 'free from non-technical risks

---

6 'Nord Stream Reaches Average Utilisation of 93% in 2017–51 bcm delivered to the European Union' (Nord Stream, press release 16 January 2018) < https://www.nord-stream.com/press-info/press-releases/nord-stream-reaches-average-utilisation-of-93-in-2017–51-bcm-delivered-to-the-european-union-500/> accessed 23 June 2019.

7 T Koivurova and I Pölönen, 'Transboundary Environmental Impact Assessment in the Case of the Baltic Sea Gas Pipeline' (2010) 25 IJMCL 151, 156.

8 'Who We Are' <www.nord-stream.com/about-us/> accessed 17 August 2016. The company was originally called 'North European Gas Pipeline Company' and had a slightly different ownership. 'Nord Stream Environmental Impact Assessment Documentation for Consultation under the Espoo Convention', Nord Stream Espoo Report, February 2009, Volume II: Chapter 1–8, 21.

9 'The Nord Stream Pipeline Project', Fact Sheet, February 2013.

10 On this matter, see further D Langlet, 'Transboundary Transit Pipelines: Reflections on the Balancing of Rights and Interests in Light of the Nord Stream Project' (2014) 63 ICLQ 977, 980.

and free of interference of a commercial or non-commercial nature by third parties'.[11]

As has been discussed elsewhere, Nord Stream has met with much opposition, not least from States such as Poland, which view it as a way to circumvent them as transit States.[12] However, this dimension of the Nord stream project is only very indirectly linked to competing uses of the seabed and will not be further discussed here.[13]

In 2015 Gazprom, BASF, E.ON, ENGIE, OMV and Royal Dutch Shell signed a Shareholders' Agreement on implementation of a 'Nord Stream 2' pipeline project to be developed by a new company.[14] Like the original Nord Stream, the Nord Stream 2 project comprises the construction of two offshore pipelines with an aggregate annual capacity of 55 bcm of gas to be transported from Russia to Germany through the Baltic Sea, largely along the same route as the two existing pipelines. The new company is, like Nord Stream AG, established in Switzerland, and continues the preparatory and planning activities initiated by Nord Stream AG.

In August 2016 the consortium behind Nord Stream 2 crumbled after a decision by the Polish anti-trust office not to approve the notification in Poland of a joint venture to construct and operate the new pipelines. The approval for a Polish joint venture was needed despite the planned pipelines not entering Polish waters because of the EU-based partners in the consortium being active in Poland. The notification was declined with reference to Nord Stream 2 restricting competition in gas supplies.[15] This lead to the new company, Nord

11 'Nord Stream Extension Project Information Document (PID)', March 2013, Document No. N-GE-PER-REP-000-PID00000-A, 15.

12 The then Polish defence minister Sikorski even said in 2006 that the project echoed the 1939 Molotov-Ribbentrop Pact (in which the territories of several European countries, among them Poland, were divided into Nazi-German and Soviet spheres of influence just before the outbreak of WWII), Nord Stream 'a waste of money' says Poland (EURACTIV, 11 January 2010, updated 31 August 2011) <https://www.euractiv.com/section/central-europe/news/nord-stream-a-waste-of-money-says-poland/> accessed 28 May 2019.

13 See instead Langlet (n 4) and Solum Whist (n 4).

14 Gazprom export, 'Nord Stream 2' at <http://www.gazpromexport.ru/en/projects/5/> accessed 28 May 2019.

15 A Rettman, 'Russia to build Nord Stream 2 despite Polish objection' (Euroserver, 22 August 2016) <https://euobserver.com/economic/134694> and 'Nord Stream 2 partners withdraw amid Poland pressure' (Financial Times, 12 August 2016) <http://www.ft.com/fastft/2016/08/12/nord-stream-2-partners-withdraw-amid-poland-pressure/> both accessed 19 May 2019.

Stream 2 AG, being wholly owned by Gazprom but financing agreements for the project have been signed with ENGIE, OMV, Shell, Uniper and Wintershall.[16]

The transit States Finland and Sweden granted permission for laying of the new pipelines in April and June 2018, respectively.[17] As of June 2019, the third transit State, Denmark, had yet to make a final decision on Nord Stream 2 AG's permit application. The legal developments in Denmark regarding Nord Stream 2 are further discussed in Section 4 below.

2.3 *Legal Status*

As concluded above, both Nord Stream AG and Nord Stream 2 AG (in the following referred to collectively as 'Nord Stream' when no distinction is necessary) are Swiss companies and are with respect to the application of international law to be regarded as nationals of Switzerland and subject to Swiss jurisdiction.[18] As a consequence of State sovereignty and according to the maxim *pacta tertiis* – the meaning of which is that no obligations can follow from an international agreement for non-parties to such agreement without the consent of the non-party in question – any obligation imposed on Nord Stream should be compatible with the legal position of Switzerland under international law.[19] Switzerland became a party to the United Nations Convention on the Law of the Sea (UNCLOS)[20] in May 2009, just a few months before Denmark, Finland and Sweden decided on Nord Stream AG's permit applications regarding routes on their respective CS. However, Switzerland is not a member of the European Union (EU) and hence not subject to EU law obligations. Customary international law and international agreements to which the EU is party, such as UNCLOS, have an elevated position in the EU legal system.[21] But this does not rule out the possibility of conflicts with

16 Nord Stream 2, 'Shareholder and Financial Investors' at <https://www.nord-stream2.com/company/shareholder-and-financial-investors/> accessed 28 May 2019.

17 'Nord Stream 2 Receives Full Set of Permits in Finland' (Nord Stream 2, press release, 12 April 2018) at < https://www.nord-stream2.com/media-info/news-events/nord-stream-2-receives-full-set-of-permits-in-finland-92/> accessed 28 May 2019; 'Decision on application from Nord Stream 2 AG' (Swedish government, press release, 7 June 2018) at <https://www.government.se/press-releases/2018/06/decision-on-application-from-nord-stream-2-ag/> accessed 28 May 2019.

18 See father Langlet (n 10) 980.

19 This customary principle has been codified in the Vienna Convention on the Law of Treaties (Vienna, 23 May 1969) 1155 UNTS 331 arts 34–36.

20 UN Convention on the Law of the Sea (Montego Bay, 10 December 1982) (UNCLOS).

21 For a succinct account of the relationship between EU law and public international law, particularly in the field of environmental protection, see D Langlet and S Mahmoudi, *EU Environmental Law and Policy* (Oxford University Press 2016) 124 et seq.

respect to the application of EU standards to non-EU nationals, such as Nord Stream AG.[22]

However, the potential imposition of an obligation on Nord Stream, as a private legal subject, that is not consistent with international law, would not be a violation of any right of Nord Stream's, but rather a transgression of Switzerland's rights under international law and could justify the exercise by Switzerland of diplomatic protection (*ius protectionis*) with respect to the company.[23]

3 Framework for Managing Competing Interests

3.1 *Mapping of Interests*

There are two sets of interests, widely construed, that are directly linked to the pipelines, their routing and operation. One is the general interest of the concerned States to exercise authority, understood as legislative and executive jurisdiction, over activities on their CS and in their EEZ which may somehow affect them. Potential conflicts are thus a matter of competing claims of jurisdiction in relation to the pipelines and their operation.

The second set of interests covers substantive interests, such as protection of the environment, unimpeded access to natural resources in or on the seabed, and the freedom to lay and operate pipelines for economic or other purposes. These sets of interests are functionally linked since whoever gets to exercise legal authority in a certain case to some extent thereby becomes the arbiter between competing substantive interests. At the same time, this competence is contingent on the specific circumstances since the right to exercise authority, understood as competence to regulate and enforce rules and decisions, over a certain area (i.e. maritime zone) varies depending on what substantive interests are at issue. Although a distinction between jurisdictional interests and substantive interests may have an analytical value they are, from a legal perspective, often indissociably linked.

---

22 For a comprehensive analysis of the relationship between EU law and international law in the related field of maritime safety, see H Ringbom, *The EU Maritime Safety Policy and International Law* (Brill – Nijhoff, 2008).

23 On diplomatic protection, see further The International Law Commission, ILC's Draft Articles on Diplomatic Protection', Official Records of the General Assembly, Sixty-first Session, Supplement No. 10 (A/61/10) art 1.

3.2 *General Rules on Jurisdiction*

One of the main purposes of UNCLOS is clearly to allocate competencies in relation to areas and activities between different actors, primarily States acting in different capacities such as flag State, coastal State, or port State. This chapter focuses on those rules most directly relevant to the use of the seabed on the CS for energy purposes, particularly the laying and operation of submarine pipelines.[24]

The maritime zones of interest here are the EEZ and the CS, and to a lesser extent the high seas. The territorial sea is in this respect less interesting since in this zone the coastal State has virtually unrestricted jurisdiction with respect to energy infrastructure.[25] There is thus little room for competing claims regarding the construction and operation of, for example, pipelines in this area.

While there is no high seas – i.e. parts of the sea not included in the EEZ, the territorial sea, or the internal waters of a State[26] – in the Baltic Sea, it is still relevant to note that the so-called freedom of the high seas comprises not only a freedom of navigation and of overflight but also, inter alia, the freedom to lay submarine pipelines and cables.[27] And, as we shall see, the freedoms of the high seas apply also, although subject to additional restrictions, in the EEZ.

---

24 For a general presentation of the jurisdictional system set out in UNCLOS, see e.g. D R Rothwell, A G Oude Elferink, K N Scott, and T Stephens (eds), *The Oxford Handbook of the Law of the Sea* (Oxford University Press 2015); and Y Tanaka, *The International Law of the Sea* (2nd edn, Cambridge University Press 2015).

25 Where geography allows, the coastal State may claim a territorial sea that stretches 12 nm from the baseline. Here the coastal State in most respects exercises sovereignty in a way similar to what applies on its land territory. UNCLOS (n 20) arts 2–3. However, as the Danish example shows, limitations on the ability to exercise this right can follow from domestic law. Until 2017 Danish law lacked a provision enabling the restriction of the laying of pipelines within the territorial waters based on for example foreign policy considerations. In December that year, an executive order was issued to the effect that approval by the Foreign Minister is required for the granting of a permit for the laying of pipelines within the Danish territorial sea. Bekendtgørelse om visse rørledningsanlæg på søterritoriet og kontinentalsoklen, BEK nr 1520 af 15/12/2017 (Executive order on certain pipelines in the territorial sea and on the continental shelf). Although drafted so as to be of general applicability, this legislative action was clearly aimed at the ongoing Nord Stream 2 process.

26 UNCLOS (n 20), art 86. To be exact, the archipelagic waters of an archipelagic State are also not part of the high seas.

27 UNCLOS (n 20) art 87(1). Additional freedoms are the freedom to construct artificial islands and other installations permitted under international law, subject to UNCLOS (n 20) Part VI; the freedom of fishing, subject to the conditions laid down in UNCLOS Part VII, section 2; and the freedom of scientific research subject to UNCLOS Parts VI and XIII.

Pipelines are normally laid on the seabed, which, at least within 200 nautical miles from the baseline,[28] forms part of the CS.[29] The CS does not have to be claimed by the coastal State but exists *ipso facto*.[30] The Baltic Sea is not wide enough for there to be any seabed beyond the CS. Also beyond this region energy transport infrastructure is for practical reasons mostly restricted to the CS.

Without affecting the legal status of the superjacent waters, the coastal State exercises sovereign rights over the CS for the purpose of exploring it and exploiting its natural resources.[31] This means, inter alia, that all extraction of oil and gas, as well as utilization of living organisms belonging to sedentary species,[32] are the exclusive prerogatives of the coastal State.

In the Baltic Sea there is no CS not covered by a water column constituting the EEZ of the pertinent coastal State. Unlike the CS, the EEZ has to be claimed by the coastal State but all such States affected by the laying of the Nord Stream pipelines have, like most coastal States, claimed an EEZ.

In many respects the legal regime for the EEZ supplements that for the CS.[33] In the EEZ the coastal State enjoys sovereign rights for the purpose of exploring and exploiting, conserving and managing the living as well as non-living natural resources. This applies to the waters superjacent to the seabed as well as to the seabed and its subsoil. The coastal State also has sovereign rights with regard to other activities for the economic exploration and exploitation of the zone.[34] Furthermore, the coastal State has jurisdiction, as provided for in relevant provisions of UNCLOS, with regard to the protection and preservation of the marine environment, marine scientific research, and the establishment and use of artificial islands, installations and structures.[35] Pipelines are not, however, installations or structures in this regard but are subject to their own regulatory structure.[36]

---

28 Baselines are defined in UNCLOS (n 20), arts 5 and 7.

29 UNCLOS (n 20) art 76 (1).

30 UNCLOS (n 20) art 77(3).

31 UNCLOS (n 20) art 78.

32 Sedentary species are those organisms which, at the harvestable stage, either are immobile on or under the seabed or are unable to move except in constant physical contact with the seabed or the subsoil. UNCLOS (n 20) art 77.

33 The relationship between the two regimes is discussed in section 4.3.

34 UNCLOS (n 20) art 56(1)(a).

35 UNCLOS (n 20) art 56(1)(b).

36 R Lagoni, 'Pipelines' in R Wolfrum (ed), *Max Planck Encyclopedia of Public International Law* (e-resource, Oxford University Press 2008, updated April 2011) para 10.

3.3 *Jurisdiction with Respect to Pipelines*

Turning now to the specific legal conditions applying to submarine pipelines it should first be noted that the regulation of pipelines in UNCLOS is characterized by a delicate, and at times rather vague, balancing of the interests of different actors, primarily coastal States and those wanting to lay and operate pipelines, and between objectives, including protection of the marine environment and the right to lay and operate pipelines as part of the freedom of the seas. However, this is far from specific to pipelines. Rather, UNCLOS is replete with provisions that balance potentially competing interests.[37]

As a point of departure, all States, and indirectly their citizens,[38] enjoy, with some exceptions, the freedom of the high seas in the EEZ. This freedom comprises, inter alia, freedom of navigation and overflight and of the laying of submarine cables and pipelines, but also other internationally lawful uses of the sea related to these freedoms, such as those associated with the operation of ships, aircraft and submarine cables and pipelines. The activities carried out under this freedom must be compatible with other relevant provisions of UNCLOS.[39] Also, with respect to the seabed and subsoil the coastal State's sovereign rights and jurisdiction are to be exercised in accordance with Part VI of UNCLOS concerning the rules on the CS.

According to UNCLOS Part VI, all States are entitled to lay submarine cables and pipelines on the CS. However, whereas the coastal State may not otherwise impede the laying or maintenance of such cables or pipelines on the CS it has a right to take 'reasonable measures for the exploration of the continental shelf, the exploitation of its natural resources and the prevention, reduction and control of pollution from pipelines'.[40] This also means that UNCLOS does not allow for restrictions on the laying or operation of submarine pipelines for other reasons, such as security or energy policy considerations.[41]

37 The most explicit reference to such weighing up of interests is perhaps found in art 59 but it is also implicit in many other provisions or combinations of provisions.

38 The freedom pertains to States, not individuals. But in practice the activities covered by the freedom of the high seas are overwhelmingly exercised by private parties. W Wiese, *Grenzüberschreitende Landrohrleitungen und Seeverlegte Rohrleitungen im Völkerrecht* (Duncker & Humblot GmbH 1997) 210.

39 UNCLOS (n 20) art 58(1). When exercising these freedoms in the EEZ other States must also comply with the laws and regulations adopted by the coastal State in accordance with the provisions of UNCLOS and other rules of international law in so far as they are not incompatible with the provisions on the EEZ in UNCLOS. ibid art 58(3).

40 UNCLOS (n 20) art 79(1) and (2).

41 Koivurova and Pölönen (7) 179. However, in the Nord Stream case 'the security of energy supply' was listed in the Finnish permit decision as one of the impacts of the project

As regards protection of the marine environment, coastal States not only have a right to control the laying of pipelines on their CS. They are also, like other States, under a general obligation to take 'all measures necessary to ensure that activities under their jurisdiction or control are so conducted as not to cause damage by pollution to other States and their environment'.[42] They are also subject to a more specific requirement to adopt laws and regulations to prevent, reduce and control pollution of the marine environment arising from or in connection with seabed activities under their jurisdiction.[43] To the extent that for example the design or routing of a pipeline is subject to the coastal State's control, it thus has a corresponding obligation to take reasonable measures to prevent the pipeline from causing pollution.

Whereas the requirement that measures must be 'reasonable' is inherently vague it is clear that the assessment by the coastal State of what is reasonable must be carried out in good faith.[44] However, nothing precludes in principle the imposition by the coastal State of conditions that make the laying of a pipeline economically or technically unviable as long as those conditions are genuinely prompted by and needed for the protection of one of the interests recognized by UNCLOS. Whether these conditions are met can be a contentious matter but anyone seeking to legally challenge an allegedly unjustified restriction imposed by the coastal State will have the burden of showing at least a prima facie breach of that State's international obligations.

Although it is not for the coastal State to question the motive or need for a pipeline, it is wholly conceivable that the importance of the pipeline to other States than the coastal State could become a factor in the assessment of whether a measure by the coastal State is reasonable.[45]

---

that were to be weighed up in the assessment. 'Consent to Exploit Finland's Exclusive Economic Zone' (5 November 2009) 678/601/2009 (Unofficial translation provided by the Ministry of Employment and the Economy) 28. But from this cannot be inferred that the Finnish government was of the view that a negative impact in this regard would in itself justify a rejection of the application. In the end, no negative impact on Finnish security of energy supply was identified.

42 UNCLOS (n 20) art 194(2). There are also more specific obligations imposed upon coastal States in this regard both by UNCLOS and numerous regional conventions. See e.g., UNCLOS art 208 and the Convention on the Protection of the Marine Environment of the Baltic Sea Area, 'Helsinki Convention' (Helsinki, 9 April 1992) 1507 UNTS 167.

43 UNCLOS (n 20) art 208.

44 This follows generally from the Vienna Convention on the Law of Treaties (n 19) art 26, but also from UNCLOS (n 20) art 300.

45 For a more extensive analysis of the meaning of 'reasonable' in this context, see Langlet (n 10) 990–3.

Another important right also pertains to the coastal State: the delineation of the course for the laying of submarine pipelines on the CS is subject to its consent.[46] It is thus not the laying as such that requires consent but the pipeline's delineation. However, for practical purposes the delineation and the laying are indissociable since any laying entails a particular delineation. This consent requirement also gives the coastal State reason to put in place a permit procedure which can also be used to decide on reasonable measures to protect legally recognized interests.

Importantly, the waters above the CS remain governed solely by the EEZ-regime. This means that structures above the seabed, such as the maintenance platform initially planned for the original Nord Stream pipelines, are subject to additional jurisdictional rights of the coastal State. Under the regime for the EEZ the coastal State has, as previously noted, sovereign rights inter alia with regard to all activities for the economic exploration and exploitation of the zone and, subject to other provisions in UNCLOS, jurisdiction with regard to the establishment and use of installations and structures.

In sum, a coastal State has rather far-reaching authority to influence and restrict the laying of submarine pipelines on its CS. However, the drafters of UNCLOS clearly did not intend to grant the coastal State a *carte blanche* for prohibiting pipelines on its CS. But the vagueness of the restrictions imposed on the coastal State in this regard does invite discrepant interpretations and potentially conflict.

In the Nord Stream case, no affected State has legally challenged any decision regarding the laying, delineation, or operation of the pipelines on the CS and the Nord Stream companies have only appealed one decision by a national authority. This is not too surprising considering that all affected States granted their consent to the two existing pipelines, although with various conditions,[47] and that two out of three transit states have also consented to Nord Stream 2. But as will be seen in the following section this does not mean that the Nord Stream project was not affected by different assessments or competing views on the weighing up of interests, sometimes resulting in legal action being taken.

46 UNCLOS (n 20) art 79(3).

47 On some of these conditions, see section 5. For a more extensive analysis, see Langlet (n 4).

4 The Impact and Handling of Competing Interests in the Nord Stream Case

The legal authority to regulate the laying and operation of submarine pipelines on the CS, and thereby act as arbiter between different competing interests, is affected by a delicate balance between on the one hand the coastal State's interest to control activities on its CS, and the related obligation to take measures for the protection of the marine environment, and on the other hand the interest of transport and communication as manifested in the right to lay submarine cables and pipelines on the CS of any State.

The limited and rather vaguely defined right of the coastal State to restrict pipeline operations is potentially problematic for a number of reasons. One fairly obvious risk is that coastal States will be tempted to circumvent the restrictions on their right to regulate pipelines by using a legitimate objective, such as protection of the marine environment, as a pretext for pursuing objectives not recognised by UNCLOS. The suspicion that coastal States will act in this manner could also make (prospective) pipeline operators, and potentially their home States, sceptical towards the environmental concerns raised by coastal States. On the other hand, legally extraneous factors, such as political pressure felt by small countries when big States invest heavily in the pursuance of a large energy project, could conceivably lead to less than diligent assessment and consideration of effects to the marine environment. So, to what extent did these risks materialize in the Nord Stream case? And was the legal framework appropriate for managing conflicting interests in a constructive manner?

Highly relevant in terms of potential conflicts is that applications by Nord Stream AG to survey the seabed in preparation for a potential routing through Estonian waters were rejected by the Estonian government in 2007 and 2012.[48] Since surveying the seabed is an indispensable part of the planning and

48 'The Government did not agree to issue a permit for the survey application' (Government Communication Unit, 20 September 2007) <valitsus.ee/et/uudised/pressiteated/ keskkonnaministeerium/13572> accessed 15 August 2013; and 'Cabinet meeting decides to deny Nord Stream AG's request to conduct marine investigations in Estonia's exclusive economic zone' (Government Communication Unit, 6 December 2012) <http://valitsus.ee/et/uudised/pressiteated/majandus-ja-kommunikatsiooniministeerium/73717> accessed 15 August 2013. On the Estonian opposition to the Nord Stream project, see Solum Whist (n 4) 71; R Götz, 'The Nord Stream Pipeline: The Energy Policy Background' (2009) 52 GYIL 233; and S Vinogradov, 'Challenges of Nord Stream: Streamlining International Legal Frameworks and Regimes for Submarine Pipelines' (2009) 52 GYIL 30.

construction of a submarine pipeline this meant that the Estonian CS was unavailable to Nord Stream AG.[49] The primary reasons cited by Estonia were not environmental but national interests in the EEZ and that surveys would give information about Estonia's natural resources and their possible use.[50] Estonia seemed to view the intended surveys as marine scientific research in the terminology of UNCLOS and thereby subject to coastal State consent.[51] Although that classification, and thereby the coastal State's right to restrict surveys on this ground, could be disputed it is nonetheless clear that the surveys, which were to involve drilling in the seabed, did require consent by the coastal State.[52] It is namely the exclusive prerogative of the coastal State to authorize drilling in the seabed and there are no circumstances in which it is explicitly required to give such consent.[53] There would thus hardly have been any prospect for successfully challenging the Estonian decision, had any affected actor desired to do so.

A somewhat similar development was seen with respect to the maintenance platform that Nord Stream initially planned to build in the Swedish EEZ. In this case no application was ever rejected but Nord Stream withdrew the platform application in 2008 and opted instead for another technical maintenance solution, based on deployment of so-called intelligent pipeline inspection gauges, or 'pigs'. The decision to withdraw the application, which according to the company was made possible by technical progress, came in response to a very critical debate in Sweden regarding perceived environmental, fisheries and security concerns relating to the platform and to some extent the pipelines as such.[54] Considering that the platform would have been governed by

49 The main purpose of such surveys is to identify a suitable pipeline route, i.e. a route that will minimize the risk for harm to the future pipeline, and indirectly harm to the environment and human activities in the vicinity of the pipeline, and avoid conflicts with other uses of the seabed. R Lagoni, 'Cable and Pipeline Surveys at Sea', in HP Hestermeyer and others (eds), *Coexistence, Cooperation and Solidarity: Liber Amicorum Rüdiger Wolfrum*, Vol. 1 (Brill 2012) 933, 933.

50 'The Government did not agree ...' (n 48).

51 On the legal preconditions for marine scientific research, see UNCLOS (n 20) Part XIII; M Gorina Ysern, *An International Regime for Marine Scientific Research* (Brill-Nijhoff 2004), and more specifically in relation to pipeline surveys, Langlet (n 10) 986.

52 On the role and legal status of seabed surveys in preparation for the laying of submarine pipelines, see Langlet (n 10) 986; Vinogradov (n 49) 284; and S Wolf, *Unterseeische Rohrleitungen und Meeresumweltschutz* (Springer 2011) 86.

53 UNCLOS (n 20) art 81.

54 'Maintenance of Nord Stream Pipelines Feasible without a Service Platform' (Press Release, 8 April 2008) < http://www.nord-stream.com/press-info/press-releases/maintenance-of-nord-stream-pipelines-feasible-without-a-service-platform-130/ > accessed 19 May 2019.

the EEZ regime, it is fairly clear that Nord Stream could not have been successful in legally challenging a potential rejection of its application by the Swedish authorities.

Whereas no part of the Nord Stream project has been challenged directly under international law, there has been some appeals to domestic courts and other appellate bodies, the most significant of which perhaps being an appeal by Nord Stream 2 AG of a decision by the Danish Energy Agency. In April 2017 the company behind Nord Stream 2 submitted an application for the laying of two pipelines through the Danish territorial sea South of the island of Bornholm, in parallel with the two existing Nord Stream pipelines. In August 2018 it submitted another application, this time for an alternative route North of Bornholm that would pass through the Danish EEZ outside of territorial waters.[55] The option of laying the pipelines beyond the territorial sea south of Bornholm was not available due to the relevant area being subject to a legal dispute between Denmark and Poland. However, in November 2018 the two countries reached an agreement, subject to ratification, on the delimitation of their respective EEZ and CS in the area south of Bornholm, thereby making it possible for Denmark to authorise the laying of pipelines on the CS in this area.[56] In early 2019 Nord Stream 2 AG was informed that it would be required to submit an application, including an environmental impact assessment, for such a routing so as to enable the Danish authorities to select the routing with the least negative impact on the marine environment and marine safety. Once a formal decision to request the assessment of a new route had been made in March 2019, Nord Stream 2 AG appealed to the Danish Energy Board of Appeal, arguing inter alia that the Danish Energy Agency lacked the right to request the full investigation of a new routing option at such a late stage in the process.[57] However, despite its appeal, the company submitted an application, with an attendant impact assessment, for a southern route through the Danish EEZ in April the same year.[58] At the same time, it argued in its appeal that any

55 Nord Stream 2, 'Permitting Process in Denmark' < https://www.nord-stream2.com/permitting-denmark/danish-permitting-process/> accessed 26 June 2019.

56 'Polen og Danmark indgår aftale om den maritime grænse i Østersøen' (Poland and Denmark reach agreement on maritime delimitation in the Baltic Sea), Ministry of foreign affairs of Denmark, press statement, 1 November 2018.

57 Klage over Energistyrelsens afgørelse af 26. marts 2019 om en sydlig rute på kontinentalsoklen (Appeal of the decision of 26 March 2019 by the Danish Energy Agency concerning a southern route on the continental shelf), 17 April 2019, Doc no 21148344.1.

58 Reportedly, the companies view was that it was 'forced to submit this third application as a mitigation measure'. Reuters, 'Nord Stream 2 says Denmark tries to delay pipeline as it seeks third route option', 15 April 2019, < https://www.reuters.com/article/us-gazprom

environmental advantage of such a route would be too marginal to justify the additional costs and delays suffered by the company. At the time of writing, the Danish Energy Board of Appeal has not decided on the matter. Depending on the outcome this could become a case in which environmental reasons have been used, and accepted by a domestic legal system, to at least partly further other interests, namely to delay the need for making a final decision on what has become a highly controversial issue in Denmark. A challenging circumstance for Denmark is of course that is has already found the laying of gas pipelines, i.e. the two existing Nord Stream pipelines, through its territorial sea south of Bornholm not to entail significant environmental harm.

During the process for the original Nord Stream pipelines the Danish permit was appealed by an Estonian NGO claiming, inter alia, that the pipelines would harm the integrity of internationally protected areas, such as Natura 2000 sites,[59] and areas protected under the Ramsar Convention.[60] However, the Energy Board of Appeal found neither evidence nor probability of such harm and upheld the original permit decision.[61]

The construction permit for the Finnish part of the original Nord Stream pipelines was appealed by Estonian environmental non-governmental organizations (NGOs). They asserted, inter alia, that the permit was based on insufficient and incorrect information and called for an independent expert assessment. They were granted standing but the permit was upheld by the Finnish Supreme Administrative Court.[62]

Neither of these are internationally authoritative assessments of the concerned States' application of international and EU law obligations pertaining to protection of the environment. They do, however, speak against the idea that the coastal States in question to some extent disregarded their obligations with respect to the marine environment as a result of the political context. The same can be said for the fact that the Swedish authorities rejected the documentation initially provided by Nord Stream AG and required a detailed

-nordstream-2-denmark/nord-stream-2-says-denmark-tries-to-delay-pipeline-as-it-seeks-third-route-option-idUSKCN1RR15F> accessed 26 June 2019.

59 On the Natura 2000 network of protected natural areas and its regulation in EU law, see Langlet and Mahmoudi (n 21) 356.

60 Convention on Wetlands of International Importance Especially as Waterfowl Habitat (Ramsar, 2 February 1971) 996 UNTS 245 (Ramsar Convention).

61 Decision by the Energy Board of Appeal (Energiklagenævnet), 31 May 2010, Eksp.nr.: 49725 (translation).

62 Decision by the Supreme Administrative Court, 22 June 2011, No 4324/1/10.

description of alternative routes for the entire route of the pipelines as well as a no-action alternative.[63]

Environmental concerns raised by coastal States have also resulted in significant amendments compared to the original plans. The initially proposed routing of the existing pipelines was changed on several occasions, including during the Finnish, the Swedish and the Danish assessments, as a result of concerns raised in consultations. Adjustments to the routing were primarily prompted by the prevalence of dumped munitions and mines and the proximity of the planned pipeline to protected areas, notably those designated as Natura 2000.[64]

That rules on the protection of Natura 2000 sites, based as they are in EU law, were applied so directly to the project may give rise to some questions, considering that the applicant is a Swiss company. However, since 'pollution of the marine environment' is quite broadly defined in UNCLOS,[65] measures for the protection of sites of particular natural value, or sites that are important for the preservation of particular species, should be consistent with the right to take reasonable measures to prevent pollution from pipelines. The obligation to take all measures necessary to prevent, reduce and control pollution of the marine environment has even been construed to encompass 'measures focussed primarily on conservation and the preservation of ecosystems.'[66] There are hence good grounds for seeing most obligations stemming from EU law on Natura 2000, as well as other nature protection provisions, as consistent with UNCLOS, as long as they are not applied arbitrarily or without a scientific basis.[67] Nord Stream AG has also accepted modifications to the proposed

63 Request for a supplement to the application for a permit for a pipeline system under the Continental Shelf Act (1966:314) and the application for a permit to build and use a service platform under the Swedish Exclusive Economic Zone Act (1992:1140), 12 February 2008, M2007/5568/F/M.

64 See, e.g., Bundesamt für Seeschiffahrt und Hydrographie, Genehmigungsbescheid, 28 December 2009, 63 et seq. For a more detailed discussion about the national permit procedures and the changes prompted by them, see Langlet (n 4).

65 'Pollution of the marine environment' is defined as 'the introduction by man, directly or indirectly, of substances or energy into the marine environment, including estuaries, which results or is likely to result in such deleterious effects as harm to living resources and marine life, hazards to human health, hindrance to marine activities, including fishing and other legitimate uses of the sea, impairment of quality for use of sea water and reduction of amenities.' UNCLOS (n 20) art 1(1)(4).

66 Chagos Marine Protected Area Arbitration (Mauritius v United Kingdom), Arbitral Award of 18 March 2015, para 538.

67 However, considering the still disputed status of precaution in international law (Y Tanaka, Principles of international marine environmental law, in R Rayfuse (ed), Research Handbook on International Marine Environmental Law (Edward Elgar 2015) 31

route based on considerations for Natura 2000 sites without raising any objections as to the legitimacy of these requirements. Nor has Russia, as the sending State and a non-EU member raised such concerns. It should also be noted that obligations to 'conserve natural habitats and biological diversity and to protect ecological processes' as well as to apply the precautionary principle, including by taking preventive measures when there is reason to assume that substances or energy introduced into the marine environment may harm living resources and marine ecosystems can be found in the Helsinki Convention to which both Russia and Switzerland are parties.[68]

Unlike the different economic and security dimensions of the projects, the potential environmental effects of the Nord Stream pipelines have been subject to formal impact assessment under both international and EU law.[69] Notably, the project was subject to an international environmental impact assessment (EIA) in accordance with the Espoo Convention,[70] which was carried out more thoroughly than the States concerned were obliged to by the Convention's minimum requirements.[71] Also Russia, a non-party to the Espoo Convention, agreed to apply the Convention to the extent permitted by its national legislation.

The Finnish government referred to its participation in the EIA under the Espoo Convention as its way to dispose of the obligation to cooperate that pertains to States bordering a semi-enclosed sea like the Baltic Sea.[72] Such States are expected to cooperate with each other in the exercise of their rights and in the performance of their duties under UNCLOS, for example by endeavouring to coordinate the implementation of their rights and duties with respect to the protection and preservation of the marine environment.[73] However, that this provision would entail a binding obligation of any level of substance

at 43) the strict version of the precautionary principle that characterises some parts of the Natura 2000 regime (Langlet and Mahmoudi, n 21, 359) are unlikely to be fully consistent with UNCLOS.

68 Helsinki Convention (n 42) arts 3 and 15.

69 See e.g. 'Nord Stream Environmental Impact Assessment (EIA) Documentation for Consultation under the Espoo Convention', Nord Stream Espoo Report: Non-Technical Summary, February 2009.

70 Convention on Environmental Impact Assessment in a Transboundary Context (Espoo, 25 February 1991) 1989 UNTS 310 (Espoo Convention).

71 Koivurova and Pölönen (n 7) 174.

72 'Consent to Exploit Finland's Exclusive Economic Zone' (n 41) 23.

73 UNCLOS (n 20) art 122. The Baltic Sea qualifies as semi-enclosed both since it is 'surrounded by two or more States and connected to another sea or the ocean by a narrow outlet' and since it consists 'entirely or primarily of the territorial seas and exclusive economic zones of two or more coastal States' ibid.

is disputed and no conclusion to the contrary can be drawn from the Nord Stream case.[74]

As far as competition for space on the seabed between communication and transport installations, i.e. primarily cables and pipelines, is concerned, the Nord Stream case indicates that this need not be a major problem. Typically there are technical solutions to handle intersections between submarine pipelines and/or cables so that they can coexist, provided that the regulatory context facilitates that.[75]

5 Conclusions

The Nord Stream projects have affected many interests and entailed complex permit procedures involving five States and an even wider transboundary EIA. Some of the main bones of contention affecting the projects have not been directly linked to the physical presence of the pipelines on a particular stretch of the seabed, and therefore not within the primary purview of this analysis. We did, however, identify a risk of issues not recognised by UNCLOS as legitimate grounds for restricting pipeline operations nonetheless influencing decisions. This stems not least from the legal framework for the laying and operation of submarine pipelines on the CS being characterised by a level of uncertainty due to the imprecise manner in which the authority of the coastal State to regulate such activities is worded.

This has not result in any legal action under international law. Since all required permits for the existing two pipelines were eventually granted, with the exception of the one for surveying the seabed in the Estonian EEZ which was in fact not essential to the project, there was no real incentive to take issue with the result of the decision making processes.

The permit procedures of the transit States also appear generally to have been characterised by a desire to construe and apply the applicable law diligently and in good faith, despite heavy criticism based on considerations not recognized by UNCLOS being directed against the project in some of these States.[76] The decision by a Danish authority to request the investigation of

74 E Franckx and M Benatar, 'The "Duty" to Co-Operate for States Bordering Enclosed or Semi-Enclosed Seas' (2013) 31 Chinese (Taiwan) YBInt'lL&Aff 66.

75 See L.O. Askheim 'Commercial arrangements and liability for crossing pipelines, power cables and telecommunication cables on the seabed', Chapter 23 of this book.

76 On this criticism, see e.g. Solum Whist (n 4) 30; and R L Larsson, 'Nord Stream, Sweden and Baltic Sea Security', Swedish Defence Research Agency (FOI), March 2007, FOI-R – 2251-SE 35–7.

a third routing option about two years after Nord Stream 2 AG's submission of its first application, one that essentially corresponds to a pipeline routing that has previously been approved by Denmark, gives rise to the suspicion that environmental and maritime safety concerns may here have been used for aims not recognized by UNCLOS. However, the final outcome is still unclear and there are no indications that Denmark intends to not approve any route.

That relevant international provisions overall seem to have been applied in good faith may have been facilitated by the projects being subject to an ambitious international impact assessment in accordance with the Espoo Convention, involving all the States concerned. Despite proposals to that effect,[77] the EIA did not consider broader policy considerations pertaining to for example energy security and climate policy, but focused on the environmental impact of the pipelines as such. That may have helped to focus the subsequent permit procedures on the legally relevant issues, including protection of the marine environment and utilization of the resources of the seabed.

One legal issue that did come to the fore in this case was the legal significance of the need to drill in the seabed in preparation for pipeline laying. Since the coastal State can reject drilling in the CS at its discretion the need for surveys involving drilling can de facto provide coastal States with additional leverage beyond the powers granted in the provisions of UNCLOS dealing with pipelines. This does not seem to increase the legal uncertainty or the propensity to engage legal remedies. Rather, since the right exercised by the coastal State in this case is set out in absolute terms it has made the coastal State's competence somewhat easier to define.

As regards the concerns that coastal States could feel pressured not to take sufficient action for the protection of the marine environment, there is in fact much to indicate that environmental concerns played a central role in the assessments made by most coastal States. They clearly had a significant impact on the final routing of the pipelines.

Overall, the experience so far from the Nord Stream projects indicate that coastal States, including those that are merely transit States, have relatively strong instruments for influencing how pipelines are laid. With one possible exception, the fear that restricting coastal State authority to the protection of a few legitimate interests would result in excessive interpretations of those interests, or otherwise in actions that would upset the balance between competing interests have not materialized.

77 Koivurova and Pölönen (n 7) 176.

Whether limiting transit State competence to reasonable measures for the protection of a few listed interests is the optimal way to regulate pipeline laying on the CS may of course be debated. Expanding that right to also include wider security or other policy considerations could, however, give such States a de facto veto over pipeline laying, something that was not intended by the drafters of UNCLOS.

CHAPTER 25

Liability and Compensation for Activities in the Area

Kristoffer Svendsen

1 Introduction

All States have a duty to prevent harm to the environment including the marine environment. States also have '… the responsibility to ensure that activities within their jurisdiction or control do not cause damage to the environment of other States or of areas beyond the limits of national jurisdiction.'[1] Similar language is found in the United Nations Convention on the Law of the Sea (UNCLOS), which provides that 'States have the obligation to protect and preserve the marine environment.'[2] UNCLOS sets forth a governance and enforcement system of our global ocean where part of the ocean is under national jurisdiction and sovereignty, after which the remaining part of the ocean is located beyond national jurisdiction (the Area), in which the freedom of the high seas applies.[3] No State has sovereignty or sovereign rights to the Area, and the Area's minerals can only be extracted according to Part XI in the UNCLOS. The International Seabed Authority (ISA) was created to organise, control, and carry out exploration exploitation activities in the Area.[4] The ISA has thereafter developed, and continues to develop, the 'Mining Code', which is a '… comprehensive set of rules, regulations and procedures issued by the International Seabed Authority to regulate prospecting, exploration and exploitation of marine minerals in the international seabed Area (defined as the seabed and subsoil beyond the limits of national jurisdiction).'[5] The current three key regulations of the 'Mining Code' are ISA Regulations on Prospecting

1 Declaration of the United Nations Conference on the Human Environment, 5–16 June 1972, Stockholm, Art. 21.

2 UNCLOS Art. 192.

3 See UNCLOS Part XI and Art. 87.

4 Id. at. Art. 153.

5 ISA, 'The Mining Code' at <https://www.isa.org.jm/mining-code>.

DOI:10.1163/9789004391567_027

and Exploration for Polymetallic Nodules,[6] Polymetallic Sulphides,[7] and Cobalt-Rich Crusts,[8] in the Area.

The interest in exploration and exploitation of seabed minerals has increased. The ISA has in total entered into 15-year contracts for exploration for polymetallic nodules, polymetallic sulphides and cobalt-rich ferromanganese crusts in the deep seabed with 29 contractors.[9] 17 of these 29 contracts are for exploration for polymetallic nodules: 16 contracts in the Clarion-Clipperton Fracture Zone, which is a geological submarine fracture zone of the Pacific Ocean, and one contract in the Central Indian Ocean Basin.[10] There are seven contracts for exploration for polymetallic sulphides in the South West Indian Ridge, Central Indian Ridge and the Mid-Atlantic Ridge and five contracts for exploration for cobalt-rich crusts in the Western Pacific Ocean.[11] These contracts grant the contractors the exclusive right to explore an initial area of up to 150,000 square kilometres of specified parts of the deep oceans outside national jurisdiction.[12]

As a result of the transition of deep seabed mining in the Area from the exploration phase to the exploitation phase, the Legal and Technical Commission of the ISA started working on drafting regulations and standard contract terms on exploitation for mineral resources in the Area in 2015.[13] The draft exploitation regulations are currently under development. The first draft was published in July 2016 and were still under discussion at the time of the 25th Annual Session

6 International Seabed Authority Council, Decision of the Council of the International Seabed Authority relating to amendments to the Regulations on Prospecting and Exploration for Polymetallic Nodules in the Area and related matters, ISBA/19/C/17, 22 July 2013, at <https://www.isa.org.jm/documents/isba19c17>.

7 International Seabed Authority Assembly, Regulations on prospecting and exploration for polymetallic sulphides in the Area, ISBA/16/A/12/Rev.1, 7 May 2010, at <https://www.isa.org.jm/documents/isba16a12-rev-1>.

8 International Seabed Authority Assembly, Regulations on Prospecting and Exploration for Cobalt-rich Ferromanganese Crusts in the Area, ISBA/18/A/11, 27 July 2012, at <https://www.isa.org.jm/documents/isba18a11>.

9 ISA, 'Deep Seabed Minerals Contractors' at <https://www.isa.org.jm/deep-seabed-minerals-contractors>.

10 Ibid.

11 Ibid.

12 See above n 6. Reg. 24.1 and 25.1. For a complete review of the applicable general regime, see J. Dingwall, Chapter 7 of this book, 'Commercial Mining Activities in the Deep Seabed Beyond National Jurisdiction: The International Legal Framework.'

13 The Legal and Technical Commission of the International Seabed Authority (2016) Working Draft Regulations and Standard Contract Terms on Exploitation for Mineral Resources in the Area at https://www.isa.org.jm/files/documents/EN/Regs/DraftExpl/Draft_ExplReg_SCT.pdf.

of ISA in July 2019. In principle, the draft exploitation regulations should be finalised by 2020.

The current draft regulations contain sections on: form of applications; fee for applications; processing of applications; consideration of applications by the Commission and Council; exploitation contracts; plan of work for exploitation; annual fees; royalties; returns, payments and refunds; record, inspection and audit; anti-avoidance measures; interests and penalties; suspension or termination of contract; disputes and review of payments; information gathering and handling; inspections; enforcement and penalties; dispute settlement; and review of the Authority's regulations. The draft regulations however do not construct a model for liability and compensation for damage as a result of these activities.

The current chapter is a contribution to the discussion on how to compensate damage caused by activities in the Area. The chapter sketches the current parallel system of the sponsoring State's responsibility for damage caused in the Area and the sponsored contractor's liability for damage caused in the Area (Section 2). The chapter attempts to draw a model for an improved liability and compensation system for damage caused by deep-sea mining in the Area (Section 3). The chapter ends with some concluding remarks (Section 4).

2 The Current 'System' of Responsibility and Liability for Pollution Damage Caused in the Area

2.1 *Prelude*

To ensure compensation of environmental damage, UNCLOS places responsibility on the sponsoring State of the company applying for exploration and exploitation in the Area to ensure compliance with the applicable parts of UNCLOS, a breach of which places liability on the supporting State.[14] The International Tribunal for the Law of the Sea (ITLOS) gave an Advisory Opinion 1 February 2011 on the Responsibilities and Obligations of States Sponsoring Persons and Entities with Respect to Activities in the Area[15] based on Nauru's submission of an application for approval of a work plan for exploratory seabed mining activities to the ISA, 'but had become concerned that the potential

14 See UNCLOS Part XI and art. 139.

15 ITLOS, Advisory Opinion of 1 February 2011, Responsibilities and Obligations of States Sponsoring Persons and Entities with Respect to Activities in the Area, at <https://www.itlos.org/fileadmin/itlos/documents/cases/case_no_17/17_adv_op_010211_en.pdf>.

liabilities or costs arising from its sponsorship of a mining entity might exceed its financial capacities as a developing country.'[16]

2.2 *Sponsorship*

The notion of sponsorship is a key element in the exploration and exploitation system of resources in the Area.[17] Enterprises and, in association with the Authority, States Parties, or state enterprises or natural or juridical persons can engage in activities in the Area.[18] Natural and juridical persons must however satisfy two requirements to be eligible to engage in activities in the Area: 1) 'they must be either nationals of a State Party or effectively controlled by it or its nationals',[19] and 2) 'they must be "sponsored by such States".'[20] The requirement of sponsorship similarly identically applies to state enterprises.[21] States Parties themselves engaged in deep seabed mining are directly bound by the obligations set forth in UNCLOS and does not need sponsorship.[22]

The sponsorship requirement is crucial, creating the necessary nexus between the international legal treaty only binding on the States Parties and the domestic legal systems, of which the Enterprises are subjects.[23] The nexus between States Parties and subjects of domestic law consists of the nationality and effective control, which requires all contractors and applicants for contracts to 'secure and maintain the sponsorship of the State or States of which they are nationals. If another State or its nationals exercises effective control, the sponsorship of that State is also necessary.'[24] All sponsoring States are in such situations jointly and severally liable, unless otherwise provided in ISA regulations.[25]

16 Gunther Handl, 'Responsibilities and Obligations of States Sponsoring Persons and Entities with respect to Activities in the Area: the International Tribunal of the Law of the Sea's recent Contribution to International Environmental Law (2011) *Review of European Community & International Environmental Law*, 20, pp. 208–213.

17 ITLOS, Advisory Opinion of 1 February 2011, *supra* note 15 at p. 32.

18 UNCLOS Art. 153 2.

19 ITLOS, Advisory Opinion of 1 February 2011, *supra* note 15 at p. 32 and UNCLOS Art. 153 2(b).

20 Ibid.

21 Ibid.

22 Id at p. 33, UNCLOS Art. 153 2(b), and Annex III Art. 4 5.

23 ITLOS, Advisory Opinion of 1 February 2011, *supra* note 15 at p. 33.

24 Id at p. 33 and UNCLOS Annex III Art. 4 3.

25 Id at p. 62.

2.3 *Liability in General*

In its advisory opinion, the Tribunal sets forth the system and different sources of liability as stated in UNCLOS:

1) rules concerning the liability of State Parties (article 139, paragraph 2, first sentence),[26]
2) rules concerning sponsoring State liability (article 139, paragraph 2, second sentence),[27] and
3) rules concerning the liability of the contractor and the Authority (referred to in Annex III, article 22).

Paragraph 2 of article 139 attach liability on a sponsoring State from its failure to carry out its own responsibilities, while not being liable for the failure of the sponsored contractor to meet its obligations. 'There is, however, a link between the liability of the sponsoring State and the failure of the sponsored contractor to comply with its obligations, thereby causing damage.'[28] This chapter discusses these rules more in-depth and towards the end offers some suggestions on improvements.

2.4 *State Fault-Based Responsibility for Failure to Properly Perform Due Diligence – Secondary Fault-Based Liability*

The State(s) sponsoring contractors or applicants for contracts for the exploration and exploitation of resources in the Area have responsibilities and obligations under the UNCLOS. These obligations are characterised as 'direct obligations'.[29] The main direct obligations incumbent on the sponsoring States are:

> the obligation to assist the Authority in the exercise of control over activities in the Area; the obligation to apply a precautionary approach; the obligation to apply best environmental practices; the obligation to take measures to ensure the provision of guarantees in the event of an emergency order by the Authority for protection of the marine environment;

26 First sentence states: 'Without prejudice to the rules of international law and Annex III, article 22, damage caused by the failure of a State Party or international organization to carry out its responsibilities under this Part shall entail liability; States Parties or international organizations acting together shall bear joint and several liability.'

27 Second sentence states: 'A State Party shall not however be liable for damage caused by any failure to comply with this Part by a person whom it has sponsored under article 153, paragraph 2(b), if the State Party has taken all necessary and appropriate measures to secure effective compliance under article 153, paragraph 4, and Annex III, article 4, paragraph 4.'

28 ITLOS, Advisory Opinion of 1 February 2011, *supra* note 15 at p. 57.

29 Id at p. 44.

> the obligation to ensure the availability of recourse for compensation in respect of damage caused by pollution; and the obligation to conduct environmental impact assessments.[30]

The sponsoring State's liability for failure to meet its direct obligations is governed exclusively by the first sentence of paragraph 2 of article 139, while a sponsoring State's liability for a failure to meet its obligations in relation to damage caused by a sponsored contractor is covered by both the first and second sentences of article 139.[31] The nature of these obligations obviously does define and determine the scope of liability.

In its advisory opinion ITLOS relied on UNCLOS article 139 paragraph 1,[32] article 153 paragraph 4,[33] and Annex III article 4 paragraph 4,[34] when confirming that 'the obligation (responsibility) of the sponsoring State is "to ensure" that the "activities in the Area" conducted by the sponsored contractor are "in conformity" or in "compliance" with the rules to which they refer.'[35] The Tribunal highlighted that one of the sponsoring State's obligations under international law expressed as a 'responsibility to ensure' in UNCLOS establishes a mechanism where UNCLOS rules concerning activities in the Area become

30 Ibid.

31 Id at p. 58.

32 The paragraph states: 'States Parties shall have the responsibility to ensure that activities in the Area, whether carried out by States Parties, or state enterprises or natural or juridical persons which possess the nationality of States Parties or are effectively controlled by them or their nationals, shall be carried out in conformity with this Part. The same responsibility applies to international organizations for activities in the Area carried out by such organizations.'

33 The paragraph states: 'The Authority shall exercise such control over activities in the Area as is necessary for the purpose of securing compliance with the relevant provisions of this Part and the Annexes relating thereto, and the rules, regulations and procedures of the Authority, and the plans of work approved in accordance with paragraph 3. States Parties shall assist the Authority by taking all measures necessary to ensure such compliance in accordance with article 139.'

34 The paragraph states: 'The sponsoring State or States shall, pursuant to article 139, have the responsibility to ensure, within their legal systems, that a contractor so sponsored shall carry out activities in the Area in conformity with the terms of its contract and its obligations under this Convention. A sponsoring State shall not, however, be liable for damage caused by any failure of a contractor sponsored by it to comply with its obligations if that State Party has adopted laws and regulations and taken administrative measures which are, within the framework of its legal system, reasonably appropriate for securing compliance by persons under its jurisdiction.'

35 ITLOS, Advisory Opinion of 1 February 2011, *supra* note 15 at p. 40.

effective for sponsored contractors, which are bound by domestic law and not UNCLOS *per se*.[36]

A violation by the sponsoring state of this obligation 'to ensure' entails liability for the actual amount of damage.[37] The sponsoring State remains liable for damage also after the completion of the exploration phase.[38] Paragraph 2 of article 139 requires two conditions for liability to arise: 1) the failure of the sponsoring State to carry out its responsibilities and 2) the occurrence of damage.[39] Such failure may be 'act or an omission that is contrary to that State's responsibilities under the deep seabed mining regime.'[40] That said, a sponsoring State is only liable for a failure to carry out its responsibilities when damage has been inflicted. Thus, no matter the situation, no damage equals no liability for the sponsoring State.[41] Therefore, 'in order for the sponsoring State's liability to arise, there must be a causal link between the failure of that State and the damage caused by the sponsored contractor.'[42] This causal link cannot be presumed and must be proven.[43] Strict liability is therefore not the standard of liability for a sponsoring state.[44]

'However, not every violation of an obligation by a sponsored contractor automatically gives rise to the liability of the sponsoring State. Such liability is limited to the State's failure to meet its obligation to "ensure" compliance by the sponsored contractor.'[45] As such, a private entity's failure does not trigger liability for the sponsoring State, but is a mere trigger mechanism for

36 Id at p. 41.

37 UNCLOS Art. 139, paragraph 2, first sentence and ITLOS, Advisory Opinion of 1 February 2011, *supra* note 15 at p. 62.

38 ITLOS, Advisory Opinion of 1 February 2011, *supra* note 15 at p. 63.

39 Id at. p. 58.

40 Ibid.

41 This is an exception from the customary international law rule on liability where 'a State may be held liable under customary international law even if no material damage results from its failure to meet its international obligations.' ITLOS, Advisory Opinion of 1 February 2011, *supra* note 15 at p. 58, relying on Rainbow Warrior Arbitration at paragraph 110 (*Case concerning the difference between New Zealand and France concerning the interpretation or application of two agreements, concluded on 9 July 1986 between the two States and which related to the problems arising from the Rainbow Warrior Affair*, UNRIAA, 1990, vol. XX, p. 215) and paragraph 9 of the Commentary to article 2 of the International Law Commission 'Draft articles on Responsibility of States for Internationally Wrongful Acts, with commentaries' (2001) II *Yearbook of the International Law Commission*, Part Two.

42 ITLOS, Advisory Opinion of 1 February 2011, *supra* note 15 at p. 59.

43 Id at p. 60.

44 Id at p. 61.

45 Id at p. 41.

potentially giving rise to liability. Only the sponsoring State's own failure to carry out its own responsibilities give rise to liability for the sponsoring State.[46] The Tribunal importantly pointed out that:

> The sponsoring State's obligation "to ensure" is not an obligation to achieve, in each and every case, the result that the sponsored contractor complies with the aforementioned obligations. Rather, it is an obligation to deploy adequate means, to exercise best possible efforts, to do the utmost, to obtain this result. To utilize the terminology current in international law, this obligation may be characterized as an obligation "of conduct" and not "of result", and as an obligation of "due diligence".[47]

In explaining the obligation to act with due diligence, the Tribunal quoted the International Court of Justice in *Pulp Mills on the River Uruguay* (*Argentina v. Uruguay*):

> It is an obligation which entails not only the adoption of appropriate rules and measures, but also a certain level of vigilance in their enforcement and the exercise of administrative control applicable to public and private operators, such as the monitoring of activities undertaken by such operators ...[48]

The International Law Commission expressed the same view.[49] Thus, the Tribunal confirmed that sponsoring States are only liable for a failure to not properly perform their due diligence. The content of this due diligence obligation is not precise partly due to diligence as 'a variable concept'.[50] Sufficiently diligent measures can be insufficient with time as a result of for example 'new scientific or technological knowledge'.[51] The level of due diligence also increases according to the level of risk. The Tribunal stated that 'the standard of due diligence has to be more severe for the riskier activities.'[52] The sponsoring

46 Id at p. 60.
47 Id at p. 41.
48 Pulp Mills on the River Uruguay (Argentina v. Uruguay), Judgment, I.C.J. Reports 2010, p. 14, paragraph 197.
49 ITLOS, Advisory Opinion of 1 February 2011, *supra* note 15 at p. 42 quoting the Prevention of Transboundary Harm from Hazardous Activities, 2001, Official Records of the General Assembly, Fifty-sixth Session, Supplement No. 10 (A/56/10).
50 ITLOS, Advisory Opinion of 1 February 2011, *supra* note 15 at p. 43.
51 Ibid.
52 Ibid.

State must adopt within its legal system 'reasonably appropriate' measures to comply with the standard of due diligence appropriate to the level of risk, for which the measures are adopted.[53] Compliance with the above-listed direct obligations 'can also be seen as a relevant factor in meeting the due diligence "obligation to ensure" and that the said obligations are in most cases couched as obligations to ensure compliance with a specific rule.'[54]

Sponsoring States also have another direct obligation, which 'gives substance to the sponsoring State's obligation to adopt laws and regulations within the framework of its legal system.'[55] Article 235, paragraph 2, states as follows:

> States shall ensure that recourse is available in accordance with their legal systems for prompt and adequate compensation or other relief in respect of damage caused by pollution of the marine environment by natural or juridical persons under their jurisdiction.

Article 235, paragraph 2, applies to sponsoring States as 'the State with jurisdiction over the persons that caused the damage.'[56] The Tribunal also points out that States may also need to 'establish substantive rules governing claims for damages before its domestic courts ... ensuring that the sponsored contractor meets its obligation under Annex III, article 22, of the Convention to provide reparation for damages caused by wrongful acts committed in the course of its activities in the Area.'[57] Additionally, the sponsoring State's laws, regulations, and administrative measures must be in force during the whole period of the contract between ISA and the contractor.[58] The existence of such laws, regulations, and administrative measures are not a condition precedent for concluding a contract with ISA, but 'it is a necessary requirement for compliance with the obligation of due diligence of the sponsoring State and for its exemption from liability.'[59] If ISA regulations are passed after already existing sponsoring State legislation, sponsoring States are under an obligation to make the necessary amendments to such legislation to keep with the provisions of

53 Ibid.
54 Id at p. 44.
55 Id at p. 49.
56 Ibid.
57 Ibid.
58 Id at p. 68.
59 Ibid.

the ISA regulations.[60] Importantly, State laws, regulations, and administrative measures cannot be in perpetuity.[61]

Examples of domestic law rules are provisions concerning 'financial viability and technical capacity of sponsored contractors, conditions for issuing a certificate of sponsorship and penalties for non-compliance by such contractors',[62] and enforcement mechanisms of ITLOS decisions. Example of administrative measures are 'the establishment of enforcement mechanisms for active supervision of the activities of the sponsored contractor',[63] and to coordinate better efficiency between the sponsored contractor and ISA to remove activities leading to duplicate work.[64]

If the sponsoring State has taken the 'reasonably appropriate' measures and thus complied with its due diligence standard under the UNLOS, the sponsoring State is exempt from liability for damage inflicted by the sponsored contractor.[65] This can result in situations, in which a sponsoring State has fulfilled its due diligence standard, while, at the same time, damage has been inflicted in the Area and has not been compensated. The question becomes on whom to place liability for harm inflicted beyond national jurisdiction without any fault by a sponsoring State in exercising its obligation to prevent harm because the standard of care required is met? The Tribunal and the UNCLOS are clear on the fact that liability cannot be placed on the sponsoring State in such a situation.

2.5 *Simultaneous Contractor Liability for Pollution Damage – National Legislation*

UNCLOS requires a contractor that wants to engage in deep seabed mining to first secure and thereafter maintain the sponsorship of a State, as mentioned above. Sponsorship is the mechanism for domestic legal entities to comply with the international legal obligations of UNCLOS only binding on State Parties, and ISA's regulations and instruments applicable to these domestic legal entities.[66] As stated in Annex III, article 4, paragraph 4:

---

60 Id at p. 69.
61 Ibid.
62 Id at p. 72.
63 Id at p. 68.
64 Ibid.
65 Id at p. 43 and UNCLOS Art. 139 paragraph 2 last sentence, and UNCLOS Annex III art. 4 paragraph 4.
66 Id at pp. 32 and 40–41.

> The sponsoring State or States shall, pursuant to article 139, have the responsibility to ensure, within their legal systems, that a contractor so sponsored shall carry out activities in the Area in conformity with the terms of its contract and its obligations under this Convention. A sponsoring State shall not, however, be liable for damage caused by any failure of a contractor sponsored by it to comply with its obligations if that State Party has adopted laws and regulations and taken administrative measures which are, within the framework of its legal system, reasonably appropriate for securing compliance by persons under its jurisdiction.

A sponsoring State is obligated to adopt laws and regulations and to take the administrative measures necessary, as the contractor's obligations cannot all be enforced through administrative measures and contractual arrangements.[67] This is partly because contractual obligations cannot be invoked by other entities, than the parties to the contract, against the sponsoring State, and that contracts generally lacks transparency as it is difficult for the public to measure the sponsoring State's success in meeting its requirements.[68] Annex III of UNCLOS and ISA regulations do not require a sponsorship agreement between the contractor and the sponsoring State, nor its submission to ISA or public publication if in existence.[69] The sponsoring State is only required to submit a certificate of sponsorship to ISA stating its assumption responsibility according to article 139, article 153, paragraph 4, and Annex III, article 4, paragraph 4, of UNCLOS.[70]

That said, the sponsored contractor is obviously obligated to conduct its activities in the Area according to the terms of the contract. Even though 29 contracts have been entered into, ISA's Legal and Technical Commission are now developing standard contractual terms 'on exploitation for Mineral Resources in the Area for consideration by the Members of the Authority and all stakeholders.'[71] If a sponsored contractor fails to comply with rules in UNCLOS, in the ISA seabed-mining regime, or in the contract, which results in damage the sponsoring State is not liable.[72]

67 Id at pp. 68–69.
68 Id at p. 69.
69 Ibid.
70 Cobalt-rich regulations, *supra* note 8 at reg. 11 para. 3(f); Sulphides regulations, *supra* note 7 at reg. 11 para. 3(f); and Nodules regulations, *supra* note 6 at reg. 11 para. 3(f).
71 Working Draft Regulations and Standard Contract Terms, *supra* note 13 at p. 4.
72 UNCLOS Art. 139, paragraph 2, and ITLOS, Advisory Opinion of 1 February 2011, *supra* note 15 at p. 60.

UNCLOS Annex III, article 22, places liability on the contractor for inflicted damage in the Area for its activities:

> The contractor shall have responsibility or liability for any damage arising out of wrongful acts in the conduct of its operations, account being taken of contributory acts or omissions by the Authority. Similarly, the Authority shall have responsibility or liability for any damage arising out of wrongful acts in the exercise of its powers and functions, including violations under article 168, paragraph 2, account being taken of contributory acts or omissions by the contractor. Liability in every case shall be for the actual amount of damage.

The contractor also remains liable for damage after the completion of the exploration phase.[73] The contractor and the ISA are mainly liable for wrongful acts conducted performing contractor's activities and ISA's powers and fictions respectfully, and not the sponsoring State.[74] As previously mentioned the sponsoring State is liable for its own failure to carry out its responsibilities, while the contractor is liable for its own non-compliance.[75]

As previously mentioned, the sponsoring State has direct obligations to ensure and adopt appropriate rules and measures, and also enforce these rules and measures at a certain level of vigilance.[76] One of these direct obligations require the implementation of rules and measures for prompt and adequate compensation or other relief in respect of damage caused by pollution of the marine environment.

2.5.1 So What Rules and Measures Have Been Implemented by Sponsoring States?

In 2014, Tonga 'became the first country in the world to put in place a law that manages seabed mineral activities within its national marine space and under its sponsorship in international waters.'[77] Whether this is in fact correct

73 Cobalt-rich regulations, *supra* note 8 at reg. 32; Sulphides regulations, *supra* note 7 at reg. 32; and Nodules regulations, *supra* note 6 at reg. 30.

74 ITLOS, Advisory Opinion of 1 February 2011, *supra* note 15 at p. 63.

75 Id at p. 64.

76 Id at p. 42 and *Pulp Mills on the River Uruguay* case, *supra* 34 at paragraph 197.

77 Pacific Community – the Geoscience Devision, 'Tonga a world leader in seabed minerals law' (1 September 2014) at http://gsd.spc.int/media-releases/1-latest-news/583–tonga-a-world-leader-in-seabed-minerals-law.

is debatable. The Seabed Minerals Act 2014[78] established the Tonga Seabed Minerals Authority, regulations of seabed minerals activities within Tonga's national jurisdiction, duties and responsibilities of individuals, prospecting permits and licensing within national jurisdiction, sponsorship of activities in the Area, certain fiscal arrangements, marine scientific research, and some miscellaneous sections.[79] The Deep Sea Minerals Project, a partnership between the EU, the Secretariat of the Pacific Community and 15 Pacific Island countries assisted in the preparation of the Seabed Minerals Act.[80] Chapter 7 of the Act is titled 'Sponsorship of Activities in the Area'; within it section 84 is titled 'Liability of Sponsored Party':

> (1) The Sponsored Party shall be responsible for the performance of all Seabed Mineral Activities carried out within the Contract Area, and their compliance with the Rules of the ISA; and will be liable for the actual amount of any compensation or damage or penalties arising out of its failure so to comply, or out of any wrongful acts or omissions and those of its employees, officers, subcontractors, and agents in the conduct of the Seabed Mineral Activities.
> (2) Any obligations which are to be observed and performed by the Sponsored Party shall at any time at which the Sponsored Party is more than one person be joint and several obligations.
> (3) A Sponsored Party shall at all times keep the Kingdom indemnified against all actions, proceedings, costs, charges, claims and demands which may be made or brought by any third party in relation to its Seabed Mineral Activities.

Other Pacific Island countries, such as Tuvalu, have implemented the same wording the same year.[81] Nauru made some smaller amendments to

78 The Seabed Minerals Act 2014 (the Kingdom of Tonga), at http://www.eisourcebook.org/cms/February%202016/Tonga%20Seabed%20Minerals%20Act%202014.pdf.

79 Ibid.

80 Pacific Community, *supra* note 77.

81 Seabed Minerals Act 2014 (Tuvalu), s. 93.

the same text,[82] following the wording of Fiji's 2013 enactment.[83] ISA has also made a list of countries' legislation/reciprocating legislation and national initiatives with respect to the activities in the Area.[84] Singapore's deep sea mining legislation for the Area is less specific:

> (1) Where a licensee is responsible or liable for any wrongful act under Annex III, Article 22 of the Convention, the Court may —
> (a) order the licensee to pay to a person such compensation for the wrongful act as may be ordered to be paid to that person pursuant to that Article; and
> (b) grant to a person such other remedy for the licensee's wrongful act as may be granted to that person pursuant to that Article.
> (2) In this section, "licensee" includes a Singapore company which has ceased to hold a licence.[85]

It is apparent from a brief look at some national legislation that sponsored parties are under fault-based liability. One problem with fault-based liability enacted on national level is the more obvious fact that the level of fault that is required to trigger liability will be different between sponsored parties sponsored by different countries, as fault is defined differently among countries. Similarly, rules of evidence various between countries too, which will impact the "hurdle" of attaching fault-based liability to a sponsored party. As these enactments show, some countries have also included an indemnity clause. Taking into account the abovementioned State's responsibility for due diligence, which lack of government oversight contributing to damage attach

82 International Seabed Minerals Act 2015 (Nauru), s. 29: '(1) A Sponsored Party shall be responsible for the performance of all Seabed Mineral Activities carried out within the Contract Area, and their compliance with the Rules of the ISA and shall be liable for the actual amount of any compensation, damage or penalties arising out of its failure so to comply, or out of any wrongful acts or omissions in the conduct of the Seabed Mineral Activities. (2) By operation of this section, Nauru shall be indemnified against all actions, proceedings, costs, charges, claims and demands which may be made or brought by any third party in relation to a Sponsored Party's Seabed Mineral Activities.'

83 International Seabed Mineral Management Decree 2013 (Fiji), s. 33. For more information on Pacific Islands' legislation, see Blue Ocean Law and Pacific Network on Globilization, 'Resource Roulette: How Deep Sea Mining and Inadequate Regulatory Frameworks Imperil the Pacific and its Peoples' (2016) at http://cer.org.za/wp-content/uploads/2016/08/Resource-Roulette-Deep-sea-Mining-and-Inadequate-Regulatory-Frameworks.pdf, Ch. 5.

84 ISA, 'Laws, regulations and administrative measures adopted by sponsoring States and other members of the International Seabed Authority with respect to the activities in the Area' (13 June 2016), ISBA /22/C/8, at https://www.isa.org.jm/sites/default/files/files/documents/isba-22c-8_1.pdf.

85 Deep Seabed Mining Act 2015 (Singapore), art. 17.

liability, and situations where a State is part of a joint venture engaged in deep seabed mining, these indemnity clauses appear unenforceable in certain situations.

2.5.2 Compensable Damage

UNCLOS does specify that the amount of damages should be for 'the actual amount of damages'.[86] This terminology 'actual amount of damages' is repeated word for word in ISA regulation Standard Clauses for exploration contracts[87] and abovementioned national law regulating activities in the Area. The Tribunal confirmed that the form of reparation depends on actual damages and the technical feasibility of restoring the pollution damage to *ex ante*,[88] relying on article 34 of the ILC Articles on State Responsibility when determining the form of reparation, which states:

> Full reparation for the injury caused by the internationally wrongful act shall take the form of restitution, compensation and satisfaction, either singly or in combination, in accordance with the provisions of this chapter.[89]

Neither UNCLOS nor the Regulations define or specify compensable damage, or which subjects are entitled to damages.[90] However, the Tribunal lists a couple of possible types of damage: damage to the Area, damage to the Area's resources considered the common heritage of mankind, and damage to the marine environment.[91] The lack of further clarity on what damage categories are compensable and how to actually calculate this 'actual amount of damages', place a large responsibility on the sponsoring States and national legislation to properly do this. It also opens up for forum shopping by contractors when choosing from with jurisdiction to conduct activities in the Area. A sense of such bias by the contractor could presumably initiate sponsoring States facilitating for more lenient national liability and compensation legislation as well

---

86 UNCLOS Annex III, art. 22.

87 Cobalt-rich regulations, *supra* note 8 at reg. 32 in conjunction with Annex IV section 16.1; Sulphides regulations, *supra* note 7 at reg. 32 in conjunction with Annex 4, section 16.1; and Nodules regulations, *supra* note 6 at reg. 30 in conjunction with Annex IV section 16.1.

88 Id at p. 63.

89 International Law Commission 'Responsibility of States for Internationally Wrongful Acts' (2001) II *Yearbook of the International Law Commission*, Part Two, art. 34.

90 Cobalt-rich regulations, *supra* note 8 at reg. 32; Sulphides regulations, *supra* note 7 at reg. 32; and Nodules regulations, *supra* note 6 at reg. 30.

91 ITLOS, Advisory Opinion of 1 February 2011, *supra* note 15 at p. 59.

as environmental legislation to attract the financial benefits of sponsoring a contractor.

3 The Need for a Proper System of Strict Liability for Pollution Damage Caused in the Area

3.1 *The Lack of a Proper Liability and Compensation Scheme for Deep Seabed Mining*

ISA, entities engaged in deep seabed mining, other users of the sea, and Coastal States are some examples the Tribunal explicitly states as subjects that may claim compensation for pollution damage in the Area.[92] Claims against the contractor (sponsored party) may either be paid or not paid. If the contractor has paid the actual amount of damages, 'there is no room for reparation by the sponsoring State.'[93] This presumes that the concept of 'actual amount of damages' corresponds with compensation in full, which should be full environmental restitution. This is hardly the case.

In certain situations, such as blameless actions of the contractor or simply contractor's bankruptcy, the contractor may end up not paying for pollution damage in the Area. As stated above, strict liability is not the standard of liability applicable to the sponsoring State. There is also no residual liability for the sponsoring State,[94] and the sponsoring State and the contractor does not bear joint and several liability, which is a standard applicable where different entities have contributed to the same damage so that compensation in full can be achieved.[95] As a result, certain 'liability gaps' may occur in situations where the contractor does not pay the actual amount of damage. The fault-based limitation on liability for sponsoring States leaves at least three liability gaps:

- where a state takes all necessary and/or appropriate measures required by international law and the blameless actions of the contractor nevertheless cause environmental harm;
- where a state takes the requisite necessary and/or appropriate measures and the private operator is blameworthy, but insolvent or its assets are beyond the reach of the sponsoring state; and
- where the sponsoring state has failed to take the required measures but there is no causal link with the environmental harm.[96]

92 ITLOS, Advisory Opinion of 1 February 2011, *supra* note 15 at p. 59.
93 Id at p. 64.
94 Ibid.
95 Id at pp. 63–64.
96 Donald K. Anton (11 November 2011) 'The Principle of Residual Liability in the Seabed Disputes Chamber of the International Tribunal for the Law of the Sea: The Advisory

These liability gaps are currently not protected under international law, and compensable. Article 304 of UNCLOS allows further developments of international law to be implemented into the deep seabed liability regime, and also further developments of the deep seabed liability regime specifically.[97]

3.2 *An Alternative Model of Liability and Compensation*

3.2.1 Sponsored Contractor's Strict Liability for Pollution Damage in the Area

ITLOS states that the sponsoring States may apply more stringent standards as far as the protection of the marine environment is concerned to their contractors.[98] Sponsoring States should implement strict liability for their contractors for pollution damage from activities in the Area, as such activities are hazardous, to ensure prompt and adequate compensation.[99] There is no good reason why injured parties and the environment should take the risk of contractors adhering to a negligence standard for pollution damage in the Area. The cost of such liability can be internalised and offset to some extent by insurance.[100] The Norwegian Petroleum Act[101] is a good example of how strict liability on the licensee for pollution damage caused by petroleum does not limit the interest from national and international companies to explore and produce oil and gas on the Norwegian Continental Shelf.

3.2.2 Pollution Damage Fund under the ISA

ITLOS suggested that ISA should consider the option of setting up a damage fund as a means to cover damages that are not otherwise covered by the deep seabed mining liability scheme, the liability gap.[102] A pollution damage fund would be a good way of creating a second layer of safety to ensure compensation of pollution damage in the Area, for situations where the sponsored contractor become delinquent and claims are beyond the scope of its insurance.[103] Such

Opinion on Responsibility and Liability for International Seabed Mining (ITLOS Case No. 17)' *McGill International Journal of Sustainable Development Law and Policy*, at https://ssrn.com/abstract=1957907, pp. 12–13 referring to Written Statement of the International Union for the Conservation of Nature, at http://www.itlos.org/fileadmin/itlos/documents/cases/case_no_17/StatementIUCN.pdf, pp. 28–29.

97 ITLOS, Advisory Opinion of 1 February 2011, *supra* note 15 at p. 66.

98 Id at p. 73.

99 ILC, 'Draft principles on the allocation of loss in the case of transboundary harm arising out of hazardous activities' (2006) International Law Commission, 58th session, art. 4(2).

100 Working Draft Regulations and Standard Contract Terms, *supra* note 13 at sec. 8.

101 Lov om petroleumsvirksomhet 1996 nr. 72.

102 ITLOS, Advisory Opinion of 1 February 2011, *supra* note 15 at pp. 65–66.

103 ILC, 'Draft principles on the allocation of loss in the case of transboundary harm arising out of hazardous activities' (2006) International Law Commission, 58th session, art. 4(4).

a fund should be under the ISA, as the international regulatory body and collector of certain deep seabed related fees. ISA has created draft regulation to establish a Environmental Liability Trust Fund.[104]

3.2.3 Residual Liability for the Sponsoring State

The sponsoring State is not residually liable for a contractor's non-compliance.[105] However, sponsoring States should be liable in last instance as deep seabed mining is a significant hazardous activity.[106] The sponsoring States set the financial requirements of the contractor and can increase financial security requirements of the contractor to decrease its own risk. The environment in the Area, the common heritage of human kind, should not be put at risk and suffer when this could be avoided.

4 Conclusion

The sponsoring State's liability arises from its own failure to carry out its responsibilities, whereas the sponsored contractor's liability arises from its own non-compliance. These two forms of liability exist simultaneously and in parallel. The only connection between the two is that the sponsoring State's liability depends upon the sponsored contractor's damage resulting from activities or omissions.[107] This system facilitates for liability gaps, which could and should be removed by enacting on a national level strict liability for pollution damage in the Area on the sponsored contractor, an international seabed pollution damage fund, and residual liability on sponsoring States. This is stricter than the current international legal situation, but international law is constantly evolving and should evolve accordingly.

---

104 ISA, 'Developing a Regulatory Framework for Mineral Exploitation in the Area: a Discussion Paper on the development and drafting of Regulations on Exploitation for Mineral Resources in the Area (Environmental Matters)' (January 2017) at https://www.isa.org.jm/files/documents/EN/Regs/DraftExpl/DP-EnvRegsDraft25117.pdf, pp. 72–74.

105 Position put before ITLOS by amongst others R.A. Makgill et al., Written Statement of International Union for Conservation of Nature and Natural Resources, Commission on Environmental Law, Oceans, Coastal and Coral Reefs Special Group (19 August 2010) at https://www.itlos.org/fileadmin/itlos/documents/cases/case_no_17/StatementIUCN.pdf, pp. 29–32.

106 ILC, 'Draft principles on the allocation of loss in the case of transboundary harm arising out of hazardous activities' (2006) International Law Commission, 58th session, art. 4(5).

107 ITLOS, Advisory Opinion of 1 February 2011, *supra* note 15 at p. 64.

Index

www.ingramcontent.com/pod-product-compliance
Lightning Source LLC
LaVergne TN
LVHW010637300926
849161LV00005B/964
* 9 7 8 9 0 0 4 3 9 1 5 5 0 *